Biotechnology in Agriculture and Forestry

Edited by
T. Nagata (Managing Editor)
H. Lörz
J. M. Widholm

Volumes already published
Volume 1: Trees 1 (1986)
Volume 2: Crops 1 (1986)
Volume 3: Potato (1987)
Volume 4: Medicinal and Aromatic Plants 1 (1988)
Volume 5: Trees 11 (1989)
Volume 6: Crops 11 (1988)
Volume 7: Medicinal and Aromatic Plants 11 (1989)
Volume 8: Plant Protoplasts and Genetic Engineering 1 (1989)
Volume 9: Plant Protoplasts and Genetic Engineering 11 (1989)
Volume 10: Legumes and Oilseed Crops 1 (1990)
Volume 11: Somaclonal Variation in Crop Improvement 1 (1990)
Volume 12: Haploids in Crop Improvement 1 (1990)
Volume 13: Wheat (1990)
Volume 14: Rice (1991)
Volume 15: Medicinal and Aromatic Plants Ill (1991)
Volume 16: Trees 111 (1991)
Volume 17: High-Tech and Micropropagation 1 (1991)
Volume 18: High-Tech and Micropropagation 11 (1992)
Volume 19: High-Tech and Micropropagation Ill (1992)
Volume 20: High-Tech and Micropropagation IV (1992)
Volume 21: Medicinal and Aromatic Plants IV (1993)
Volume 22: Plant Protoplasts and Genetic Engineering 111 (1993)
Volume 23: Plant Protoplasts and Genetic Engineering IV (1993)
Volume 24: Medicinal and Aromatic Plants V (1993)
Volume 25: Maize (1994)
Volume 26: Medicinal and Aromatic Plants VI (1994)
Volume 27: Somatic Hybridization in Crop Improvement 1 (1994)
Volume 28: Medicinal and Aromatic Plants VII (1994)
Volume 29: Plant Protoplasts and Genetic Engineering V (1994)
Volume 30: Somatic Embryogenesis and Synthetic Seed 1 (1995)
Volume 31: Somatic Embryogenesis and Synthetic Seed 11 (1995)
Volume 32: Cryopreservation of Plant Germplasm 1 (1995)
Volume 33: Medicinal and Aromatic Plants VIII (1995)
Volume 34: Plant Protoplasts and Genetic Engineering VI (1995)
Volume 35: Trees IV (1996)
Volume 36: Somaclonal Variation in Crop Improvement 11 (1996)
Volume 37: Medicinal and Aromatic Plants IX (1996)
Volume 38: Plant Protoplasts and Genetic Engineering VII (1996)
Volume 39: High-Tech and Microprogation V (1997)
Volume 40: High-Tech and Microprogation VI (1997)
Volume 41: Medicinal and Aromatic Plants X (1998)
Volume 42: Cotton (1998)
Volume 43: Medicinal and Aromatic Plants XI (1999)
Volume 44: Transgenic Trees (1999)
Volume 45: Transgenic Medicinal Plants (1999)
Volume 46: Transgenic Crops 1 (1999)
Volume 47: Transgenic Crops 11 (2001)
Volume 48: Transgenic Crops Ill (2001)
Volume 49: Somatic Hybridization in Crop Improvement 11 (2001)
Volume 50: Cryopreservation of Plant Germplasm 11 (2002)
Volume 51: Medicinal and Aromatic Plants XII (2002)
Volume 52: Brassicas and Legumes: From Genome Structure to Breeding (2003)
Volume 53: Tobacco BY-2 Cells (2004)
Volume 54: *Brassica* (2004)
Volume 55: Molecular Marker Systems in Plant Breeding and Crop Improvement (2004)

Volumes in preparation
Volume 56: Haploids in Crop Improvement II
Tropical Crops I
Tropical Crops II

Biotechnology in Agriculture and Forestry 55

*Molecular Marker Systems
in Plant Breeding and Crop Improvement*

Edited by
H. Lörz and G. Wenzel

With 42 Figures, 5 in Color, and 48 Tables

Springer

Series Editors

Professor Dr. TOSHIYUKI NAGATA
University of Tokyo
Gradulate School of Science
Department of Biological Sciences
7-3-1 Hongo, Bunkyo-ku
Tokyo 113-0033, Japan

Professor Dr. HORST LÖRZ
Universität Hamburg
Biozentrum Klein Flottbek
Zentrum für Angewandte Molekularbiologie
der Pflanzen (AMP II)
Ohnhorststraße 18
22609 Hamburg, Germany

Professor Dr. JACK WIDHOLM
University of Illinois
285A E.R. Madigan Laboratory
Department of Crop Sciences
1201 W. Gregory
Urbana, IL 61801, USA

Volume Editors

Professor Dr. HORST LÖRZ (*address see above*)

Professor Dr. GERHARD WENZEL
Technische Universität München
Lehrstuhl für Pflanzenbau und Pflanzenzüchtung
Alte Akademie 12
85250 Freising-Weihenstephan, Germany

ISSN 0934-943X
ISBN 3-540-20689-2 Springer-Verlag Berlin Heidelberg New York

Library of Congress Control Number: 2004103886

This work is subject to copyright. All rights reserved, whether the whole or part of the material is concerned, specifically the rights of translation, reprinting, reuse of illustrations, recitation, broadcasting, reproduction on microfilm or in any other way, and storage in data banks. Duplication of this publication or parts thereof is permitted only under the provisions of the German Copyright Law of September 9, 1965, in its current version, and permission for use must always be obtained from Springer-Verlag. Violations are liable to prosecution under the German Copyright Law.

Springer-Verlag is a part of Springer Science+Business Media
springeronline.com

© Springer-Verlag Berlin Heidelberg 2005
Printed in Germany

The use of general descriptive names, registered names, trademarks, etc. in this publication does not imply, even in the absence of a specific statement, that such names are exempt from the relevant protective laws and regulations and therefore free for general use.

Cover Design: Design & Production, Heidelberg
Typesetting: Mitterweger & Partner GmbH, Plankstadt
31/3150-WI – 5 4 3 2 1 0 – Printed on acid-free paper

Dedicated to
Professor Dr. Dr. h.c. Gerhard Fischbeck

Preface

More food and feed have to be grown on less land. At the same time, agronomy and agrochemistry have reached a level where further progress is neutral in relation to yield. Scientific progress helps environmental needs. Genetics alone may close the yield gap by a combined application of classical techniques and molecular knowledge. Large-scale genome analysis and related technologies provide access to a refined understanding of the genome. Fortunately, molecular markers are a genetic tool accepted by the public which often judges gene technology critically. There is even a tendency for classical combination breeding, together with marker-assisted selection, to be more effective in achieving complex breeding goals.

We are grateful to the authors of this book, who contributed new information about molecular marker collections and molecular gene maps, elucidating the usefulness of these tools: such as higher efficiency in parent selection; controlled combination of heterotic parents; uncovering a rare, but desired genotype in a large segregating population; or pyramiding single traits to result in more complex characters, e.g., durable resistance. We restricted the selection of topics to functioning marker approaches rather than to all possible strategies. Nevertheless, the spectrum of crop plants and trees covered is quite broad with a close connection to application.

Successful breeders need to have green fingers and this hopefully might in future be confirmed by molecular markers. Soon, it will seem unbelievable that breeders have been successful without using such a technique. We would like to dedicate this book to one such breeding artist, Prof. Dr. Dr. h.c. Gerhard Fischbeck. He has bred wheat varieties giving optimal quality under suboptimal European climatic conditions and has found numerous new resistance genes for powdery mildew in cereals. He held the Chair of Agronomy and Plant Breeding at the Technical University in Munich, Germany, and, although he is approaching his 80th birthday, he is still very active and curious to discover the secrets of nature and always full of life and spirit. In 1989, he started a project in barley genomics, thereby giving an impressive example of moving from classical to molecular breeding.

Even most sophisticated new marker techniques will not replace a sharp eye, green fingers, or a bright mind – and certainly not the field experiment. Nevertheless, such techniques do allow more reproducible and faster development of new and better cultivars. Today, a breeder has to know how such tools work and how to use them in an intelligent manner. Hopefully, this book will help to transform most recent marker theories into many superior varieties of crops.

Horst Lörz
Hamburg, Germany

Gerhard Wenzel
München, Germany

May 2004

Contents

Section I Basics

I.1	The Principle: Identification and Application of Molecular Markers	3
	P. Langridge and K. Chalmers	
1	Introduction	3
2	Status	4
3	Molecular Markers	5
4	Identifying Marker/Trait Associations	6
5	Application of Molecular Markers	12
6	Directions	17
7	Conclusions	20
	References	20
I.2	Genotyping Tools in Plant Breeding: From Restriction Fragment Length Polymorphisms to Single Nucleotide Polymorphisms	23
	V. Mohler and G. Schwarz	
1	Introduction	23
2	Restriction Fragment Length Polymorphisms	23
3	Microsatellites	25
4	Random Amplified Polymorphic DNAs	25
5	Amplified Fragment Length Polymorphisms	26
6	Single Nucleotide Polymorphisms	28
7	Conclusions	33
	References	34
I.3	A Model Crop Species: Molecular Markers in Rice	39
	D.J. Mackill and K.L. McNally	
1	Introduction	39
2	Gene Mapping with Molecular Markers in Rice	41
3	Positional Cloning	42
4	Array-Based Markers	45
5	Candidate Genes as Markers	47
6	Conclusions	48
	References	49

I.4	From Markers to Cloned Genes: Map-Based Cloning	55
	W.-R. SCHEIBLE, O. TÖRJEK, and T. ALTMANN	
1	Introduction	55
2	Outline of the General Map-Based Cloning Strategy	56
3	Map-Based Cloning in a Model Species with a Fully Sequenced Genome (*Arabidopsis thaliana*)	70
4	Map-Based Cloning in Crop Species	76
5	Conclusions	80
	References	81

Section II Specific Crops

II.1	DNA Markers in *Brassica*: Use of Genetic Information from *Arabidopsis* and Development of Sequence Tagged Site Markers	89
	T. SAKAI, H. FUJIMOTO, R. IMAI, and J. IMAMURA	
1	Introduction	89
2	Characterization of DNA Markers in *Brassica*	90
3	Application of *Arabidopsis* Genome Information to *Brassica* DNA Markers	96
4	Developing Random Amplified Polymorphic DNA Sequence Tagged Site Markers Linked to the Radish-Derived Fertility Restoration (*Rf*) Locus in *B. napus*	97
5	Conclusions	100
	References	101

II.2	Genomics as Efficient Tools: Example Sunflower Breeding	107
	A. SARRAFI and L. GENTZBITTEL	
1	Introduction	107
2	Linkage Mapping	107
3	Quantitative Trait Loci Identification	111
4	Sunflower Genomics: Towards Genes and Functions	114
5	Conclusions	116
	References	116

II.3	Genome Analysis: Mapping in Sugar Beet	121
	C. JUNG	
1	Introduction: The Species of the Genus *Beta*	121
2	Resources and Techniques	121
3	Repetitive DNA Classes	126
4	Genetic Relationships Between Species of the Genus *Beta*	128
5	Genetic Mapping of Mendelian Traits	129
6	Mapping of Quantitative Trait Loci	132
7	Conclusions	133
	References	133

II.4	Molecular Markers in Genetics and Breeding: Improvement of Alfalfa (*Medicago sativa* L.) 139
	I.J. MAUREIRA and T.C. OSBORN
1	Introduction 139
2	Characterization of Alfalfa Germplasms 140
3	Development of Genetic Maps and Identification of Regions Affecting Traits of Interest 142
4	Additional Uses of Molecular Markers in Alfalfa Breeding 148
5	Conclusions 149
	References 150

II.5	Localization of Important Traits: The Example Pea (*Pisum sativum* L.) 155
	W.K. SWIECICKI and G. TIMMERMAN-VAUGHAN
1	Introduction 155
2	The *Pisum* Genetic Map and Loci for Agronomic Characters 156
3	Resistance to Powdery Mildew (*Erysiphe pisi* Syd.) 158
4	Resistance to Fusarium Wilt (*Fusarium oxysporum* f. sp. *pisi* (van Hall) Snyd & Hans) 159
5	Plant Virus Resistance 159
6	Flowering Genetics 161
7	Quantitative Trait Loci 162
8	Conclusions 166
	References 166

II.6	Molecular Markers in *Vigna* Improvement: Understanding and Using Gene Pools 171
	A. KAGA, D.A. VAUGHAN, and N. TOMOOKA
1	Introduction 171
2	Application of Molecular Markers to Understand the *Vigna* Crop Gene Pools 171
3	Linkage Maps 174
4	Synteny 176
5	Gene Mapping 178
6	Transformation Systems 181
7	Conclusions 183
	References 183

II.7	Molecular Markers for Genetics and Breeding: Development and Use in Pepper (*Capsicum* spp.) 189
	V. LEFEBVRE
1	Introduction 189
2	Molecular Markers 190
3	Genetic Diversity 190
4	Variety Identification and Hybrid Purity 191

5	Genetic Maps	192
6	Maps Position and Markers of Loci Governing Traits of Interest	198
7	Genomic Resources	206
8	Marker-Assisted Selection	208
9	Conclusion	209
	References	210

II.8 Potato Genetics: Molecular Maps and More ... 215
C. Gebhardt

1	Introduction: The Potato as an Object of Genetic Analysis	215
2	Reference Molecular Maps of Potato	216
3	Synteny of Potato With Other Plant Genomes	216
4	Potato Function Map for Pathogen Resistance	218
5	Potato Function Map for Tuber Traits	221
6	Potato Function Maps as Basis for Innovative Approaches to Breeding	223
7	Conclusions	224
	References	224

II.9 Molecular Marker Maps of Barley:
A Resource for Intra- and Interspecific Genomics ... 229
R.K. Varshney, M. Prasad, and A. Graner

1	Introduction	229
2	Molecular Markers	230
3	Construction of Molecular Maps	230
4	Comparative Mapping and Synteny	236
5	Conclusions	238
	References	239

II.10 Genomics in Rice: Markers as a Tool for Breeding ... 245
Y. Kishima, K. Onishi, and Y. Sano

1	Introduction	245
2	Conventional Markers	246
3	The Beginnings of Molecular Marker Analysis	246
4	Molecular Markers Currently Used in Rice Breeding	247
5	Amplified Fragment Length Polymorphism Analysis in Rice	248
6	MITE-Transposon Display in Rice	248
7	Transposable Elements as Markers for Major Genes	250
8	Conclusions: Quantitative Trait Loci and Future Prospects	251
	References	252

II.11 Wheat Microsatellites: Potential and Implications ... 255
M.S. Röder, X.-Q. Huang, and M.W. Ganal

1	Introduction	255
2	Development of Microsatellite Markers	255

3	The Bridge to Practical Applications	257
4	Diagnostic Markers for Traits of Interest	257
5	Analysis of Genetic Diversity	261
6	Conclusions	261
	References	262

II.12 **Comparative Genetic Mapping in Trees: The Group of Conifers** .. 267
D.B. NEALE and K.V. KRUTOVSKY

1	Introduction	267
2	Conifer Genomes	268
3	Loblolly Pine Reference Genetic Map	269
4	Genetic Markers for Comparative Mapping in Conifers	269
5	Comparative Mapping in *Pinus*	270
6	Comparative Mapping in Pinaceae	273
7	Conclusions: Needs of Linking the Genetic Map to Chromosomes	275
	References	275

II.13 **Markers in Fruit Tree Breeding: Improvement of Peach** 279
E. DIRLEWANGER and P. ARÚS

1	Introduction	279
2	Use of Molecular Markers for Fruit Quality Improvement	285
3	Use of Molecular Markers for Disease Resistance	289
4	Marker-Assisted Selection for Tree Architecture Characters	291
5	Synteny Among *Prunus* Species	292
6	Development of Peach Molecular Markers, Their Use for Fingerprinting and for the Evaluation of Genetic Resources	296
7	Conclusions	297
	References	297

Section III Breeding Strategies and Silviculture Based on Markers

III.1 **General Considerations: Marker-Assisted Selection** 305
V. MOHLER and C. SINGRÜN

1	Introduction	305
2	Requirements of Markers for Marker-Assisted Selection	305
3	Present Status of Validated Molecular Markers for Molecular Breeding of Important Crops	307
4	Marker-Assisted Selection for Quantitative Trait Loci	307
5	Marker-Assisted Selection in Gene Pyramiding	310
6	Marker-Assisted Selection in Backcross Breeding	311
7	Conclusions	313
	References	313

III.2	Breeding Strategies: Optimum Design of Marker-Assisted Backcross Programs 319
	M. FRISCH
1	Introduction 319
2	Introgression of One Dominant Gene 319
3	Introgression of a Recessive Gene 327
4	Introgression of Two Dominant Genes 330
5	Length of the Intact Donor Chromosome Segment Around the Target Gene 331
	References 333

III.3	From Theory to Practice: Marker-Assisted Selection in Maize 335
	D.A. HOISINGTON and A.E. MELCHINGER
1	Introduction 335
2	Practical Examples of Marker-Assisted Selection 342
3	Economics of Marker-Assisted Selection 348
4	New Marker-Assisted Selection Strategies 349
5	Conclusions 349
	References 350

III.4	Molecular Markers for Disease Resistance: The Example Wheat .. 353
	C. FEUILLET and B. KELLER
1	Introduction 353
2	Development of Molecular Markers 354
3	Use of Molecular Markers in Marker-Assisted Selection for Disease Resistance 362
4	Conclusions 363
	References 364

III.5	Application of DNA Markers: Soybean Improvement 371
	M.J. IQBAL and D.A. LIGHTFOOT
1	Introduction 371
2	Choice of Markers 373
3	Identification of Polymorphism 378
4	Marker-Assisted Recovery of Recurrent Parent Genome 378
5	Marker-Assisted Selection in Recurrent Cross Populations 379
6	Marker-Assisted Selection for Targeted Genes/Traits 379
7	Methods for Marker-Assisted Selection 380
8	Conclusions 382
	References 383

III.6	Forest Management and Conservation Using Microsatellite Markers: The Example of *Fagus* 387
	Y. Tsumura, M. Takahashi, T. Takahashi, N. Tani, Y. Asuka, and N. Tomaru
1	Introduction 387
2	Development and Evaluation of Microsatellite Markers in *Fagus* 388
3	Spatial Analysis of Genetic Structure Within Forests by Microsatellite Markers 390
4	Genetic Management of *Fagus* Forests for Conservation and Sustainable Use 393
5	Conclusions 395
	References 395

III.7	Molecular Markers in Tree Improvement: Characterisation and Use in *Eucalyptus* 399
	M. Shepherd and M.E. Jones
1	Introduction 399
2	Base Population Characterisation – *Eucalyptus globulus* is geographically structured with chloroplast haplotypes coincident with quantitative genetic variation 400
3	Effect of Utilisation on the Base Population Resource – Influence of Silvicultural and Harvesting on *Eucalyptus sieberi* Genetic Diversity 401
4	Defining the Gene Pool for Breeding – Hybridisation 402
5	Direct Measures of Gene Flow – Implications for Orchard Design 403
6	Genetic Architecture of Commercial Traits – Quantitative Trait Loci and Candidate Gene Mapping 405
	References 409

III.8	DNA Markers for Identification and Evaluation of Genetic Resources in Forest Trees: Case Studies in *Abies*, *Picea* and *Populus* 413
	B. Ziegenhagen and M. Fladung
1	Introduction 413
2	Which DNA Marker at Which Scale and for Which Purpose? 415
3	Case Studies with Fir (*Abies* sp.), Norway Spruce [*Picea abies* (Karst.) L.] and Poplar (*Populus* sp.) 417
4	Conclusions 425
	References 426

Section IV Legal Aspects

IV.1	Intellectual Property Rights in the Field of Molecular Marker Analysis 433	
	P. Jorasch	
1	Introduction 433	
2	What is a Patent? 434	
3	Microsatellite or Simple Sequence Repeat Markers 434	
4	The Selection of Microsatellite Primers and the PCR Reaction ... 436	
5	Analysis of PCR Products................................. 437	
6	Marker-Assisted Breeding Methods 437	
7	Conclusions 438	
	References 471	

Subject Index 473

List of Contributors

T. ALTMANN
Institute of Biochemistry and Biology – Genetics, University of Potsdam, 14415 Potsdam, Germany

P. ARÚS
Laboratori CSIC-IRTA de Genètica Molecular Vegetal, Departament de Genética Vegetal, Carretera de Cabrils s/n; 08348 Cabrils (Barcelona), Spain

Y. ASUKA
Laboratory of Forest Ecology and Physiology, Graduate School of Bioagricultural Sciences, Nagoya University, Nagoya 464-8601, Japan

K. CHALMERS
Australian Centre for Plant Functional Genomics, University of Adelaide, Waite Campus, Glen Osmond, SA 5064, Australia

E. DIRLEWANGER
Unité de Recherches sur les Espèces Fruitières et la Vigne, INRA, BP 81, 33883 Villenave d'Ornon, France

C. FEUILLET
Institute of Plant Biology, University of Zürich, Zollikerstrasse 107, 8008 Zürich, Switzerland

M. FLADUNG
Federal Research Centre for Forestry and Forest Products, Institute for Forest Genetics and Forest Tree Breeding, Sieker Landstrasse 2, 22927 Grosshansdorf, Germany

M. FRISCH
Institute of Plant Breeding, Seed Science and Population Genetics, University of Hohenheim, 70593 Stuttgart, Germany

H. FUJIMOTO
Plantech Research Institute, 1000 Kamosida-cho Aoba-ku Yokohama, Japan 227-0033

M.W. GANAL
TraitGenetics GmbH, Am Schwabeplan 1b, 06466 Gatersleben, Germany

C. GEBHARDT
Max-Planck Institute for Plant Breeding Research, Carl-von-Linne-Weg 10, 50829 Köln, Germany

L. GENTZBITTEL
Department of Biotechnology and Plant Breeding, Pôle de Biotechnologie Végétale, ENSAT-BAP, 18 Chemin de Borde Rouge, BP 107, 31326 Castanet, France

A. GRANER
Institute of Plant Genetics and Crop Plant Research (IPK), Corrensstrasse 3, 06466 Gatersleben, Germany

D.A. HOISINGTON
International Maize and Wheat Improvement Center, Applied Biotechnology Center and Bioinformatics, Apdo Postal 6–641, COL. Juárez, 06600 Mexico, D.F. Mexico

X.-Q. HUANG
Institut für Pflanzengenetik und Kulturpflanzenforschung (IPK), Corrensstr. 3, 06466 Gatersleben, Germany

R. IMAI
Plantech Research Institute, 1000 Kamosida-cho, Aoba-ku Yokohama, Japan 227-0033

J. IMAMURA
Plantech Research Institute, 1000 Kamosida-cho, Aoba-ku Yokohama, Japan 227-0033

M.J. IQBAL
Center of Excellence in Soybean Research, Teaching and Outreach, Department of Plant, Soil and General Systems, 176 Ag. Building, MC 4415, Southern Illinois University at Carbondale, Carbondale, Illinois 62901-4415, USA

M.E. JONES
Centre for Plant Conservation Genetics, Southern Cross University, P.O. Box 157, Lismore, NSW 2480 Australia

List of Contributors

P. Jorasch
Gesellschaft für Erwerb und Verwertung von Schutzrechten GVS mbH,
Kaufmannstr. 71–73, 53115 Bonn, Germany

C. Jung
Plant Breeding Institute, Christian-Albrechts-University of Kiel,
Olshausenstr. 40, 24098 Kiel, Germany

A. Kaga
National Institute of Agrobiological Sciences, Kannondai 2-1-2, Tsukuba,
Ibaraki, 305-8602, Japan

B. Keller
Institute of Plant Biology, University of Zürich, Zollikerstrasse 107,
8008 Zürich, Switzerland

Y. Kishima
Laboratory of Plant Breeding, Graduate School of Agriculture, Hokkaido
University, Kita-9, Nishi-9, Kita-ku, Sapporo 060-8589, Japan

K.V. Krutovsky
Institute of Forest Genetics, Pacific Southwest Research Station,
USDA Forest Service, Davis, California 95616, USA

P. Langridge
Australian Centre for Plant Functional Genomics, University of Adelaide,
Waite Campus, Glen Osmond, SA 5064, Australia

V. Lefebvre
Institut National de la Recherche Agronomique, Unité de Génétique et
Amélioration des Fruits et Légumes, Domaine Saint Maurice, BP94,
84143 Montfavet cedex, France

D.A. Lightfoot
Center of Excellence in Soybean Research, Teaching and Outreach, Department of Plant, Soil and General Systems, 176 Ag. Building, MC 4415,
Southern Illinois University at Carbondale, Carbondale, Illinois 62901–4415,
USA

D.J. Mackill
International Rice Research Institute (IRRI), DAPO Box 7777, Metro
Manila, Philippines

I.J. Maureira
Department of Agronomy, University of Wisconsin–Madison, 1575 Linden
Drive, Madison, Wisconsin 53706–1597, USA

K.L. McNally
International Rice Research Institute (IRRI), DAPO Box 7777, Metro Manila, Philippines

A.E. Melchinger
Institute of Plant Breeding, Seed Science and Population Genetics, University of Hohenheim, 70593 Stuttgart, Germany

V. Mohler
Chair of Agronomy and Plant Breeding, Department of Plant Sciences, Center of Life and Food Sciences Weihenstephan, Technical University Munich, 85350 Freising-Weihenstephan, Germany

D.B. Neale
Institute of Forest Genetics, Pacific Southwest Research Station, USDA Forest Service, Davis, California 95616, USA

K. Onishi
Laboratory of Plant Breeding, Graduate School of Agriculture, Hokkaido University, Kita-9, Nishi-9, Kita-ku, Sapporo 060-8589, Japan

T.C. Osborn
Department of Agronomy, University of Wisconsin–Madison, 1575 Linden Drive, Madison, Wisconsin 53706-1597, USA

M. Prasad
Institute of Plant Genetics and Crop Plant Research (IPK), Corrensstrasse 3, 06466 Gatersleben, Germany

M.S. Röder
Institut für Pflanzengenetik und Kulturpflanzenforschung (IPK), Corrensstr. 3, 06466 Gatersleben, Germany

T. Sakai
Plantech Research Institute, 1000 Kamosida-cho, Aoba-ku Yokohama, Japan 227-0033

Y. Sano
Laboratory of Plant Breeding, Graduate School of Agriculture, Hokkaido University, Kita-9, Nishi-9, Kita-ku, Sapporo 060-8589, Japan

A. Sarrafi
Department of Biotechnology and Plant Breeding, Pôle de Biotechnologie Végétale, ENSAT-BAP, 18 Chemin de Borde Rouge, BP 107, 31326 Castanet, France

List of Contributors

W.-R. SCHEIBLE
Max-Planck-Institute of Molecular Plant Physiology, 14424 Potsdam, Germany

G. SCHWARZ
EpiGene GmbH, Biotechnology in Plant Protection, 85354 Freising, Germany

M. SHEPHERD
Centre for Plant Conservation Genetics, Southern Cross University, PO Box 157, Lismore, NSW 2480, Australia

C. SINGRÜN
Department of Plant Sciences, Center of Life and Food Sciences Weihenstephan, Technical University Munich, 85350 Freising-Weihenstephan, Germany

W.K. SWIECICKI
Institute of Plant Genetics, Polish Academy of Sciences, 60–479 Strzeszynska 34, Poznan, Poland

M. TAKAHASHI
Forest Tree Breeding Center, Juo, Ibaraki 319-1301, Japan

T. TAKAHASHI
Graduate School of Science and Technology, Niigata University, Niigata 950-2181, Japan

N. TANI
Department of Forest Genetics, Forestry and Forest Products Research Institute, Tsukuba, Ibaraki 305-8687, Japan

G. TIMMERMAN-VAUGHAN
Crop and Food Research, PO Box 4704, Christchurch, New Zealand

O. TÖRJEK
Institute of Biochemistry and Biology – Genetics, University of Potsdam, 14415 Potsdam, Germany

N. TOMARU
Laboratory of Forest Ecology and Physiology, Graduate School of Bioagricultural Sciences, Nagoya University, Nagoya 464-8601, Japan

N. TOMOOKA
National Institute of Agrobiological Sciences, Kannondai 2-1-2, Tsukuba, Ibaraki 305-8602, Japan

Y. TSUMURA
Department of Forest Genetics, Forestry and Forest Products Research Institute, Tsukuba, Ibaraki 305-8687, Japan

R.K. VARSHNEY
Institute of Plant Genetics and Crop Plant Research (IPK), Corrensstrasse 3, 06466 Gatersleben, Germany

D.A. VAUGHAN
National Institute of Agrobiological Sciences, Kannondai 2-1-2, Tsukuba, Ibaraki 305-8602, Japan

B. ZIEGENHAGEN
Philipps-University of Marburg, Faculty of Biology, Conservation Biology, Karl-von-Frisch-Strasse, 35032 Marburg, Germany

Section I Basics

I.1 The Principle: Identification and Application of Molecular Markers

P. LANGRIDGE and K. CHALMERS[1]

1 Introduction

Plant breeding is based around the identification and utilisation of genetic variation. The breeder makes decisions at several key points in the process. First in deciding on the most appropriate parents to use for the initial cross or crosses and then in the selection strategy used in identifying the most desirable individuals amongst the progeny of the cross. The efficiency of the breeding and selection process can be assessed in many different ways including the ultimate success of the varieties released and the frequency with which new varieties are produced. A major cost and logistical issue in plant breeding are the actual number of lines that need to be carried through the evaluation and selection phases of a program. Large breeding programs for annual crops may carry hundreds of thousands of lines to produce a new variety only once every few years. Field trials can be expensive and evaluation of some traits, such as quality and yield stability can be expensive to assess. Molecular markers have proved to be a powerful tool in replacing bioassays and there are now many examples available to show the efficacy of such markers.

The use of molecular markers to track loci and genome regions in crop plants is now routinely applied in many breeding programs. The location of major loci is now known for many disease resistance genes, tolerances to abiotic stresses and quality traits. Improvements in marker screening techniques have also been important in facilitating the tracking of genes. For markers to be effective, they must be closely linked to the target locus and be able to detect polymorphisms in material likely to be used in a breeding program. The prime applications of markers in most breeding programs have been in backcross breeding where loci are tracked to eliminate specific genetic defects in elite germplasm, for the introgression of recessive traits and in the selection of lines with a genome make-up close to the recurrent parent. In progeny breeding, markers have proved valuable in building crucial parents and in enriching F_1s from complex crosses. Markers have also improved the strategies for gene deployment and enhanced the understanding of the genetic control of complex traits such as components of quality and broad adaptation.

---

[1] Australian Centre for Plant Functional Genomics, University of Adelaide. Waite Campus, Glen Osmond, SA 5064, Australia

2 Status

Recent developments that have occurred in molecular markers for many crop species have major implications for the future of the technology. There are three key components that are particularly significant. First, for many species, we now have markers closely linked to many traits of importance in the breeding programs. Indeed, for major crop species, we have markers for more loci than can be screened in a conventional breeding program. Second, we have tools that allow marker scanning of the whole genome. Of particular importance has been the development of microsatellite or SSR markers that now form the basis for analysis and allow highly multiplexed SSR screens. This trend will continue as newer, cheaper marker screening based on SNPs become available. The technological advances have improved our capacity for whole genome screens. Third, through association mapping projects we have, or are in the process of developing, whole genome fingerprints for many key lines and varieties of importance in breeding programs. These studies are developing large databases of historic germplasm that should, over the next few years, start to reveal the ways in which breeding programs have selected for and against specific regions of the genome. We can see these developments, particularly in crops such as maize and barley, where markers for most of the major disease resistance clusters, for key components of feed or processing quality and for many loci conditioning tolerance to abiotic stresses are available.

The new marker systems have several important implications for the future of marker-assisted selection (MAS) and breeding strategies in general. Existing strategies for MAS were initiated with a view of markers as providing a rapid and cheap alternative to bioassays and they have largely been used in this role. While highly successful, this strategy does not fully exploit the technology. The key limitation to an expansion of the scale and complexity of marker use is the size of the populations that would be required if one were to try and select for alleles at a large number of loci simultaneously. A further important feature of recent advances has been related to how we best take advantage of the genome information that has been generated for major crop species. We know, for example that chromosome 2H in barley and group 7 chromosomes of wheat, carry clusters of genes, often in repulsion that we would like to break up. Again conventional use of markers has not been very effective in utilising such genome regions. Conversely, we know that there are some chromosomes where there is little allelic variation between lines and it is a waste of effort to try and break these up in a breeding program.

The key challenge of new work is to investigate strategies for whole genome breeding: how we can use genome-wide information in the form of graphical genotypes and known locations of key loci and marker tags for both desirable and undesirable alleles, to design optimal breeding strategies that integrate as much of the available information as possible.

3 Molecular Markers

Molecular markers have been taken, in recent years, to refer to assays that allow the detection of specific sequence differences between two or more individuals. However, it should be recognized that isoenzyme and other protein-based marker systems also represent molecular markers and were in wide use long before DNA markers became popular. One of the earliest type of DNA-based molecular markers, restriction fragment length polymorphisms (RFLPs), were based around the detection of variation in restriction fragment length detected by Southern hybridisation. The types of sequence variation detected by this procedure could be caused by single base changes that led to the creation or removal of a restriction endonuclease recognition site or through insertions or deletions of sufficient size to lead to a detectable shift in fragment size. This technique has been largely superceded by microsatellite or simple-sequence repeat (SSR) markers and is now rarely used in screening material for breeding programs, but it remains an important research tool. SSR markers detect variation in the number of short repeat sequences, usually two or three base repeats. The number of such repeat units has been found to change at a high frequency and allows the detection of multiple alleles. The large expansion of DNA, particularly EST, sequence databases has now opened the opportunity for the identification of single nucleotide polymorphisms, SNPs. These occur at varying frequencies depending on the species and genome region being considered. In Arabidopsis SNP frequencies of 0.007–0.0104 have been measured (Kawabe et al. 1997; Purugganan and Suddith 1998) while in maize a range of 0.00047–0.0037 has been measured (Hilton and Gaut 1998; Wang et al. 1999). SNPs are widely seen as providing the key advantage of multiple detection systems many of which, such as mass spectroscopy, offer high throughput at low detection cost. Importantly, new array based screening methods, such as DArT (Jaccoud et al. 2001) appear to offer still cheaper assays due to their very high multiplexing capability. Interestingly, molecular markers may be coming full circle with protein markers again being proposed as viable genetic markers for MAS. Mass spectrometric methods for mass fingerprinting of proteins and for the analysis of low molecular weight proteins, again opens the option for high throughput protein screening. In these cases, single amino acid changes in protein sequence can often be detected and this provides a means for revealing variation in the corresponding DNA coding sequence.

In each method, DNA sequence variation is being detected. However, each method analyses different aspects of DNA sequence variation and different regions of the genome. For example, RFLPs were detected using cDNA clones, namely coding sequence, but frequently detected variation that lay in regions flanking the genes. SSR markers have generally been from non-coding regions although the recent move to three-base repeats and the use of ESTs as the source of SSR markers is changing this. Other markers such as RAPD and

AFLP markers appear to frequently target repetitive regions of the genome. The stability of the sequence difference may also be an issue in some cases. SSRs are seen as being too unstable for some applications since the mutation rate may in some cases be high.

The decision about the most appropriate marker system to use will vary greatly depending on the species, the objective of the marker work and resources available.

4 Identifying Marker/Trait Associations

The most widely used methods for identifying marker/trait associations are based around the construction, phenotyping and genotyping, with molecular markers, of special populations. The steps in identifying marker/trait associations and developing the markers through to application are summarised in Table 1. The populations are generally constructed from two varieties that show a major difference in the traits targeted for mapping. The genetic structure of the segregating populations can be immortalised by producing double haploids or recombinant inbred lines. The populations produced then become a major resource for a wide range of studies. Many such populations have become international resources used by researchers around the world. The ITMI population used for wheat research is an example of this. The population made from a cross between the wheat variety Opata 85 × W7984 a synthetic wheat, has become the international reference for wheat genetic research (Langridge et al. 2001). New markers, such as SSR and SNP, are being placed on the population continually and the population has been screened for a wide range of disease, abiotic stress tolerance, physiological and quality traits. The beauty of these populations is that they continue to grow in value as they are more and more widely used. Such reference populations are now available for several crop and model plant species.

However, there are also problems with the use of such structured populations. Many of the reference populations were constructed to facilitate marker screening and were based on highly diverse parents, this was the case for the ITMI reference population

There are three important issues that will frequently impact on the most appropriate procedure to be used in finding marker/trait associations:

- There is a major cost in phenotyping. This clearly varies depending on the trait being analysed, but usually the more complex and expensive the phenotypic screening is, the more valuable will be markers for the trait. Costs of phenotyping can be particularly important for traits that require extensive field trials, such as yield or tolerance to some stresses, or require large amounts of material for analysis, such as malting quality in barley or animal feeding trials. Due to costs, the number of replicates and sites is often limited, reducing the sensitivity of some of the analyses.

Table 1. Steps in identifying marker/trait association

1. Defining the target
Decision about marker development
- Is the trait of importance to breeding program or to biological research?
- Is a molecular marker needed?
  - What is the cost of the bioassay relative to marker assay?
  - Is the trait dominant versus recessive? – recessive traits may be hard to identify in a bioassay and will be a prime target for marker development
  - Perhaps there is no alternative to marker use:
    - Quarantine trait – e.g., resistance to a disease not present in the country
    - Pyramiding resistances – accumulating multiple genes for resistance to protect against resistance breakdown
    - Map-based cloning of genes – high resolution map is needed to minimize region that needs to assessed
    - Gene deployment – where desirable alleles are available to several loci, but only one is really needed. How does one decide on the best one to use?

2. Identify germplasm for marker development
Available germplasm, with and without the trait

3. Population structures
Deciding on the best material to use for identifying the marker trait associations
- Knowledge of genetics
  - Is the trait simply inherited or multigenic?
  - What is the heritability?
  - If this information is not available, a trial experiment may be needed

  A simple cross can be constructed to measure segregation ratio and heritability
  Complex traits and traits of low heritability are often prime targets for marker development as they are hard to assay otherwise
- Decide on best population structure
  The structure of the population will be related to the trait and purpose
  Populations structure will differ between:
    - in-bred versus out-breeding species
    - long generation versus short generation plants
    - perennial versus annual plants
  - Doubled haploids – one meiotic event per line
  - F_2s – two-meiotic events per plant
  - Recombinant inbreds or single seed descent
  - Complex crosses between highly heterozygous parents
  Population size
  - For single gene 50 F_2s may be adequate
  - Map-based cloning over 1000 required
  Is an existing population already available?
  - Screen parents of existing crosses and mapped populations

4. Phenotypic evaluation
- Is phenotypic evaluation possible for single plants?
- For some traits a large number of seeds or plants may be required or field trials at multiple sites, e.g., quality and yield and traits of low heritability
- For association mapping phenotypic information can be collected from existing programs or lines pooled that have a common phenotypes, e.g., lines adapted to a common environment or of common quality ranking

Table 1. (Continue)

5. Genotyping
- Identify markers that detect polymorphisms between parents
- Screen population

Marker density will depend on objective
- Full maps
  Screen with sufficient markers to give good genome coverage, usually around 1 marker every 10 cM
- Bulked segregant analysis
  Bulk two extremes of phenotype and screen with markers
- Map-based cloning
  - high resolution in small region
  - Screen population with markers flanking target region
  - Identify recombination events between the flanking markers
  - Use the recombinant lines for the high-density marker screening
    Note: only the recombinant lines need to be phenotyped
- Association mapping
  Genotype multiple lines, for example in pedigree

6. Identifying marker trait associations
- Full maps
  Construct linkage map based around molecular markers
  Locate trait loci by regression analysis, interval mapping or related technique
- Bulked segregant analysis
  Screen pools of lines with and without trait with molecular markers
- Association mapping
  Measure rate of linkage disequilibrium between traits and markers
  Deviations from the expected frequency of alleles

7. Developing markers for application – Marker validation
- Test marker trait association in alternative populations and estimate reliability of marker in predicting phenotype
- Identify polymorphisms between lines used in breeding program
- Develop a palette of suitable markers with associated polymorphism data
  Usually around 10 markers within 10 cM of trait
  Provide protocols and polymorphism data to breeding programs

- The lines (varieties) used to construct the populations are often out-of-date by the time the marker/trait information is available. For example, many mapping programs are using populations constructed over a decade ago. This reduces the value of the information gathered and slows its implementation into active breeding programs.
- The structure of the populations limits the types of traits that can be mapped and many of the subtleties of adaptation can only be analysed with special populations.
- Generally, mapping is restricted to known traits for which a well-defined bioassay is available.

4.1 Full Linkage Maps

Complete linkage maps generated from screening the progeny of a cross have provided the basis for most early marker development work. However, this is difficult and labour-intensive, particularly in species with large numbers of linkage groups such wheat, where linkage maps must be constructed for 21 chromosomes or where chromosome numbers are large or variable, such as kiwi fruit (*Actinidia* sp.) or sugarcane. Usually, 10 to 20 markers are required for each chromosome to give reasonable genome coverage. For many species where the germplasm base is small, the level of polymorphism may be low so a large number of markers are required to detect sufficient polymorphisms for mapping. The key feature of such maps is that they provide considerable information on the genome structure of an organism and provide major resource for researchers even though their application to practical plant breeding is now becoming increasingly limited. As mentioned above, once established, a well-mapped population can be used for a wide range of genetic studies provided the individual lines can be maintained.

4.2 Bulked Segregant Analysis

Bulked segregant analysis (BSA) was described by Michelmore et al. (1991). It involves pooling individuals from the two phenotypic extremes of a segregating F_2, doubled haploid or similar population. DNA isolated from the two pools is then screened with DNA markers, usually SSR or AFLP, and polymorphic bands identified. Clear polymorphisms seen between the two pools will be derived from regions of the genome that are common between the individuals that made up the pools. The remainder of the genome will be randomly contributed by the parents and should show no polymorphisms between the pools.

This technique offers the important advantage of identifying markers associated with the trait without the need for full map construction. However, it requires markers that can be easily screened in mixed DNA preparation. This essentially means PCR-based marker assays. AFLP markers have proved particularly suitable for these assays. The method also allows the use of a wide variety of population structures and can often be applied to material produced within a breeding program.

The key disadvantage is that one is not provided with a genetic distance between the marker and the trait. Indeed, it is always necessary to confirm the marker/trait association by screening individuals from the population to confirm that the marker is a reliable predictor of the trait. This can usually be done by taking the individual plants used to construct the bulks and determining just how many of the lines actually show the expected marker pattern.

BSA is now widely used and in many marker development programs has become the major method for marker identification.

4.3 Association Mapping

Association mapping is becoming an increasingly important tool for marker analysis and application. Particular emphasis will be placed on this technique here since we believe it will replace alternative marker development procedures, will greatly facilitate marker delivery to breeding programs and will provide valuable insights into the genetics and evolution of crop plants.

Molecular markers offer an easily quantifiable measure of genetic variation within crop species. However, many species, such as wheat, display a low level of polymorphism (Chao et al. 1989; Lui et al. 1990) hampering the identification of markers linked to agronomically important traits and complicating the differentiation of varieties and the analysis of genetic variability.

While the low level of polymorphism may be problematic, the level of genome conservation observed between varieties offers an opportunity to identify markers associated with traits of interest (Paull et al. 1998). The rationale of the approach is based on 'linkage drag' (Hanson 1959; Stam and Zeven 1981), a feature of chromosome behaviour whereby flanking DNA surrounding the target gene diminishes at a much slower rate than unlinked regions. Since varieties of a particular crop species are often closely related, differences between related accessions may reflect differences in important agronomic traits. A comparison of the molecular marker profiles of accessions demonstrating a particular trait with those lacking the trait, facilitates the identification of linked markers. Loci linked to the trait of interest will show the same marker phenotype within each group while unlinked loci will show a random distribution of marker alleles. This approach is analogous to bulk segregant analysis (Michelmore et al. 1991), but is dependent on traits remaining as part of a larger linkage block during crossing and selection.

A further problem with the identification of marker/trait associations using defined populations such as F_2 populations, doubled haploids derived from F_1s or recombinant inbreds, is that they are usually built from only two parents and often do not reflect germplasm in an active breeding program. By the time the populations are ready for mapping they often involve germplasm that is out of date and no longer optimal for a pragmatic breeding program.

These limitations in existing mapping strategies can be addressed through association or linkage disequilibrium (LD) mapping. LD mapping is based on seeking associations between phenotype and allele frequencies. It is the basis for gene mapping in species where large mapping populations cannot be readily produced such as mapping in farm animals and humans. There are three advantages of this approach in mapping in crop species. Firstly, it provides a new perspective for trait mapping. This is because it uses different population structures (based largely around pedigree) and it uses a different set of phenotypic data. Consequently, we can expect to see new marker/trait associations, but more importantly, this technique will help identify targets for more detailed analysis. For example, we can expect to find genetic associations with lines or varieties that have performed particularly well at certain sites. In many cases, we will not be able to associate this with a specific aspect of the site environment, but we will have an indication of where to look for an environmental factor.

Secondly, LD mapping also provides detailed fingerprinting information on a large number of lines and varieties and this information will be valuable in several of the breeding strategies outlined below.

Thirdly, the LD method uses real breeding populations, the material is diverse and relevant and the most important genes (for example, for adaptation) should be segregating in such populations. The breeder is also integrally involved in the process and this may lead to improved rate and efficiency of validation and adoption. Many breeding programs are reluctant to grow and assess a huge number of lines with little or no potential for a direct commercial outcome. The advantage of LD mapping to the breeder is that mapping and commercial variety development can be conducted simultaneously.

When a novel mutation occurs at a locus determining the expression of a QTL, all other alleles of that locus are considered to be in complete linkage disequilibrium with the new mutant. However, as time goes by, the level of observed linkage disequilibrium will deteriorate as recombination between the mutant allele and other loci occurs. The level of LD observed will be in direct relationship to the distance from the mutant allele and also a function of the number of generations that have passed since the original mutation event. Several factors will also mediate this effect. In the case of outcrossing species there will be a relatively rapid breakdown of LD and in outcrossing species it is expected that LD will only be detected over a relatively short distance that may be measured in the region of a few tens of kb, although the actual extent of LD in maize still appears controversial (Remington et al. 2001; Ching et al. 2002). In these species it may be necessary to use direct DNA sequencing to identify and track linkage disequilibrium. However, for in-breeding species, LD breaks down relatively more slowly and is clearly detectable at the centiMorgan level. This offers the opportunity to use more conventional marker-based assay detection systems such as SSRs as an appropriate detection system. The existence of detectable levels of LD in in-breeding populations then offers the possibility of carrying out association mapping. This may be done by systematically screening molecular marker loci at defined intervals across the genome. In human and animal systems, this has readily been achieved using the abundant SNP resources available. However, in most crop species, such a resource is unlikely to become available in the near future. The question then is will the available marker systems, in particular SSRs provide sufficient genome coverage? This question is currently being addressed for several species and the results should start appearing in publications over the next few years.

5 Application of Molecular Markers

The following section aims to provide an overview of the current and predicted applications of molecular markers within plant breeding programs. The application of molecular markers to pedigree/progeny and backcross breeding is summarised in Table 2 for each stage in the breeding process. Although the focus here is on application of markers to practical breeding programs, it is important to remember that molecular markers have also become a critical tool for a wide range of genetic studies and are widely used in many aspects of genetic research from map-based cloning of genes through to the study of genome structure, organisation and behaviour.

5.1 Trait-Based Selection

One of the earliest demonstrations of the power of molecular markers was provided by Beckman and Soller (1986) for the indirect selection of genes in a breeding program. Molecular markers offer several key advantages over many bioassay systems:

- DNA can be extracted from tissue sampled from growing plants at very early stages of development. This allows sufficient time to use linked markers to identify heterozygous, backcross or topcross F_1 individuals prior to anthesis and further crossing. In contrast, optimum expression of many important phenotypes in a bio-assay, such as disease resistances, often occurs at development stages close to or after anthesis when crossing should take place. Plants in the bio-assay system are also frequently grown under suboptimum conditions for crossing and seed set.
- Markers can be used accurately on a single plant. For most bio-assays, many individual plants must usually be screened although this will vary with the bioassay. It is rare that a single plant assay will give a reliable phenotypic assessment. In contrast, the proportion of single plants incorrectly scored from a molecular marker assay is related to the closeness of linkage between the marker and the trait. For markers within 10 cM, the error is therefore less than 10%. The confidence level of single plant selection would increase even further with the use of flanking markers (Beckmann and Soler 1986).

An example of this application can be seen in the South Australian Barley Improvement Program. In a breeding strategy commencing in the spring of 1994, a single dominant resistance gene to cereal cyst nematode (*Ha2*) was transferred from a resistant to susceptible variety through three cycles of marker-assisted backcrossing using a single molecular marker. At no stage were more than four BC_xF_1 plants backcrossed. One hundred and twenty doubled haploid lines were produced from marker selected BC_3F_1 and 66% of the

Table 2. Role of marker in pedigree/progeny and backcross breeding for inbred annual crops

Phase	Pedigree/progeny	Backcrossing		Marker roles
Parental choice	A×B			Understand the genetic relationships between current and future members of the germplasm pool. In hybrids, estimate likely heterosis through diversity analysis
				Choose parents with small genetic distance to reduce number of backcrosses.
				Develop strategy for introgressing and selection of "quarantine traits"
				Transfer transgenes into elite lines
Crossing		A×B		Identify progeny
				• Carrier of trait
		BC1	F$_1$×A	• Low percentage donor genome
		BC2	F$_1$×A	• Small introgression segment
		BCn	F$_1$×A	Save one generation per backcross for recessive traits
				Enrichment of F$_1$s – markers used to characterise germplasm
				Enrichment of F$_1$s – markers used to characterise germplasm
				Enables more complex crossing strategies
Segregating generations	A×B			High throughput markers required to select desired alleles
	(A×B)×C			Markers used to identify desired progeny in parent building schemes
	(A×B)×(C×D)			Use markers to choose lines close to recurrent parent to "fast-track" line to release
	F$_2$ to F$_n$	Fixation		by reducing evaluation requirements as before
				Limited role until QTL for adaptation/quality validated
Evaluation of fixed lines	Year 1 Limited sites, limited replications			
	Year 2 More sites, more replications	Year 1 limited sites and seasons		
	Year 3 Regional trials	Year 2 regional trials		
	Year 4 National list year 1	Year 3 national list		
	Year 5 National list year 2			
Pure seed				Whole genome marker analysis can identify individuals close to recurrent parent, thereby saving expensive yield and quality testing
				Random genome survey plus key economic traits
				Markers used to compare reselections for bulking
Commercialisation				Use markers to provide evidence in "essentially derived" discussions
				Markers used to identify or compare new varieties against other varieties of "common knowledge"

regenerants were classed as resistant by phenotypic assay. A line selected from among these was commercially released 7 years from the time of the first cross. This compares to 14 years for a conventional breeding strategy for malting barley.

5.1.1 Enrichment of Complex F_1s

Plant breeders often use three- and four-way crosses since they allow an increase in the range of traits that can be simultaneously incorporated into elite progeny. However, the frequency of elite progeny from this type of cross is usually very low and this has reduced the application of this strategy. Breeders have tended to take the longer route of making simple crosses, fixing desirable alleles and then intercrossing selected, fixed lines. MAS offers a powerful alternative to increase the desirable allele frequency for each locus contributed from a quarter parent from 25% of progeny to 50% by screening the top cross F_1 or four-way cross F_1. This application has become the most common application of molecular markers in both wheat and barley breeding in southern Australia and its application continues to grow. Importantly, it is expected to increase in power through the development of new whole genome screening and selection strategies.

5.1.2 Early Generation Selection

The selection theory required to implement MAS in early generations is similar to other forms of selection although MAS is closer to 'simultaneous' rather than 'tandem' (or stepwise) selection, which is often a feature of early generation, phenotypic selection. In early generations breeders usually visually select traits of high heritability since complex traits such as yield cannot be effectively selected in rows or small plots. However, MAS is more effective than phenotypic selection when population sizes are large and heritability is low (Lande and Thompson 1990; Whittaker et al. 1997). Therefore, breeders and geneticists need to design and implement MAS strategies that allow selection of complex traits in early generations. The combination of MAS with techniques such as single seed descent and doubled haploid generation offers the option to address some of these difficulties. The testing of these strategies is now well advanced in many breeding programs, but assessment of the full impact of this strategy will, as with gene enrichment, only be possible in later generations. However, the efficient handling and management of single plants, a pre-requisite for single seed descent (and doubled haploid production), lend them to MAS.

5.2 Whole Genome Selection

5.2.1 Choice of Donor Parent in Backcrossing

Marker-based genetic diversity studies have generated large data sets that can be used to select from a number of possible donor parents for a desired trait. The objective is to identify the donor parent that is the minimum genetic distance from the recurrent parent. This should reduce the number of backcrosses required to recover the recurrent parent phenotype. Genetic distance estimates may also assist in assessing the suitability of prospective donor parents of unknown or diverse pedigree or where limited phenotypic information is available. Information gained from the routine DNA fingerprinting of potential parents could contribute to the broader information base used for selection of suitable donor parents including the selection of more diverse parents for speculative crosses.

5.2.2 Recovery of Recurrent Parent Genotype in Backcrossing

The key objective of a backcrossing strategy is to reduce the proportion of the donor parent genome by about 50% at each generation of backcrossing. Until recently, most backcrossing has focused on this principle and ignored the variation in the proportion of the donor parent genome that exists around the expected mean. Molecular markers allow selection for the desired donor allele, and also for recombinant individuals that have a genome composition closer to the recurrent parent than would be predicted from theoretical expectations (Tanksley and Rick 1980). MAS against donor parent or for recurrent parent genome, provides a means of reducing both the time and number of generations required to adequately recover the recurrent parent genotype.

This strategy has been applied in the South Australian wheat and barley breeding programs. In general, BC_1 derived lines have been identified that carried a proportion of donor parent genome not significantly different from the mean of the BC_3 generation. Selecting these individuals saves two cycles of backcrossing. Similar results were found in simulation studies conducted by Hospital et al. (1992), Visscher et al. (1996), and Frisch et al. (1999) where they also found that at least two generations of backcrossing could be saved.

A major constraint to the adoption of recurrent parent background selection in a practical backcrossing program is the large number of polymorphic marker alleles required to cover the entire genome. New marker developments, such as DArT and SNP-based markers will help address this limitation. In their simulation study, Frisch et al. (1999) showed that the use of marker loci of known location was more efficient than random marker alleles. This means that markers that have not been localised to the genome, such as AFLP, are of less value in this strategy than SSR markers or DNA micro-array technologies.

Stam and Zevens (1981) estimated that the typical segment length of donor parent DNA retained after three backcrosses was surprisingly high at 51 cM in a 100-cM chromosome. There are now many examples of deleterious linkage drag in plant breeding to support this theoretically derived conclusion particularly for alien segment (Paull et al. 1994). Therefore, a more focused selection strategy may be more useful. The methods would be as follows:

- use flanking markers at 10–20 cM around the estimated position of the gene, to maintain the donor allele frequency during later generations of backcrossing
- use more distant (30–40 cM) flanking markers to select for small donor segment around the desired gene
- use carefully selected markers spaced 30–40 cM over the remainder of the genome to select against donor parent alleles.

5.2.3 Linkage Block Analysis and Selection

It has been clear to most breeders that certain chromosomal regions carry key clusters of genes which have been highly conserved through selection. The high degree of success of conservative breeding strategies around the world in the past and the apparent poor combining ability found between many germplasm pools suggests that major linkage blocks may also be important to breeding programs for most major crop species. For example, a study in barley indicated that conserved regions of the genome derived from a North African landrace introduction in the early 1970s appears to have been crucial for the improved adaptation of successful South Australian varieties released from that time. Regions on chromosome 2H appeared to be particularly strongly associated with improved adaptation. This region also showed significant associations with grain yield, grain yield stability, grain size and flowering time (all across a number of environments; Atmodjo et al., pers. comm.). The syntenous region in wheat also appears to carry a block of genes associated with adaptation including the photoperiod sensitivity genes, *Ppd1*, *Ppd2*, and *Ppd3* on chromosomes 2D, 2B and 2A respectively (Borner et al. 1998) and QTL associated with flag leaf area, flag leaf length, leaf width, plant height at stem elongation and anthesis, number of heads per square meter, and grain weight (Coleman et al. 2001). This type of information could lead to the development of specific targeted breeding strategies, where either allelic variation in these regions is actively sought or linkage blocks from superior adapted genotypes are actively conserved through marker assisted selection.

5.2.4 Key Recombination Events

The presence of major linkage blocks containing groups of genes that have been important for adaptation has several implications for breeding programs. For most species recombination is not evenly spread across the genome nor are target traits evenly distributed. This logically leads to the suggestion that specific recombinational events may be important in making major advances in breeding. This is particularly important for species where wild relatives are used as sources for new alleles, such as disease resistances. For example, the wild relatives of wheat have proved valuable sources of useful traits yet these alien segments have shown very low recombination with wheat even when derived from a close relative. Molecular markers are valuable in identifying the rare lines where recombination has occurred and in characterising the recombination events (Langridge et al. 2001). A major benefit arising from the various mapping initiatives has been the increased knowledge of the structure and behaviour of the genomes of crop species and the physical and genetic control of important traits. In very well studied species, such as barley, a comprehensive understanding of the genome has facilitated the design and recent adoption of complex crossing and key recombination event selection strategies using markers. As our understanding of the genetic control of important traits in crop species improves, so will the potential for us to apply specific targeted marker-assisted breeding strategies to compliment the broader, traditional approaches.

6 Directions

6.1 Quantitative Trait Loci

The agronomic performance of crop varieties is mainly influenced by complex quantitative traits, for example, components of yield and quality. Since the development of molecular markers, it has become feasible to identify and genetically localise the contributing genetic factors as quantitative trait loci (QTLs) and to utilise these QTLs for crop improvement. This has led to an increasing number of QTL studies, involving most agronomically important crop species.

Despite successes in mapping QTLs, the relevance of this information for breeding new varieties is limited. In most cases, the QTL analysis has been carried out in crosses utilising parents drawn from elite germplasm sources. Hence, in most cases the studies have been able to identify a limited number of alleles that are already present in the mainstream breeding programs and offer little opportunity for variety improvement.

6.2 Diversity

The risk most frequently raised for MAS is the temptation to use only parents for which either markers and/or polymorphic markers exist, thus further narrowing genetic diversity within breeding programs. In particular, this may concentrate the use of a few, well-characterised disease resistance genes to the exclusion of less well documented sources. This risk can be minimised by breeder discretion allocating a proportion of the program to 'new' or 'uncharacterised' sources as has already happened in some breeding programs. It might be more useful to think of this problem as a challenge of how marker technology be used to expand the useful gene pool.

The primary gene pools of many crop plants are so depleted in genetic variability that breeders have relied upon wild relatives for sources of disease resistance and other traits. Although crop germplasm collections contain many thousands of potentially useful wild accessions, their utilisation is sometimes hindered by hybridisation barriers preventing interspecific crosses and/or by undesirable characteristics inherent in exotic germplasm. Breeders have used exotic germplasm almost exclusively as a source of major genes for disease and insect resistances, and have mostly relied on repeated intercrossing of adapted elite genotypes for the improvement of quantitative traits, like yield and quality.

Tanksley and Nelson (1996) presented a novel alternative to the limitations inherent in conventional approaches of utilising exotic germplasm. By combining the introgression of novel QTL alleles from exotic sources of germplasm with QTL analysis and discovery, they have been able to demonstrate significant variation in the expression of a number of agronomically important traits. This procedure has been termed advanced-backcross-QTL analysis (AB-QTL). This utilises exotic germplasm as the genetic donor for the improvement of quantitative agronomic traits and combines marker and phenotype analysis in advanced backcross generations such as BC_2, BC_3 or more recently, crosses such as BC_2F_2.

To date, several reports on the application of the AB-QTL strategy are available for tomato, rice, wheat and barley. In each case, favourable exotic QTL alleles for important agronomic traits have been identified. It is proposed that the introgression of new exotic QTL alleles, will contribute to an increased level of genetic diversity in a range of cultivated species. An example of this has been demonstrated in fruit yield in cultivated tomato. Through the introgression of wild-species alleles from *Lycopersicon pimpinellifolium* and *L. peruvianum*, fruit yield was increased by up to 17 and 34%, respectively (Tanksley and Nelson 1996; Fulton et al. 1997). Similar results were seen from AB-QTL studies in rice. In this case, two wild-species QTL alleles have been associated with an increase of yield by 17 and 18% on rice chromosomes 1 and 11, respectively (Xiao et al. 1996, 1998). More recently, *Hordeum spontaneum* has been used as a source of novel alleles in barley and in wheat, synthetic hexaploids created from *Triticum dicoccides* and *Aegilops taushii*. In both examples, alleles from the exotic germplasm were associated with a positive effect on agronomic traits.

An alternative approach to that of AB-QTL for utilising genetic variation within exotic germplasm is to attempt to identify alternative alleles by undertaking germplasm screens. The use of wild barley (*Hordeum vulgare* ssp. – *spontaneum*) as a source of novel malting quality alleles has been reported by Eglinton et al. (1998). One hundred and fifty four accessions of *H. spontaneum* were screened for β-amylase polymorphism and three novel alleles (*Bmy*1-Sd3, -Sd4, and -Sd5) were identified in addition to those detected in cultivated barley. The corresponding Sd4 and Sd5 enzymes exhibit intermediate levels of thermostability, similar to the Sd1 β-amylase. The Sd3 β-amylase from wild barley exhibits thermostability significantly greater than the other five allelic forms of β-amylase and as such provides for improved fermentability during processing.

6.3 Whole Genome Breeding

As a result of work carried out on many crop species over the past decade, markers are becoming available for a large number of important traits. We have good strategies in place to use these markers to accelerate a number of breeding techniques, in particular backcrossing. However, we do not have strategies to manage the introgression of more than about five loci simultaneously, whether by direct crossing or by merging crossing streams. A major problem is that our existing breeding and marker implementation strategies are based around the selection and monitoring of individual loci. However, we now have high-throughput marker screening techniques that allow us to monitor the entire genome. Indeed, since the major cost of marker screening is the DNA isolation, the more marker information that is gathered for each line, the greater the cost/benefit ratio. Can we move from a trait-based selection system to a recombination-based strategy where we manage the entire genome and select individuals with a particular genome configuration (based on recombination events)? This approach would offer major gains in the efficiency of breeding programs because it would allow a dramatic reduction in the sizes of populations needed to achieve a specific outcome. However, to do this effectively we need to understand and analyse the behaviour of the entire genome in a breeding program and the major genomic events that have led to adaptation to our environment.

7 Conclusions

Molecular markers are now well established as tools in plant breeding and genetics. They have also provided a major new impetus to plant breeding programs offering considerable improvements in the efficiency and sophistication of breeding. Their use as research tools is also well developed and they have played a key role in improving our understanding of genome organisation, structure and behaviour for many of our major crops. However, the application of molecular markers in practical plant breeding has been patchy. Marker resources and capabilities for marker implementation are largely unavailable for minor crops. Even for some of the major crops, such as wheat and rice, markers are not widely used in public breeding programs with a few notable exceptions. Given their huge potential, this slow acceptance and implementation is disappointing and is probably related to a lack of flexibility by many public breeding programs to make the structural and strategic changes needed for effective marker implementation. It may also be partly due to the lack of active participation of breeders in the marker development programs in some countries. As the results of marker application become more apparent and move from theory to released varieties, this attitude may change.

The key developmental challenges for molecular markers now lies in developing new breeding strategies where the objectives will be increasing the germplasm base and increasing the number of traits that can be effectively selected simultaneously. The new marker technologies that offer greatly reduced costs in marker screening and high multiplexing capabilities will be central to these developments. Essentially we will move to whole genome-based selection strategies where specific recombinational events are sought and changes will be assessed on a genome-wide scale. In this way we can look to better manage chromosome regions that may come from wild relatives or land races, track several traits at once and keep the population sizes as small as possible.

References

Beckmann JS, Soller M (1986) Restriction fragment length polymorphisms in plant genetic improvement of agricultural species. Euphytica 35:111–124

Borner A, Korzun V, Worland AJ (1998) Comparative genetic mapping of loci affecting plant height and development in cereals. Euphytica 100:245–248

Chao S, Sharp PJ, Worland AJ, Warham EJ, Koebner RMD, Gale MD (1989) RFLP-based genetic linkage maps of wheat homologous group 7 chromosomes. Theor Appl Genet 78:495–504

Ching A, Caldwell KS, Jung M, Dolan M, Smith OS, Tingey SV, Morgante M, Rafalski AJ (2002) SNP frequency, haplotype structure and linkage disequilibrium in elite maize inbred lines. BMC Genet 7:3–19

Coleman RK, Gill GS, Rebetzke GJ (2001) Identification of quantitative trait loci for traits conferring weed competitiveness in wheat (*Triticum aestivum*). Aust J Agric Res 52:1235–1246

Eglinton JK, Langridge P, Evans DE (1998) Thermostability variation in alleles of barley *beta-amylase*. J Cereal Sci 28:301–309

Frisch M, Bohm M, Melchinger AE (1999) Comparison of selection strategies for marker assisted backcrossing of a gene. Crop Sci 39:1295–1301

Fulton TM, Nelson JC, Tanksley SD (1997) Introgression and DNA marker analysis of *Lycopersicum peruvianum*, a wild relative of the cultivated tomato into *Lycopersicum esculentum*, followed through three successive backcross generations. Theor Appl Genet 95:895–902

Hanson WD (1959) Early generation analysis of lengths of heterozygous chromosome segments around a locus held heterozygous with backcrossing and selfing. Genetics 44:833–837

Hilton H, Gaut BS (1998) Speciation and domestication in maize and its wild relatives: evidence from the globulin-1 gene. Genetics 150:863–872

Hospital F, Chevalet C, Mulsant P (1992) Using markers in gene introgression breeding programs. Genetics 132:1199–1210

Jaccoud D, Peng K, Feinstein D, Kilian A (2001) Diversity Arrays: a solid state technology for sequence information independent genotyping. Nucleic Acids Res 29(4):E25

Kawabe A, Innan H, Terauchi R, Miyashita NT (1997) Nucleotide polymorphism in the acid chitinase locus (*ChiA*) region of the wild plant *Arabidopsis thaliana*. Mol Biol Evol 14:1303–1315

Lande R, Thompson R (1990) Efficiency of marker-assisted selection in the improvement of quantitative traits. Genetics 124:743–756

Langridge P, Lagudah E, Holton T, Appels R, Sharp P, Chalmers K. (2001) Trends in genetic and genome analyses in wheat: a review. Aust J Agric Res 52:1043–1077

Lui YG, Mori N, Tsunewaki K (1990) Restriction fragment length polymorphism (RFLP) analysis in wheat. I. Genomic DNA library construction and RFLP analysis in common wheat. Jpn J Genet 65:367–380

Michelmore RW, Paran I, Kesseli RV (1991) Identification of markers linked to disease-resistance genes by bulked segregant analysis: a rapid method to detect markers in specific genomic regions by using segregating populations. Proc Natl Acad Sci USA 88:9828–9832

Paull JG, Pallotta MA, Langridge P, The TT (1994) RFLP markers associated with *Sr22* and recombination between chromosome 7A of bread wheat and the diploid species *Triticum boeoticum*. Theor Appl Genet 89:1039–1045

Paull JG, Chalmers KJ, Karakousis A, Kretschmer JM, Manning S, Langridge P (1998) Genetic diversity in Australian wheat varieties and breeding material based on RFLP data. Theor Appl Genet 96:435–446

Purugganan MD, Suddith JI (1998) Molecular population genetics of the *Arabidopsis* CAULIFLOWER regulatory gene: nonneutral evolution and naturally occurring variation in floral function. Proc Natl Acad Sci USA 95:8130–8134

Remington DL, Thornsberry JM, Matsuoka Y, Wilson LM, Whitt SR, Doebley J, Kresvick S, Goodman MM, Buckler ES (2001) Structure of linkage disequilibrium and phenotypic associations in the maize genome. Proc Natl Acad Sci USA 98:11479–11484

Stam P, Zeven AC (1981). The theoretical proportion of the donor genome in near-isogenic lines of self-fertilizers bred by backcrossing. Euphytica 30:227–238

Tanksley SD, Nelson JC (1996) Advanced backcross QTL analysis – a method for the simultaneous discovery and transfer of valuable QTLs from unadapted germplasm into elite breeding lines. Theor Appl Genet 92:191–203

Tanksley SD, Rick CM (1980) Isozymic gene linkage map of tomato: applications in genetics and breeding. Theor Appl Genet 57:161–170

Visscher PM, Haley CS, Thompson R (1996) Marker-assisted introgression in backcross breeding programs. Genetics 144:1923–1932

Wang RL, Stec A, Hey J, Lukens L, Doebley J (1999) The limits of selection during maize domestication. Nature 398:236–239

Whittaker JC, Haley CS, Thompson R (1997) Optimal weighting of information in marker assisted selection. Gen Res Cam 69:137–144

Xiao J, Li J, Grandillo S, Ahn S, McCouch SR, Tanksley SD, Yuan L (1996) Genes from wild rice improve yield. Nature 384:223–224

Xiao J, Li J, Grandillo S, Ahn S, Yuan L, Tanksley SD, McCouch SR (1998) Identification of trait-improving QTL alleles from a wild rice relative, *Oryza rufipogon*. Genetics 150:899–909

Young DA (1999) A cautiously optimistic vision for marker assisted breeding. Mol Breeding 5:505–510

I.2 Genotyping Tools in Plant Breeding: From Restriction Fragment Length Polymorphisms to Single Nucleotide Polymorphisms

V. Mohler[1] and G. Schwarz[2]

1 Introduction

A series of molecular marker types are available for plant genotyping, but no single technique is generally applicable to the wide range of questions in plant genome analysis; each available technique exhibits both assets and drawbacks, thus, a decision is needed which marker system should be used for which research aim. The marker types differ in information content, number of scorable polymorphism per reaction, and degree of automation. In addition, the choice of method often depends on the genetic resolution needed as well as on technological and financial constraints. Here, we review the basic principles of the most important techniques and their suitability for molecular plant breeding.

2 Restriction Fragment Length Polymorphisms

Both the basis and techniques for restriction fragment length polymorphisms (RFLPs; Botstein et al. 1980) in plant genome mapping have been extensively reviewed (Tanksley et al. 1988) and as such will only be briefly outlined here. In complex plant genomes, RFLPs are detected upon hybridization of (mainly) single-copy DNA probes to restriction enzyme-hydrolyzed, agarose gel electrophoresis-separated and nylon membrane-bound genomic DNA. For the detection of polymorphisms, the DNAs of the genotypes to be surveyed are digested, usually with various restriction enzymes, and marker alleles are identified by size differences of the restriction fragments to which the probes hybridize. Thus, RFLP markers are specified by a clone/restriction enzyme combination. Major sources of RFLP probes are species-specific genomic DNA and cDNA sequences. The use of heterologous probes for cross-genome RFLP analysis will be discussed later in this section.

---

[1] Chair of Agronomy and Plant Breeding, Department of Plant Sciences, Center of Life and Food Sciences Weihenstephan, Technical University Munich, 85350 Freising, Germany
[2] EpiGene GmbH, Biotechnology in Plant Protection, 85354 Freising, Germany

RFLPs allowed the construction of whole-genome linkage maps in plants for the first time (Bernatzky and Tanksley 1986; Helentjaris et al. 1986) and, thus, the localization of any genetically inherited trait. Despite being 'old-fashioned' and the most time-consuming molecular marker technique, RFLP analysis still displays high efficiency for the generation of high-density transcript genetic maps. The major reason is the availability of diverse, marker-saturated mapping populations for most crop species allowing simultaneous parental survey and, thus, the mapping of as many EST (expressed sequence tag) polymorphisms as possible. In the German GABI genome project, three barley doubled haploid populations (involving populations Igri × Franka with 71 lines, Steptoe × Morex with 150 and Oregon Wolfe Barley Dom × Oregon Wolfe Barley Rec with 94) are being exploited to generate a transcript map comprising 1200 ESTs (http://pgrc.ipk-gatersleben.de/barley–proj–2001.pdf). In addition to denaturing high-performance liquid chromatography mapping of single nucleotide polymorphisms (SNPs) between EST alleles (Kota et al. 2001), the major part of EST mapping is managed through DNA gel blot hybridizations.

RFLP markers initiated a rapid evolution in the field of comparative genomics by linking plant genomes through comparative genetic maps. Especially for species from the same plant family, the use of a common set of heterologous, single-copy probes in DNA gel blot experiments allowed the detection of orthologous marker loci and, thus, the prediction of genomic localization of both qualitative and quantitative traits across related species (Gale and Devos 1998; Doganlar et al. 2002). However, for plant species belonging to different families, comparative mapping via common probes is difficult because gene similarities are reduced due to a greater evolutionary divergence time, for which reason gene probes that have remained relatively stable in both sequence and copy number will be required. Fulton et al. (2002) identified more than 1000 conserved genes which were referred to as conserved ortholog set (COS) markers by computationally comparing more than 20,000 tomato ESTs with the *Arabidopsis thaliana* genome sequence. ESTs having a single match in the *Arabidopsis* genome (criterion to avoid problems with multigene families for which orthology and paralogy cannot be differentiated) were probed against tomato DNA to ensure that these COS markers were truly single- or low-copy. The evolutionary stability of some of the COS markers was shown by the consecutive hybridization of 'garden blots' (which were composed of DNA from a wide range of plant species) with tomato ESTs and their counterpart *Arabidopsis* probes with both sets detecting (nearly) identical restriction fragments.

Due to the time-consuming multi-step protocol and the requirement of radioactivity for satisfying fragment detection, RFLPs lost their importance in marker-assisted breeding. However, the development of simple PCR markers from sequenced RFLP probes provides an opportunity to maintain useful polymorphism found in previous gene mapping studies.

3 Microsatellites

Microsatellites also termed simple sequence repeats (SSRs), short tandem repeats (STRs) or sequence-tagged microsatellite sites (STMS) are tandem repeats of short nucleotide sequence motifs (mono-, di-, tri-, tetra- or pentanucleotide units). Microsatellites are abundant and are relatively evenly spaced throughout eukaryotic genomes (Tautz and Renz 1984). SSR loci are extremely variable in the number of repeat units among individuals of a given species (mutation rates for di- and tetranucleotide repeats: 0.001 mutations per generation, Goldstein and Pollock 1997) and can be easily typed via STS (sequence-tagged site)-PCR due to unique sequences bracketing individual microsatellites (Weber and May 1989). The opportunity to tag a fast mutating – repetitive – sequence motif as a simple PCR marker represents the ideal resource for the development of a valuable, if not the most important, marker class to be applied in practical plant breeding, particularly in species otherwise characterized by low levels of genetic diversity. As a consequence, for the most agriculturally important crops, a huge number of SSR markers are already publicly available for research (Milbourne et al. 1998; Röder et al. 1998; Cregan et al. 1999; Ramsay et al. 2000; Temnykh et al. 2000; Sharopova et al. 2002; Song et al. 2002). Furthermore, this type of repetitive sequence is also occurring within genes as has been demonstrated by searching EST databases for the presence of microsatellites (Eujayl et al. 2002; Hackauf and Wehling 2002; Thiel et al. 2003). EST-derived SSRs are expected to display slightly less polymorphisms than genomic library-derived SSRs, as there is pressure for sequence conservation in coding regions (Scott 2001). However, the availability of SSR markers from the expressed portion of the genome might facilitate their transferability across genera, compared to the low efficiency of SSR markers that have been retrieved from gene-poor areas (Peakall et al. 1998). This approach would benefit plant species with minimal resource and research expenditure.

For an efficient exploitation of SSR polymorphisms, particularly for dinucleotide repeats, in which alleles may differ by only two base pairs, analysis on a DNA sequencing instrument is recommended. This further simplifies scoring of SSR loci showing characteristic sub-banding, the so-called stutter bands.

4 Random Amplified Polymorphic DNAs

In 1990, Williams et al. and Welsh and McClelland independently described the utilization of a single, random-sequence oligonucleotide primer in a low-stringency PCR (35–45 °C) for the simultaneous amplification of several discrete DNA fragments referred to as random amplified polymorphic DNA

(RAPD) and arbitrarily primed PCR (AP-PCR), respectively. Usually, ten-base oligomers of varying GC content (ranging from 40 to 100%) are applied. RAPDs are qualified as genetic markers, since short and inverted repeats on complementary strands to which the primer hybridize, are more or less regularly spaced throughout the genome. Polymorphism between individuals results primarily from sequence differences in one or both primer sites or from indels (insertions-deletions) exceeding the PCR-amplifiable distance, and are visible in conventional agarose gel electrophoresis as presence or absence of a particular RAPD band. RAPDs predominantly provide dominant markers, but homologous alleles can sometimes be identified with the help of detailed pedigree information.

Most advantageous for this molecular marker technique is the use of universal primers enabling the cost-effective accomplishment of various genetic analyses in a short period of time (for review: Tingey and del Tufo 1993). Nevertheless, due to frequently observed problems with reproducibility of overall RAPD profiles and specific bands, this marker class is often treated with reserve (Ellsworth et al. 1993; Muralidharan and Wakeland 1993; Schierwater and Ender 1993). In replication studies by Pérez et al. (1998), mispriming error amounted up to 60%. To overcome this problem, Paran and Michelmore (1993) converted RAPD fragments to simple and robust PCR markers termed sequence characterized amplified regions (SCARs). Primer pairs are deduced from cloned RAPD fragments by usually extending the original decamer primer sequence with 10–15 bases.

A modified approach of the RAPD technique, DNA amplification fingerprinting (DAF; Caetano-Anollés et al. 1991) employs one or more primers as short as five nucleotides in length to produce complex banding patterns that are resolved by polyacrylamide gel electrophoresis. A further enhancement in detection of polymorphic DNA was achieved by profiling endonuclease-digested DNA (Caetano-Anollés et al. 1993). Advances in the fingerprinting of small templates up to 250 kb (e.g., BACs) were achieved with primers harboring an arbitrary sequence of only three nucleotides if a stable mini-hairpin was attached to their 5'-ends (Caetano-Anollés and Gresshoff 1994).

5 Amplified Fragment Length Polymorphisms

Like RAPD, amplified fragment length polymorphism (AFLP; Vos et al. 1995) is a universal, multi-locus marker technique that can be applied to genomes of any source or complexity. The method is based on the selective PCR amplification of restriction fragments from a total double-digest of genomic DNA under high stringency conditions. Oligonucleotide adaptors are ligated to the restricted DNA, which serve with the restriction site sequences as target sites for primer annealing. By using primers having extensions into the restriction fragments, the specific amplification of only those fragments matching the primers is achieved. The option to permute the order of the selective bases and to recombine the primers with each other theoretically discloses the chance of the gradual collection of all restriction fragments from a particular enzyme combination showing a suitable size for DNA fragment analysis from a genotype. AFLP products need to be separated in high-resolution electrophoresis systems. Thus, a well-balanced number of amplified restriction fragments ranges from 50–150. A major improvement has been made by switching from radioactive to fluorescent dye-labeled primers for the detection of fragments in gel-based or capillary DNA sequencers in which fluorescently labeled fragments pass the detector near the bottom of the gel/the end of the capillary, resulting in a linear spacing of DNA fragments and, therefore, increasing the resolution over the whole size range (Schwarz et al. 2000).

A two-step amplification strategy was developed for complex genomes (10^9–10^{10} bp) using six- (*Eco*RI, *Pst*I, *Hin*dIII) and four-base (*Mse*I, *Taq*I) cutters for AFLP template preparation and three selective nucleotides on both primers. The first PCR amplification step utilizes primers each having a single selective nucleotide and reduces the overall complexity of the mixture up to 16-fold, allowing the target sequence to become the predominant species. PCR products from preamplification are diluted and used as templates for a second amplification round using primers with full base extensions. Although the complexity is reduced with each additional selective nucleotide, the selectivity is maintained with nucleotide extensions to a maximum of three selective bases. For small plant genomes (10^7–10^8 bp), two to three selective nucleotides on the 3'-end of each primer are sufficient to reveal polymorphism. Another widespread enzyme system substitutes six-base for eight-base cutting enzymes, such as *Sse*8387I or its isoschizomer *Sda*I, and usually employs two to three selective bases on both primers (Law et al. 1998; Hartl et al. 1999).

Due to its capacity to reveal many polymorphic bands in a single reaction, the AFLP technique has been extensively used with plant DNA for the construction of both whole-genome (Becker et al. 1995; Keim et al. 1997; Haanstra et al. 1999; Vuylsteke et al. 1999; Chalmers et al. 2001) and high-resolution maps around loci that control agronomically important traits (Thomas et al. 1995; Büschges et al. 1997; Simons et al. 1998; Schwarz et al. 1999; Ballvora et al. 2001). Furthermore, AFLP markers have been proven to be reliable and reproducible not only within, but also among mapping populations, for which reason allele-specificity of comigrating AFLP markers has been used for aligning genetic maps from different genotypes (Rouppe van der Voort et al. 1997; Waugh et al. 1997a; Groh et al. 2001; Singrün et al. 2003). Consequently, AFLP provided an ideal tool for estimating genetic variation of both cultivated and natural populations (Travis et al. 1996; Barrett and Kidwell 1998).

The AFLP technique can be modified so that one primer is obtained from a known multi-copy sequence to detect sequence-specific amplification polymorphisms (S-SAP). This approach has been used successfully to generate genome-wide *Bare-1* retrotransposon-like markers in barley (Waugh et al. 1997b) and diploid *Avena* (Yu and Wise 2000) as well as in alfalfa by making use of consensus sequences from long terminal repeats (LTRs) of *Tms1* retrotransposon (Porceddu et al. 2002). Based on the LTR regions of retrotransposable elements *Tps12* and *Tps19* of pea (Pearce et al. 2000) and *Ty1-copia* of sweetpotato (Berenyi et al. 2002) S-SAP systems were established for their use in genetic diversity studies.

The cDNA-AFLP technique (Bachem et al. 1996), which applies the standard AFLP protocol on a cDNA template, was used to display transcripts whose expression was rapidly altered during race-specific resistance reaction (Durrant et al. 2000), for the isolation of differentially expressed genes from a specific chromosome region using aneuploids (Money et al. 1996; Kojima et al. 2000) and for the construction of genome-wide transcriptome maps (Brugmans et al. 2002).

6 Single Nucleotide Polymorphisms

The next generation of genetic markers is based on SNPs, which are defined as single nucleotide positions in a given DNA stretch at which there are variations between different individuals within a species. Thus, single base insertion/deletion (indels) variants would not formally be considered to be SNPs (Brookes 1999). In general, SNPs are the most common form of DNA sequence polymorphisms and, therefore, sufficiently abundant for comprehensive haplotype analyses. In the human genome SNPs occur once every 100–300 bases, but up to now there is not much information about the frequency of SNPs in plant genomes. Only for *Arabidopsis thaliana*, a collection of 37,344 predicted SNPs between the whole genome sequences of two ecotypes is documented, and estimated to occur at one out of every 3.3 kb (http://www.arabidopsis.org/cereon/). In addition, the binary (di-allelic) character and stability from generation to generation make SNPs amenable to automated, high-throughput genotyping and, therefore, an attractive tool for quantitative trait loci (QTL) mapping studies and marker-assisted selection in plant breeding programs.

The basic method for determination of SNP genotypes has been Sanger dideoxysequencing. Since sequencing generates more information than necessary, misses SNPs when the DNA template is heterozygous and, thus, is time-consuming and very expensive, other gel-based assays were applied for SNP genotype detection. A widely used agarose gel-based SNP detection technique is PCR-RFLP (Fig. 1A), also referred to as cleaved amplified polymorphic sequence (CAPS; Konieczny and Ausubel 1993). A target DNA segment containing a SNP is amplified by PCR and the product is incubated with an appropriate restriction enzyme, which cuts the molecule if it contains the SNP variant creating the recognition site, but not if it contains the other.

Though CAPS analysis is performed routinely in many laboratories, the time-consuming post-PCR digestion step limits its application for small-scale experiments. Allele-specific PCR (AS-PCR, Fig. 1B) is a further agarose gel-based method to assay SNPs by using oligonucleotide primers with 3'-nucleotides complementary to the SNP site (Ugozzoli and Wallace 1991). Allele-specific primers perfectly match only with the corresponding alleles, whereas they have 3'-mismatches with the alternative alleles. Since in many cases a single base pair change at the 3'-end of alleles is not sufficient for robust allele discrimination, AS-PCR procedure was improved by utilizing competing polymorphism-detecting primers (Mohler and Jahoor 1996; See et al. 2000; Ellis et al. 2002) or by introducing artificial mismatches at 3'-subterminal positions of the allele-specific primers (Drenkard et al. 2000). Another important technology for SNP scoring is single-stranded conformation polymorphism analysis (SSCP; Orita et al. 1989; Fig. 1C). Allele discrimination relies on the secondary structure being different for single-stranded DNA derived from PCR products that differ by one or more internal nucleotides. PCR products are denatured and electrophoretically separated in native polyacrylamide gels. Differences in electrophoretic mobility between amplicons from the wild type and mutant genotype, respectively, will suggest the presence of SNP(s). Even though SSCP analysis is simple, accurate and relatively inexpensive, the polyacrylamide gel-based assay format prevents its use for high-throughput allele typing.

The limitations of the above-mentioned techniques for high-throughput SNP genotyping forced the development of new technology platforms utilizing the reaction principles of minisequencing, heteroduplex analysis and allele-specific hybridization.

The robust detection of known mutations using minisequencing (Fig. 1D) bases on oligonucleotides which anneal immediately upstream of the query SNP to be extended by single dideoxynucleotides (ddNTPs) in cycle sequencing reactions. The fidelity of thermostable proof-reading DNA polymerases guarantees that only the complementary ddNTP is incorporated. Several detection methods have been described for the discrimination of primer extension (PEX) products. Most popular is the use of dideoxynucleotide terminators which are labeled with different fluorescent dyes. The differentially dye-labeled PEX products can readily be detected on CCD camera-based DNA sequencing instruments. For minisequencing in parallel, solid-phase assays with detection by fluorescence scanning, also denoted arrayed primer extension (APEX) were developed, in which one detection primer per SNP is covalently immobilized on glass slides. Prior to analysis, single-stranded templates are prepared through degrading one strand of PCR products by digestion with an exonuclease enzyme (Shumaker et al. 1996). Both gel- and array-based PEX approaches were applied for SNP analysis in barley (Kanazin et al. 2002). Matrix-assisted laser desorption/ionization time-of-flight (MALDI-TOF) mass spectrometry is another well-suited method for detection and discrimination of small DNA molecules. Due to the inherent molec-

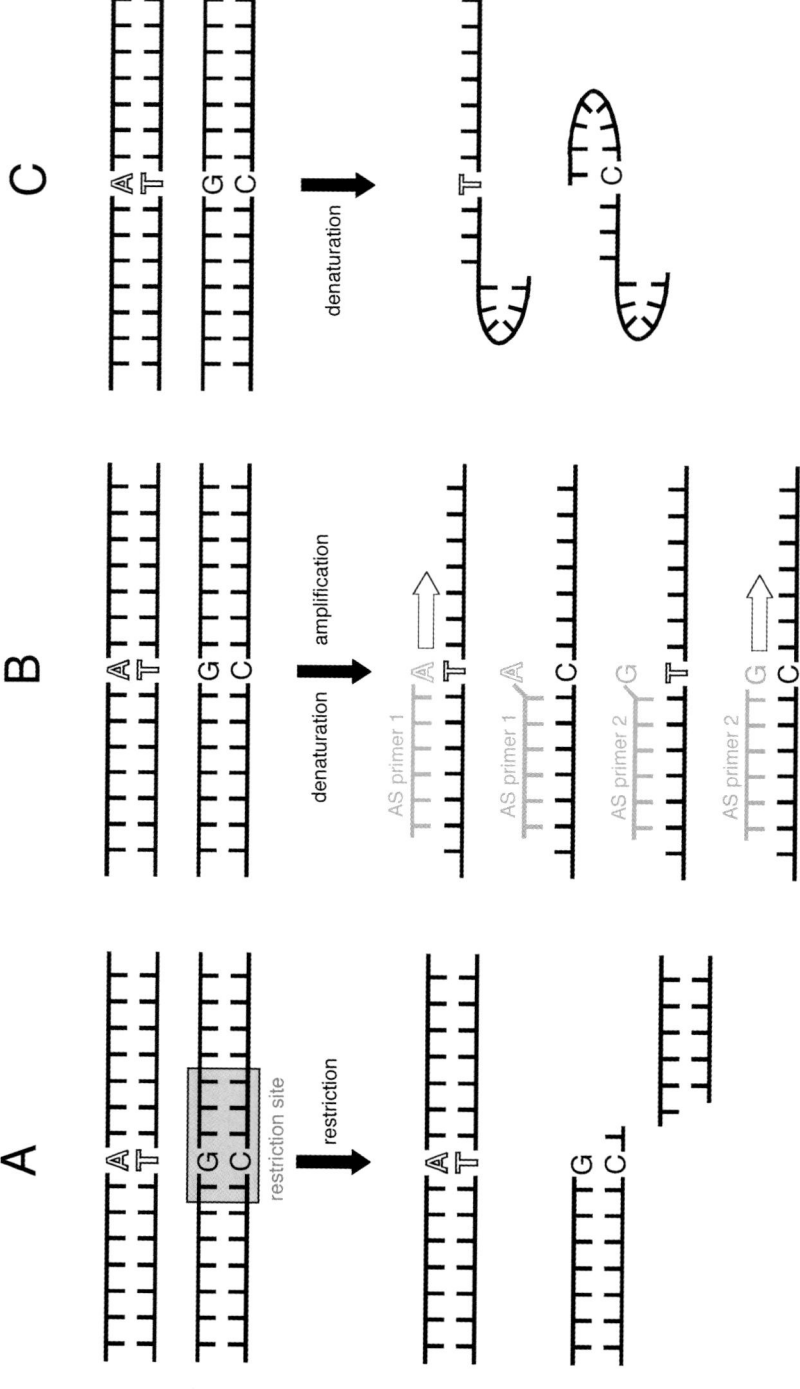

Fig. 1. Biochemical reaction principles for SNP genotype determination: A restriction endonuclease digestion, B allele-specific primer PCR, C single-stranded conformation polymorphism, D minisequencing, E heteroduplex analysis, F allele-specific oligonucleotide probe hybridization

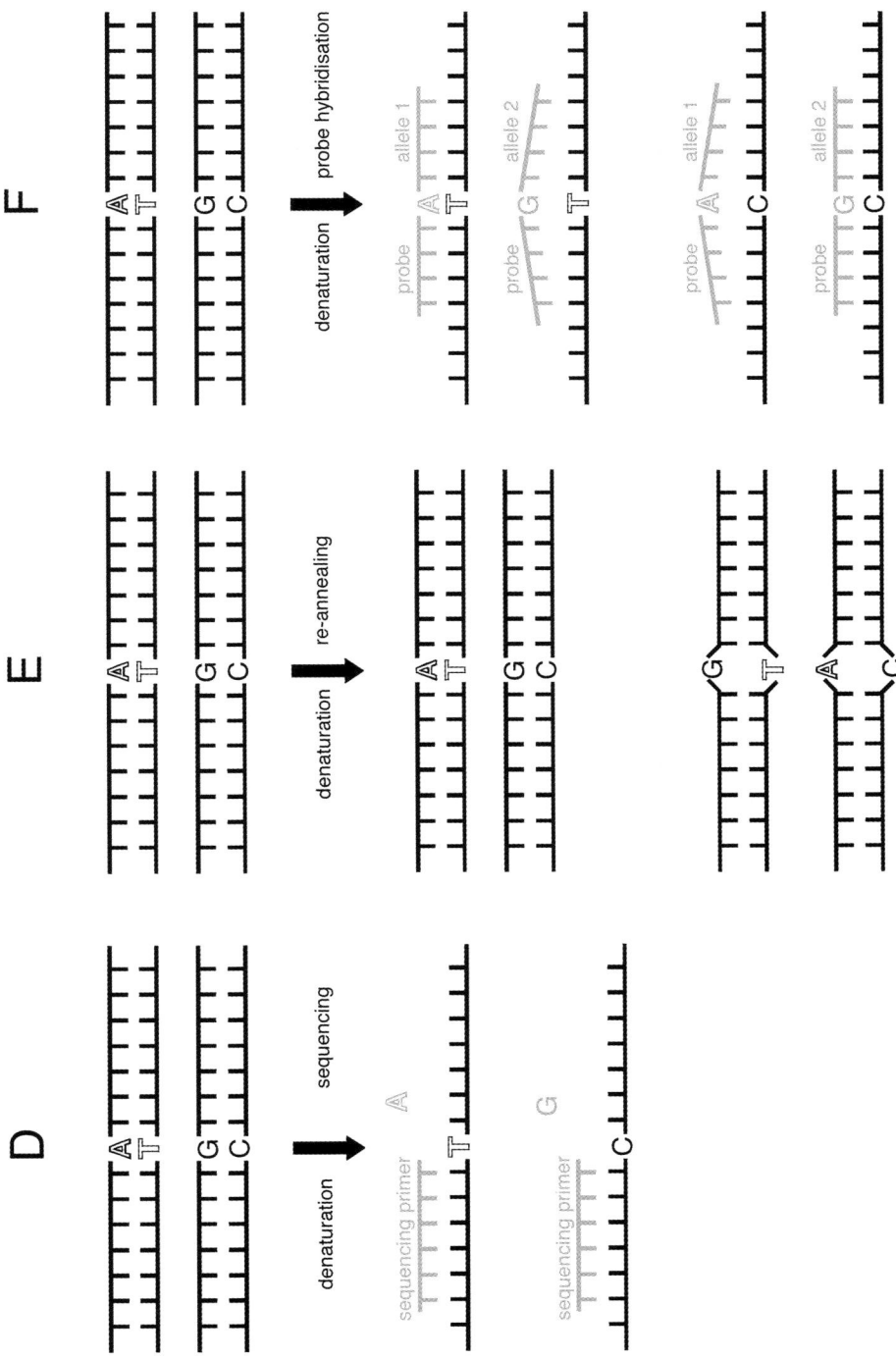

ular weight difference of DNA bases, incorporated nucleotides are identified by the increase in mass of extended primers. The MALDI-TOF method is particularly advantageous for detection of PEX products in multiplex (Ross et al. 1998; Paris et al. 2002). In the pyrosequencing format, real-time monitoring of PEX relies on the bioluminometric detection of inorganic pyrophosphate released upon incorporation of deoxynucleotide triphosphate (Ahmadian et al. 2000). The genotype of a SNP is determined by the sequential addition (and degradation) of nucleotides. The yielded light is proportional to the amount of incorporated nucleotides, for which reason pyrosequencing is qualified for the quantitative estimation of allele frequencies in pooled DNA samples. Furthermore, pyrosequencing proved to be an appropriate method for genotyping SNPs in polyploid plant genomes, such as potato, because all possible allelic states of binary SNPs could be accurately distinguished (Rickert et al. 2002).

Heteroduplex analysis (Fig. 1E) uses the different electrophoretic mobility behaviors of homoduplex and heteroduplex DNA molecules. The most straightforward method is denaturing high-performance liquid chromatography (DHPLC), an ion-pair reversed-phase chromatography method, which allows the detection of single mismatches in PCR products by injecting them into an adequately preheated mobile phase that results in partial denaturation of the DNA molecules (Oefner and Underhill 1998). Owing to reduced melting temperatures, heteroduplexes will be eluted earlier from the column than the homoduplexes. Besides scoring of known SNPs, DHPLC can be applied for the rapid and cost-effective scanning of unknown mutations in DNA amplicons (Cho et al. 1999). DHPLC was reported to have greater than 99% sensitivity in detecting unknown SNPs and mutations between two *Arabidopsis thaliana* ecotypes (Spiegelman et al. 2000).

Oligonucleotide hybridization probes (Fig. 1F) are also an attractive tool for SNP genotyping. Under optimized assay conditions, the SNP discrimination is solely based on differences in melting temperature (T_m) of the two probe–template hybrid classes. For high reliability of SNP genotype calling, T_m differences have to be maximized by using probes as short as possible. Originally, allele-specific hybridization (ASH) used the dot blot format where probes hybridized to membrane-bound genomic DNA or PCR fragments. Meanwhile the advanced, PCR-based dynamic allele-specific hybridization (DASH) method in a microtiter plate format is available (Howell et al. 1999). Since one of the PCR primers is biotinylated at the 5'-end, the PCR products can be bound to streptavidin-coated wells and denatured under alkaline conditions. An oligonucleotide probe, complementary to one allele, is added to the single-stranded target DNA molecules. The differences in melting curves are measured by slowly heating and observing the change in fluorescence of a double-strand-specific, intercalating dye. Large-scale scanning of SNPs in a huge number of loci by allele-specific hybridization can be performed on high-density oligonucleotide chips (Wang et al. 1998). Another interesting option for SNP scoring by hybridization is provided by the use of fluorescently labeled allele-specific oligonucleotides such as Taqman (Livak et al. 1995) and Molecular Beacon (Tyagi and Kramer 1996) DNA probes. Both systems use two fluorogenic probes for allele discrimination, one for each allele. Typically, the probes are labeled with different reporter dyes at the 5'- and a quencher dye at the 3'-end. In the intact Taqman probe, fluorescence of the reporter dye is quenched by the close proximity of the 3'-dye label due to fluorescence resonance energy transfer (FRET). After hybridization of the probes to the target alleles, degradation of probes by the 5'3' exonuclease activity of the processing *Taq* DNA polymerase during PCR extension steps interrupts the interaction of the dye molecules resulting in a fluorescent signal. Contrary to linear Taqman probes, Molecular Beacons form a hairpin structure due to complementary 5'- and 3'-ends which brings the reporter and quencher into immediate vicinity of one another. Hybridization of the probe to the target sequence during PCR annealing steps results in linearization and subsequent emission of fluorescence signals.

Besides Taqman and Molecular Beacon probes, other allele-specific hybridization assays were developed for SNP genotype determination: Scorpion assays combine forward primer and probe in a single molecule (Thelwell et al. 2000), Padlock assays use oligonucleotide probes to be ligated into circles upon target recognition and isothermal rolling circle amplification (Nilsson et al. 1997) and invasive cleavage assays involve a FLAP 5'-endonuclease that is specific for a three-dimensional structure formed by two (the Invader and the primary SNP detection probe) overlapping oligonucleotides (Lyamichev et al. 1999).

7 Conclusions

A wide range of marker techniques is now available for genotyping plant genomes. Meanwhile, markers are employed not only in plant breeding research, but also in practical plant breeding. In order to exploit the full power of each marker class, one has to make its special character consistent with the desired application. Unfortunately, highly informative marker types like SSRs and SNPs have been elaborated for only a few well-studied crop plants. Due to the lack of sequencing and mapping data, genotyping in 'undiscovered' plant genomes still has to be performed using universal marker techniques like RAPD and AFLP. However, the strong synteny between closely related species will allow, to a certain extent, the transfer of marker information thereby increasing the molecular marker pool in genomes of plant families. Finally, reducing genotyping costs for high-throughput techniques, e.g., microarrays, is a major challenge for the comprehensive integration of markers into plant breeding programs.

References

Ahmadian A, Gharizadeh B, Gustafsson AC, Sterky F, Nyren P, Uhlen M, Lundeberg J (2000) Single nucleotide polymorphism analysis by pyrosequencing. Anal Biochem 280:103–110

Bachem CWB, van der Hoeven RS, de Bruijn SM, Vreugdenhil D, Zabeau M, Visser RGF (1996) Visualization of differential gene expression using a novel method of RNA fingerprinting based on AFLP: analysis of gene expression during potato tuber development. Plant J 9:745–753

Ballvora A, Schornack S, Baker B, Ganal M, Bonas U, Lahaye T (2001) Chromosome landing at the tomato *Bs4* locus. Mol Gen Genom 266:639–645

Barrett BA, Kidwell KK (1998) AFLP-based genetic diversity assessment among wheat cultivars from the Pacific Northwest. Crop Sci 38:1261–1271

Becker J, Vos P, Kuiper M, Salamini F, Heun M (1995) Combined mapping of AFLP and RFLP markers in barley. Mol Gen Genet 249:65–73

Berenyi M, Gichuki ST, Schmidt J, Burg K (2002) *Ty1-copia* retrotransposon-based S-SAP (sequence-specific amplified polymorphism) for genetic analysis of sweetpotato. Theor Appl Genet 105:862–869

Bernatzky R, Tanksley SD (1986) Toward a saturated linkage map in tomato based on isozymes and random cDNA sequences. Genetics 112:887–898

Botstein D, White RL, Skolnick M, Davis RW (1980) Construction of a genetic linkage map in man using restriction fragment length polymorphisms. Am J Hum Genet 32:314–331

Brookes AJ (1999) The essence of SNPs. Gene 234:177–186

Brugmans B, del Carmen AF, Bachem CWB, van Os H, van Eck HJ, Visser RGF (2002) A novel method for the construction of genome wide transcriptome maps. Plant J 31:211–222

Büschges R, Hollricher K, Panstruga R, Simons G, Wolter M, Frijters A, van Daelen R, van der Lee T, Diergaarde P, Groenendijk J, Töpsch S, Vos P, Salamini F, Schulze-Lefert P (1997) The barley *mlo* gene: a novel control element of plant pathogen resistance. Cell 88:695–705

Caetano-Anollés G, Gresshoff PM (1994) DNA amplification fingerprinting using arbitrary mini-hairpin oligonucleotide primers. Biotechnology 12:619–623

Caetano-Anollés G, Bassam BJ, Gresshoff PM (1991) DNA amplification fingerprinting using very short arbitrary oligonucleotide primers. Biotechnology 9:553–557

Caetano-Anollés G, Bassam BJ, Gresshoff PM (1993) Enhanced detection of polymorphic DNA by multiple arbitrary amplicon profiling of endonuclease-digested DNA – identification of markers tightly linked to the supernodulation locus in soybean. Mol Gen Genet 241:57–64

Chalmers KJ, Campbell AW, Kretschmer J, Karakousis A, Henschke P, Pierens S, Harker N, Palotta M, Cornish GB, Shariflou MR, Rampling LR, McLauchlan A, Daggard G, Sharp PJ, Holton TA, Sutherland MW, Appels R, Langridge P (2001) Construction of three linkage maps in bread wheat, *Triticum aestivum*. Aust J Agric Res 52:1089–1119

Cho RJ, Mindrinos M, Richards DR, Sapolsky RJ, Anderson M, Drenkard E, Dewdney J, Reuber TL, Stammers M, Federspiel N, Theologis A, Yang W-H, Hubbell E, Au M, Chung EY, Lashkari D, Lemieux B, Dean C, Lipshutz RJ, Ausubel FM, Davis RW, Oefner PJ (1999) Genome-wide mapping with biallelic markers in *Arabidopsis thaliana*. Nat Genet 23:203–207

Cregan PB, Jarvik T, Bush AL, Shoemaker RC, Lark KG, Kahler AL, Kaya N, van Toai TT, Lohnes DG, Chung J, Specht JE (1999) An integrated genetic linkage map of the soybean genome. Crop Sci 39:1464–1490

Doganlar S, Frary A, Daunay M-C, Lester RN, Tanksley SD (2002) Conservation of gene function in the Solanaceae as revealed by comparative mapping of domestication traits in eggplant. Genetics 161:1713–1726

Drenkard E, Richter BG, Rozen S, Stutius LM, Angell NA, Mindrinos M, Cho RJ, Oefner PJ, Davis RW, Ausubel FM (2000) A simple procedure for the analysis of single nucleotide polymorphisms facilitates map-based cloning in *Arabidopsis*. Plant Physiol 124:1483–1492

Durrant WE, Rowland O, Piedras P, Hammond-Kosack KE, Jones JDG (2000) cDNA-AFLP reveals a striking overlap in race-specific resistance and wound response gene expression profiles. Plant Cell 12:963–977

Ellis MH, Spielmeyer W, Gale KR, Rebetzke GJ, Richards RA (2002) "Perfect" markers for the *Rht-B1b* and *Rht-D1b* dwarfing genes in wheat. Theor Appl Genet 105:1038–1042

Ellsworth DL, Rittenhouse KD, Honeycutt RL (1993) Artifactual variation in randomly amplified polymorphic DNA banding patterns. Biotechniques 14:214–216

Eujayl I, Sorrells ME, Baum M, Wolters P, Powell W (2002) Isolation of EST-derived microsatellite markers for genotyping the A and B genomes of wheat. Theor Appl Genet 104:399–407

Fulton TM, van der Hoeven R, Eannetta NT, Tanksley SD (2002) Identification, analysis, and utilization of conserved ortholog set markers for comparative genomics in higher plants. Plant Cell 14:1457–1467

Gale M, Devos K (1998) Comparative genetics in the grasses. Proc Natl Acad Sci USA 95:1971–1974

Goldstein DB, Pollock DD (1997) Launching microsatellites: a review of mutation processes and methods of phylogenetic inference. J Hered 88:335–342

Groh S, Zacharias A, Kianian SF, Penner GA, Chong J, Rines HW, Phillips RL (2001) Comparative AFLP mapping in two hexaploid oat populations. Theor Appl Genet 102:876–884

Haanstra JPW, Wye C, Verbakel H, Meijer-Dekens F, van den Berg P, Odinot P, van Heusden AW, Tanksley S, Lindhout P, Peleman J (1999) An integrated high-density RFLP-AFLP map of tomato based on two *Lycopersicon esculentum* × *L. pennellii* F_2 populations. Theor Appl Genet 99:254–271

Hackauf B, Wehling P (2002) Identification of microsatellite polymorphisms in an expressed portion of the rye genome. Plant Breed 121:17–25

Hartl L, Mohler V, Zeller FJ, Hsam SLK, Schweizer G (1999) Identification of AFLP markers closely linked to the powdery mildew resistance genes *Pm1c* and *Pm4a* in common wheat (*Triticum aestivum* L.). Genome 42:322–329

Helentjaris T, Slocum M, Wright S, Schaefer A, Nienhuis J (1986) Construction of genetic-linkage maps in maize and tomato using restriction-fragment-length-polymorphisms. Theor Appl Genet 72:761–769

Howell WM, Jobs M, Gyllensten U, Brookes V (1999) Dynamic allele-specific hybridization. A new method for scoring single nucleotide polymorphisms. Nat Biotechnol 17:87–88

Kanazin V, Talbert H, See D, DeCamp P, Nevo E, Blake T (2002) Discovery and assay of single-nucleotide polymorphisms in barley (*Hordeum vulgare*). Plant Mol Biol 48:529–537

Keim P, Schupp JM, Travis SE, Clayton K, Zhu T, Shi L, Ferreira A, Webb DM (1997) A high-density soybean genetic map based on AFLP markers. Crop Sci 37:537–543

Kojima T, Habu Y, Iida S, Ogihara Y (2000) Direct isolation of differentially expressed genes from a specific chromosome region of common wheat: application of the amplified fragment length polymorphism-based mRNA fingerprinting (AMF) method in combination with a deletion line of wheat. Mol Gen Genet 263:635–641

Konieczny A, Ausubel F (1993) A procedure for mapping *Arabidopsis* mutations using co-dominant ecotype-specific PCR-based markers. Plant J 4:403–410

Kota R, Wolf M, Michalek W, Graner A (2001) Application of denaturing high-performance liquid chromatography for mapping of single nucleotide polymorphisms in barley (*Hordeum vulgare* L.). Genome 44:523–528

Law JR, Donini P, Koebner RMD, Reeves JC, Cooke RJ (1998) DNA profiling and plant variety registration III: the statistical assessment of distinctness in wheat using amplified fragment length polymorphisms. Euphytica 102:335–342

Livak KJ, Flood SJ, Marmaro J, Giusti W, Deetz K (1995) Oligonucleotides with fluorescent dyes at opposite ends provide a quenched probe system useful for detecting PCR product and nucleic acid hybridization. PCR Methods Appl 4:357–362

Lyamichev V, Mast AL, Hall JG, Prudent JR, Kaiser MW, Takova T, Kwiatkowski RW, Sander TJ, de Arruda M, Arco DA, Neri BP, Brow MAD (1999) Polymorphism identification and quantitative detection of genomic DNA by invasive cleavage of oligonucleotide probes. Nat Biotechnol 17:292–296

Milbourne D, Meyer RC, Collins AJ, Ramsay LD, Gebhardt C, Waugh R (1998) Isolation, characterisation and mapping of simple sequence repeat loci in potato. Mol Gen Genet 259:233–245

Mohler V, Jahoor A (1996) Allele-specific amplification of polymorphic sites for the detection of powdery mildew resistance loci in cereals. Theor Appl Genet 93:1078–1082

Money T, Reader S, Qu LJ, Dunford RP, Moore G (1996) AFLP-based mRNA fingerprinting. Nucleic Acids Res 24: 2616–2617

Muralidharan K, Wakeland EK (1993) Concentration of primer and template qualitatively affects products in random-amplified polymorphic DNA PCR. Biotechniques 14:362–364

Nilsson M, Krejci K, Koch J, Kwiatkowski M, Gustavsson P, Landegren U (1997) Padlock probes reveal single-nucleotide differences, parent of origin and *in situ* distribution of centromeric sequences in human chromosomes 13 and 21. Nat Genet 16:252–255

Oefner PJ, Underhill PA (1998) DNA mutation detection using denaturing high-performance liquid chromatography (DHPLC). In: Dracopoli NC, Haines JL, Korf BR, Moir DT, Morton CC, Seidman CE (eds) Current protocols in human genetics [Suppl 19]. Wiley, New York, 7.10.1–7.10.12

Orita M, Iwahana H, Kanazawa H, Hayashi K, Sekiya T (1989) Detection of polymorphisms of human DNA by gel electrophoresis as single-strand conformation polymorphisms. Proc Natl Acad Sci USA 86:2766–2770

Paran I, Michelmore RW (1993) Development of reliable PCR-based markers linked to downy mildew resistance genes in lettuce. Theor Appl Genet 85:985–993

Paris M, Jones MGK, Eglinton JK (2002) Genotyping single nucleotide polymorphisms for selection of barley beta-amylase alleles. Plant Mol Biol Rep 20:149–159

Peakall R, Gilmore S, Keys W, Morgante M, Rafalski A (1998) Cross-species amplification of soybean (*Glycine max*) simple sequence repeats (SSRs) within the genus and other legume genera: implications for the transferability of SSRs in plants. Mol Biol Evol 15:1275–1287

Pearce SR, Knox M, Ellis THN, Flavel AJ, Kumar A (2000) Pea *Ty1-copia* group retrotransposons: transpositional activity and use as markers to study genetic diversity in *Pisum*. Mol Gen Genet 263:898–907

Pérez T, Albornoz J, Domínguez A (1998) An evaluation of RAPD fragment reproducibility and nature. Mol Ecol 7:1347–1358

Porceddu A, Albertini E, Barcaccia G, Marconi G, Bertoli F, Veronesi F (2002) Development of S-SAP markers based on an LTR-like sequence from *Medicago sativa* L. Mol Gen Genomics 267:107–114

Ramsay L, Macaulay M, Degli Ivanissevich S, MacLean K, Cardle L, Fuller J, Edwards KJ, Tuvesson S, Morgante M, Massari A, Maestri E, Marmiroli N, Sjakste T, Ganal M, Powell W, Waugh R (2000) A simple sequence repeat-based linkage map of barley. Genetics 156:1997–2005

Rickert AM, Premstaller A, Gebhardt C, Oefner PJ (2002) Genotyping of SNPs in a polyploid genome by pyrosequencing™. Biotechniques 32:592–603

Röder MS, Korzun V, Wendehake K, Plaschke J, Tixier MH, Leroy P, Ganal MW (1998) A microsatellite map of wheat. Genetics 149:2007–2023

Ross P, Hall L, Smirnov I, Haff L (1998) High level multiplex genotyping by MALDI-TOF mass spectrometry. Nat Biotechnol 16:1347–1351

Rouppe van der Voort JNAM, van Zandvoort P, van Eck HJ, Folkertsma RT, Hutten RCB, Draaistra J, Gommers FJ, Jacobsen E, Helder J, Bakker J (1997) Use of allele specificity of comigrating AFLP markers to align genetic maps from different potato genotypes. Mol Gen Genet 255:438–447

Schierwater B, Ender A (1993) Different thermostable DNA polymerases may amplify different RAPD products. Nucleic Acids Res 21:4647–4648

Schwarz G, Michalek W, Mohler V, Wenzel G, Jahoor A (1999) Chromosome landing at the *Mla* locus in barley (*Hordeum vulgare* L.) by means of high resolution mapping with AFLP markers. Theor Appl Genet 98:521–530

Schwarz G, Herz M, Huang XQ, Michalek W, Jahoor A, Wenzel G, Mohler V (2000) Application of fluorescence-based semi-automated AFLP analysis in barley and wheat. Theor Appl Genet 100:545–551

Scott KD (2001) Microsatellites derived from ESTs, and their comparison with those derived by other methods. In: Henry RJ (ed) Plant genotyping: the DNA fingerprinting of plants. CABI Publishing, Oxford, pp 225–237

See D, Kanazin V, Talbert H, Blake TK (2000) Electrophoretic detection of single nucleotide polymorphisms. Biotechniques 28:710–716

Sharopova N, McMullen MD, Schultz L, Schroeder S, Sanchez-Villeda H, Gardiner J, Bergstrom D, Houchins K, Melia-Hancock S, Musket T, Duru N, Polacco M, Edwards K, Ruff T, Register JC, Brouwer C, Thompson R, Velasco R, Chin E, Lee M, Woodman-Clikeman W, Long MJ, Liscum E, Cone K, Davis G, Coe EH Jr (2002) Development and mapping of SSR markers for maize. Plant Mol Biol 48:483–499

Shumaker JM, Metspalu A, Caskey CT (1996) Mutation detection by solid-phase primer extension. Hum Mutat 7:346–354

Simons G, Groenendijk J, Wijbrandi J, Reijans M, Groenen J, Diergaarde P, van der Lee T, Bleeker M, Onstenk J, de Both M, Haring M, Mes J, Cornelissen B, Zabeau M, Vos P (1998) Dissection of the *Fusarium I2* gene cluster in tomato reveals six homologs and one active gene copy. Plant Cell 10:1055–1068

Singrün Ch, Hsam SLK, Hartl L, Zeller FJ, Mohler V (2003) Powdery mildew resistance gene *Pm22* in cultivar Virest is a member of the complex *Pm1* locus in common wheat (*Triticum aestivum* L. em Thell.). Theor Appl Genet 106:1420–1424

Song QJ, Fickus EW, Cregan PB (2002) Characterization of trinucleotide SSR motifs in wheat. Theor Appl Genet 104:286–293

Spiegelman JI, Mindrinos MN, Oefner PJ (2000) High-accuracy DNA sequence variation screening by DHPLC. Biotechniques 29:1084–1092

Tanksley SD, Miller J, Paterson A, Bernatzky R (1988) Molecular mapping of plant chromosomes. In: Gustafson JP, Appels R (eds) Chromosome structure and function – impact of new concepts. Proceedings of the 18th Stadler genetics symposium. Plenum Press, New York, pp 157–173

Tautz D, Renz M (1984) Simple sequences are ubiquitous repetitive components of eukaryotic genomes. Nucleic Acids Res 12:4127–4138

Temnykh S, Park WD, Ayres N, Cartinhour S, Hauck N, Lipovich L, Cho YG, Ishii T, McCouch SR (2000) Mapping and genome organization of microsatellite sequences in rice (*Oryza sativa* L.). Theor Appl Genet 100:697–712

Thelwell N, Millington S, Solinas A, Booth J, Brown T (2000) Mode of action and application of Scorpion primers to mutation detection. Nucleic Acids Res 28:3752–3761

Thiel T, Michalek W, Varshney RK, Graner A (2003) Exploiting EST databases for the development and characterization of gene-derived SSR-markers in barley (*Hordeum vulgare* L.). Theor Appl Genet 106:411–422

Thomas CM, Vos P, Zabeau M, Jones DA, Norcott KA, Chadwick BP, Jones JDG (1995) Identification of amplified restriction fragment polymorphism (AFLP) markers tightly linked to the tomato *Cf-9* gene for resistance to *Cladosporium fulvum*. Plant J 8:785–794

Tingey SV, del Tufo JP (1993) Genetic analysis with RAPD markers. Plant Physiol 101:349–352

Travis SE, Maschinski J, Keim P (1996) An analysis of genetic variation in *Astragalus cremnophylax* var. *cremnophylax*, a critically endangered plant, using AFLP markers. Mol Ecol 5:735–745

Tyagi S, Kramer FR (1996) Molecular beacons: probes that fluoresce upon hybridization. Nat Biotechnol 14:303–308

Ugozzoli L, Wallace RB (1991) Allele-specific polymerase chain reaction. Methods Enzymol 2:42–48

Vos P, Hogers R, Bleeker M, Reijans M, Van De Lee T, Hornes M, Frijters A, Pot J, Peleman J, Kuiper M, Zabeau M (1995) AFLP: a new technique for DNA fingerprinting. Nucleic Acids Res 23:4407–4414

Vuylsteke M, Mank R, Antonise R, Bastiaans E, Senior ML, Stuber CW, Melchinger AE, Lübberstedt T, Xia XC, Stam P, Zabeau M, Kuiper M (1999) Two high density AFLP linkage maps of *Zea mays* L.: analysis of distribution of AFLP markers. Theor Appl Genet 99:921–935

Wang DG, Fan J-B, Siao C-J, Berno A, Young P, Sapolsky R, Ghandour G, Perkins N, Winchester E, Spencer J, Kruglyak L, Stein L, Hsie L, Topaloglou T, Hubbell E, Robinson E, Mittmann M, Morris MS, Shen N, Kilburn D, Rioux J, Nusbaum C, Rozen S, Hudson TJ, Lipshutz R, Chee

M, Lander ES (1998) Large-scale identification, mapping, and genotyping of single-nucleotide polymorphisms in the human genome. Science 280:1077–1082

Waugh R, Bonar N, Baird E, Thomas B, Graner A, Hayes P, Powell W (1997a) Homology of AFLP products in three mapping populations of barley. Mol Gen Genet 255:311–321

Waugh R, Mclean K, Flavell AJ, Pearce SR, Kumar A, Thomas BBT (1997b) Genetic distribution of *Bare-1*-like retrotransposable elements in the barley genome revealed by sequence-specific amplification polymorphism (S-SAP). Mol Gen Genet 253:687–694

Weber JL, May PE (1989) Abundant class of human DNA polymorphism which can be typed using the polymerase chain reaction. Am J Hum Genet 44:388–396

Welsh J, McClelland M (1990) Fingerprinting genomes using PCR with arbitrary primers. Nucleic Acids Res 18:7213–7218

Williams JGK, Kubelik AR, Livak KJ, Rafalski JA, Tingey SV (1990) DNA polymorphisms amplified by arbitrary primers are useful as genetic markers. Nucleic Acids Res 18:6531–6535

Yu G-X, Wise RP (2000) An anchored AFLP- and retrotransposon-based map of diploid *Avena*. Genome 43:736–749

I.3 A Model Crop Species: Molecular Markers in Rice

D.J. MACKILL and K.L. MCNALLY[1]

1 Introduction

Molecular markers were being used in the study of rice genetics even before the emergence of techniques for easy manipulation of DNA. Morphological markers had relatively limited applications, but isozyme markers were used extensively to study rice systematics (Second 1982; Glaszmann 1987). The development of restriction fragment length polymorphism (RFLP) markers and, subsequently, random amplified polymorphic DNA (RAPD), amplified fragment length polymorphism (AFLP), and microsatellite or simple sequence repeat (SSR) markers allowed the genetic mapping of many important traits. Use of these markers for rice has been recently reviewed (Mackill and Ni 2001; Temnykh et al. 2001; Xu 2002).

Model species have been used extensively in biology. The first model plant species chosen was *Arabidopsis*, and its complete genome sequence has been published recently (Kaul et al. 2000). Rice is the second model plant species, and it is a representative of monocot plants, in addition to its immense agricultural importance. Rice has many practical advantages for use in molecular genetics research, which include its small genome size and relatively low amount of repetitive DNA, its diploid nature, and its ease of manipulation in tissue culture. Table 1 lists the major milestones in the development of rice as a model crop species. The completion of a high-quality draft of the rice genome sequence by the International Rice Genome Project was announced on 18 December 2002 (http://rgp.dna.affrc.go.jp/rgp/Dec18–NEWS.html).

The term "molecular markers" usually signifies the use of DNA fragments from locations in the genome to map and follow the segregation of these fragments or observe underlying genetic variation. A useful feature of molecular markers is that the fragments themselves need have no function; their utility lies in the ease with which they can be assayed and the amount of information relative to genetic variation they impart. In this sense, a major value of markers is that they can be applied easily in any crop. Nevertheless, the enormity of the accumulated genetic information on rice offers unique opportunities for the development and deployment of molecular markers for breeding applications and advanced biological studies.

---

[1] International Rice Research Institute (IRRI), DAPO Box 7777, Metro Manila, Philippines

Table 1. Milestones in the molecular genetic analysis of rice

Milestone	References
First RFLP map	McCouch et al. (1988)
Transgenic japonica rice	Toriyama et al. (1988); Zhang et al. (1988); Zhang and Wu (1988)
Transgenic indica rice	Datta et al. (1988)
Major gene mapping	Yu et al. (1991)
RAPD markers	Zheng et al. (1991)
Microsatellite markers (SSR)	Zhao and Kochert (1992, 1993); Wu and Tanksley (1993)
QTL mapping	Ahn et al. (1993); Wang et al. (1994)
Agrobacterium transformation	Hiei et al. (1994)
Positional cloning	Song et al. (1995)
AFLP markers	Cho et al. (1996); Mackill et al. (1996)
Rice YAC library	Umehara et al. (1996)
Rice BAC library	Jiang et al. (1995); Wang et al. (1995)
Rice genome draft	Goff et al. (2002); Yu et al. (2002)

Several useful applications of molecular markers are facilitated or enhanced by the availability of the rice genome sequence, and these will be greatly augmented with the discovery of functionally important genes. These advantages are already being realized in *Arabidopsis*. The sequence information itself can be used to identify new microsatellite markers in particular regions for saturation mapping at high resolution (Casacuberta et al. 2000). This approach has been used to identify 2537 of the 2740 experimentally verified SSR primer pairs now available (McCouch et al. 2002). Orthologous genomic sequences from two or more sources can also be used to develop single nucleotide polymorphisms (SNPs) that can be used to map or identify candidate genes through association mapping (Buckler and Thornsberry 2002; Rafalski 2002a, b). High-throughput genetic mapping using multiplexed SSRs and small mapping populations can be used to rapidly map important genes (Ponce et al. 1999) and determine their sequence in relatively small positional cloning experiments (Lukowitz et al. 2000).

The identification of the function of known genes will follow from the annotation of the sequence of the entire rice genome. Those genes having only "hypothetical" as the rationale for their annotation will need to be identified through the efforts of functional genomics. Following assignment of function, the most difficult and important part of this process will be the discovery of useful allelic variation for genes that affect economically important traits. Techniques that allow the mining of these useful alleles will produce the most useful molecular markers. These represent DNA sequence changes that confer improved phenotype in plants, and they can be used directly to follow segregation of these genes in segregating populations.

2 Gene Mapping with Molecular Markers in Rice

2.1 Molecular Maps

Rice benefited early from the development of RFLP maps, largely because of the coordinated rice biotechnology program of the Rockefeller Foundation and later the Rice Genome Program of Japan. The first genetic map was published in 1988 (McCouch et al. 1988), and this was followed by much denser maps (Causse et al. 1994; Harushima et al. 1998). RFLP markers were used to map several important traits in rice. These markers were very useful because of their reliability and well-defined map location. However, PCR-based markers such as RAPDs, AFLPs, and SSRs became more popular because of their ease of use. AFLP markers were particularly useful because of the large number of markers that could be determined with few reactions (Cho et al. 1998). They have also been useful in fine-scale mapping because of the potentially large number of markers available (Xu et al. 2000). However, SSR markers have proven to be the most popular because of their high polymorphism and codominant inheritance (Chen et al. 1997; Temnykh et al. 2000). Primers and map positions for thousands of SSR markers are now available with an experimentally verified SSR placed on average every 157 kb of sequence (McCouch et al. 2002), and researchers can develop their own markers from the sequence data.

Early mapping studies relied on F_2 or F_2-derived F_3 ($F_{2:3}$) populations, which are still useful for major gene traits because they are easy to develop. However, these populations have limited use beyond the initial study and they cannot be easily regenerated or maintained. Fixed populations developed by repeated self-pollination (recombinant inbred lines, RILs), anther culture (doubled haploids, DHs), or backcrossing (near-isogenic lines, NILs; substitution lines, near-isogenic introgression lines, NIILs) are preferred because they can be replicated indefinitely and used in many studies. Xu (2002) listed 14 permanent mapping populations being used in rice, but undoubtedly more have been developed. Molecular mapping data are available for these populations, and it is therefore easy to add new traits, assuming that genetic variation exists for these traits.

2.2 Mapping Useful Genes

The mapping of important traits, including those controlled by major genes or quantitative trait loci (QTLs), has been reviewed previously (Yano and Sasaki 1997; Mackill 1999; Mackill and Ni 2001; Xu 2002). Genetic mapping has resulted in much useful information on important major genes such as those for disease resistance and morphological traits. This information is particularly helpful in clarifying the allelism of genes conferring similar phenotypes. Linked markers are already being used in marker-assisted selection (MAS) programs for developing improved rice cultivars (Chen et al. 2000; Hittalmani et al. 2000; Sanchez et al. 2000; Gregorio et al. 2002).

The accumulated information is particularly interesting for identifying QTLs. One problem of mapping QTLs is that those with a relatively small effect (i.e., most of them) are difficult to accurately identify and map with good resolution. However, when several traits are being mapped in different populations and/or different environments, the nature of QTL variation in different germplasm can be assessed. Xu (2002) has provided examples of QTLs mapped across different populations. These comparisons indicate that some QTLs are important in diverse varieties and environments, while others are specific to a particular cultivar or location.

One of the implications of this finding is that the results from a single QTL study do not reveal a complete picture of the genetic control of a trait. Most QTL studies have used a parent with a high expression of the trait of interest. QTLs with the largest effects are the most interesting to breeders. However, different QTLs may show a strong effect in different donors, or under different conditions. Molecular approaches for identifying this type of variation (i.e., allele mining) will be described below. A sequential backcrossing approach, in which up to 200 "donors" representing maximum genetic diversity are crossed to one or a few elite cultivars, was described by Li (2001). Selection is practiced during the backcrossing stage to maximize the occurrence of desirable genotypes and allow the identification of chromosomal segments associated with improved phenotype. This approach will sample a larger number of alleles than standard QTL analysis, and will also allow the identification of alleles that express well in a desirable genetic background.

3 Positional Cloning

3.1 Examples of Positional Cloning in Rice

The positional cloning approach is appropriate for situations in which a gene is identified based on phenotype, but its function is unknown (forward genetics approach). This is the situation for most genes affecting agronomic traits. This approach has been used extensively in *Arabidopsis* and increasingly in rice.

The first gene cloned by this approach in rice (and in any monocot) was the *Xa21* gene for resistance to bacterial blight disease (Song et al. 1995). Several additional genes have been cloned by a map-based approach (Table 2), and many ongoing projects will result in new genes in the near future. While most of these projects involve major genes, two QTLs that control heading date, *Hd1* and *Hd6*, have been cloned. *Hd1* is an allele of the photoperiod-sensitive gene *Se1*, which is inherited as a major gene in many cultivars, but

Table 2. Positional cloning of genes in rice reported in the literature

Gene	Trait	Source of allele	Description of gene	Population size for map	Description of how gene was isolated	Reference
Xa21	Bacterial blight resistance	O. longistaminata	Receptor kinase with LRR	386	RFLP marker RG103 cosegregating with Xa21 was used to probe BAC subclones which were transformed into TP309	Song et al. (1995)
Xa1	Bacterial blight resistance	IRBB1, Kogyoku	NBS-LRR	4225	Selected by homology from 7 cDNA sequences that cosegregated with Xa1	Yoshimura et al. (1998)
d1	Dwarf	FL2 (mutant)	α subunit of GTP binding protein	13,000	Identified from a cDNA fragment cosegregating with d1	Ashikari et al. (1999)
Pib	Blast resistance	Tohoku IL9 (from source Engkatek)	NBS-LRR resistance gene	3305	Transcribed gene with NBS was found in 80-kb region determined by recombination	Wang et al. (1999)
Hd1	Se1, photoperiod sensitivity	Niponbare-Kasalath	Transcription factor homologous to CONSTANS	>9000	Region limited to 12 kb by recombination found to contain the CONSTANS homolog	Yano et al. (2000)
Pi-ta	Blast resistance	Tadukan	Cytoplasmic receptor, NBS	Not reported	Candidate with NBS was identified in sequences of BAC clones spanning about 850 kb	Bryan et al. (2000)
Hd6	Heading date QTL	Niponbare-Kasalath	α subunit of protein kinase CK2	2807	Only one EST found in the 26.4-kb region delimited by recombination	Takahashi et al. (2001)
Spl7	Spotted leaf lesion mimic	KL210 mutant line	Heat stress transcription factor (HSF)	2944	Gene prediction from 3-kb region delimited by recombination	Yamanouchi et al. (2002)
Sd1	Semidwarf gene	DGWG	Gibberellin 20 oxidase	3477	1 ORF identified in 6-kb interval	Monna et al. (2002)
Hd3	Heading date QTL	Niponbare-Kasalath	Protein similar to flowering locus (FT) of Arabidopsis	2207	Region limited to 20 kb by recombination found to contain the gene similar to FT	Kojima et al. (2002)

not in the genetic background (Nipponbare/Kasalath) from which it was cloned (Yano et al. 2000).

The usual approach for positional cloning is to develop a low-resolution map, followed by a high-resolution map with around 3000 or more F_2 progeny. Bryan et al. (2000), however, did low-coverage sequencing through a bacterial artificial chromosome (BAC) contig of about 1.5 cM representing about 850 kb of DNA to identify the location of the *Pi-ta* resistance gene for rice blast. For most traits, the large population is screened for the trait of interest. DNA pools can be used to reduce the number of marker assays (Ahn et al. 2002). Selected progeny can be screened for flanking markers obtained from the low-resolution map. However, phenotyping of the entire large population is not necessary. When measuring the trait is difficult, flanking markers from the low-resolution map can be used to screen the large population before plants are phenotyped (Xu et al. 2000). A subset of plants showing recombination between markers that flank the gene of interest is then screened for more markers. In some cases, AFLP markers are screened on these progeny. Other types of markers can be used, such as CAPS (cleaved amplified polymorphic sequence) or EST (expressed sequence tag) markers. A physical map is created using large-insert clones such as YACs (yeast artificial chromosomes) or BACs. One or two of the clones can be sequenced, and there is an attempt to narrow the location of the gene to as small a fragment as possible by determining where the closest recombination events occur. Candidate genes from this segment are ultimately evaluated through transformation. However, if a strong candidate can be identified in a small interval flanked by recombination events around the gene, sequence differences characteristic of the particular phenotype would be a good indication that the candidate is the gene of interest, and supporting evidence could be provided by expression data.

3.2 Future Use of Positional Cloning

With a small genome size and good transformation protocols, rice has been a good system for positional cloning. The relatively few cases reported so far (Table 2) should be augmented greatly in the near future. These projects have not been trivial undertakings, and they usually involve several years of intensive work. It would be expected that the genome sequence information would greatly accelerate the process of positional cloning as has been the case with *Arabidopsis* (Jander et al. 2002).

A formula was devised for determining the number of progeny required to identify crossovers on either side of a candidate gene in *Arabidopsis* (Durrett et al. 2002):

$$P = 1 - [1 + NT/(100R)]e^{-NT/(100R)}$$

where R is the kb/cM ratio for the region, N is the number of gametes to sample (number of testcross progeny or twice the number of F_2 plants, and P is the probability of finding a minimum of two crossovers, one on each side of the gene, at a physical distance less than Durrett et al. (2002) point out that, assuming that a probability of 0.95 is needed for crossovers within a BAC clone, the population size would be about 600 F_2 plants. This was computed for a region where 250 kb corresponded to 1 cM genetic distance, which is approximately the average for the rice genome. However, the formula assumes constant recombination within the target region, and this would not be the case over the entire rice genome.

Positional cloning efforts in rice typically rely on large populations of more than 3000 F_2 individuals for fine-scale mapping (Table 2). By the above formula, using a population of 3000 F_2 plants, an interval of 10 kb could be delimited by recombination with a probability of 0.69 assuming that $R=250$ kb/cM, or a probability of about 0.34 if a more conservative R of 500 is used. In the future, however, such large population sizes should not be needed. Annotation of the sequence will be improved markedly with data from large-scale expression studies using microarrays and large mutant collections. The strategy of map-based cloning will merge with the other methods of functional genomics. This will be particularly important for identifying genes underlying QTLs, for which map position is more difficult to determine (Wayne and McIntyre 2002). Molecular markers are used to define the position of the genes controlling a particular trait and to detect selection for a trait in artificial or natural populations, and potential candidate genes can be identified by reference to genes of known function and expression data. Even in such a strategy, it will be helpful to have good resolution of the genes or QTLs. On average, a 1-cM genetic distance would correspond to nearly 30 potential candidates.

4 Array-Based Markers

To date, most efforts using microarrays in rice have focused on expression profiling. Studies have been published for expression analyses using rice ESTs for responses during salt stress (Kawasaki et al. 2001) and resistance to blast disease (Rao et al. 2002), oligonucleotides for 21,000 rice genes for nutrient partitioning during grain filling (Zhu et al. 2003), and full-length cDNA clones (Kikuchi et al. 2002) for monitoring chemical induction of disease resistance (Shimono et al. 2003). The full-length cDNA microarray developed by Kikuchi and collaborators has also been used to analyze gene expression during Fe-deficiency stress in barley (Negishi et al. 2002). Kikuchi and collaborators have recently published their rice expression database (Yazaki et al. 2002; http://red.dna.affrc.go.jp/RED/), which will serve as the entry point for microarray analyses from a consortium of more than 50 laboratories.

Nevertheless, microarrays promise to be a convenient route for genotyping and mapping. Current applications of array-based markers involve the use of either genomic clones as in the diversity array technology or DArT (Jaccoud et al. 2001) or oligonucleotides as probes. The latter technique has centered around the use of GeneChips, such as those produced by Affymetrix (Fodor et al. 1991, 1993; Pease et al. 1994), wherein oligonucleotides are synthesized directly on the array substrate. Such arrays typically involve tens if not hundreds of thousands of oligonucleotide "features".

DArT was designed as a means to produce diversity fingerprints via DNA/DNA hybridization between reduced-complexity genomic clones and target germplasm. This is a sequence-independent method that does not rely on sequence information to identify the clones on the array. Hence, this approach is amenable to any organism for which little or no sequence data exist. Jaccoud et al. (2001) developed this technique using a panel of nine *Oryza sativa* lines as proof of concept for the procedure. If the clones on the diversity array are anchored to a genetic map, the arrays can then be used not only for fingerprinting diversity, but also to provide molecular markers for mapping. We are currently implementing DArT at IRRI to characterize rice genetic diversity, while, at CIAT, diversity arrays are under development for *Phaseolus* spp. (J. Tohme, pers. comm.).

For oligonucleotide arrays, several genotyping examples as well as their use in mapping have now been published on nonhuman model species. These studies include work on yeast (Winzeler et al. 1998, 2003; Steinmetz et al. 2002), *Arabidopsis* (Cho et al. 1999; Spiegelman et al. 2000; Borevitz et al. 2003), mouse (Lindblad-Toh et al. 2000), and zebrafish (Stickney et al. 2002). Two approaches have been used for the design of the oligonucleotide arrays in these studies. The first approach is the variation detector array, in which 16 features are routinely synthesized for each locus (the coding and noncoding strands for two alleles with all four combinations of bases for the polymorphic site of each allele). The advantage of this approach is the accurate identification of the SNP at the locus, whereas the main disadvantages are the need to know the base constitution for the alleles to be queried and the high production costs for the design, optimization, and synthesis of the arrays. Other approaches for using SNPs as markers are described in the next section.

The second approach uses existing oligonucleotide arrays developed for expression analyses to identify genetic variation as measured by differential hybridization intensities to features on the array (Borevitz et al. 2003; Winzeler et al. 2003). The polymorphisms discovered in this manner have been termed "single-feature polymorphisms" or SFPs since the specific identity of the nucleotides leading to the differential hybridization may not be known. A recent development in oligonucleotide array technology is the maskless array system pioneered by NimbleGen Systems, Inc. (Singh-Gasson et al. 1999; Nuwaysir et al. 2002). This method of array synthesis uses a digital-micromirror system to direct light at specific elements during each round of synthesis, thus allowing for quick turnaround in array design and optimization. A pilot project to develop rice oligonucleotide arrays for profiling abiotic and biotic stress-related genetic diversity and expression using the

NimbleGen approach has begun between David Frisch at the Genome Center of the University of Wisconsin and IRRI.

5 Candidate Genes as Markers

A dense microsatellite map has been developed in rice, and most mapping studies now rely on these markers. They have also proven to be popular for applications in marker-assisted breeding. The many advantages of SSR markers have been described, and foremost is their high level of polymorphism, which allows them to be used in a wide range of germplasm. Yet, they can still be problematic to use in closely related germplasm. For example, in temperate japonica cultivars, genetic diversity can be relatively low despite marked differences in phenotypes (Mackill et al. 1996; Ni et al. 2002). However, with such an abundance of microsatellite markers, estimated to exist at one per 8000 bp on average from sequence data (Goff et al. 2002), even these limitations could be overcome to develop markers at specific locations in the genome.

The outcome of functional genomics research should allow the identification of gene function for all the rice genes, and this will provide a means of manipulating these genes directly for cultivar improvement. However, this is seen as a long-term objective, and the development of useful products will depend on many factors, such as how a useful phenotype can be produced by manipulating the genes. In the shorter term, these candidate genes can be used directly in identifying favorable alleles and following their inheritance in segregating populations. The use of gene sequences as selectable markers has several advantages over the use of SSR or other markers that are obtained from linkage maps. Linked markers will always have the problem of recombination, often necessitating the use of flanking markers for selection. Furthermore, identification of the sequence change that imparts a desirable phenotype will allow the development of a marker specific for the favorable allele. In addition to avoiding any problem with recombination, this would also allow the use of the marker in nearly any population, and also as a general screen of germplasm or elite breeding lines for genes of interest.

Markers specific for alleles of a gene are most likely those that detect single nucleotide polymorphisms. Alignment of genomic sequence from the *japonica* and *indica* subspecies as well as cDNA or EST sequences from other varieties will allow SNPs to be located and primers designed for the target alleles or intervals. Numerous experimental techniques are available for SNP detection (for recent reviews see Gut 2001, Kwok 2001, Syvänen 2001 and Kirk et al. 2002). Basically, existing SNP detection methods rely on four reaction principles: hybridization with allele-specific probes, oligonucleotide ligation, single nucleotide primer extension, or enzymatic cleavage. The separation step(s) of the assays can occur: (1) on a solid support (microarray, microtiter plate, or microspheres); (2) in liquid phase by electrophoresis, flow cytometry, or denaturing high-performance liquid chromatography; or (3) in liquid phase with no requirement on separation. Products are then detected by indirect colorimetry, mass spectrometry, fluorescence, fluorescence resonance energy transfer, fluorescence polarization, or chemiluminescence. The choice of any particular method will be determined by the ease of automation, sensitivity of the assay, feasibility of multiplexing, or whether de novo sequencing is a prerequisite. Recently, Nasu et al. (2002) identified 2800 SNPs located in 417 regions from three *Oryza sativa* subsp. *japonica* cultivars, two *indica* cultivars, and one wild *O. rufipogon* accession by sequencing and aligning about 250 kb. From these SNPs, they established a set of 213 codominant SNP markers suitable for use in molecular breeding.

The targeted local lesions in genomes (TILLING) approach developed for reverse genetics (Colbert et al. 2001; Till et al. 2003) is also applicable for SNP genotyping and discovery. Genomic loci are chosen for querying, and differentially labeled primers are produced for this locus. These primers are used to amplify PCR products from pools of chemically mutagenized plant lines. Following denaturation/renaturation, the PCR products are treated with the enzyme CEL-I that only cuts mismatched base pairs as small as a single base. If, for the target locus, a chemically induced mutation occurred in one or more lines, cleaved products can be visualized on automated genotypers as new bands. If, instead of pools of chemically mutagenized plants, heteroduplexes are formed between a reference cultivar and a query cultivar, genetic variation in the form of SNPs or indels can be detected, and this application has been termed "EcoTILLING" (Comai et al. 2004).

SSR markers could still be used after the identification of candidate genes. Their convenience, codominant inheritance, and high polymorphism may make SSRs preferable over SNPs identified from the favorable allele. SSR markers adjacent to or within genes can serve for this purpose. An example is the *waxy* gene, which contains a microsatellite within it (Bligh et al. 1995). For this SSR locus, differences in amylose content are associated with repeat length (Ayres et al. 1997).

6 Conclusions

The impact of the DNA sequence of rice is just being felt and undoubtedly many functional genes will be identified in the next few years. In addition to using this information to improve rice and other crops through a transgenic approach, there will be a need for mining alleles of important genes from the largely underused germplasm collections. As an entry point for this effort, a core collection of 11,200 accessions from the International Rice Genebank Collection (IRGC) has recently been developed at IRRI. The set of accessions chosen for the core collection was based on species, variety group, ecocultural type, source location, estimates of possible deployment based on tracing pedigrees in the International Rice Information System crop information database, and characterization data.

The level of coverage of accessions in the IRGC is about 11% for the cultivated species *O. sativa* and *glaberrima* and from 100 to 20% for the wild species (depending on the number of accessions per species in the IRGC). These levels of coverage seem adequate such that the core collection will encompass most of the diversity contained in the entire collection. This set of germplasm is currently being processed to produce an archive of lyophilized tissue and the genomic DNA. SSR and/or DArT fingerprinting of the core collection materials will be carried out in the near future to define population structure as a prerequisite for association mapping. In addition, candidate genes for a wide variety of biotic and abiotic stresses and nutritional factors are being identified, and markers for these target loci established. PCR-based methods such as EcoTILLING, locus-specific SNP detection, or length polymorphisms will be used to identify alleles in the core collection. Pooling strategies enabling the detection of infrequent alleles will be used to increase throughput as much as possible.

This process seems likely to identify numerous novel alleles; accessions carrying these alleles will be phenotyped for traits appropriate to the physiological function of the candidate genes. The analysis of these data by association mapping and/or linkage disequilibrium will identify those alleles that make a positive contribution to the phenotype of interest. Marker-assisted breeding programs can then benefit by an infusion of new alleles with markers that are the underlying genes or are located within the haplotype block determined by the extent of linkage disequilibrium at that locus. Furthermore, the identification of novel alleles from a wide range of germplasm will lead to the pyramiding of alleles in favorable genetic backgrounds by breeding schemes using multiple donors that previously would have been difficult to devise. Through allele mining, the products of functional genomics will be delivered to the ultimate end-users, farmers in the developing world, in the form of new rice varieties with enhanced nutrition and improved tolerance for biotic and abiotic stresses.

References

Ahn BO, Miyoshi K, Itoh JI, Nagato Y, Kurata N (2002) A genetic and physical map of the region containing *plastochron1*, a heterochronic gene, in rice (*Oryza sativa* L.). Theor Appl Genet 105:654–659

Ahn SN, Bollich CN, McClung AM, Tanksley SD (1993) RFLP analysis of genomic regions associated with cooked-kernel elongation in rice. Theor Appl Genet 87:27–32

Ashikari M, Wu JZ, Yano M, Sasaki T, Yoshimura A (1999) Rice gibberellin-insensitive dwarf mutant gene *Dwarf 1* encodes the alpha-subunit of GTP-binding protein. Proc Natl Acad Sci USA 96:10284–10289

Ayres NM, McClung AM, Larkin PD, Bligh HFJ, Jones CA, Park WD (1997) Microsatellites and a single-nucleotide polymorphism differentiate apparent amylose classes in an extended pedigree of US rice germ plasm. Theor Appl Genet 94:773–781

Bligh HFJ, Till RI, Jones CA (1995) A microsatellite sequence closely linked to the *waxy* gene of *Oryza sativa*. Euphytica 86:83–85

Borevitz JO, Liang D, Plouffe D, Chang H-S, Zhu T, Weigel D, Berry CC, Winzeler E, Chory J (2003) Large-scale identification of single-feature polymorphisms in complex genomes. Genome Res 13:513–523

Bryan GT, Wu KS, Farrall L, Jia YL, Hershey HP, McAdams SA, Faulk KN, Donaldson GK, Tarchini R, Valent B (2000) A single amino acid difference distinguishes resistant and susceptible alleles of the rice blast resistance gene *Pi-ta*. Plant Cell 12:2033–2045

Buckler ES, Thornsberry JM (2002) Plant molecular diversity and applications to genomics. Curr Opin Plant Biol 5:107–111

Casacuberta E, Puigdomenech P, Monfort A (2000) Distribution of microsatellites in relation to coding sequences within the *Arabidopsis thaliana* genome. Plant Sci 157:97–104

Causse MA, Fulton TM, Cho YG, Ahn SN, Chunwongse J, Wu KS, Xiao JH, Yu ZH, Ronald PC, Harrington SE, Second G, McCouch SR, Tanksley SD (1994) Saturated molecular map of the rice genome based on an interspecific backcross population. Genetics 138:1251–1274

Chen S, Lin XH, Xu CG, Zhang QF (2000) Improvement of bacterial blight resistance of 'Minghui 63', an elite restorer line of hybrid rice, by molecular marker-assisted selection. Crop Sci 40:239–244

Chen X, Temnykh S, Xu Y, Cho YG, McCouch SR (1997) Development of a microsatellite framework map providing genome-wide coverage in rice (*Oryza sativa* L.). Theor Appl Genet 95:553–567

Cho RJ, Mindrinos M, Richards DR, Sapolsky RJ, Anderson M, Drenkard E, Dewdney L, Reuber TL, Stammers M, Federspiel N, Theologis A, Yang WH, Hubbell E, Au M, Chung EY, Lashkari D, Lemieux B, Dean C, Lipshutz RJ, Ausubel FM, Davis RW, Oefner PJ (1999) Genome-wide mapping with biallelic markers in *Arabidopsis thaliana*. Nat Genet 23:203–207

Cho YG, Blair MW, Panaud O, McCouch SR (1996) Cloning and mapping of variety-specific rice genomic DNA sequences: amplified fragment length polymorphisms (AFLP) from silver-stained polyacrylamide gels. Genome 39:373–378

Cho YG, McCouch SR, Kuiper M, Kang MR, Pot J, Groenen JTM, Eun MY (1998) Integrated map of AFLP, SSLP and RFLP markers using a recombinant inbred population of rice (*Oryza sativa* L.). Theor Appl Genet 97:370–380

Colbert T, Till BJ, Tompa R, Reynolds S, Steine MN, Yeung AT, McCallum CM, Comai L, Henikoff S (2001) High-throughput screening for induced point mutations. Plant Physiol 126:480–484

Comai L, Young K, Reynolds SH, Codomo C, Enns L, Johnson J, Burtner C, Henikoff JG, Greene EA, Till BJ, Henikoff S (2004) Efficient discovery of nucleotide polymorphisms in populations by ecotilling. Plant J (in press)

Datta SK, Peterhaus A, Datta K, Potrykus I (1988) Genetically engineered fertile Indica-rice recovered from protoplasts. Bio/Technology 8:736–740

Durrett RT, Chen KY, Tanksley SD (2002) A simple formula useful for positional cloning. Genetics 160:353–355

Fodor SP, Read JL, Pirrung MC, Stryer L, Lu AT, Solas D (1991) Light-directed, spatially addressable parallel chemical synthesis. Science 251:767–773

Fodor SP, Rava RP, Huang XC, Pease AC, Holmes CP, Adams CL (1993) Multiplexed biochemical assays with biological chips. Nature 364:555–556

Glaszmann JC (1987) Isozymes and classification of Asian rice varieties. Theor Appl Genet 74:21–30

Goff SA, Ricke D, Lan TH, Presting G, Wang RL, Dunn M, Glazebrook J, Sessions A, Oeller P, Varma H, Hadley D, Hutchinson D, Martin C, Katagiri F, Lange BM, Moughamer T, Xia Y, Budworth P, Zhong JP, Miguel T, Paszkowski U, Zhang SP, Colbert M, Sun WL, Chen LL, Cooper B, Park S, Wood TC, Mao L, Quail P, Wing R, Dean R, Yu YS, Zharkikh A, Shen R, Sahasrabudhe S, Thomas A, Cannings R, Gutin A, Pruss D, Reid J, Tavtigian S, Mitchell J, Eldredge G, Scholl T, Miller RM, Bhatnagar S, Adey N, Rubano T, Tusneem N, Robinson R, Feldhaus J, Macalma T, Oliphant A, Briggs S (2002) A draft sequence of the rice genome (*Oryza sativa* L. ssp. *japonica*). Science 296:92–100

Gregorio GB, Senadhira D, Mendoza RD, Manigbas NL, Roxas JP, Guerta CQ (2002) Progress in breeding for salinity tolerance and associated abiotic stresses in rice. Field Crop Res 76:91–101

Gut IW (2001) Automation in genotyping of single nucleotide polymorphisms. Hum Mutat 17:475–492

Harushima Y, Yano M, Shomura P, Sato M, Shimano T, Kuboki Y, Yamamoto T, Lin SY, Antonio BA, Parco A, Kajiya H, Huang N, Yamamoto K, Nagamura Y, Kurata N, Khush GS, Sasaki T (1998) A high-density rice genetic linkage map with 2275 markers using a single F_2 population. Genetics 148:479–494

Hiei Y, Ohta S, Komari T, Kumashiro T (1994) Efficient transformation of rice (*Oryza sativa* L.) mediated by *Agrobacterium* and sequence analysis of the boundaries of the T-DNA. Plant J 6:271–282

Hittalmani S, Parco A, Mew TV, Zeigler RS, Huang N (2000) Fine mapping and DNA marker-assisted pyramiding of the three major genes for blast resistance in rice. Theor Appl Genet 100:1121–1128

Jaccoud D, Peng K, Feinstein D, Kilian A (2001) Diversity arrays: a solid state technology for sequence information independent genotyping. Nucleic Acids Res 29:E25

Jander G, Norris SR, Rounsley SD, Bush DF, Levin IM, Last RL (2002) Arabidopsis map-based cloning in the post-genome era. Plant Physiol 129:440–450

Jiang J, Gill BS, Wang GL, Ronald PC, Ward DC (1995) Metaphase and interphase fluorescence in situ hybridization mapping of the rice genome with bacterial artificial chromosomes. Proc Natl Acad Sci USA 92:4487–4491

Kaul S, Koo HL, Jenkins J, Rizzo M, Rooney T, Tallon LJ, Feldblyum T, Nierman W, Benito MI, Lin XY, Town CD, Venter JC, Fraser CM, Tabata S, Nakamura Y, Kaneko T, Sato S, Asamizu E, Kato T, Kotani H, Sasamoto S, Ecker JR, Theologis A, Federspiel NA, Palm CJ, Osborne BI, Shinn P, Conway AB, Vysotskaia VS, Dewar K, Conn L, Lenz CA, Kim CJ, Hansen NF, Liu SX, Buehler E, Altafi H, Sakano H, Dunn P, Lam B, Pham PK, Chao QM, Nguyen M, Yu GX, Chen HM, Southwick A, Lee JM, Miranda M, Toriumi MJ, Davis RW, Wambutt R, Murphy G, Dusterhoft A, Stiekema W, Pohl T, Entian KD, Terryn N, Volckaert G, Choisne N, Rieger M, Ansorge W, Unseld M, Fartmann B, Valle G, Artiguenave F, Weissenbach J, Quetier F, Wilson RK, de la Bastide M, Sekhon M, Huang E, Spiegel L, Gnoj L, Pepin K, Murray J, Johnson D, Habermann K, Dedhia N, Parnell L, Preston R, Hillier L, Chen E, Marra M, Martienssen R, McCombie WR, Mayer K, White O, Bevan M, Lemcke K, Creasy TH, Bielke C, Haas B, Haase D, Maiti R, Rudd S, Peterson J, Schoof H, Frishman D, Morgenstern B, Zaccaria P, Ermolaeva M, Pertea M, Quackenbush J, Volfovsky N, Wu DY, Lowe TM, Salzberg SL, Mewes HW, Rounsley S, Bush D, Subramaniam S, Levin I, Norris S, Schmidt R, Acarkan A, Bancroft I, Brennicke A, Eisen JA, Bureau T, Legault BA, Le QH, Agrawal N, Yu Z, Copenhaver GP, Luo S, Pikaard CS, Preuss D, Paulsen IT, Sussman M, Britt AB, Selinger DA, Pandey R, Mount DW, Chandler VL, Jorgensen RA, Pikaard C, Juergens G, Meyerowitz EM, Dangl J, Jones JDG, Chen M, Chory J, Somerville C (2000) Analysis of the genome sequence of the flowering plant *Arabidopsis thaliana*. Nature 408:796–815

Kawasaki S, Borchert C, Deyholos M, Wang H, Brazille S, Kawai K, Galbraith D, Bohnert HJ (2001) Gene expression profiles during the initial phase of salt stress in rice. Plant Cell 13:889–905

Kikuchi S, Yazaki J, Kishimoto N, Ishikawa M, Kojima K, Namiki T, Shimbo K, Fujii F, Ohata T, Shimatani Z, Hashimoto A, Nagata Y, Honda S, Toyoshima K, Sakata K, Yamamoto K, Sasaki T (2002) Rice functional genomics via cDNA microarray: systems for the microarray analysis and the expression profiles of stress responsible gene expression. JIRCAS Working Rep 23:93–98

Kirk BW, Feinsod M, Favis R, Kliman RM, Barany F (2002) Single nucleotide polymorphism seeking long term association with complex disease. Nucleic Acids Res 30:3295-3311

Kojima S, Takahashi Y, Kobayashi Y, Monna L, Sasaki T, Araki T, Yano M (2002) Hd3a, a rice ortholog of the *Arabidopsis FT* gene, promotes transition to flowering downstream of *Hd1* under short-day conditions. Plant Cell Physiol 43:1096-1105

Kwok PY (2001) Methods for genotyping single nucleotide polymorphisms. Annu Rev Genomics Hum Genet 2:235-258

Li Z (2001) QTL mapping in rice: a few critical considerations. In: Khush GS, Brar DS, Hardy B (eds) Rice genetics IV. International Rice Research Institute, Los Baños, Philippines, pp 153-171

Lindblad-Toh K, Winchester E, Daly MJ, Wang DG, Hirschhorn JN, Laviolette JP, Ardlie K, Reich DE, Robinson E, Sklar P, Shah N, Thomas D, Fan JB, Gingeras T, Warrington J, Patil N, Hudson TJ, Lander ES (2000) Large-scale discovery and genotyping of single-nucleotide polymorphisms in the mouse. Nat Genet 24:381-386

Lukowitz W, Gillmor CS, Scheible WR (2000) Positional cloning in Arabidopsis. Why it feels good to have a genome initiative working for you. Plant Physiol 123:795-805

Mackill DJ (1999) Genome analysis and breeding. In: Shimamoto K (ed) Molecular biology of rice. Springer, Berlin Heidelberg New York, pp 17-41

Mackill DJ, Ni J (2001) Molecular mapping and marker-assisted selection for major-gene traits in rice. In: Khush GS, Brar DS, Hardy B (eds) Rice genetics IV. Proceedings of the 4th international rice genetics symposium, 22-27 Oct 2000. International Rice Research Institute, Los Baños, Philippines, pp 137-151

Mackill DJ, Zhang Z, Redona ED, Colowit PM (1996) Level of polymorphism and genetic mapping of AFLP markers in rice. Genome 39:969-977

McCouch SR, Kochert G, Yu ZH, Wang ZY, Khush GS, Coffman WR, Tanksley SD (1988) Molecular mapping of rice chromosomes. Theor Appl Genet 76:815-829

McCouch SR, Teytelman L, Xu Y, Lobos KB, Clare K, Walton M, Fu B, Maghirang R, Li Z, Zing Y, Zhang Q, Kono I, Yano M, Fjellstrom R, DeClerck G, Schneider D, Cartinhour S, Ware D, Stein L (2002) Development and mapping of 2240 new SSR markers for rice (*Oryza sativa* L.). DNA Res 9:199-207

Monna L, Kitazawa N, Yoshino R, Suzuki J, Masuda H, Maehara Y, Tanji M, Sato M, Nasu S, Minobe Y (2002) Positional cloning of rice semidwarfing gene, *sd-1*: rice "green revolution gene" encodes a mutant enzyme involved in gibberellin synthesis. DNA Res 9:11-17

Nasu S, Suzuki J, Hasegawa K, Yui R, Kitawa N, Monna L, Minobe Y (2002) Search for and analysis of single nucleotide polymorphisms in rice (*Oryza sativa, Oryza rufipogon*) and establishment of SNP markers. DNA Res 9:163-171

Negishi T, Nakanishi H, Yazaki J, Kishimoto N, Fujii F, Shimbo K, Yamamoto K, Sakata K, Sasaki T, Kikuchi S, Mori S, Nishizawa NK (2002) cDNA microarray analysis of gene expression during Fe-deficiency stress in barley suggests that polar transport of vesicles is implicated in phytosiderophore secretion in Fe-deficient barley roots. Plant J 30:83-94

Ni J, Colowit PM, Mackill DJ (2002) Evaluation of genetic diversity in rice subspecies using microsatellite markers. Crop Sci 42:601-607

Nuwaysir EF, Huang W, Albert TJ, Singh J, Nuwaysir K, Pitas A, Richmond T, Gorski T, Berg JP, Ballin J, McCormick M, Norton J, Pollock T, Sumwalt T, Butcher L, Porter D, Molla M, Hall C, Blattner F, Sussman MR, Wallace RL, Cerrina F, Green RD (2002) Gene expression analysis using oligonucleotide arrays produced by maskless photolithography. Genome Res 12:1749-1755

Pease AC, Solas D, Sullivan EJ, Cronin MT, Holmes CP, Fodor SP (1994) Light-generated oligonucleotide arrays for rapid DNA sequence analysis. Proc Natl Acad Sci USA 91:5022-5026

Ponce MR, Robles P, Micol JL (1999) High-throughput genetic mapping in *Arabidopsis thaliana*. Mol Gen Genet 261:408-415

Rafalski A (2002a) Applications of single nucleotide polymorphisms in crop genetics. Curr Opin Plant Biol 5:94-100

Rafalski JA (2002b) Novel genetic mapping tools in plants: SNPs and LD-based approaches. Plant Sci 162:329–333

Rao ZM, Dong HT, Zhuang JY, Chai RY, Fan YY, Li DB, Zheng KL (2002) Analysis of gene expression profiles during host *Magnaporthe grisea* interactions in a pair of near isogenic lines of rice (in Chinese). Yi Chuan Xue Bao 29:887–893

Sanchez AC, Brar DS, Huang N, Li Z, Khush GS (2000) Sequence tagged site marker-assisted selection for three bacterial blight resistance genes in rice. Crop Sci 40:792–797

Second G (1982) Origin of the genic diversity of cultivated rice (*Oryza* spp.): study of the polymorphism scored at 40 isozyme loci. Jpn J Genet 57:25–57

Shimono M, Yazaki J, Nakamura K, Kishimoto N, Kikuchi S, Iwano M, Yomamoto K, Sakata K, Sasaki T, Nishiguchi M (2003) cDNA microarray analysis of gene expression in rice plants treated with probenazole, a chemical inducer of disease resistance. J Gen Plant Pathol 69:76–82

Singh-Gasson S, Green RD, Yue Y, Nelson C, Blattner F, Sussman MR, Cerrina F (1999) Maskless fabrication of light-directed oligonucleotide microarrays using a digital micromirror array. Nat Biotechnol 17:974–978

Song WY, Wang GL, Chen LL, Kim HS, Pi LY, Holsten T, Gardner J, Wang B, Zhai WX, Zhu LH, Fauquet C, Ronald P (1995) A receptor kinase-like protein encoded by the rice disease resistance gene, *Xa21*. Science 270:1804–1806

Spiegelman JI, Mindrinos MN, Fankhauser C, Richards D, Lutes J, Chory J, Oefner PJ (2000) Cloning of the Arabidopsis *RSF1* gene by using a mapping strategy based on high-density DNA arrays and denaturing high-performance liquid chromatography. Plant Cell 12:2485–2498

Steinmetz LM, Sinha H, Richards DR, Spiegleman JI, Oefner PJ, McCusker JH, Davis RW (2002) Dissecting the architecture of a quantitative trait locus in yeast. Nature 416:326–330

Stickney HL, Schmutz J, Woods IG, Holtzer CC, Dickson MC, Kelly PD, Myers RM, Talbot WS (2002) Rapid mapping of zebrafish mutations with SNPs and oligonucleotide microarrays. Genome Res 12:1929–1934

Syvänen AC (2001) Accessing genetic variation: genotyping single nucleotide polymorphisms. Nat Rev Genet 2:930–942

Takahashi Y, Shomura A, Sasaki T, Yano M (2001) *Hd6*, a rice quantitative trait locus involved in photoperiod sensitivity, encodes the alpha subunit of protein kinase CK2. Proc Natl Acad Sci USA 98:7922–7927

Temnykh S, Park WD, Ayres N, Cartinhour S, Hauck N, Lipovich L, Cho YG, Ishii T, McCouch SR (2000) Mapping and genome organization of microsatellite sequences in rice (*Oryza sativa* L.). Theor Appl Genet 100:697–712

Temnykh S, DeClerck G, Lukashova A, Lipovich L, Cartinhour S, McCouch S (2001) Computational and experimental analysis of microsatellites in rice (*Oryza sativa* L.): frequency, length variation, transposon associations, and genetic marker potential. Genome Res 11:1441–1452

Till BJ, Reynolds SH, Greene EA, Codomo CA, Enns LC, Johnson JE, Burtner C, Odden AR, Young K, Taylor NE, Henikoff JG, Comai L, Henikoff S (2003) Large-scale discovery of induced point mutations with high-throughput TILLING. Genome Res 13:524–530

Toriyama K, Arimoto Y, Uchimiya H, Hinata K (1988) Transgenic rice plants after direct gene transfer into protoplasts. Bio/Technology 6:1072–1074

Umehara Y, Tanoue H, Kurata N, Ashikawa I, Minobe Y, Sasaki T (1996) An ordered yeast artificial chromosome library covering over half of rice chromosome 6. Genome Res 6:935–942

Wang GL, Mackill DJ, Bonman JM, McCouch SR, Champoux MC, Nelson RJ (1994) RFLP mapping of genes conferring complete and partial resistance to blast in a durably resistant rice cultivar. Genetics 136:1421–1434

Wang GL, Holsten TE, Song WY, Wang HP, Ronald PC (1995) Construction of a rice bacterial artificial chromosome library and identification of clones linked to the *Xa-21* disease resistance locus. Plant J 7:525–533

Wang ZX, Yano M, Yamanouchi U, Iwamoto M, Monna L, Hayasaka H, Katayose Y, Sasaki T (1999) The *Pib* gene for rice blast resistance belongs to the nucleotide binding and leucine-rich repeat class of plant disease resistance genes. Plant J 19:55–64

Wayne ML, McIntyre LM (2002) Combining mapping and arraying: An approach to candidate gene identification. Proc Natl Acad Sci USA 99:14903–14906

Winzeler EA, Richards DR, Conway AR, Goldstein AL, Kalman S, McCullough MJ, McCusker JH, Stevens DA, Wodicka L, Lockhart DJ, Davis RW (1998) Direct allelic variation scanning of the yeast genome. Science 281:1194–1197

Winzeler EA, Castillo-Davis CI, Oshiro G, Liang D, Richards DR, Zhou Y, Hartl DL (2003) Genetic diversity in yeast assessed with whole-genome oligonucleotide arrays. Genetics 163:79–89

Wu KS, Tanksley SD (1993) Abundance, polymorphism and genetic mapping of microsatellites in rice. Mol Gen Genet 241:225–235

Xu K, Xu X, Ronald PC, Mackill DJ (2000) A high-resolution linkage map in the vicinity of the rice submergence tolerance locus *Sub1*. Mol Gen Genet 263:681–689

Xu Y (2002) Global view of QTL: rice as a model. In: Kang MS (ed) Quantitative genetics, genomics and plant breeding. CAB International, Wallingford, pp 109–134

Yamanouchi U, Yano M, Lin H, Ashikari M, Yamada K (2002) A rice spotted leaf gene, *Spl7*, encodes a heat stress transcription factor protein. Proc Natl Acad Sci USA 99:7530–7535

Yano M, Sasaki T (1997) Genetic and molecular dissection of quantitative traits in rice. Plant Mol Biol 35:145–153

Yano M, Katayose Y, Ashikari M, Yamanouchi U, Monna L, Fuse T, Baba T, Yamamoto K, Umehara Y, Nagamura Y, Sasaki T (2000) *Hd1*, a major photoperiod sensitivity quantitative trait locus in rice, is closely related to the Arabidopsis flowering time gene *CONSTANS*. Plant Cell 12:2473–2483

Yazaki J, Kishimoto N, Ishikawa M, Kikuchi S (2002) Rice Expression Database: the gateway to rice functional genomics. Trends Plant Sci 7:563–564

Yoshimura S, Yamanouchi U, Katayose Y, Toki S, Wang ZX, Kono I, Kurata N, Yano M, Iwata N, Sasaki T (1998) Expression of *Xa1*, a bacterial blight-resistance gene in rice, is induced by bacterial inoculation. Proc Natl Acad Sci USA 95:1663–1668

Yu J, Hu SN, Wang J, Wong GKS, Li SG, Liu B, Deng YJ, Dai L, Zhou Y, Zhang XQ, Cao ML, Liu J, Sun JD, Tang JB, Chen YJ, Huang XB, Lin W, Ye C, Tong W, Cong LJ, Geng JN, Han YJ, Li L, Li W, Hu GQ, Huang XG, Li WJ, Li J, Liu ZW, Liu JP, Qi QH, Liu JS, Li T, Wang XG, Lu H, Wu TT, Zhu M, Ni PX, Han H, Dong W, Ren XY, Feng XL, Cui P, Li XR, Wang H, Xu X, Zhai WX, Xu Z, Zhang JS, He SJ, Zhang JG, Xu JC, Zhang KL, Zheng XW, Dong JH, Zeng WY, Tao L, Ye J, Tan J, Ren XD, Chen XW, He J, Liu DF, Tian W, Tian CG, Xia HG, Bao QY, Li G, Gao H, Cao T, Zhao WM, Li P, Chen W, Wang XD, Zhang Y, Hu JF, Liu S, Yang J, Zhang GY, Xiong YQ, Li ZJ, Mao L, Zhou CS, Zhu Z, Chen RS, Hao BL, Zheng WM, Chen SY, Guo W, Li GJ, Liu SQ, Tao M, Zhu LH, Yuan LP, Yang HM (2002) A draft sequence of the rice genome (*Oryza sativa* L. ssp *indica*). Science 296:79–92

Yu ZH, Mackill DJ, Bonman JM, Tanksley SD (1991) Tagging genes for blast resistance in rice via linkage to RFLP markers. Theor Appl Genet 81:471–476

Zhang HM, Yang H, Rech EL, Golds TJ, Davis AS, Mulligan BJ, Cocking EC, Davey MR (1988) Transgenic rice plants produced by electroporation-mediated plasmid uptake into protoplasts. Plant Cell Rep 7:379–384

Zhang W, Wu R (1988) Efficient regeneration of transgenic plants from rice protoplasts and correctly regulated expression of the foreign gene in the plants. Theor Appl Genet 76:835–840

Zhao X, Kochert G (1992) Characterization and genetic mapping of a short, highly repeated, interspersed DNA sequence from rice (*Oryza sativa* L.). Mol Gen Genet 231:353–359

Zhao XP, Kochert G (1993) Phylogenetic distribution and genetic mapping of a $(GGC)_n$ microsatellite from rice (*Oryza sativa* L.). Plant Mol Biol 21:607–614

Zheng K-L, Shen B, Qian H-R (1991) DNA polymorphisms generated by arbitrary primed PCR in rice. Rice Genet Newslett 8:134–136

Zhu T, Budworth P, Chen W, Provart N, Chang H-S, Guimil S, Su W, Estes B, Zou G, Wang X (2003) Transcriptional control of nutrient partitioning during rice grain filling. Plant Biotech J 1:59–70

I.4 From Markers to Cloned Genes: Map-Based Cloning

W.-R. Scheible[1], O. Törjek[2], and T. Altmann[2]

1 Introduction

The term map-based cloning, also called positional cloning or recombinational mapping, refers to a universally applicable technique for the isolation of genes characterized by a phenotypic alteration (either visible or detectable through a specific analytical procedure) usually caused by variation in its DNA sequence. This sequence polymorphism may either have originated from a mutation induced in the course of a mutagenesis experiment or may be the result of natural genetic variation. The central procedure used in map-based cloning is the genetic mapping of the gene of interest at extremely high resolution. A prerequisite for the application of this procedure is the availability of at least one line sufficiently genetically deviant from the line carrying the mutant allele of the gene of interest, which can be used to create the necessary mapping population. The high resolution mapping results in the identification of a small interval harboring the gene, which is defined by the two flanking markers most closely linked to the gene. In the case of (model) organisms such as *Arabidopsis thaliana* or *Oryza sativa* (rice), the genomes of which have been physically mapped and largely or fully sequenced, both the identification of genetic markers at high density and the characterization of the DNA segment defined by the molecular markers is very much facilitated by the available sequence information. In the case of species, the genomes of which have not been characterized to the same extent as those of *Arabidopsis* or rice (essentially all other crop species), further intermediate steps such as the establishment of clone-based physical maps, the identification of expressed regions and extensive sequencing and sequence annotation may be necessary. Finally, several approaches may be used for the verification of the identified gene harboring a sequence polymorphism. These include (1) sequence analysis of several independently isolated alleles, (2) the introduction of a wild-type copy of the gene into the mutant or (in the case of dominant mutations) a mutant copy of the gene into the wild-type (in the case of natural variation: introduction of a copy of the alternative allele into one of

---

[1] Max-Planck-Institute of Molecular Plant Physiology, 14424 Potsdam, Germany
[2] Institute of Biochemistry and Biology – Genetics, University of Potsdam, 14415 Potsdam, Germany

the initially selected genetic variants), (3) the introduction of an antisense/ RNAi construct or an over-expression construct of the identified gene into the wild type (or one of the initially selected genetic variants).

Map-based cloning is the principal procedure that can be used to isolate genes controlling traits, for which very little or no information is available on the underlying molecular mechanism. Furthermore, this approach is superior to other molecular genetic analyses (e.g., reversed genetic approaches) for the isolation of essential genes that cause lethality upon loss-of-function mutations or for the identification of genes that cause a detectable phenotypic alteration only upon very specific sequence changes.

2 Outline of the General Map-Based Cloning Strategy

Map-based cloning involves very high-resolution mapping of a gene characterized by at least two different alleles giving rise to discernable phenotypes. This is usually performed in a step-wise fashion initially done at low resolution followed by the enrichment of recombinants around the gene, which are then used for high-resolution mapping. The alleles of the gene studied are either a wild-type and a mutant allele, the latter usually has been induced and identified during the course of a mutagenesis experiment, or they are available as natural diversity represented in different strains or accessions of the species studied. The induced mutations usually segregate in a Mendelian fashion and the mapping is usually done using an F_2 population derived from a cross of the mutant to a wild type of a different strain that can be distinguished from the mutant through DNA-polymorphisms (Fig. 1). In contrast, natural variation is often characterized by polygenic traits and individual loci contributing to the phenotypic variation are identified as QTLs (quantitative trait loci; e.g., Tanksley 1993; Alonso-Blanco and Koornneef 2000; Yano 2001; Asins 2002; Morgante and Salamini 2003). QTLs can be detected in F_2 populations, but the use of recombinant inbred lines (RILs), advanced backcross lines, or nearly isogenic lines (NILs; see Fig. 2) is much more powerful. Map-based cloning of a gene characterized by such a QTL usually is more tedious than in the case of a mutant allele due to several potential complications: (1) the phenotypic alteration caused by the QTL may be much more subtle than that caused by an induced mutation. This poses difficulties in unequivocal genotype determination based on the phenotype, and in general, only strong-effect QTLs can readily be used for map-based identification of the corresponding gene. (2) The QTL may be in epistatic relation to one or more other genomic loci, different alleles of which may strongly affect the phenotypic expression of the QTL. (3) Finally, a QTL may be a complex locus composed of more than one gene affecting the analyzed trait. In the course of the fine mapping, these may be separated and the effects of the individual loci may be fairly weak and thus difficult to detect and pinpoint. Except for situations

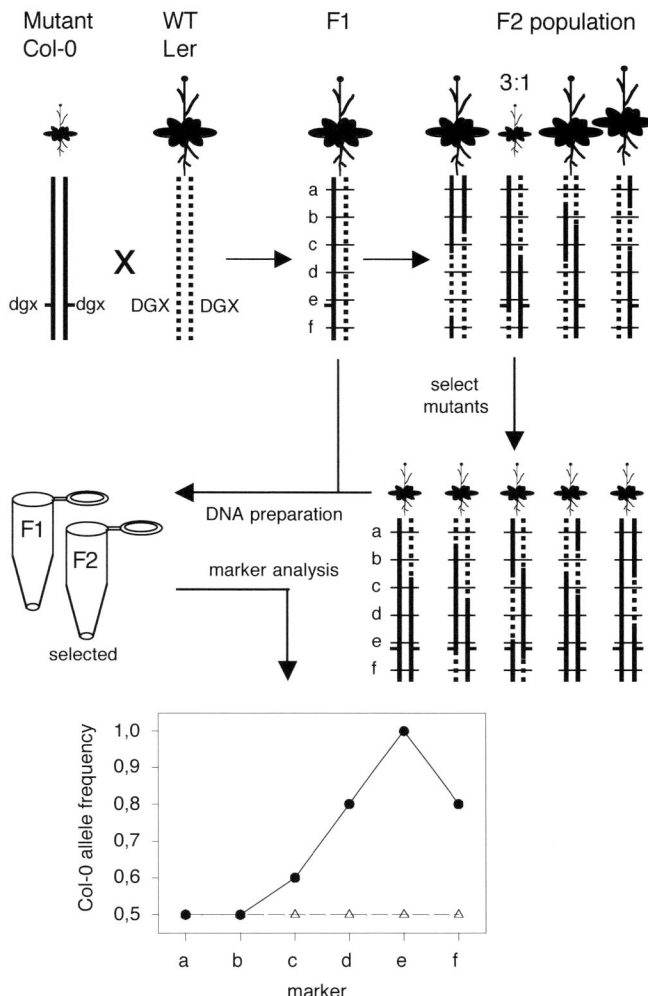

Fig. 1. Schematic representation of bulked segregant analysis for low-resolution mapping. A dwarfed, recessive *Arabidopsis* mutant (the yet unknown mutated *dwarf gene x, dgx*, is marked with a *tick*), selected in a Columbia-0 (Col-0) genome background (*solid lines*), is first outcrossed to another *Arabidopsis* wild type (*WT*) like *Landsberg erecta* (Ler, the diploid genome is depicted as a pair of *dashed lines*). A resulting F_1 plant, which is heterozygous at every locus of the genome, is allowed to self-pollinate, yielding an F_2 population that segregates at every genomic locus, due to meiotic recombination. The WT phenotype and the recessive mutant dwarf phenotype hence segregate in a 3:1 ratio. Mutant F_2 plants are selected based on their dwarf phenotype and tissue samples are pooled (bulked segregant material). DNA is prepared from the selected F_2 pool, as well as from the F_1 plant. Subsequently, molecular markers (*a–f*), which are ideally equally spread over the length of the genome, are analyzed with the F_2 DNA sample (*filled circle*) and the F_1 control DNA sample (*open triangle*). Markers located in proximity to the mutant (Col-0) allele of *dgx* yield a high frequency of the Col-0 marker allele, i.e., they show linkage, whereas markers distant to gene *dgx* yield an equal frequency of the Col-0 and the Ler marker alleles. The heterozygous F_1 control sample will yield equal allele frequencies for all markers interrogated. The conclusion of this bulked segregant analysis is that *dgx* must be located between markers *d* and *f*, in close proximity to marker *e*

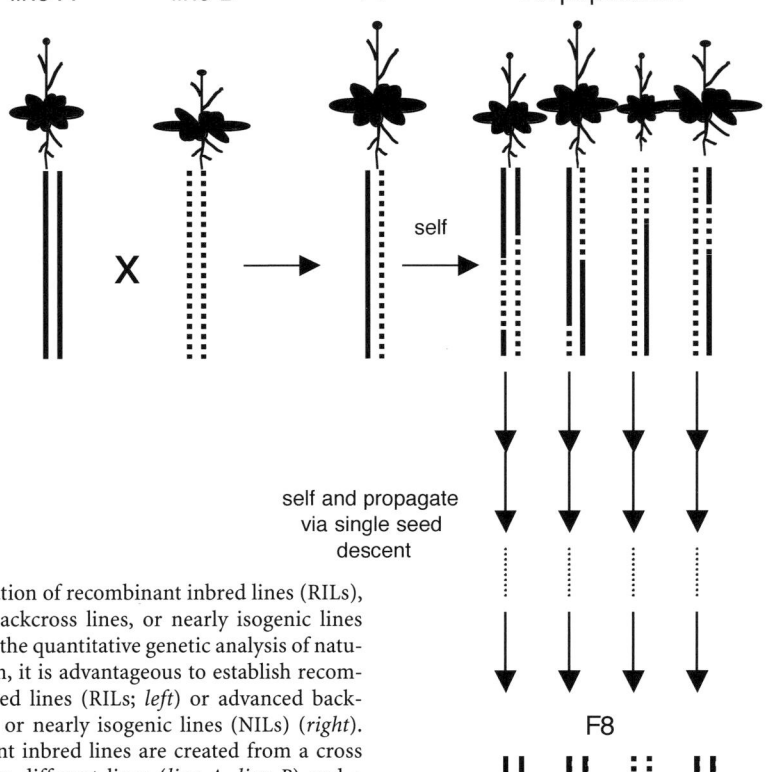

Fig. 2. Creation of recombinant inbred lines (RILs), advanced backcross lines, or nearly isogenic lines (NILs). For the quantitative genetic analysis of natural variation, it is advantageous to establish recombinant inbred lines (RILs; *left*) or advanced backcross lines, or nearly isogenic lines (NILs) (*right*). Recombinant inbred lines are created from a cross between two different lines (*line A, line B*) and a series of randomly selected F_2 plants (usually several hundred) is propagated to the F_8 via single seed descent (for every initial F_2 plant, only one individual per progeny is propagated to yield the next generation). After progression to the F_8, on average more than 99% homozygosity is achieved in the obtained plants (and their amplified progeny that each represent a recombinant inbred line), each of which, however, contains a unique combination of genome segments derived from the two parental lines that is determined by marker analysis. Advanced backcross lines (*right*) are obtained upon repeated backcrossing of progeny derived from the initial cross of the two parental lines A and B. This results in introgression of genome segments from a donor line (*line B in the left half of the scheme*) into the genome of the recurrent parent line (*line A in the left half of the scheme*). On average, the fraction of donor genome is reduced by 50% with every backcross (50% heterozygosity in BC1F1, 25% heterozygosity in BC2F1, etc.). Upon selfing of backcross progeny (e.g. BC2F1) individual populations segregating for the introgressed donor genome segment(s) (e.g. BC2F2) are obtained, which can be used for QTL mapping. Through marker analysis, lines with very few or single introgressed donor genome segments can be selected that are nearly isogenic to the recurrent parent (except for the introgression) and thus constitute nearly isogenic lines (NILs). In this way, series of NILs may be established that constitute a complete collection of introgression lines (a 'genetic library'; Zamir 2001), in which all donor genome segments are represented

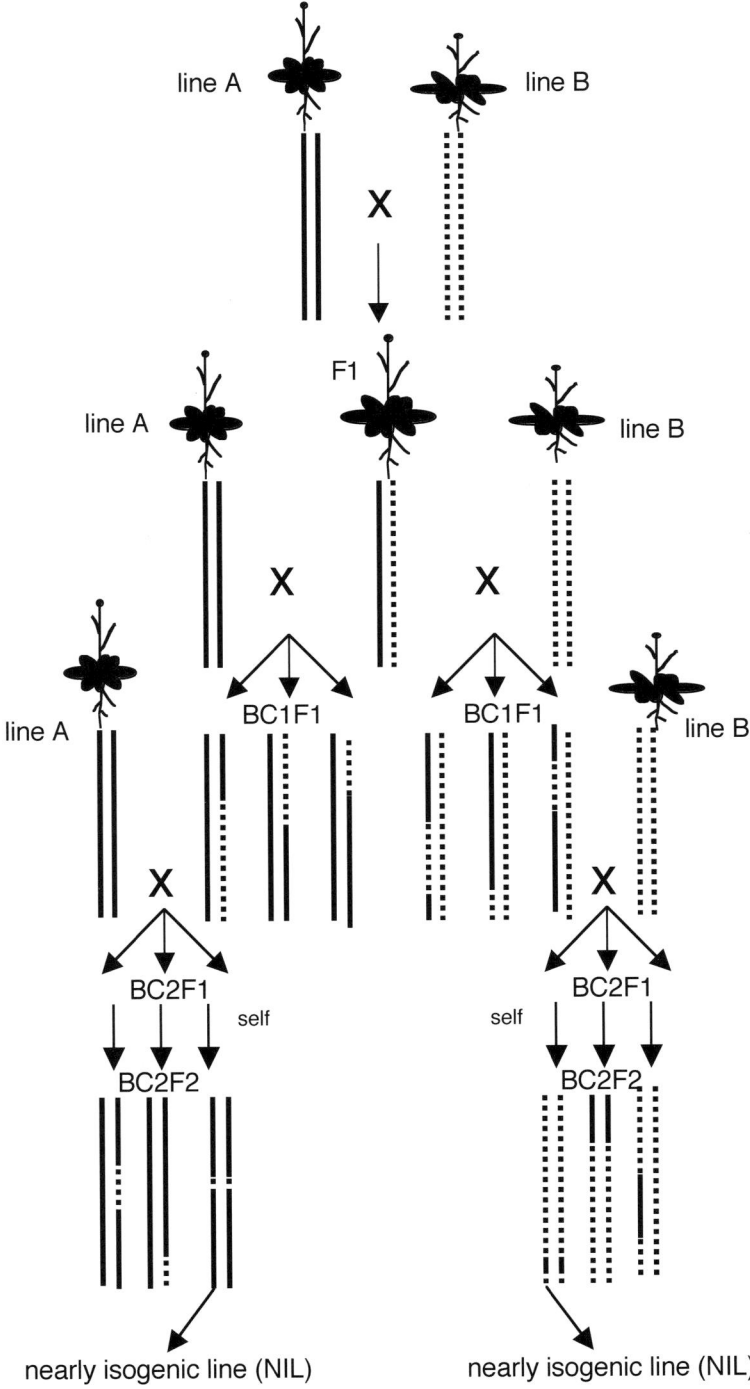

Fig. 2. (continued)

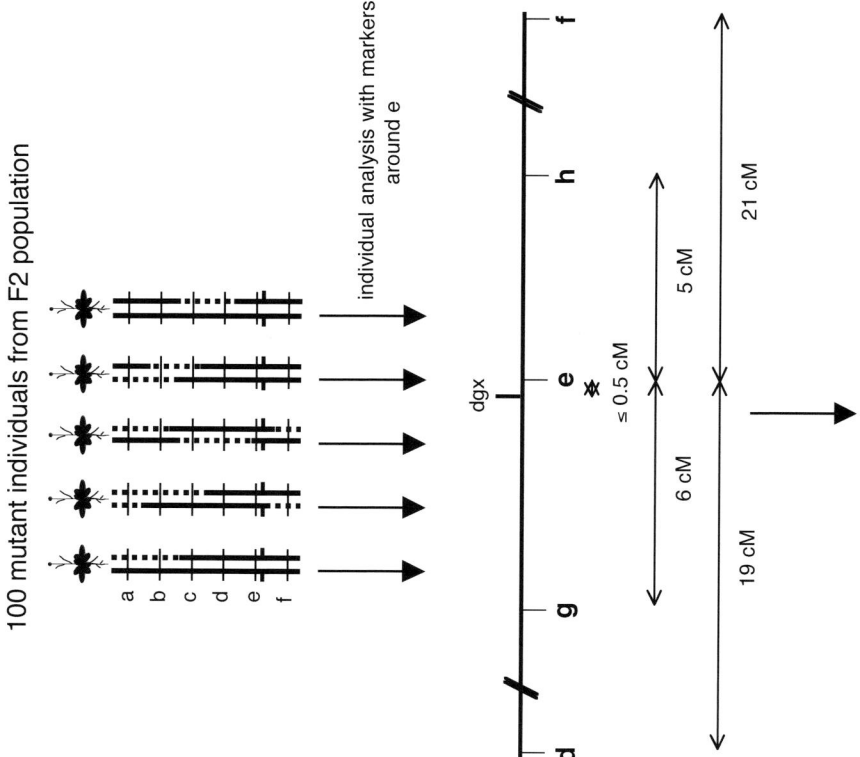

From Markers to Cloned Genes: Map-Based Cloning

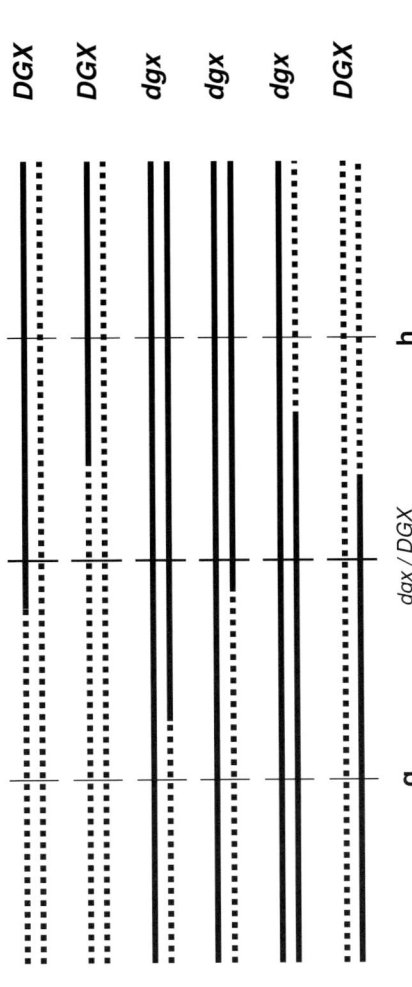

Fig. 3. Schematic outline of the enrichment procedure for recombinants in the genomic region of interest, which are used for high-resolution mapping of the gene to be isolated. Using bulked segregant analysis (cf. Fig. 1), the *dwarf gene x*, *dgx*, was mapped to the region between markers *d* and *f*, in proximity to marker *e*. Robust flanking markers (*g* and *h*) are now selected or developed that border a segment of ~ 10 cM or less including *dgx*. These markers are used to identify recombinants in the vicinity of *dgx*, and the positions of the recombination breakpoints are mapped into the region to the left of *dgx* (between marker *g* and *dgx*) or into the region to the right of *dgx* (between *dgx* and marker *h*). This is either achieved by testing 3000 to 4000 unselected F₂ individuals for the genotypes of markers *g* and *h* and by phenotyping (*DGX*: wild type, *dgx*: mutant phenotype) recombinant F₂ individuals (and, if appropriate their F₃ progenies) or by genotyping about 1000 *dgx* mutant F₂ individuals with both markers (in the case of the easily scorable *dgx* mutant phenotype, the latter is preferable)

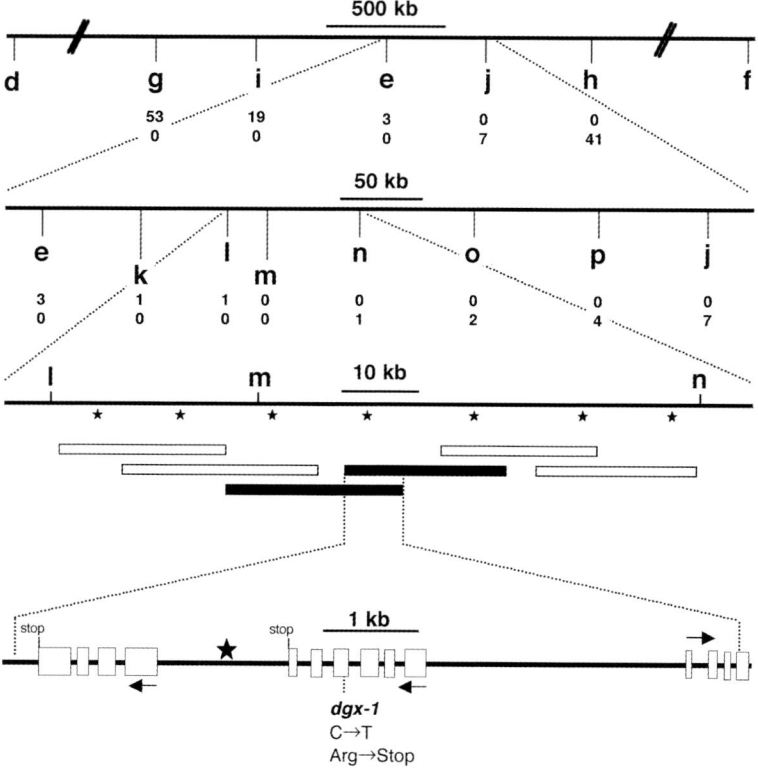

Fig. 4. Schematic depiction of gene identification in *Arabidopsis* by fine mapping and complementation. The evaluation of about 1000 *dgx* mutant F_2 individuals led to the identification of, e.g., 53 homozygous *dgx* plants that are heterozygous at *g* (and thus carry recombination breakpoints between *g* and *dgx*) as well as 41 mutant plants that are heterozygous at *h* (with recombination breakpoints between *dgx* and *h*). The DNAs of these recombinants are subsequently analyzed with additional, interjacent markers (*e*, *i*, *j*), and the remaining recombinants (three that are heterozygous at *e*, but homozygous at *j* and *h*; and seven that are heterozygous at *j*, but homozygous at *e* and *i*) confine the *dgx* gene to the region between *e* and *j*. Subsequently, the interval is further restricted by the analysis of as many markers (*k–p*) as possible, until only one recombinant remains on either side. These two last recombinants, hence, define the final mapping interval between markers *l* and *n* (here approximately 90 kb in size). If no apparent suspect genes are present in this interval and if only one mutant allele is available, the depicted complementation approach with genomic fragments is a powerful alternative to sequencing or mutation detection. To that effect, sequence-specific DNA probes (*asterisks*), equally spread over the length of the interval are PCR-synthesized, labeled and used to screen a genomic Arabidopsis library (15–30 kb fragment size) inserted in a binary plant transformation vector (Meyer et al. 1994) by colony-hybridization. Identified clones (*horizontal bars*) can be sorted by their ability to serve as templates for probe amplification. Overlapping clones are individually transformed into *dgx* mutants via *Agrobacterium* and complementing clones (*black bars*), which reconstitute wild-type growth, are identified in the T_1 or T_2 generation. Size determination of the clones by end-sequencing reveals that a region <10 kb, carrying only two genes is responsible for complementation. The point mutation (e.g., C→T, leading to the replacement of an arginine codon into a stop codon) in the mutant *dgx* gene is finally identified by sequencing

where naturally occurring monogenic traits are studied (e.g., in the case of resistance genes acting in a gene-for-gene fashion (Flor 1971; Dangl and Jones 2001), the QTL under investigation is usually converted into a Mendelian factor by the selection of an appropriate genetic substitution line (NIL). The genome of such a NIL is uniformly composed of only one genotype except for the region containing the QTL, hence segregating in a monogenic fashion (see Sect. 4.2).

After the initial mapping and, in the case of QTLs, the creation/selection of appropriate NILs, the fine-mapping procedures applied to induced or natural genetic variation are very similar (Figs. 3, 4), but the efficiency with which they can be performed are very strongly dependent on the extent of available genome information and characteristics of the species analyzed, such as its suitability for genetic transformation or the extent of sequence polymorphism between the two chosen lines (i.e., ecotypes, accessions, or cultivars) of the species considered.

2.1 Creation of a Mapping Population

The first step in a map-based cloning project is the creation of an appropriate mapping population. In the case of an induced mutant, it is usually crossed to a corresponding wild-type of another, genetically deviant strain. It is most beneficial if the latter (wild-type) strain shows a high frequency of DNA polymorphisms, but has very similar phenotypic characteristics as the (true) wild type in whose background the mutant was obtained. Furthermore, care has to be taken that no major genetic interactions occur in combinations of the two wild-types (e.g., Lee et al. 1994; Narang and Altmann 2001) that would result in enhanced phenotypic variation in a segregating population (e.g., an F_2), or may even suppress the phenotypic expression of the mutation and would compromise phenotyping and thus genotyping of individuals of the mapping population with respect to the mutant locus. If crosses of the strains to be used have not previously been analyzed for the trait of interest, it is advisable to perform crosses between the mutant and several different (genetically deviant) wild-type strains and to check in parallel crosses of these strains with the (true) wild-type of the mutant strain. As mentioned above, in the case of mutants with high phenotypic expression, usually an F_2 population is generated and used for low- and high-resolution mapping (Figs. 1, 3, 4). Natural variation is usually studied in populations derived from crosses between different strains, e.g., elite breeding material of different origin, wild relatives to cultivated lines, or even related (sub) species (which can be crossbred), which may have been pre-selected to show differences in the expression of the trait of interest. However, even in cases where the parental strains show little difference in trait expression, the crossbred progeny may show considerable variation useful to study the underlying genetic factors (e.g., Alonso-Blanco et al. 1998). This variation may be due to different mechanisms established in the parental strains resulting in the same trait expression. However, these different processes may be combined in the progeny, giving rise to transgressive segregation. Furthermore, compensatory interactions between multiple loci may be established in the parental strains, which may be broken up in the crossbred progeny through genetic segregation. (e.g., Alonso-Blanco et al. 1998). Except for cases of monogenic inheritance, loci responsible for or contributing to the natural genetic diversity are most frequently identified through QTL mapping in advanced lines such as RILs, advanced backcross lines, or NILs derived from crosses (Fig. 2). A RIL population created by single seed descent of crossbred progeny over several generations (frequently advanced to the F8-generations yielding about 99% homozygosity) is a highly versatile and generally applicable resource, because any genotype created by a certain combination of genome segments of the parental strains is almost identically replicated among the individuals of a given RIL. In contrast to an F_2 population, where every individual has a unique genetic constitution (which nevertheless could also be used for QTL mapping, e.g., Hayashi and Ukai 1999), RILs provide the means to perform replicated phenotypic analyses of each of the genotypes represented by a given line (with a correspondingly high precision). For the analysis of agriculturally important traits, it is particularly advantageous that analyses can be replicated in different environments and thus QTLs can be identified that exert their effects under a broad range of conditions. Furthermore, a RIL population essentially constitutes an immortalized mapping population, which needs to be genotyped only once and can then be used for QTL mapping of all traits that segregate among the lines. The prerequisites for successful QTL mapping are the availability of sufficiently large populations of lines (e.g., Utz et al. 2000), a decent number of genetic markers showing polymorphisms between the strains used, and sufficiently strong effects of individual loci to be detectable. The QTL mapping procedures are detailed in other chapters (e.g., Chap. II.3, III.7) and are therefore not outlined here.

A strategy that integrates the detection and the introgression of beneficial genomic segments is the advanced backcross QTL analysis (Tanksley and Nelson 1996). In this approach, which is most useful for efficient introgression of valuable QTL alleles from unadapted donor lines (e.g., land races, wild species), QTL analysis is performed in the BC2 or BC3 generation using lines that have undergone selection to reduce the presence of deleterious donor alleles.

2.2 Initial (Low Resolution) Mapping

Map-based cloning is initiated through a first-pass, low-resolution mapping. For monogenically inherited traits this is usually done by phenotyping a limited set of individuals of the F_2 mapping population and the selection of a relatively small number of individuals expressing the recessive phenotype, which are genotyped for a series of genetic markers. Depending on the genetic size of the genome and the distribution of the markers, the number of individuals and the range of markers necessary to detect linkage vary. In the case of Arabidopsis the number of individuals may be as low as 21 and the minimum number of markers necessary to find at least one linked to the mutant locus may be only 15 (Ponce et al. 1999; Swann et al. 2002; Törjék et al. 2003). Usually, the initial goal is to map the mutant locus into an interval of about 20–25 cM. Using an increased number of mutant F_2 individuals (100–150), further markers located in the region of interest are interrogated and a pair of markers flanking the mutant locus on either side at a distance of ca. 5 cM is sought. As an alternative to this stepwise genotyping of increasing numbers of individuals with successive marker sets, bulked segregant analysis can be used as an effective way to identify markers that are genetically linked to a mutation (Michelmore et al. 1991; Lukowitz et al. 2000). This procedure involves the analysis of pools of DNA rather than many individual samples, which substantially reduces the number of individual marker assays required to establish linkage (Fig. 1). Typically, two DNA pools are established: The first pool serves as control and contains DNA that represents a 1:1 mixture of the two alleles of all loci. This can either be achieved by extracting DNA from F_1 individuals used to create the mapping population, or by pooling plant material and then extracting DNA from randomly picked (i.e., not phenotypically selected) F_2 individuals. The latter has the advantage that unequal allele transmission (certation) can be detected and controlled for. The second DNA pool is created only from phenotypically selected F_2 individuals, e.g., displaying the recessive mutant phenotype. Typically, both pools are composed of DNA from about 100 individual F_2 plants. As it is important that a similar amount of DNA from each individual is present in the pool, it is advisable to combine similar amounts of fresh weight of each individual plant prior to DNA extraction. If this is not easily achieved, DNA should first be prepared from each individual plant and then combined in similar amounts to constitute the pool. Due to linkage in coupling phase, markers located in the vicinity of the mutant locus will show in the mutant pool an overrepresentation of the allele derived from the strain that carried the mutation, while nonlinked markers will show the same ratio of the two alleles as in the control pool (expected: 1:1). Allele overrepresentation will subsequently be preserved during PCR-based marker analysis, unless a given marker assay is strongly biased towards one of the two polymorphic sequences with respect to PCR amplification.

For species with (very) well-analyzed genomes such as *Arabidopsis*, rice, tomato/potato, barley, maize, *Lotus* or *Medicago*, appropriate markers usually are readily available, if standard combinations of mapping lines with known polymorphisms are used. In cases of species with fully sequenced genomes such as Arabidopsis, very large numbers of polymorphisms have been identified between various accessions, especially Columbia-0 and Landsberg *erecta* (Cho et al. 1999; Jander et al. 2002; Schmid et al. 2003) and highly efficient marker analysis tools have been implemented (Cho et al. 1999; Drenkard et al. 2000; Törjék et al. 2003) that can be used at high throughput (see Sect. 3). The use of these resources provides the means to map almost every mutant locus to a sufficiently small genetic interval in a single step. In sharp contrast, in species with less well-studied genomes or upon use of nonstandard lines, additional work may be necessary to establish appropriate markers for mapping the gene of interest into the desired interval. In such cases, the amplified length polymorphism (AFLP) marker system (Vos et al. 1995) is a highly useful tool as it provides very large numbers of markers that can be simultaneously mapped without any prior knowledge except for an estimate of the genome complexity. This marker system, which is described in detail in Chapter I.2, has therefore been used very widely, in particular in crop plants. It can also be applied to bulked segregant analysis. In this way, favorable primer combinations detecting linked polymorphic sites can be pre-selected efficiently. These markers are essentially anonymous and are characterized by a specific primer combination and a certain fragment size.

QTL mapping results usually define relatively large intervals, the sizes of which are determined by the number of lines and markers analyzed, the effect of the QTL and its heritability. Any further delineation of the genomic segment containing the QTL is usually done by the use of NILs that contain single introgressions of the region of interest (Fig. 2). Present as the only heterozygous region in the (otherwise homozygous = nonpolymorphic) recipient genomes of the progeny of a test cross, the QTL acts here as a single Mendelian factor. If the QTL alleles cause sufficient phenotypic differences (i.e., the QTL has a strong effect), recombinants within the introgressed region can effectively be used to define smaller and smaller segments in which the locus resides. The fine mapping of a QTL can therefore be done essentially in the same way as that of a mutant locus.

2.3 Enrichment of Recombinants, High-Resolution Mapping and Linking Genetic and Physical Maps

As soon as closely linked genetic markers have been identified that flank the locus of interest at either side, they can effectively be used to identify chromosomes with recombination events within the interval they border (Fig. 3). In the mapping population (either F_2 or segregating progeny of a test cross, TCF2), the large majority of the plants will have the same genotype at both flanking markers, but a small subset of the individuals will show different genotypes at the two flanking markers. The fraction of recombinants (i.e., the plants that show a recombination event) depends on the genetic distance of the two markers (e.g., ca. 10% if each of the two markers are about 5 cM apart from the gene/QTL). These recombinants in the vicinity of the locus to be fine-mapped are highly informative and are the essential resource to determine the position of the locus relative to further markers within the region. To gain sufficient resolution, typically a large number of recombinants has to be selected. The necessary number of recombinants is determined by the ratio of genetic to physical distance and a simple formula has been derived to estimate the number of gametes to sample, which is equivalent to the number of test cross progeny or twice the number of F_2 progeny (Durrett et al. 2002). However, the gene density also has to be considered. Both measures vary between different species and between different genomic regions within a certain genome: in Arabidopsis 1 cM on average corresponds to 200 kb, but this varies between 100 and 400 kb/cM throughout the genome, and the average gene density is 1/4.5 kb (Schmidt et al. 1995; Lin et al. 1999; The Arabidopsis Genome Initiative 2000). An exception to this represents the gene-poor centromeric regions, where 1 cM appears to correspond to 1000 kb and more (Copenhaver et al. 1998; The Arabidopsis Genome Initiative 2000). Because of the low gene-density in those regions, probably just a very small percentage of *Arabidopsis* genes are not amenable to a map-based cloning approach.

The average relations of genetic to physical distances vary dramatically between different plant species: 244 kb/cM for rice (Chen et al. 2002); 700 kb/cM for the diploid *Brassica* species *B. rapa* (Sadowski et al. 1996); 750 kb/cM for tomato (Tanksley et al. 1992); 1180 kb/cM for *Lotus japonicus* (Pedrosa et al. 2002); 1460 kb/cM for maize (Civardi et al. 1994); ca. 4400 kb/cM for barley and wheat (Faris et al. 2000; Künzel et al. 2000). Estimated average gene densities range from 1/20 for rice to 1/250 kb for wheat (Keller and Feuillet 2000 and references therein). In principle therefore, much higher genetic map-resolution with markers much more closely linked to the gene/QTL to be isolated have to be achieved in these plant species compared to *Arabidopsis*. However, in several large plant genomes uneven gene distribution has been observed with gene-dense regions showing gene densities as high as 1/15 in barley, 1/6 in maize, and 1/5 kb in wheat (Keller and Feuillet 2000 and references therein). Furthermore, recombination appears to be suppressed in gene-poor regions mostly consisting of repetitive sequences and occur more frequently in the gene-rich regions (Gill et al. 1996; Künzel et al. 2000; Sandhu and Gill 2002; Weng and Lazar 2002), which thus appear genetically expanded, and many genetic markers have been established from ESTs (expressed sequence tags) representing genes. Therefore, the map resolution required for a positional cloning experiment in such a species may not be as high as indicated by the genome-wide average ratios of genetic to physical distances and average gene densities. In any case, to collect the necessary number of recombinants, a large set of individuals of the fine-mapping population (usually a few thousand; Durrett et al. 2002; Lukowitz et al. 2000) has to be genotyped for the flanking markers. Therefore, technically very robust markers should be selected for this purpose, which ideally are well suited for high-throughput analysis of low-quality DNA. Most frequently, molecular markers (e.g., single nucleotide polymorphisms (SNP)-based markers or simple sequence length polymorphisms (SSLP) or microsatellites) are used for this purpose, but in several cases convenient visible or selectable markers (such as T-DNA or transposon insertions carrying antibiotic or herbicide resistance genes, which have been genetically mapped or those positions have been determined through flanking DNA-sequence information), may be available that could effectively be used for this purpose (van Lisjebettens et al. 1996). Whether or not the use of visible or selectable markers is advantageous depends very much on circumstances and several possible pitfalls have to be considered: The genetic background of marker strains is often difficult to reconstruct and in many cases visible and molecular markers have been mapped using different mapping populations. Consequently, the map position of visible markers with respect to molecular markers is often not exactly known and needs to be confirmed or determined from first principles before they can be used in a fine-mapping experiment. Furthermore, in most cases new mapping populations need to be established using appropriate marker lines, which could cause considerable delays.

The recombinants thus selected are then phenotypically characterized for the locus of interest in order to determine the position of the recombination relative to the locus and the two flanking markers. Most informative individuals are homozygous for the marker allele linked in coupling to the recessive allele at the locus of interest of one of the flanking markers and are heterozygous for the other. Depending on the phenotype of the individual and the marker constitution, the position of the recombination (either between the left marker and the mapped locus, or between the right marker and the mapped locus) can be determined. With the recombinants at hand, the final mapping steps are directed towards the definition of a minimal interval of molecular markers that includes the locus of interest (Fig. 4). To this end, as many markers as possible are genotyped in the recombinants and the markers are ordered according to the number of recombination breakpoints positioned between them and the mapped locus and (conversely) the recombinations are positioned accordingly.

The efficiency with which the final high-resolution fine mapping step can be performed and (candidate) genes can be identified in the delineated region is most strongly influenced by the degree of genome characterization achieved in the species under investigation and the resources that thus are available: in the case of species such as *Arabidopsis* or rice with physically completely mapped and fully or largely sequenced genomes (Mozo et al. 1999; The Arabidopsis Genome Initiative 2000), this is highly straightforward as the necessary information and the required resources are readily available. The necessary markers can easily be established based on the available sequence: either by identification of polymorphisms via sequencing of PCR products amplified from the two parental DNAs or by testing such amplicons for restriction site polymorphisms (to be monitored as cleaved amplified polymorphic sequences (CAPS), Konieczny and Ausubel 1993). If the region of interest has not been (fully) sequenced, usually sufficient sequence information for marker development is available from end sequences of the clones used to set up the physical map (e.g., Chen et al. 2002). For species with only partially characterized genomes, much stronger efforts may be necessary: depending on the status of the genome analysis, large insert libraries (usually yeast artificial chromosomes, YACs, or bacterial artificial chromosomes, BACs; Shizuya et al. 1992; Ramsay 1994) may have to be generated, contigs spanning the region defined by the two closest flanking markers may need to be established via chromosome walking or other approaches (Fairweather 1997; Horrigan and Westbrook 1997; Bancroft 2000), and (end) sequences of the clones included in the contigs have to be generated. Finally, the region defined by the recombinations/markers closest to the locus of interest has to be fully sequenced and analyzed for the presence of genes.

In principle, every gene that is completely or partially included in the region bordered by the two closest flanking markers is a candidate for the genetic factor in search. Only in rare cases has the extent of the interval been narrowed down to such a small size that only a single candidate gene is present or the site of the mutation/allelic variation has been assigned to a subsegment of the gene (e.g., Fridman et al. 2000). Usually, however, the genetically defined genomic segment contains several (potential) genes and different criteria may be used to define the most promising candidate(s): (1) the gene responsible for the trait of interest should be expressed in the tissue/organ that exhibits the phenotypic differences; (2) the gene may show differences in expression in the two genotypes; (3) the gene is expected to show (allelic) sequence diversity among the two genotypes. In opportune cases, one or a few of the candidate genes encode functions that can be directly related to the investigated trait and that may thus be prioritized for further analysis.

2.4 Verification of Identified Genes

The final step in a map-based cloning project is to provide proof for the identity of the isolated gene. Depending on the particular situation, this may be achieved in different ways and different efforts may be required.

One of the most straightforward approaches is the characterization of multiple independently isolated alleles (especially in cases where the observed phenotype is caused by loss-of-function mutations). The demonstration that several independently isolated allelic mutants all show mutations in the same gene is very strong (and generally sufficient) evidence for its correct identification. This is true for induced mutations as the probability that mutations in the true gene are always accompanied by mutations in a second gene (the one assayed) is extremely low. The situation may be much more equivocal, if natural diversity is analyzed. Here, sequence divergence frequently does not cause loss of gene function and rational links to the phenotypic variation cannot be drawn easily. Furthermore, the probability is very high that genetically divergent lines exhibit sequence variation in any given gene that has no relation to the characterized trait.

Another way of verification that the correct gene has been identified is by genetic transformation: if a recessive mutation is studied, the wild-type (dominant) allele is transferred into the homozygous mutant and genetic complementation is demonstrated by documentation of a wild-type phenotype for several independently created transgenic lines (alternatively, heterozygous material may be used for transformation and progeny homozygous for the mutant locus is selected or transgenic wild-type plants are generated and transgenic mutant segregants are selected after crossing to the mutant). For dominant mutations, the reverse configuration is used (the mutant allele is introduced into the wild-type). For some genes, it may be possible to use for transformation a cDNA driven by a standard, largely constitutive promoter, but other genes may require properly controlled expression (in terms of expression level and spatial, temporal, or developmental regulation) and their native regulatory elements need to be used. Even then, only a subset of the transgenic lines may show full (or only partial) complementation. Alternatively, antisense or dsRNAi constructs (Turner and Schuch 2000) may be created and introduced into the wild-type to demonstrate the phenotypic consequences of reduced gene expression. In some cases, overexpression of the identified gene may also cause characteristic phenotypic changes suitable to demonstrate the role of the gene.

Transgenic approaches can also be used to demonstrate that a gene corresponding to a QTL has been correctly identified. This may, however, be more difficult to achieve, because the demonstration of significance of the (weak or moderate) phenotypic changes elicited by the transgene may require very extensive replication, proper experimental design and quantitative analysis with appropriate statistical analysis. Also, considerable complications can occur if the effect of the superior allele is influenced by the presence of the inferior allele (e.g., in cases of semidominance). In most cases, the inferior allele is not a null allele and gene (allele) replacement (e.g., through homologous recombination) can usually not be achieved. Phenotypic expression may, furthermore, be subject to gene dosage effects that could cause considerable complications (e.g., in cases where an additional gene copy may be detrimental rather than beneficial, even the addition of the superior allele may cause inferior trait expression).

3 Map-Based Cloning in a Model Species with a Fully Sequenced Genome (*Arabidopsis thaliana*)

A number of different model species have been used to isolate genes via the map-based approach. These include *Lotus japonicus* and *Medicago truncatula*, which serve as model legumes and from which genes involved in symbiosis have been identified through positional cloning (*NORK:* Endre et al. 2002; *SYMRK:* Stracke et al. 2002). The most favorable situation for map-based cloning, however, has been established in the model system *Arabidopsis thaliana* and this plant species has been used for this procedure most frequently by far. Map-based cloning projects are very strongly supported by the availability of the excellently annotated full genome sequence (The Arabidopsis Genome Initiative 2000). Using the genome sequence as a basis, large collections of sequence polymorphisms and many new molecular markers have recently been established, which are another key prerequisite for map-based cloning. The Arabidopsis Information Resource Website (http://www.arabidopsis.org) currently provides information about 2074 "classical" molecular markers (such as restriction fragment length polymorphism (RFLPs), random amplified polymorphic DNA (RAPDs), SSLPs/microsatellites, CAPS/dCAPS, and AFLPs) that can readily be used as standard molecular markers for first pass mapping at low resolution if standard accessions such as Col-0 and L*er* are studied. Furthermore, multiple marker sets have been assembled, which were specifically designed to limit the number of marker assays necessary for initial mapping (Ponce et al. 1999; Drenkard et al. 2000; Lukowitz et al. 2000; Baumbusch et al. 2001; Törjék et al. 2003). Very large sets of markers have been developed through comparative sequencing (Cho et al. 1999; Jander et al. 2002; Schmid et al. 2003; http://www.arabidopsis.org/servlets/Search?action=new_search&type=polyallele, M. Piercy et al., unpubl.; http://www.tigr.org/tdb/e2k1/ath1/atgenome/Ler.shtml) and more than 55,000 SNPs have been detected, the vast majority of them between the accessions Col-0 and L*er* with the largest contribution provided by a twofold coverage whole genome shotgun sequencing effort performed for L*er* (Jander et al. 2002). Thus, for these two standard accessions, an average density of polymorphisms of about 1/2.5 kb has been achieved. Most of these polymorphisms can be easily exploited to create co-dominant molecular PCR-markers (SSLP, CAPS, dCAPS), and a web-based program is available to facilitate marker design (http://helix.wustl.edu/dcaps/dcaps.html; Neff et al. 2002). When mutations induced in either of these two accessions are mapped, this collection of polymorphisms together with novel, high efficiency genotyping systems (see below) provides the means to advance the fine mapping steps very rapidly.

The support that map-based cloning projects in Arabidopsis receive through the annotated genome sequence, through large collections of molecular markers and other readily available resources (e.g., genomic and cDNA clone libraries) as well as refined protocols for high-throughput work has been outlined recently (Lukowitz et al. 2000; Jander et al. 2002), with one of the most important take-home messages being that the amount of effort required for map-based cloning of Arabidopsis genes has dropped dramatically in recent years. While only 6–10 years ago it was a time-consuming (3–5 years) specialist approach, it can now be considered a standard technique that can be set up in almost any plant lab and carried out by a single student within 1 year or so, depending on circumstances. Map-based cloning of the gene affected in an *Arabidopsis* mutant nowadays merely involves a set of consecutive steps as outlined in detail in Section 2: (1) production of an F_2 mapping population with the mutant of interest; (2) bulked F_2 segregant analysis using established marker sets to determine a rough genome position (5–10 cM); (3) search for or establish robust flanking markers based on publicly available database information; (4) genotyping of about 1000 individual phenotypically chosen F_2 plants (e.g., plants showing the recessive mutant phenotype) with flanking markers and collection of recombinants; (5) use of recombinants and additional markers to restrict the genetic interval as much as possible; (6) identification and verification of the gene within the annotated interval by sequencing and/or genetic complementation.

While it is possible, but not very likely, that mapping projects eventually fail during the course of this process (see Lukowitz et al. 2000 for a discussion of potential problems), it is more likely that the remaining mapping interval, as determined by the last recombinant on either side, is still too large to unambiguously identify the sought-after gene. This is especially true when the full potential of map-based cloning as an unbiased forward genetic approach, is finally highlighted by the identification of a gene of previously unknown function, a characteristic of ~40% of the Arabidopsis genes at this time (The Arabidopsis Genome Initiative 2000). The size of the smallest genetic interval containing the gene of interest is much dependent on parameters like the size of the mapping population, the number of collected recombinants, and hence the recombination frequency in the genomic region to which the gene of interest maps. Small mapping intervals (<10 kb) containing a single or very few genes have been established in only a few cases (e.g., Sakai et al. 1995; Lukowitz et al. 2001), making the identification of the gene responsible for the mutant phenotype quite easy. A typical mapping experiment, involving 1000–2000 plants from the F_2 mapping population, will, however, frequently end up with intervals between 30 and 100 kb (Lukowitz et al. 2000; Durrett et al. 2002). Because of the high gene-density in the Arabidopsis genome, intervals of this size may contain dozens of genes, and predicting the right one is frequently not possible, especially when the interval contains only genes that are not immediately attributable to the mutant phenotype. Since it is usually too laborious and inefficient to further reduce the size of the mapping interval by identifying additional recombinants and testing new molecular markers, more goal-oriented approaches are typically employed at this level to identify the correct gene. When several point mutation alleles are available, the approaches to choose from include systematic searches for polymorphisms between wild-type and mutant DNA by high-quality sequencing or mutation detection methods (Cotton et al. 1998) including enzymatic or chemical cleavage of mismatched bases (Taylor 1999), analysis of single-strand conformational polymorphisms (Nataraj et al. 1999; Berger and Altmann 2000), heteroduplex analysis (Hauser et al. 1998) or denaturing HPLC (O'Donovan et al. 1998; Spiegelman et al. 2000). Another less laborious, but potentially somewhat slower approach, which is highly recommended when only one point mutation allele is available, is functional complementation of a recessive mutant by transformation with a wild-type copy of the yet unknown gene, by using overlapping genomic fragments (e.g., 15–30 kb in length) cloned into a binary plant transformation vector (Meyer et al. 1994), covering the entire mapping interval (exemplified in Scheible et al. 2001, 2003; Fig. 4). This latter approach can reduce the genetic interval to less than 10 kb, and has the potential to directly pinpoint the correct gene.

3.1 High-Efficiency Genetic Marker Systems

As has been pointed out above, a significant part of the work involved in a map-based cloning project is covered by marker development and performance of marker analyses. The development of molecular markers began less than two decades ago. During this time a series of different marker techniques has been developed among which initially the RFLP and RAPD, later on the SSLP and AFLP were the most widely preferred techniques for gene mapping by linkage analysis in plants. In the more recent past, SNPs have emerged as the new generation of molecular markers. Their high and equal density in genomes and the appearance of highly automated methodologies for parallel SNP-genotyping make the SNPs the fastest and cheapest marker system available for map-based cloning. In addition to these advantages, this marker system is also the best choice for association mapping (Risch 2000; Buckler and Thornsberry 2002) due to the slow mutation rate per generation ($\sim 10^{-8}$ to 10^{-9}) in comparison to the microsatellites ($\sim 10^{-4}$; Brumfield et al. 2003). The SNPs are co-dominant and mostly biallelic, because the frequency of multiple mutations at a single site is very low. This makes the evaluation and interpretation of the obtained SNP data particularly straight-forward and well suited to automated analysis (Jander et al. 2002).

A large number of very diverse methods for SNP genotyping are now available. These procedures can be divided into different classes according to the discrimination methods and the detection systems applied (for extensive reviews, see Syvänen 1999; ; Gupta et al. 2001; Gut 2001; Jenkins and Gibson 2002; Kwok 2001). Broadly, the sequence-specific discrimination procedures can be grouped into five main classes: allele-specific amplification, allele-specific hybridization; allele-specific nucleotide incorporation, allele-specific oligonucleotide ligation and allele-specific invasive cleavage. The detection systems are based on fluorescence detection, mass spectrometry or light detection. The advantages/disadvantages for specific applications of the different techniques are determined by the following features:

1. SNP setup: how many SNPs can on average be assayed? How much effort is needed to assay an SNP? How much are the costs for the setup of a new SNP?

2. SNP genotyping: accuracy, robustness, analysis cost per one data point, possibility and degree of multiplexing, flexibility, possibility of automation, possibility to measure allele frequencies in pools, DNA amount and quality needed for the analysis, throughput.
3. Data evaluation: possibility of automation, speed, error rate.

Hitherto, no ideal genotyping method has been available, which satisfies all these desires. Among other aspects, the accuracy and the robustness is always less than 100% (Gut 2001). Finding the method of choice, therefore requires some consideration of the features most relevant for the specific application. One of the most decisive factors is the number of SNPs and the number of individuals to be analyzed. Previous studies have shown that high-density DNA arrays are very powerful when hundreds or thousand SNPs are typed in a few individuals (Wang et al. 1998). In some cases, such as in the first rough mapping stage of a map-based cloning or association mapping project, the opportunity to precisely quantify allele frequencies using only a few pooled DNA samples could provide a highly efficient way to localize the chromosomal region of interest. Previous studies showed that allelic frequencies can be accurately estimated from pools using Pyrosequencing (Ronaghi et al. 1996, 1999), TaqMan (Applied Biosystems; Breen et al. 2000), primer extension followed by dHPLC (Hoogendoorn et al. 2000), SNaPshot (Applied Biosystems; Le Hellard et al. 2002; Norton et al. 2002), Maldi-Tof mass spectrometry (Ross et al. 2000), MassARRAY (Sequenom; Buetow et al. 2001; Le Hellard et al. 2002) and kinetic PCR (Germer et al. 2000). Furthermore, the possibility to analyze multiple loci (multiplexing) is advantageous and a high flexibility to assemble different sets of markers for multiplexed analysis without the need of extensive optimization is highly desirable. In this respect, the SNaPshot assay (outlined in Fig. 5) proved to be highly useful for low to medium throughput analyses, e.g., successive genotyping of a limited number of individuals with several different marker sets (Törjék et al. 2003). Other methods (e.g., TaqMan and Maldi-Tof assays) are preferred when a small number of SNPs is tested in a large population, e.g., for the selection of recombinants. In many cases, a combination of two or more different techniques provides the optimal solution for cheap and fast mapping.

Another aspect to consider is the requirement of highly specialized and expensive instrumentation for the detection methods used in many of these techniques. In addition, the probes and reagents needed for many genotyping methods can be quite expensive. For institutions which cannot afford one of these systems, a number of service providers offer easy and fast ways to make use of these techniques and to rapidly obtain genotyping information at low cost (for review, see Jenkins and Gibson 2002).

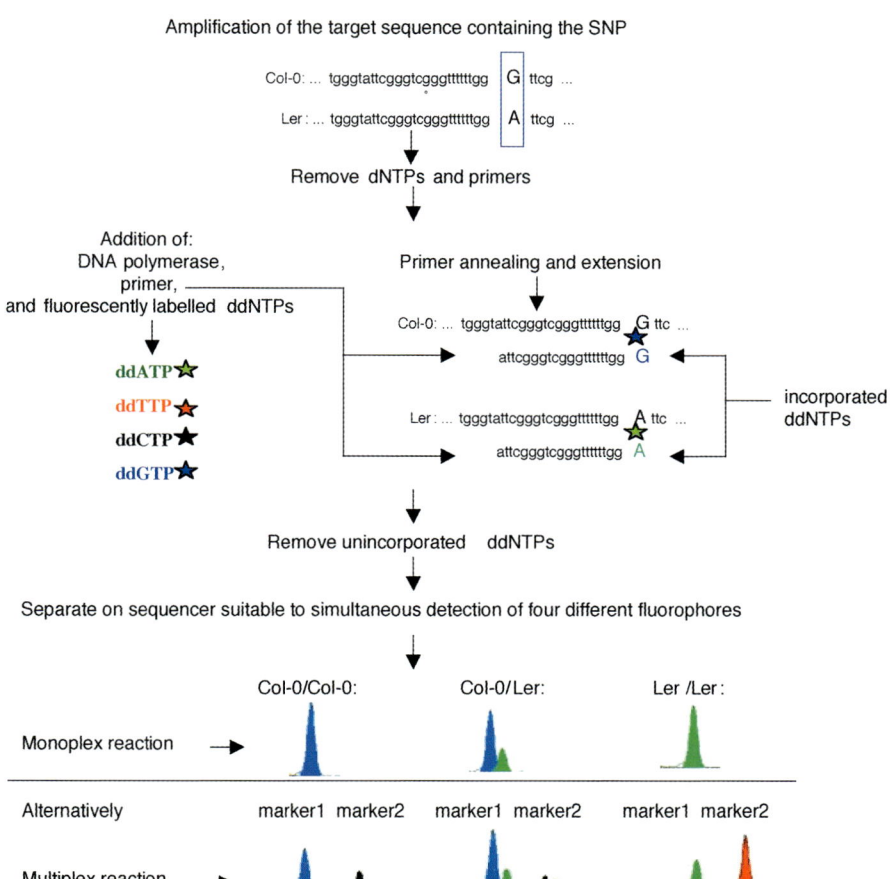

Fig. 5. Schematic representation of the SNaPshot procedure (Applied Biosystems) for rapid SNP typing. The SNaPshot procedure relies upon the 5' extension of a specific primer immediately adjacent to the SNP using four different fluorescently labeled didesoxynucleotides (ddNTPs). After amplification of the 100–300 bp target sequence the dNTPs and unincorporated primers are removed by shrimp alkaline phosphatase (SAP; Amersham Pharmacia) and exonuclease 1 (Amersham Pharmacia) treatment. The purified PCR products are used as templates in the extension reaction, in which a specific primer is annealed adjacent to the SNP and extended by incorporation of the fluorescently labeled complementary ddNTP at the polymorphic site. After removal of the unincorporated ddNTPs by SAP and denaturation of the samples, the reaction products are separated and visualized by electrophoresis on sequencers such as ABI310, ABI3100, or ABI3700, that are designed to simultaneously detect the signals of multiple fluorophores. The SNaPshot Multiplex Kit also allows multilocus interrogation in a single tube/single capillary format using locus-specific extension primers of different lengths. Major advantages of this method are the high success-rate and the high speed of SNP marker establishment, the possibility of multiplexing (parallel analysis of up to eight loci; Törjék et al. 2003), the flexibility to change the markers in multiplex sets, and the very high accuracy

4 Map-Based Cloning in Crop Species

Despite the difficulties associated with the unfavorable situation encountered in most crop species, many of which have very large genomes characterized to various limited extents, several successful map-based cloning projects have been conducted. A number of genes corresponding to traits showing simple (monogenic) inheritance with either naturally occurring allelic variation or with induced mutant alleles have been isolated through positional cloning from several different plant species (see below). Most notably, several genes corresponding to QTLs have also been identified in this way from tomato and rice (see below). The range of approaches used on these projects will be indicated in the following section using a few selected examples. They highlight the fact that the optimal strategies to isolate genes may differ very considerably and that in addition to the availability of the genomic resources mentioned above, the success of a map-based cloning project in a crop species depends a lot on the available plant material, the complexity of the genomic regions containing the genes of interest, and characteristics of the traits under investigation and the properties of the plant species/varieties:

4.1 Identification of Crop Genes Corresponding to Monogenic Traits

Several genes controlling disease resistance of tomato, rice, barley, sugar beet, and potato have been isolated via map-based cloning (*Pto*: Martin et al. 1993; *Xa21*: Song et al. 1995; *Cf-2*: Dixon et al. 1996; *Mlo*: Büschges et al. 1997; *Hs1^{pro-1}*: Cai et al. 1997; *Xa1*: Yoshimura et al. 1998; *Rar1*: Shirasu et al. 1999; *Mla1*: Zhou et al. 2001; *Rpg1*: Brueggeman et al. 2002; *Hero*: Ernst et al. 2002; *R1*: Ballvora et al. 2002) and a number of genes have been isolated by this approach from tomato that have been characterized by mutation (*Fen*: Martin et al. 1994; *Beta/Old-gold*: Ronen et al. 2000; *Tangerine*: Isaacson et al. 2002; *Fer*: Ling et al. 2002; *Chloronerva*: Ling et al. 1999; *Self-pruning*: Pnueli et al. 1998; *Lateral suppressor*: Schumacher et al. 1999; *Jointless*: Mao et al. 2000; *Blind*: Schmitz et al. 2002).

The *Pto* gene of tomato was one of the first genes ever cloned from plants through this approach and is the first plant disease resistance gene that has been identified (Martin et al. 1993). Using 251 F_2 segregating for the resistance allele, an RFLP marker was found that co-segregated with the *Pto* locus and that was used to identify a 400-kb YAC clone. End-specific probes of this YAC were placed on a high-resolution linkage map of the region and the left end probe mapped 1.8 cM from *Pto*. The demonstration that the right end probe of the YAC mapped opposite *Pto* was achieved through analysis of 1300 plants with the initial RFLP marker and the right end probe that resulted in the identification of one crossover between *Pto* and the right end probe. The YAC thus shown to span the *Pto* region was used to isolate corresponding leaf cDNA clones, one of which cosegregated with *Pto* and was shown to represent a member of a gene family. The identity of the *Pto* gene was confirmed through genetic complementation using a 2.4-kb cDNA placed under the control of the CaMV35S that conferred resistance to a susceptible cultivar upon transformation.

The isolation of the *Mlo* gene of barley, recessive alleles of which confer resistance to powdery mildew, marked another breakthrough in map-based cloning (Büschges et al. 1997): This study demonstrated the feasibility of positional cloning in a large grass genome of ca. 5.3 Gbp per haploid genome following the 'chromosome landing' approach (Tanksley et al. 1995). The AFLP marker technology was used to identify very closely linked markers by analysis of a line carrying the mutant locus on a small (8–10 cM) introgressed DNA segment in comparison to the recurrent parent line and analysis of bulks established from a segregating F_2 population of a cross between the two lines. About 1900 AFLP primer combinations were tested and 38 linked candidate AFLP markers were identified and 21 of these could be positioned to opposite sides of *Mlo*.

Using two co-dominant AFLP markers positioned on opposite sides of *Mlo*, 2022 F2 segregants were analyzed and 76 recombinants were identified whose genotype at *Mlo* was determined by testing their F_3 progeny for resistance. This analysis revealed one AFLP marker cosegregating with Mlo (without crossing over) and two flanking markers at distances of 0.24 and 0.4 cM, respectively. A YAC library was constructed and four YAC clones were isolated by screening for the cosegregating AFLP marker. Three of the four YAC clones also contained the two flanking AFLP markers and one was chosen for BAC subcloning. A 60-kb BAC containing the cosegregating AFLP marker and one of the flanking AFLP markers was selected and from this BAC new polymorphic markers were developed. By testing 25 of the aforementioned recombinants, one of these markers was positioned opposite *Mlo* showing that the BAC contained *Mlo*. A physical map of the BAC revealed that the two markers defined an interval of approximately 30 kb. Randomly chosen BAC subclones were sequenced and of the sequence contigs only one (which included the cosegregating AFLP marker) revealed regions of high coding probability. These regions corresponded to an RNA of ca 2 kb consisting of 12 exons. The correct identification of the *Mlo* gene was confirmed by sequencing six different mutants that all showed nucleotide alterations and by selection of intragenic recombinants between different mutant alleles that occurred as rare susceptible F_2 individuals.

The sugar beet $Hs1^{pro-1}$ gene that confers resistance to the beet cyst nematode, was also isolated by positional cloning (Cai et al. 1997). The strategy used here, however, differed from commonly used approaches, because the resistance gene has been transferred to cultivated sugar beet (*Beta vulgaris*) from a wild beet (*Beta procumbens*) and no recombination occurs between the sugar beet and the wild beet chromosomes. The region covering the $Hs1^{pro-1}$ gene was restricted to a small segment of the wild chromosome by mapping chromosomal break points using species-specific satellite markers. Among a panel of chromosomal mutants that were created upon transfer of the resistance gene by species hybridization and backcrossing, a translocation line with the smallest wild beet segment was identified with a *B. procumbens* genome-specific satellite marker. Three YAC clones from a library of this translocation line were isolated using the satellite marker as a probe and were used to identify three cDNAs from a library of the same translocation line. One of the three cDNAs gave a single-copy signal exclusively with DNA of the resistant line when used to probe genomic DNA and, in RNA blot analysis, detected a 1.6-kb transcript present mainly in roots. This cDNA was confirmed to be encoded by the $Hs1^{pro-1}$ gene through genetic complementation. Here, advantage was taken from the fact that compatible and incompatible reactions of susceptible or resistant roots are maintained in hairy root cultures obtained by transformation using *Agrobacterium rhizogenes*, obviating the need to create intact transgenic plants. The isolated cDNA was thus shown to confer resistance to hairy roots when expressed in sense orientation in roots of the susceptible sugar beet line and caused susceptibility to roots of the resistant line when expressed in antisense orientation.

4.2 Identification of Crop Genes Corresponding to Quantitative Trait Loci

Positional cloning of genes corresponding to QTLs has been recently achieved for six loci, three in tomato and three in rice (Frary et al. 2000; Fridman et al. 2000; Yano et al. 2000; Takahashi et al. 2001; Kojima et al. 2002; Liu et al. 2002). This marked another breakthrough in map-based cloning and demonstrated that numerous further genes controlling traits of agricultural importance will be amenable to isolation via this approach.

All three tomato genes, which all control fruit characteristics including size, contents of soluble compounds, and shape, have been isolated using nearly isogenic lines derived from a cross of the red-fruited cultivated tomato, *Lycopersicon esculentum*, and a green-fruited wild tomato species, *Lycopersicon pennellii* (Eshed and Zamir 1995).

fw2.2 was detected as a QTL controlling fruit size in a backcross 1 population derived from *L. esculentum* × *L. pimpinellifolium* and an *L. esculentum* × *L. pennellii* introgression line F_2 (Alpert et al. 1995). Using 3472 individuals of the latter population (F_2 NIL), *fw2.2* was mapped to an 0.8-cM interval bordered by two RFLP markers. Fifty-five F_2 NIL plants that contained recombinations in this interval were used to develop a high-resolution map around *fw2.2*. Using the two flanking markers and four further markers located within this segment, six YAC clones were isolated and arranged as a contig. Their ends were isolated and used as markers in addition to those previously positioned in that interval and the recombinants were phenotyped to place *fw2.2* more precisely on the high-resolution map. *fw2.2* could thus be narrowed down to an interval of less than 150 kb included within two YACs.

One of these was used to screen a *L. pennellii* cDNA library and cDNAs representing four different genes were identified, which were positioned on the high-resolution map. The four cDNAs were used to screen a *L. pennellii* cosmid library and four nonoverlapping cosmids were isolated, which were assembled into a physical contig of the *fw2.2* region and were used to transform tomato lines, that carried the partially recessive large-fruit allele of *fw2.2*. Transfer of one of the cosmids resulted in plants with significantly smaller fruits showing that this cosmid carried *fw2.2*. The sequence of the cosmid insert revealed two open reading frames (ORFs) one of which was excluded from representing *fw2.2* due to its position beyond a recombination that delimited the end of the *fw2.2* candidate region. The *fw2.2* gene thus identified showed heterochronic allelic variation in expression, which is associated with concomitant changes in mitotic activity in the early stages of fruit development (Cong et al. 2002).

The aforementioned introgression lines developed from a cross between *L. esculentum* and *L. pennellii* have been analyzed for fruit contents of total soluble solids (called brix) and 23 QTLs controlling this important property were mapped (Eshed and Zamir 1995). One of these QTLs, *Brix9-2-5*, was mapped to a 9-cM interval on tomato chromosome 9 flanked by two RFLP markers (Eshed and Zamir 1996). For map-based cloning, these two markers were used to analyze 7000 F_2 NILs segregating for the *L. esculentum* and the *L. pennellii* alleles and 145 recombinants were identified (Fridman et al. 2000). A BAC contig was established and the two ends of one BAC marked a segment that included 28 recombinations. BAC subclones were used as further markers that located *Brix9-2-5* in an 18-kb region, which was sequenced. Polymorphic products amplified with primers designed according to the sequence were mapped using the 28 recombinants and one 1-kb fragment showed complete so-segregation with the QTL. This 1-kb region was sequenced for the parental types and the recombinants and based on detected single nucleotide polymorphisms, 13 recombinations were found to reside in this segment, an apparent recombination hot spot. The phenotypes of these 13 recombinants allowed *Brix9-2-5* to be positioned in an extraordinarily small interval of 484 bp. This region was identified as part of an apoplastic invertase gene, *Lin5*, covering intron 3 and parts of exons 3 and 4. Comparison of the *L. esculentum* and *L. pennellii* sequences revealed several sequence differences including three amino acid substitutions. More elaborate sequence differences were apparent in the intron with several deletions and a small insertion found in the *L. esculentum* sequence. These differences may have consequences in the activity of the *Lin5* invertase gene, however, no clear differences in mRNA expression were detected between the *Brix9-2-5* NILs (Fridman et al. 2000).

The three isolated rice genes correspond to three of fourteen QTLs controlling flowering time (heading date) identified in a cross between the rice cultivars Nipponbare (*japonica*) and Kasalath (*indica*; Yano et al. 2001). *Hd1* was detected in the F_2 of this cross (Yano et al. 1997) and was mapped more precisely using BC3F2 plants and BC3F2 progeny tests into a 0.6-cM interval on a rice RFLP linkage map (Yamamoto et al. 1998). Approximately 9000 BC3F3 plants derived from BC3F2 plants heterozygous for the region of chromosome 6 that included *Hd1* were grown and nearly 2000 showing early heading (indicative for homozygosity of the Kasalath *Hd1* allele) were selected for fine mapping (Yano et al. 2000). Recombinants within the 0.6-cM interval were selected by analysis of the 2000 BC3F3 plants (initially as pools of five plants) for the genotypes of the two flanking RFLPs. Eleven plants were selected and analyzed for the genotypes of additional markers (two RFLPs and one CAPS). Two YAC clones and two PAC clones were selected with markers from this region and the two YACs and one of the PACs were shown to contain the *Hd1* region. The PAC clone was sequenced and a 21-kb segment was defined as the *Hd1* candidate region through the position of the two closest flanking markers. Using nine CAPS markers developed on the basis of the sequence and the recombinants, the *Hd1*-region was narrowed down to 12 kb, which contained two putative genes predicted according to the sequence. One of the two genes was further considered due to its similarity to the *Arabidopsis CONSTANS* gene known to be involved in the control of flowering. The Nipponbare and Kasalath sequences of this gene showed many differences and two rice flowering mutants were shown to carry induced mutations in this gene. Its identity as *Hd1* was finally confirmed by transformation of a 7.1-kb Nipponbare fragment containing the candidate *Hd1* gene into a nearly isogenic line (a Nipponbare recipient line homozygous for the introgressed Kasalath non-functional alleles of *Hd1* and *Hd2*). Genetic complementation was confirmed through early flowering of transgenic plants (Yano et al. 2000).

5 Conclusions

The excellent genomic resources developed and made available for the Arabidopsis model system provide the means to apply map-based cloning very efficiently and at high throughput: In comparison to the situation a few years ago, when the total effort necessary for positional cloning of a gene was three to five person-years, the work load per isolated gene has now dropped to less than one person-year (Jander et al. 2002), and many labs have implemented this approach as a routine procedure. Through several examples it was firmly demonstrated that application of positional cloning is also possible in non-model crop species with very large genomes and that even genes corresponding to QTLs can be isolated via this approach. The rapid accumulation of genomic (sequence) information and resources such as genomic libraries and highly efficient markers for these crop species will strongly support the application of map-based cloning in crops and will lead to a major increase in the number of genes of agronomic importance isolated from these species using this approach.

References

Alonso-Blanco C, Koornneef M (2000) Naturally occurring variation in *Arabidopsis*: an underexploited resource for plant genetics. Trends Plant Sci 5:22–29

Alonso-Blanco C, El-Assal SE-D, Coupland G, Koornneef M (1998) Analysis of natural allelic variation at flowering time loci in the Landsberg erecta and Cape Verde Islands ecotypes of *Arabidopsis thaliana*. Genetics 149:749–764

Alpert KB, Grandillo S, Tanksley SD (1995) FW-2.2 – a major QTL controlling fruit weight is common to both red-fruited and green-fruited tomato species. Theor Appl Genet 91:994–1000

Arabidopsis Genome Initiative (2000) Analysis of the genome sequence of the flowering plant *Arabidopsis thaliana*. Nature 408:796–815

Asins MJ (2002) Present and future of quantitative trait locus analysis in plant breeding. Plant Breed 121:281–291

Ballvora A, Ercolano MR, Weiß J, Meksem K, Bormann CA, Oberhagemann P, Salamini F, Gebhardt C (2002) The *R1* gene for potato resistance to late blight (*Phytophthora infestans*) belongs to the leucine zipper/NBS/LRR class of resistance genes. Plant J 30:361–371

Bancroft I (2000) Physical mapping: YACs, BACs, cosmids, and nucleotide sequences. In: Wilson ZA (ed) Practical approach series. Arabidopsis 223:199–224

Baumbusch LO, Sundal IK, Hughes DW, Galau GA, Jakobsen KS (2001) Efficient protocols for CAPS-based mapping in *Arabidopsis*. Plant Mol Biol Rep 19:137–149

Berger D, Altmann T (2000) A subtilisin-like serine protease involved in the regulation of stomatal density and distribution in *Arabidopsis thaliana*. Genes Dev 14:1119–1131

Breen G, Harold D, Ralston S, Shaw D, St Clair D (2000) Determining SNP allele frequencies in DNA pools. Biotechniques 28:464–470

Brueggeman R, Rostoks N, Kudrna D, Kilian A, Han F, Chen J, Druka A, Steffenson B, Kleinhofs A (2002) The barley stem rust-resistance gene *Rpg1* is a novel disease-resistance gene with homology to receptor kinases. Proc Natl Acad Sci USA 99:9328–9333

Brumfield RT, Beerli P, Nickerson DA, Edwards SV (2003) Single nucleotide polymorphisms (SNPs) as markers in phylogeography. Trends Ecol Evol 18:249–256

Buckler ES, Thornsberry JM (2002) Plant molecular diversity and application to genomics. Curr Opin Plant Biol 5:107–111

Büschges R, Hollricher K, Panstruga R, Simons G, Wolter M, Frijters A, van Daelen R, van der Lee T, Diergaarde P, Groenendijk J, Topsch S, Vos P, Salamini F, Schulze-Lefert P (1997) The barley *Mlo* gene: a novel control element of plant pathogen resistance. Cell 88:695–705

Buetow KH, Edmonson M, MacDonald R, Clifford R, Yip P, Kelley J, Little DP, Strausberg R, Koester H, Cantor CR, Braun A (2001) High-throughput development and characterization of a genome-wide collection of gene-based single nucleotide polymorphism markers by chip-based matrix-assisted laser desorption/ionization time-of-flight mass spectrometry. Proc Natl Acad Sci USA 98:581–584

Cai D, Kleine M, Kifle S, Harloff HJ, Sandal NN, Marcker KA, Klein-Lankhorst RM, Salentijn EM, Lange W, Stiekema WJ, Wyss U, Grundler FM, Jung C (1997) Positional cloning of a gene for nematode resistance in sugar beet. Science 275:832–834

Chen M, Presting G, Barbazuk WB, Goicoechea JL, Blackmon B, Fang G, Kim H, Frisch D, Yu Y, Sun S, Higingbottom S, Phimphilai J, Phimphilai D, Thurmond S, Gaudette B, Li P, Liu J, Hatfield J, Main D, Farrar K, Henderson C, Barnett L, Costa R, Williams B, Walser S, Atkins M, Hall C, Budiman MA, Tomkins JP, Luo M, Bancroft I, Salse J, Regad F, Mohapatra T, Singh NK, Tyagi AK, Soderlund C, Dean RA, Wing RA (2002) An integrated physical and genetic map of the rice genome. Plant Cell 14:537–545

Cho RJ, Mindrinos M, Richards DR, Sapolsky RJ, Anderson M, Drenkard E, Dewdney J, Reuber TL, Stammers M, Federspiel N, Theologis A, Yang W-H, Hubbell E, Au M, Chung EY, Lashkari D, Lemieux B, Dean C, Lipshutz RJ, Ausubel FM, Davis RW, Oefner PJ (1999) Genome-wide mapping with biallelic markers in *Arabidopsis thaliana*. Nat Genet 23:203–207

Civardi L, Xia Y, Edwards KJ, Schnable PS, Nikolau BJ (1994) The relationship between genetic and physical distances in the cloned a1-sh2 interval of the Zea mays L. genome. Proc Natl Acad Sci USA 1994:8268–8272

Cong B, Liu J, Tanksley SD (2002) Natural alleles at a tomato fruit size quantitative trait locus differ by heterochronic regulatory mutations. Proc Natl Acad Sci USA 99: 13606–13611

Copenhaver GP, Browne WE, Preuss D (1998) Assaying genome-wide recombination and centromere functions with Arabidopsis tetrads. Proc Natl Acad Sci USA 95:247–252

Cotton RGH, Edkins E, Forrest S (1998) Mutation detection. IRL Press at Oxford University Press, Oxford

Dangl JL, Jones JDG (2001) Plant pathogens and integrated defence responses to infection. Nature 411:826–833

Dixon MS, Jones DA, Keddie JS, Thomas CM, Hanson K, Jones JDG (1996) The tomato Cf-2 disease resistance locus comprises two functional genes encoding leucine rich repeats. Cell 84:451–459

Drenkard E, Richter BG, Rozen S, Stutius LM, Angell NA, Mindrinos M, Cho RJ, Oefner PJ, Davis RW, Ausubel FM (2000) A simple procedure for the analysis of single nucleotide polymorphisms facilitates map-based cloning in Arabidopsis. Plant Physiol 124:1483–1492

Durrett RT, Chen KY, Tanksley SD (2002) A simple formula useful for positional cloning. Genetics 160:353–355

Endre G, Kereszt A, Kevei Z, Mihacea S, Kalo P, Kiss GB (2002) A receptor kinase gene regulating symbiotic nodule development. Nature 417:962–966

Ernst K, Kumar A, Kriseleit D, Kloos DU, Phillips MS, Ganal MW (2002) The broad-spectrum potato cyst nematode resistance gene (Hero) from tomato is the only member of a large gene family of NBS-LRR genes with an unusual amino acid repeat in the LRR region. Plant J 31:127–136

Eshed Y, Zamir D (1995) An introgression line population of Lycopersicon pennellii in the cultivated tomato enables the identification and fine mapping of yield-associated QTL. Genetics 141:1147–1162

Eshed Y, Zamir D (1996) Less-than-additive epistatic interactions of quantitative trait loci in tomato. Genetics 143:1807–1817

Fairweather N (1997) Construction and use of cosmid contigs. In: Boultwood J (ed) Gene isolation and mapping protocols. Methods Mol Biol 68:137–148

Faris JD, Haen KM, Gill BS (2000) Saturation mapping of a gene-rich recombination hot spot region in wheat. Genetics 154:823–835

Flor HH (1971) Current status of the gene-for-gene concept. Annu Rev Phytopathol 9:275–296

Frary A, Nesbitt TC, Grandillo S, Knaap E, Cong B, Liu J, Meller J, Elber R, Alpert KB, Tanksley SD (2000) fw2.2: a quantitative trait locus key to the evolution of tomato fruit size. Science 289:85–88

Fridman E, Pleban T, Zamir D (2000) A recombination hotspot delimits a wild-species quantitative trait locus for tomato sugar content to 484 bp within an invertase gene. Proc Natl Acad Sci USA 97:4718–4723

Germer S, Holland MJ, Higuchi R (2000) High-throughput SNP allele-frequency determination in pooled DNA samples by kinetic PCR. Genome Res 10:258–266

Gill KS, Gill BS, Endo TR, Boyko EV (1996) Identification and high-density mapping of gene-rich regions in chromosome group 5 of wheat. Genetics 143:1001–1012

Gupta PK, Roy JK, Prasad M (2001) Single nucleotide polymorphisms: a new paradigm for molecular marker technology and DNA polymorphism detection with emphasis on their use in plants. Curr Sci 80:524–535

Gut IG (2001) Automation in genotyping of single nucleotide polymorphisms. Hum Mutat 17:475–492

Hauser M-T, Adhami F, Dorner M, Fuchs E, Glössl J (1998) Generation of co-dominant PCR-based markers by duplex analysis on high-resolution gels. Plant J 16:117–125

Hayashi T, Ukai Y (1999) Method of QTL mapping in an F_2 population using phenotypic means of F3 lines (article). Breed Sci 49(2):105–114

Hoogendoorn B, Norton N, Kirov G, Williams N, Hamshere ML, Spurlock G, Austin J, Stephens MK, Buckland PR, Owen MJ, O'Donovan MC (2000) Cheap, accurate and rapid allele frequency estimation of single nucleotide polymorphisms by primer extension and DHPLC in DNA pools. Hum Genet 107:488–493

Horrigan S, Westbrook C (1997) Construction and use of YAC contigs in disease regions. In: Boultwood J (ed) Gene isolation and mapping protocols. Methods Mol Biol 68:123–136

Isaacson T, Ronen G, Zamir D, Hirschberg J (2002) Cloning of *tangerine* from tomato reveals a carotenoid isomerase essential for the production of β-carotene and xanthophylls in plants. Plant Cell 14:333–342

Jander G, Norris SR, Rounsley SD, Bush DF, Levin IM, Last RL (2002) Arabidopsis map-based cloning in the post-genome era. Plant Physiol 129:440–450

Jenkins S, Gibson S (2002) High-throughput SNP genotyping. Comp Funct Genom 3:57–66

Keller B, Feuillet C (2000) Colinearity and gene density in grass genomes. Trends Plant Sci 5:246–251

Kojima S, Takahashi Y, Kobayashi Y, Monna L, Sasaki T, Araki T, Yano M (2002) Hd3a, a rice ortholog of the *Arabidopsis* FT gene, promotes transition to flowering downstream of Hd1 under short-day conditions. Plant Cell Physiol 43:1096–1105

Konieczny A, Ausubel FM (1993) A procedure for mapping *Arabidopsis* mutations using co-dominant ecotype-specific PCR-based markers. Plant J 4:403–410

Künzel G, Korzun L, Meister A (2000) Cytologically integrated physical restriction fragment length polymorphism maps for the barley genome based on translocation breakpoints. Genetics 154:397–412

Kwok P (2001) Methods for genotyping single nucleotide polymorphisms. Annu Rev Genom Hum Genet 2:235–258

Le Hellard S, Ballereau SJ, Visscher PM, Torrance HS, Pinson J, Morris SW, Thomson ML, Semple CAM, Muir WJ, Blackwood DHR, Porteous DJ, Evans KL (2002) SNP genotyping on pooled DNAs: comparison of genotyping technologies and a semi-automated method for data storage and analysis. Nucleic Acids Res 30:74–84

Lee I, Michaels SD, Masshardt AS, Amasino RM (1994) The late-flowering phenotype of *FRIGIDA* and mutations in *LUMINIDEPENDENS* is suppressed in the Landsberg *erecta* strain of *Arabidopsis*. Plant J 6:903–909

Lin X, Kaul S, Rounsley S, Shea TP, Benito M-I, Town CD, Fujii CY, Mason T, Bowman CL, Barnstead M, Feldblyum TV, Buell CR, Ketchum KA, Lee J, Ronning CM, Koo HL, Moffat KS, Cronin LA, Shen M, Pai G, Van Aken S, Umayam L, Tallon LJ, Gill JE, Venter JC (1999) Sequence and analysis of chromosome 2 of the plant *Arabidopsis thaliana*. Nature 402:761–768

Ling H-Q, Koch G, Bäumlein H, Ganal MW (1999) Map-based cloning of *chloronerva*, a gene involved in iron uptake of higher plants encoding nicotianamine synthase. Proc Natl Acad Sci USA 96:7098–7103

Ling H-Q, Bauer P, Bereczky Z, Keller B, Ganal M (2002). The tomato *fer* gene encoding a bHLH protein controls iron-uptake responses in roots. Proc Natl Acad Sci USA 99:13938–13943

Liu J, van Eck J, Cong B, Tanksley SD (2002) A new class of regulatory genes underlying the cause of pear-shaped tomato fruit. Proc Natl Acad Sci USA 99:13302–13306

Lukowitz W, Gillmor CS, Scheible W-R (2000) Positional cloning in Arabidopsis. Why it feels good to have a genome initiative working for you. Plant Physiol 123:795–805

Lukowitz W, Nickle TC, Meinke DW, Last RL, Conklin PL, Somerville CR (2001) *Arabidopsis cyt1* mutants are deficient in a mannose-1-phosphate guanylyltransferase and point to a requirement of N-linked glycosylation for cellulose biosynthesis. Proc Natl Acad Sci USA 98:2262–2267

Mao L, Begum D, Chuang H-W, Budiman MA, Szymkowiak EJ, Irish EE, Wing RA (2000) JOINTLESS is a MADS-box gene controlling tomato flower abscission zone development. Nature 406:910–913

Martin GB, Brommonschenkel SH, Chunwongse J, Frary A, Ganal MW, Spivey R, Wu T, Earle ED, Tanksley SD (1993) Map-based cloning of a protein kinase gene conferring disease resistance in tomato. Science 262:1432–1436

Martin GB, Frary A, Wu T, Brommonschenkel S, Chunwongse J, Earle ED, Tanksley SD (1994) A member of the tomato *Pto* gene family confers sensitivity to fenthion resulting in rapid cell death. Plant Cell 11:1543–1552

Meyer K, Leube MP, Grill E (1994) A protein phosphatase 2c involved in ABA signal transduction in *Arabidopsis thaliana*. Science 264:1452–1455

Michelmore RW, Paran I, Kesseli RV (1991) Identification of markers linked to disease-resistance genes by bulked segregant analysis: a rapid method to detect markers in specific genomic regions by using segregating populations. Proc Natl Acad Sci USA 88:9828–9832

Morgante M, Salamini F (2003) From plant genomics to breeding practice. Curr Op Biotechnol 14:214–219

Mozo T, Dewar K, Dunn P, Ecker JR, Fischer S, Kloska S, Lehrach H, Marra M, Martienssen R, Meier-Ewert S, Altmann T (1999) A complete BAC-based physical map of the *Arabidopsis thaliana* genome. Nat Genet 22:271–275

Narang RA, Bruene A, Altmann T (2001) Analysis of phosphate acquisition efficiency in different *Arabidopsis* accessions. Plant Physiol 124:1786–1799

Nataraj AJ, Olivos-Glander I, Kusukawa N, Highsmith WE (1999) Single-strand conformation polymorphism and heteroduplex analysis for gel-based mutation detection. Electrophoresis 20:1177–1185

Neff MM, Turk E, Kalishman M (2002) Web-based primer design for single nucleotide polymorphism analysis. Trends Genet 18:613–615

Norton N, Williams NM, Williams HJ, Spurlock G, Kirov G, Morris DW, Hoogendoorn B, Owen MJ, O'Donovan MC (2002) Universal, robust, highly quantitative SNP allele frequency measurement in DNA pools. Hum Genet 110:471–478

O'Donovan M, Oefner PJ, Roberts SC, Austin J, Hoogendoorn B, Guy C, Speight G, Upadhyaya M, Sommer SS, McGuffin P (1998) Blind analysis of denaturing high-performance liquid chromatography as a tool for mutation detection. Genomics 52:44–49

Pedrosa A, Sandal N, Stougaard J, Schweizer D, Bachmair A (2002) Chromosomal map of the model legume *Lotus japonicus*. Genetics 161:1661–1672

Pnueli L, Carmel-Goren L, Hareven D, Gutfinger T, Alvarez J, Ganal M, Zamir D, Lifschitz E (1998). The *SELF-PRUNING* gene of tomato regulates vegetative to reproductive switching of sympodial meristems and is the ortholog of *CEN* and *TFL1*. Development 125:1979–1989

Ponce MR, Robles P, Micol JL (1999) High-throughput mapping in *Arabidopsis thaliana*. Mol Gen Genet 261:408–415

Ramsay M (1994) Yeast artificial chromosome cloning. Mol Biotechnol 1:181–201

Risch NJ (2000) Searching for genetic determinants in the new millennium. Nature 405:847–856

Ronaghi M, Karamohamed S, Pettersson B, Uhlen M, Nyren P (1996) Real-time DNA sequencing using detection of pyrophosphate release. Anal Biochem 242:84–89

Ronaghi M, Nygren M, Lundeberg J, Nyren P (1999) Analyses of secondary structures in DNA by pyrosequencing. Anal Biochem 267:65–71

Ronen G, Carmel-Goren L, Zamir D, Hirschberg J (2000) An alternative pathway to β-carotene formation in plant chromoplasts discovered by map-based cloning of *Beta* and *old-gold* color mutations in tomato. Proc Natl Acad Sci USA 97:11102–11107

Ross P, Hall L, Haff LA (2000) Quantitative approach to single-nucleotide polymorphism analysis using MALDI-TOF mass spectrometry. Biotechniques 29:620–629

Sadowski J, Gaubier P, Delseny M, Quiros CF (1996) Genetic and physical mapping in *Brassica* diploid species of a gene cluster defined in *Arabidopsis thaliana*. Mol Gen Genet 251:298–306

Sakai H, Medrano LJ, Meyerowitz EM (1995) Role of *SUPERMAN* in maintaining *Arabidopsis* floral whorl boundaries. Nature 378:199–203

Sandhu D, Gill KS (2002) Gene-containing regions of wheat and the other grass genomes. Plant Physiol 128:803–811

Scheible W-R, Eshed R, Richmond T, Delmer D, Somerville CR (2001) Modifications of cellulose synthase confer resistance to isoxaben and thiazolidinone herbicides in *Arabidopsis Ixr1* mutants. Proc Natl Acad Sci USA 98:10079–10084

Scheible W-R, Fry B, Kochevenko A, Schindelasch D, Zimmerli L, Somerville S, Loria R, Somerville CR (2003). An *Arabidopsis* mutant resistant to thaxtomin A, a cellulose synthesis inhibitor from *Streptomyces* spp. Plant Cell 15:1781–1794

Schmid KJ, Sorensen TR, Stracke R, Törjek O, Altmann T, Mitchell-Olds T, Weisshaar B (2003) Large-scale identification and analysis of genome-wide single-nucleotide polymorphisms for mapping in *Arabidopsis thaliana*. Genome Res 13:1250–1257

Schmidt R, West J, Love K, Lenehan Z, Lister C, Thompson H, Bouchez D, Dean C (1995) Physical map and organization of *Arabidopsis thaliana* chromosome 4. Science 270:480–483

Schmitz G, Tillmann E, Carriero F, Fiore C, Cellini F, Theres K (2002) The tomato *Blind* gene encodes a MYB transcription factor that controls the formation of lateral meristems. Proc Natl Acad Sci USA 99:1064–1069

Schumacher K, Schmitt T, Rossberg M, Schmitz G, Theres K (1999) The *Lateral suppressor* (*Ls*) gene of tomato encodes a new member of the VHIID protein family. Proc Natl Acad Sci USA 96:290–295

Shirasu K, Lahaye T, Tan MW, Zhou F, Azevedo C, Schulze-Lefert P (1999) A novel class of eukaryotic zinc-binding proteins is required for disease resistance signaling in barley and development in *C. elegans*. Cell 99:355–366

Shizuya H, Birren B, Kim UJ, Mancino V, Slepak T, Tachiiri Y, Simon M (1992) Cloning and stable maintenance of 300-kilobase-pair fragments of human DNA in *Escherichia coli* using an F-factor-based vector. Proc Natl Acad Sci USA 89:8794–8797

Song W-Y, Wang G-L, Chen L-L, Kim H-S, Pi L-Y, Holsten T, Gardner J, Wang B, Zhai W-X, Zhu L-H, Fauquet C, Ronald P (1995) A receptor kinase-like protein encoded by the rice disease resistance gene, *Xa21*. Science 270:1804–1806

Spiegelman JI, Mindrinos MN, Oefner PJ (2000) High-accuracy DNA sequence variation screening by DHPLC. Biotechniques 29:1084–1092

Stracke S, Kistner C, Yoshida S, Mulder L, Sato S, Kaneko T, Tabata S, Sandal N, Stougaard J, Szczyglowski K, Parniske M (2002) A plant receptor-like kinase required for both bacterial and fungal symbiosis. Nature 417:959–962

Swann KA, Curtis DE, McKusick KB, Voinov AV, Mapa FA, Cancilla MR (2002) High-throughput gene mapping in *Caenorhabditis elegans*. Genome Res 12:1100–1105

Syvänen AC (1999) From gels to chips: "minisequencing" primer extension for analysis of point mutations and single nucleotide polymorphisms. Hum Mutat 13:1–10

Takahashi Y, Shomura A, Sasaki T, Yano M (2001) Hd6, a rice quantitative trait locus involved in photoperiod sensitivity, encodes the alpha subunit of protein kinase CK2. Proc Natl Acad Sci USA 98:7922–7927

Tanksley SD (1993) Mapping polygenes. Annu Rev Genet 27:205–233

Tanksley SD, Nelson JC (1996) Advanced backcross QTL analysis: a method for the simultaneous discovery and transfer of valuable QTLs from unadapted germplasm into elite breeding lines. Theor Appl Genet 92:191–203

Tanksley SD, Ganal MW, Prince JP, de Vicente MC, Bonierbale MW, Broun P, Fulton TM, Giovannoni JJ, Grandillo S, Martin GB, Messeguer R, Miller JC, Miller L, Paterson AH, Pineda O, Röder MS, Wing RA, Wu W, Young ND (1992) High density molecular linkage maps of the tomato and potato genomes. Genetics 132:1141–1160

Tanksley SD, Ganal MW, Martin GB (1995) Chromosome landing: a paradigm for map-based gene cloning in plants with large genomes. Trends Genet 11:63–68

Taylor GR (1999) Enzymatical and chemical cleaving methods. Electrophoresis 20:1125–1130

Törjék O, Berger D, Meyer R, Müssig C, Schmid K, Rosleff-Sörensen T, Weisshaar B, Mitchell-Olds T, Altmann T (2003) Establishment of a high-efficiency SNP-based framework marker set for *Arabidopsis*. Plant J 36:122-140

Turner M, Schuch W (2000) Post-transcriptional gene-silencing and RNA interference: genetic immunity, mechanisms and applications. J Chem Technol Biotechnol 75:869–882

Utz HF, Melchinger AE, Schoen CC (2000) Bias and sampling error of the estimated proportion of genotypic variance explained by quantitative trait loci determined from experimental data in maize using cross validation and validation with independent samples. Genetics 154:1839–1849

Van Lisjebettens M, Wang X, Cnops G, Boerjan W, Desnos T, Höfte H, van Montagu M (1996) Transgenic Arabidopsis tester lines with dominant marker genes. Mol Gen Genet 251:365–372

Vos P, Hogers R, Bleeker M, Reijans M, Vandelee T, Hornes M, Frijters A, Pot J, Peleman J, Kuiper M, Zabeau M (1995) AFLP: a new technique for DNA fingerprinting. Nucleic Acids Res 23:4407–4414

Wang DG, Fan JB, Siao CJ, Berno A, Young P, Sapolsky R, Ghandour G, Perkins N, Winchester E, Spencer J, Kruglyak L, Stein L, Hsie L, Topaloglou T, Hubbell E, Robinson E, Mittmann M, Morris MS, Shen N, Kilburn D, Rioux J, Nusbaum C, Rozen S, Hudson TJ, Lipshutz R, Chee M, Lander ES (1998) Large-scale identification, mapping, and genotyping of single-nucleotide polymorphisms in the human genome. Science 280:1077–1082

Weng Y, Lazar MD (2002) Comparison of homoeologous group-6 short arm physical maps of wheat and barley reveals a similar distribution of recombinogenic and gene-rich regions. Theor Appl Genet 104:1078–1085

Yamamoto T, Kuboki Y, Lin SY, Sasaki T, Yano M (1998) Fine mapping of quantitative trait loci Hd-1, Hd-2 and Hd-3, controlling heading date of rice, as single Mendelian factors. Theor Appl Genet 97:37–44

Yano M (2001) Genetic and molecular dissection of naturally occurring variation. Curr Opin Plant Biol 4:130–135

Yano M, Harushima Y, Nagamura Y, Kurata N, Minobe Y, Saski T (1997) Identification of quantitative trait loci controlling heading date of rice using a high-density linkage map. Theor Appl Genet 95:1025–1032

Yano M, Katayose Y, Ashikari M, Yamanouchi U, Monna L, Fuse T, Baba T, Yamamoto K, Umehara Y, Nagamura Y, Sasaki T (2000) Hd1, a major photoperiod sensitivity quantitative trait locus in rice, is closely related to the Arabidopsis flowering time gene CONSTANS. Plant Cell 12:2473–2484

Yano M, Kojima S, Takahashi Y, Lin H, Sasaki T (2001) Genetic control of flowering time in rice, a short-day plant. Plant Physiol 127:1425–1429

Yoshimura S, Yamanouchi U, Katayose Y, Toki S, Wang ZX, Kono I, Kurata N, Yano M, Iwata N, Sasaki T (1998) Expression of Xa1, a bacterial blight-resistance gene in rice, is induced by bacterial inoculation. Proc Natl Acad Sci USA 95:1663–1668

Zamir D (2001) Improving plant breeding with exotic genetic libraries. Nat Rev Genet 2:983–989

Zhou F, Kurth J, Wei F, Elliott C, Vale G, Yahiaoui N, Keller B, Somerville S, Wise R, Schulze-Lefert P (2001) Cell-autonomous expression of barley Mla1 confers race-specific resistance to the powdery mildew fungus via a Rar1-independent signaling pathway. Plant Cell 13:337–350

Section II Specific Crops

II.1 DNA Markers in *Brassica*: Use of Genetic Information from *Arabidopsis* and Development of Sequence Tagged Site Markers

T. Sakai, H. Fujimoto, R. Imai, and J. Imamura[1]

1 Introduction

The genus *Brassica* is part of a taxon grown worldwide and includes important vegetables and oil crops. Cultivation of *Brassica* is thought to have started in the Neolithic age (Kimber and McGregor 1995), and since that time, *Brassica* crops have been exposed to intensive selection throughout the world. As a result, they have diversified into many types of vegetables, oilseeds, condiments and forages (reviewed by Tsunoda 1980).

Using conventional taxonomic and cytological criteria, *Brassica* species are classified into six species, three of which are diploid: *B. rapa* = AA, *B. oleracea* = CC, *B. nigra* = BB. The remaining three species are amphidiploids (*B. napus* = AACC, *B. carinata* = BBCC, *B. juncea* = AABB), derivatives of the diploid species. The relationships between the A, B and C genomes were established by U who successfully produced artificial amphidiploids (U 1935), and the relationship became known as the "triangle of U". Subsequent interspecific hybridization among the diploid species revealed common ancestral chromosomes in *Brassica* (reviewed by Mizushima 1980). Plant breeders have exploited the high success rate of interspecific crosses to transfer numerous desirable agronomic characteristics into several *Brassica* species. In addition to the significant cross-ability, *Brassica* species possess a pollination system involving several series of self-incompatible (SI) alleles, thus creating complicated genetic backgrounds (reviewed by Hinata and Nishio 1980).

Recent progress in plant biotechnology has produced novel breeding tools such as microspore culture (Downey and Rimmer 1993), hybrid seed production (Buzza 1995), genetic transformation (Nap et al. 2003) and DNA markers. These techniques have been incorporated into *Brassica* breeding regimens and are routinely used for the efficient production of elite lines. In this chapter, we describe recent developments in the use of DNA markers in both vegetable and oilseed *Brassica* breeding.

In the early 1990s, restriction fragment length polymorphism (RFLP) analysis was utilized to detect genetic diversity and construct genetic maps in *Brassica*; linkage maps of several *Brassica* species have been demonstrated by

---

[1] Plantech Research Institute, 1000 Kamosida-cho, Aoba-ku Yokohama, Japan 227-0033

RFLP analysis (Slocum et al. 1990; Landry et al. 1991; Song et al. 1991). RFLP markers were widely used in marker-assisted selection because they were often codominant and identified low or single copy DNA sequences. However, isolation and identification of the markers often required laborious procedures including DNA cloning and DNA–DNA hybridization. Since the advent of the polymerase chain reaction (PCR), various DNA markers have been developed in plant breeding, and random amplified polymorphic DNA (RAPD), amplified fragment length polymorphism (AFLP) and simple sequence repeat (SSR) markers have been extensively integrated into marker development in *Brassica* varieties.

To develop DNA markers, the first important step is to select or produce appropriate breeding populations. Nearly isogenic lines (NILs) are frequently used to select DNA markers in self-pollinated plants such as rice and tomato (reviewed by Tanksley et al. 1995), though significant time and effort are required to develop NILs via backcrossing. A novel method termed bulked segregant analysis (BSA) was developed for efficient DNA marker isolation (Michelmore et al. 1991) and is based on a plant population comprising both dominant and recessive segregants for target traits. When comparing the bulked segregants with other breeding populations such as NILs, bulked segregant populations may be developed more quickly in the F_2 generation by first dividing them into two pools of dominant and recessive individuals. The two pools differ in their allelic content only at the chromosomal region near the target gene. We used the BSA method to develop DNA markers in populations for which NILs did not exist.

Recently, construction of genetic linkage maps in *Brassica* species has been reported, and various DNA markers have been used to elucidate the organization of *Brassica* genomes and construct genetic linkage maps to develop new DNA markers linked to genes of interest. Schmidt (2000) has recently reviewed the progress of *Brassica* genome organization and evolution. In this chapter, we summarize recent progress on DNA markers in *Brassica*, focusing mainly on marker development in the construction of genetic linkage maps and marker isolation for genes of interest. In addition, we describe the isolation of DNA markers linked with the fertility restorer gene in *B. napus*.

2 Characterization of DNA Markers in *Brassica*

Table 1 summarizes examples of DNA markers that were developed in recent years and linked to particular genes or traits. DNA markers for several disease resistance loci, fertility restoration (*Rf*) genes for cytoplasmic male sterility (CMS) and other important agronomic traits such as flowering time quantitative trait loci (QTL), self-incompatibility and fatty acid composition have been reported. In the case of *fad3* and *SLG*-loci genes, sequence polymorphisms that identified exons of genes have been used as markers. It is

Table 1. Examples of DNA markers that linked with specific genes or locus of interests in *Brassica* species

	Target genes or loci[a]	Plant material	Type of markers[b]	Mode of marker selection[c]	References[d]
1	*fad3* (linolenic acid contents)	*B. napus*	RAPD-STS	Identify polymorphism of target gene sequence	Jourdren et al. (1996)
2	*SLG* (self-incompatibility)	*B. oleracea*, *B. napus*	RAPD-STS	Mapping markers with individual DH (*B. napus*) or F_2 (*B. oleracea*) plants	Cheung et al. (1997b)
3	*Rfp1, Rfp2* (fertility restorer locus for polima CMS)	*B. napus*	RFLP, RAPD	BSA, NIL	Jean et al. (1997)
4	*pb-3, pb-4* (clubroot resistance genes)	*B. oleracea*	RFLP, AFLP	Mapping markers with individual DH plants	Voorrips et al. (1997)
5	Clubroot resistance(CR) genes	*B. rapa*	RAPD	Mapping markers with individual DH plants	Kuginuki et al. (1997)
6	Flowering time QTL	*B. napus, B. rapa*	RFLP, AFLP	Mapping markers with individual DH (*B. napus*) or RI (*B. rapa*) plants	Osborn et al. (1997)
7	*Bzh* (dwarf BREIZH)	*B. napus*	RAPD	Mapping markers with individual DH plants	Barret et al. (1998)
8	*Rfo* (fertility restorer locus for Ogu-INRA CMS-Rf system)	*B. napus*	RFLP, RAPD	BSA	Delourme et al. (1998)
9	$Ac2_1$ (white rust resistance genes)	*B. juncea*	RAPD	BSA	Prabhu et al. (1998)
10	Flowering time QTL	*B. olreacea*	RFLP	Mapping markers with individual DH plants	Bohuon et al. (1998)
11	CR gene	*B. rapa*	RAPD-STS	Mapping markers with F_2 population	Kikuchi et al. (1999)
12	Cultural efficiency of microspore	*B. rapa*	RAPD, RFLP	Mapping markers with F_2 population	Ajisaka et al. (1999)
13	*bb* (yellow seed color)	*B. juncea*	AFLP-SCAR	BSA	Negi et al. (2000)
14	QTL of sculpting curd	*B. olreacea*	EST	Mapping of *Arabidopsis* ESTs in F_2 populations	Lan and Paterson (2000)
15	*PhR2, PhR3* (phoma resistance genes)	*B. napus*	RFLP-STS, AFLP-STS	BSA	Plieske and Struss (2001b)

Table 1. (Continue)

	Target genes or loci[a]	Plant material	Type of markers[b]	Mode of marker selection[c]	References[d]
16	BoGLS-ALK (glucosinolate gene)	B. olreacea	SRAP	Mapping markers with individual RI plants	Li and Quiros (2001)
17	QTL for extreme late bolting	B. rapa	RAPD, RFLP	BSA	Ajisaka et al. (2001)
18	Ac_2V_1 (white rust resistance gene)	B. napus	AFLP	BSA	Sommers et al. (2002)
19	Rfk1 (fertility restorer locus for kos-CMS-Rf system)	B. napus	RAPD-STS, AFLP-STS	BSA	Imai et al. (2003)

[a] Genes or loci are described in the same manner as the original report
[b] Abbreviation of markers: *RAPD* random amplified polymorphic DNA, *STS* sequence-tagged site, *RFLP* restriction fragment length polymorphism, *AFLP* amplified fragment length polymorphism, *SCAR* sequence characterized amplified region, *EST* expressed sequence tag, *SRAP* sequence-related amplified polymorphism
[c] Abbreviation of populations used for marker selections: *BSA* bulked segregant analysis, *DH* doubled haploid, *NIL* near isogenic lines, *RI* recombinant inbred
[d] The table entries are ordered chronologically by date of publication

noteworthy that half of the DNA markers in Table 1 were isolated by the BSA method in combination with PCR-based methods such as RAPD and AFLP. The BSA method has frequently been used in *Brassica* because it is difficult to develop NILs in these species. Doubled haploid (DH) lines, which are genetically similar to the recombinant inbred (RI) line, are also frequently used in *Brassica*. *Arabidopsis* expressed sequence tag (EST) information has also been applied to obtain DNA markers in *Brassica*. In the following sections, we discuss several DNA markers that have been used in *Brassica* breeding.

2.1 Restriction Fragment Length Polymorphism Markers

Since Figdore et al. (1988) demonstrated a high degree of polymorphism for RFLP markers in *Brassica*, this property has been exploited for various purposes in both basic research and *Brassica* breeding, i.e., assessment of genetic diversity among varieties (Song et al. 1988; Diers and Osborn 1994; Becker et al. 1995), mapping genes of interest (see Table 1), and the study of genome organization and evolution (reviewed by Schmidt 2000). In many of these cases, PCR-based markers have been used in conjunction with RFLP markers (Table 1).

RFLP analysis has been used to explore the origin and evolution of cultivated *Brassica* species (Song et al. 1988); indeed, linkage maps of RFLPs have been developed within *B. oleracea* and *B. rapa* (Slocum et al. 1990; Song et al. 1991). The authors of these studies constructed nine and ten linkage groups which represented the corresponding haploid chromosome set of the genome by examining 188 and 197 cloned *B. rapa* and *B. oleracea* sequences, respectively. Among the isolated 188 and 197 clones, 280 and 258 loci were detected, respectively, half of which could detect a single locus and the other half of which detected two or more loci in the F_2 population (Slocum et al. 1990; Song et al. 1991). During construction of the linkage groups, the authors found an elevated level of sequence duplication among the *Brassica* genome compared to other crops. They suggested that the extensive rearrangement and chromosome restructuring resulted from frequent duplication that was an important feature in the evolution of the *Brassica* genome. In addition, they suggested that the isolated markers could be useful as tools, not only for the study of genome evolution, but also as genetic markers for selecting agricultural traits.

In contrast to the diploid species, linkage maps of amphidiploids constructed only with RFLP markers do not completely cover the entire genome (Becker et al. 1995; Cheung et al. 1997a), partly because the genome is typically larger than that of their ancestors. However, since these initial studies were conducted, PCR techniques have become more common and PCR-based DNA markers have since become deeply integrated into marker technologies.

2.2 Random Amplified Polymorphic DNA

Among PCR-based DNA markers, RAPD markers have several advantages including convenience, ease of use, the use of nonradioactive materials, and their requirement for only small amounts of DNA. Random arbitrary primers, 10–12 bp in length, that correspond to genomic DNA sequences can reveal polymorphisms, and the RAPD markers can detect single base changes in some instances. Nearly all RAPD markers are dominant and it is difficult to identify allelic relationships between each fragment on the gel, where several DNA fragments are often observed in one lane (Williams et al. 1990). For this reason, the use of RAPD markers is limited in some cases. In addition, unreliability and noncodominance are concerns with RAPD markers. Halldén et al. (1994) evaluated RFLP and RAPD markers by comparing genetic relationships in *B. napus* breeding lines and concluded that RAPD markers possessed the same resolving ability as RFLP markers.

In the last 7 years, RAPD was frequently utilized to isolate DNA markers (Table 1), and in some cases it was combined with other marker techniques. During construction of the *B. napus* linkage map, RAPD markers were used to obtain DNA markers that allow global comparison of genomes with respect to genetic distance (Foisset et al. 1996). It was mentioned that utilization of the RAPD markers was limited in breeding programs due to potential unreliability and noncodominance (Halldén et al. 1994; Iketani 1998). To overcome these drawbacks, RAPD markers may be converted into sequence tagged site (STS) markers that can be used practically for breeding purposes (Table 1).

RAPD may be used to estimate genetic distance. Potential problems with the reliability and reproducibility of RAPD markers in this respect have been discussed (Halldén et al. 1994; Iketani 1998). Ma et al. (2000) utilized differences in RAPD patterns to calculate phylogenic distances between Asian and European *B. napus* varieties. Their results agree with the geographic distribution and cultivation types of the varieties, and were also consistent with the findings of Diers and Osborn (1994) that estimated genetic diversity of European and Asian *B. napus* varieties by comparing RFLP patterns. These facts suggest that RAPD is a convenient tool to estimate genetic distance among breeding lines.

2.3 Amplified Fragment Length Polymorphism Markers

Amplified fragment length polymorphism is a relatively new method that combines the reliability of RFLP with the sensitivity of PCR. Since its development in 1995, it has become an indispensable tool in plant molecular genetics (Vos et al. 1995). AFLP is characterized by detection of sequence differences with respect to polymorphisms in restriction enzyme sites following selective amplification of the restriction fragments. Polymorphisms in selected amplified fragments are detected by differential migration in gels. The number of amplified fragments in a given sample set may be controlled by the selection of the restriction enzyme as well as the number of selective bases in PCR primers.

The advantages of AFLP markers are (1) large numbers of DNA markers may be obtained in a single experiment, (2) they are highly reproducible compared with RAPD, (3) high multiplex ratios are obtained. These advantages make AFLP suitable for the development of DNA markers to construct high-resolution maps of important genes in various species (Thomas et al. 1995; Bentolila and Hanson 2001; Klein et al. 2001; Liu et al. 2001). In *Brassica*, the *Rf* gene for radish CMS was isolated by map-based cloning, and AFLP markers were used to map both sides of the *Rf* gene and thus served as key landmarks for positional cloning using a DNA library (Imai et al. 2003; Koizuka et al. 2003).

Due to its reliability and high multiplex ratio, AFLP is often combined with other markers to develop high-density genetic maps (Pradhan et al. 2002). Linkage maps developed with AFLP and RFLP markers have been reported for *B. oleracea* (Sebastian et al. 2000), *B. juncea* (Pradhan et al. 2002), and *B. napus* (Lombard and Delourme 2001), and subsequent construction of high-density linkage maps allowed us to develop consensus linkage maps across populations derived from genetically different combinations of crosses (Sebastian et al. 2000; Lombard and Delourme 2001). These primary consensus maps also allow us to cross reference markers to identify genes or QTL positions in different varieties.

2.4 Simple Sequence Repeat Markers

Simple sequence repeats, or microsatellites, consist of short tandem base repeats (2–8 bp units). SSRs are codominant and highly efficient DNA markers for genome and pedigree analyses. SSR markers were originally developed as probes consisting of synthetic dinucleotide or trinucleotide motifs to screen numerous clones in DNA libraries. Based on the DNA sequence of screened candidates, PCR primers are designed to identify the motifs that may be dispersed throughout the genome. These PCR primers can discriminate DNA sequences that share the same motifs, but contain a different number of repeats. The detection of SSRs is both complicated and laborious and results in low throughput and high costs for marker development. However, once SSR markers are developed, their ability to identify high polymorphic loci distributed throughout the entire genome sequence give them significant advantages over other marker techniques.

The informative nature of SSR markers was reported in *B. rapa*. A total of 35 of 38 primer pairs (92%) detected 232 alleles among 19 *B. rapa* cultivars (Suwabe et al. 2002). In *B. oleracea,* Saal et al. (2001) utilized 120 primer pairs to evaluate the *B. napus* genome, 63 of which yielded clear amplification of polymorphic bands. In *B. napus*, 198 polymorphisms were detected with 61 primer pairs, 56% of which detected two to seven allelic loci (Plieske and Struss 2001a). These findings suggest that primer pairs evaluated in one *Brassica* species may be applicable to other *Brassica* crops. Any observed differences in primer efficiency may be a consequence of differences in genome organization (diploid or amphidiploid), PCR primer nature, or PCR amplification conditions.

Furthermore, SSR alleles may be distributed randomly throughout the genome, implying that a small number of PCR primer sets could potentially cover a wide range of chromosomes (Saal et al. 2001). For example, 15 strictly selected SSR primer pairs detected 18 loci covering 34% of the chromosome arms in *B. napus* (Tommasini et al. 2002). This feature lends SSR to the development of DNA markers for both the construction of genetic linkage maps and the discrimination of varietal differences (Tommasini et al. 2002). Several authors demonstrated that SSR markers combined with fluorescence-based automated technology constitute a highly efficient tool to discriminate varieties and estimate heterogeneity within varieties (Kresovich et al. 1995; Tommasini et al. 2002).

3 Application of *Arabidopsis* Genome Information to *Brassica* DNA Markers

Comparative mapping between *Arabidopsis* and *Brassica* species shows that the evolution of these genomes was punctuated with frequent rearrangements, translocations and inversions as corroborated by genome-wide comparison of linkage maps (reviewed by Schmidt 2000). Despite differences in genome size between *Brassica* crops and *Arabidopsis*, the order of the DNA markers is highly conserved. Microcolinearity of functional genes was also observed between *Brassica* species and *Arabidopsis*, but corresponding regions in *Brassica* are of considerably larger size. Lan et al. (2000) developed a comparative map of *B. oleracea* by mapping *Arabidopsis* ESTs in *Brassica* populations. They found extensive synteny between *Arabidopsis* and *B. oleracea* EST maps. Thus, the comparative data helped identify a number of candidate genes in *Brassica*.

The QTL for flowering time was assessed in *B. napus* and *B. rapa* by comparing the corresponding region of the *Arabidopsis* genome in which several candidate genes have been characterized (Bohoun et al. 1998). Further investigations identified the QTL for flowering time, disease resistance (Sillito et al. 2000), and sculpting of the curd in *B. oleracea* (Lan and Paterson 2000) utilizing information from the comparative map of the *Brassica* and *Arabidopsis* sequences.

"Amplified consensus genetic markers" (ACGM) is another type of DNA marker that utilizes *Arabidopsis* sequence information. In *B. napus*, ACGM was used to construct a gene map based on amplified DNA corresponding to various known genes (Fourmann et al. 2002). The authors selected 32 genes to develop ACGMs, 25 of which were derived from *Arabidopsis*. The corresponding sequences were then amplified in *Brassica*. A total of 37 sequences were amplified from *Arabidopsis* while 102 were amplified from *B. napus*. The number of amplified homologous sequences varied, and the number of amplified fragments were strict additions of the parental genome in most cases. Polymorphisms were primarily observed within introns of corresponding genes. Among the 102 *B. napus* sequences, 59 failed to detect the origin of the parental genome and were therefore not suitable as markers. The remaining 43 sequences identified the parental origin by comparing intron sequences. Finally, 12 genes were mapped that may be related to fatty acid synthesis and disease resistance in *B. napus* linkage groups. This method represents a rapid and efficient method to develop DNA markers for agronomic traits such as oil content QTL in *B. napus*.

4 Developing Random Amplified Polymorphic DNA-Sequence Tagged Site Markers Linked to the Radish-Derived Fertility Restoration (*Rf*) Locus in *B. napus*

4.1 Background of the Experiments

Cytoplasmic male sterility (CMS) and the nuclear-encoded *Rf* gene are among the most important agronomic traits in *Brassica*. To construct a novel CMS-Rf system in *B. napus*, we transferred CMS cytoplasm between a radish (*Raphanus sativus* L.) CMS line and *B. napus* cv. Westar via protoplast fusion, resulting in CMS *B. napus* plants (Sakai and Imamura 1992). These plants were then used to perform donor-recipient protoplast fusions with the radish Rf line that had been X-ray-irradiated (Sakai et al. 1996). Several backcrossings of the *B. napus* regenerants carrying the *Rf* locus yielded Rf lines in *B. napus*. In this CMS-Rf system, termed the Kosena system, CMS may be governed by the mitochondrial-encoded gene *orf125*, which is closely related to the ogu-CMS associated gene *orf138* (Iwabuchi et al. 1999). Expression of *orf125* may be controlled by a single nuclear gene in *B. napus*. We further investigated the genetics of the radish *Rf* locus in *B. napus* to isolate the radish *Rf* gene, and defined a single dominant radish *Rf* locus '*Rfk1*' corresponding to fertility restoration (Koizuka et al. 2000). We developed a regional high-resolution map as a first step to positional cloning of the radish *Rfk1* gene using PCR-based methods (Imai et al. 2003). In the section below, we describe details of the marker development linked to the introduction of the radish *Rfk1* into the *B. napus* Rf line.

4.2 Marker Development with Random Amplified Polymorphic DNA-Sequence Tagged Site in *B. napus*

To isolate RAPD markers that were tightly linked to the radish *Rfk1* locus in *B. napus*, we used fertile and sterile bulks, all of which were derived from the F_2 population between the *B. napus* CMS line and the homozygous *B. napus* Rf line (Koizuka et al. 2000). Ten plants were randomly selected from fertile and sterile segregants of *B. napus*, and leaf DNA was isolated from individual plants. Equal weights of DNA from individual plants were then mixed as fertile and sterile bulks. RAPD analysis was performed using ~1200 primers (10-mer, Wako Pure Chemical Industries, Japan; and 12-mer, The University of British Columbia, Vancouver, Canada). Details of the PCR conditions are described in Imai et al. (2003). We screened 14 RAPD markers that were successfully transferred into STS markers after eluting the PCR products from an agarose gel, cloning into a vector (TA cloning kit, Invitrogen) and sequencing the insert DNA. All of the fertile-specific RAPD fragments were successfully transferred into STS markers.

Table 2. Estimation of marker order by analysis of the presence or absence of RAPD-STS markers in fertile and sterile recombinant plants derived from a *B. napus* Rf population

Order of RAPD-STS markers in *B. napus* F$_2$ population[1]		D92	F51L 534	475	D51	484 313 245	Rfk 1	D43	128 602 F51S	314S	585
Number of *B. napus* recombinants among 1344 F$_2$ plants	Fertility of the recombinants[2]										
5	F					■	■	■	■	■	■
3	F		■	■	■	■	■	■	■	■	■
2	F						■	■	■	■	■
1	F							■	■	■	■
2	F							■			■
1	S	n.t.	■	■							
1	S		■								

[a] The 12 RAPD-STS markers derived from the *B. napus* F$_2$ population and the two RAPD-STS markers derived from radish F$_2$ populations (bold letters) were ordered. The most probable order was estimated by connecting flanking positive markers in fertile recombinants of *B. napus*. *Rfk1* indicates the Kosena *Rf* locus. *Shaded boxes* indicate the presence of RAPD-STS markers and *blank areas* indicate loss of amplification of the corresponding markers. Genetic distance described in Fig. 1B was calculated on the basis of the number of recombinants in 1344 F$_2$ progeny
[b] F and S indicate male fertile and male sterile, respectively. *n.t.* Not tested

A total of 1344 F$_2$ plants were derived from the cross of *B. napus* CMS and the homozygous Rf line, and these plants were screened to yield 15 fertile and sterile recombinants (Table 2). These 15 selected recombinants were assayed by PCR using the RAPD-STS markers derived from *B. napus* and radish F$_2$ populations. Due to the low frequency of recombination in this region, the positive signals were arranged in the most probable order in the recombinants. When recombination frequency was calculated in the *B. napus* F$_2$ population, we found that all of the markers mapped within a span of 1 cM (Table 2, Fig. 1). We screened bulks for *Rfk1* and *rfk1* in the radish F$_2$ population to isolate RAPD-STS markers cosegregating with *Rfk1*. After screening with the same arbitrary primer set, six RAPD-STS markers were isolated that linked to *Rfk1*, though four failed to detect the *Rfk1* locus in the *B. napus* Rf line due to nonspecific amplification.

We mapped six radish-derived RAPD-STS markers and a *B. napus* Rf line-derived RAPD-STS marker (D43) in a genetic map of radish F$_2$ populations (Fig. 1A). The most probable marker order was calculated by Mapmaker/exp (v3.0) that also estimated inter-marker distances. The order of radish-derived markers in the *B. napus* F$_2$ population was inconsistent (Table 2). For example, the analysis indicated that D51 and D43 are located on opposite sides of

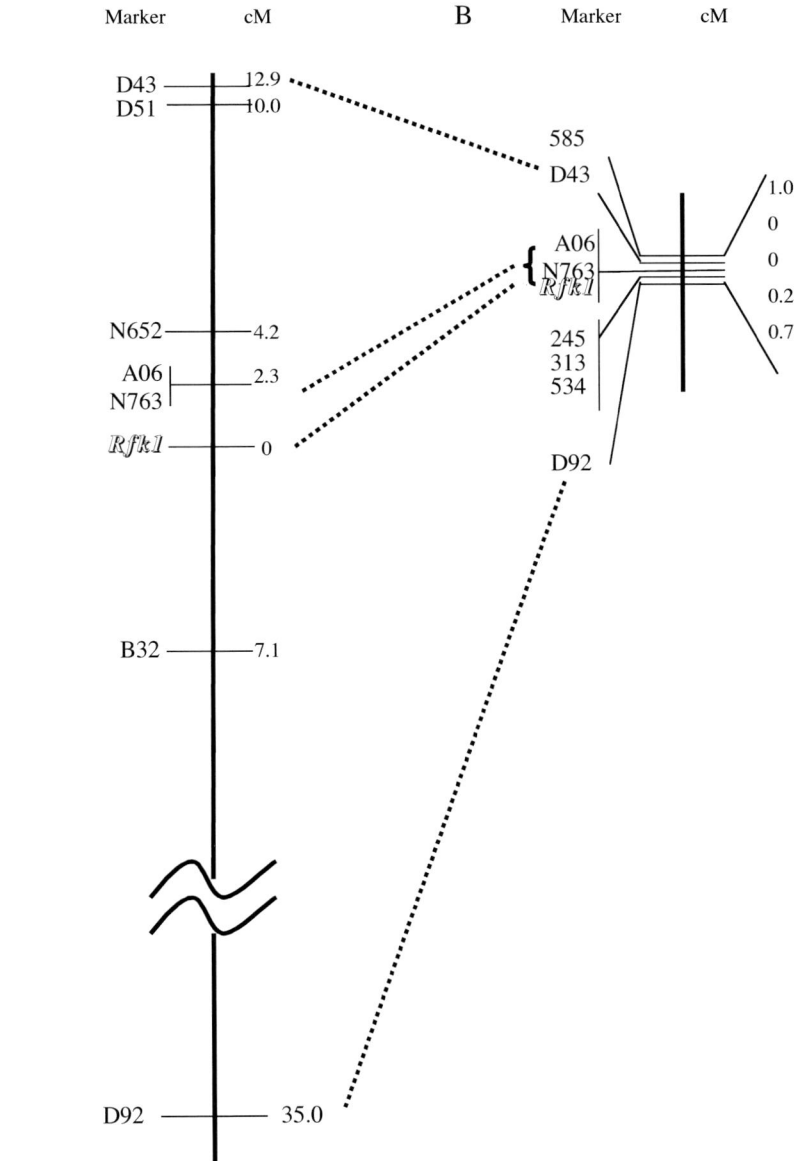

Fig. 1. Comparison of the linkage map covering the radish restorer locus *Rfk1*. **A** Linkage map including five radish-derived RAPD-STS markers and one *B. napus* Rf-derived RAPD-STS marker (D43). **B** Linkage map of the radish *Rfk1* region in *B. napus* constructed from RAPD-STS markers (A06, 763 and D92) derived from radish F_2 populations and a *B. napus* Rf plant. To compare the region, the four markers D43, A06, N763 and D92 were mapped in both populations. For *B. napus*, map distances were calculated on the basis of the number of recombinants in the *B. napus* F_2 population (Table 2). Linkage between *Rfk1* and RAPD-STS markers in radish was calculated with Mapmaker/V3.0 software. (Imai et al. 2003)

Table 3. Southern hybridization of RAPD-STS markers originating from an F_2 population of B. napus CMS × B. napus Rf line in the Kosena system

Marker name	Number of radish-specific DNA fragments in Southern hybridization[a]	Number of homologues in the B. napus genome[b] in the same experiment
585	1	6
D43	1	2
F51S	1	3
602	1	2
128	1	4
534	1	0
313	1	0
245	1	0
314S	1	0
484	1	3
F51L	1	2

[a] Individual plants segregated for male fertile and sterile were randomly selected from the population, and total DNAs were isolated by the CTAB method (Imai et al. 2003). The B. napus total DNAs were digested with three different restriction enzymes (EcoRI, EcoRV and HindIII) and separated on a 1% agarose gel. A single radish-specific fragment was observed for each of the three enzymes

[b] The number of fragments in sterile B. napus plants estimated by number of homologues in B. napus. The numbers of fragments were the same for each of the three enzymes

Rfk1 in B. napus, whereas they map to the same side of Rfk1 in radish. The comparison of the two regional maps indicated that the recombination of the Rfk1 region was highly suppressed in B. napus (Fig. 1), suggesting that a large portion of the radish chromosome may be retained in the B. napus Rf line. All of the RAPD-STS markers were detected as an additional radish-specific fragment in a Southern hybridization analysis (Table 3), indicating that all the Rfk1-related RAPD-STS markers were derived from the radish chromosome region. A detailed chromosomal analysis will be required to verify the chromosome construction.

5 Conclusions

With recent technological advances, DNA markers have been applied to various breeding regimens not only for the selection of agronomic traits, but also for quality control among seeds. In the future, DNA markers will be defined for complicated, but important QTL. For example, the QTL for flowering times have been delineated by comparative genetics between the *Arabidopsis* and *Brassica* genomes. Comparative genetics with the *Arabidopsis* genome will also be applied to identify various QTL such as yield, oil quality and quantity, pest and disease resistance, heterosis as well as other traits.

Intellectual property protection of plant breeders' rights has become an important issue in the plant breeding field. Under the agreement of the International Union for Protection of Plant Varieties (UPOV), new crop varieties should be tested for their distinctness, uniformity and stability (DUS) in UPOV countries. As UPOV member countries increase, variety collections will expand making it difficult to distinguish DUS among candidate varieties by simple comparison of standardized morphological characteristics. However, DNA technology has improved the accuracy of detecting sequence differences, and consequently, appropriate sets of DNA markers can be used to distinguish closely related varieties. DNA markers are efficient and cost-effective for assessing DUS as part of the regulation of *Brassica* varieties (Tommasini et al. 2002).

Genetically modified (GM) canola, including *B. napus* and *B. rapa*, is an example of successful genetic engineering in crop plants. Canola varieties harboring herbicide resistance or male-sterile genes have been developed and are commercially available in Canada. In the next generation of GM canola, scientists and canola breeders anticipate that the crop may produce novel and interesting traits or pharmaceutical compounds (Nap et al. 2003). The rapid development of GM varieties will be required to accommodate marketing strategies, and therefore, DNA marker technology is an essential tool that will allow the proper control of GM varieties during breeding as well as reduce the time required for varietal formulation and production.

In the last two decades, *Brassica* breeding strategies have changed drastically. Hybrids have become more common in amphidiploid oilseed varieties in addition to vegetable *Brassica*. Furthermore, novel biotechnology including DNA markers and genetic engineering has been integrated into the breeding strategies. Currently, and in the future, plant breeders must respond to the many demands of consumers. In addition, protection of superior varieties will be of great importance, and *Brassica* DNA markers will become indispensable for cost-effective breeding and the protection of breeders' rights.

Acknowledgements. We thank Prof. Masashi Hirai (Kyoto Prefectural University) and Dr. Masahiko Ishida (National Institute of Vegetable and Tea Science) for critically reading the manuscript.

References

Ajisaka H, Kuginuki Y, Shiratori M, Ishiguro K, Enomoto S, Hirai M (1999) Mapping of loci affecting the cultural efficiency of microspore culture of *Brassica rapa* L. syn. *campestris* L. using DNA polymorphism. Breed Sci 49:187–192

Ajisaka H, Kuginuki Y, Yui S, Enomoto S, Hirai M (2001) Identification and mapping of a quantitative trait locus controlling extreme late bolting in Chinese cabbage (*Brassica rapa* L. ssp. *pekinensis* syn. *campestris* L.) using bulked segregant analysis. Euphytica 118(1):75–81

Barret P, Delourme R, Foisset N, Renard M (1998) Development of a SCAR (sequence characterized amplified region) marker for molecular tagging of the dwarf BREIZH (*Bzh*) gene in *Brassica napus* L. Theor Appl Genet 97:828–833

Becker HC, Engiqvist GM, Karlsson B (1995) Comparison of rapeseed cultivars and resynthesized lines based on allozyme and RFLP markers. Theor Appl Genet 91:62–67

Bentolila S, Hanson MR (2001) Identification of a BIBAC clone that co-segregates with the petunia restorer of fertility (*Rf*) gene. Mol Genet Genom 266:223–230

Bohoun EJR, Ramsay LD, Craft JA, Arthur AE, Marshall DF, Lydiate DJ, Kearsey MJ (1998) The association of flowering time quantitative trait loci with duplicated regions and candidate loci in *Brassica oleracea*. Genetics 150:393–401

Buzza GC (1995) Plant breeding. In: Kimber D, McGregor DI (eds) *Brassica* oilseeds. Production and utilization. CAB International, Cambridge, UK, pp 153–175

Cheung WY, Friesen L, Raskow GFW, Seguin-Swartz G, Landry BS (1997a) A RFLP-based linkage map of mustard [*Brassica juncea* (L.) Czem. And Coss]. Theor Appl Genet 94:841–851

Cheung WY, Champagne G, Hubert N, Tulsieram L, Charne D, Patel J, Landry BS (1997b) Conservation of S-locus for self-incompatibility in *Brassica napus* (L.) and *Brassica oleracea* (L.). Theor Appl Genet 95:73–82

Delourme R, Foisset N, Horvais R, Barret P, Champagne G, Cheung WY, Landry BS, Renard M (1998) Characterisation of the radish introgression carrying the *Rfo* restorer gene for the Ogu-INRA cytoplasmic male sterility in rapeseed (*Brassica napus* L.). Theor Appl Genet 97:129–134

Diers BW, Osborn TC (1994) Genetic diversity of oilseed *Brassica napus* germplasm based on restriction fragment length polymorphisms. Theor Appl Genet 88:662–668

Downey RK, Rimmer R (1993) Agronomic important in oilseed *Brassicas*. In: Sparks DL (ed) Advances in agronomy, vol 50. Academic Press, New York, pp 1–66

Figdore SS, Kennard WC, Song KM, Slocum MK, Osborn TC (1988) Assessment of the degree of restriction length polymorphism in *Brassica*. Theor Appl Genet 75:833–940

Foisset N, Delourme R, Barret P, Hubert N, Landry BS, Renard M (1996) Molecular-mapping analysis in *Brassica napus* using isozyme, RAPD and RFLP markers on a doubled haploid progeny. Theor Appl Genet 93:1017–1025

Fourmann M, Barret P, Froger N, Baron C, Charlot F, Delourme R, Brunel D (2002) From *Arabidopsis thaliana* to *Brassica napus*: development of amplified consensus genetic markers (ACGM) form construction of a gene map. Theor Appl Genet 105:1196–1206

Halldén C, Nilsson N-O, Rading IM, Sall T (1994) Evaluation of RFLP and RAPD markers in a comparison of *Brassica napus* breeding lines. Theor Appl Genet 88:123–128

Hinata K, Nishio T (1980) Self-incompatibility in Cruciferes. In: Tsunoda S, Hinata K, Gómez-Campo C (eds) *Brassica* crop and wild allies. Japan Scientific Society Press, Tokyo, Japan, pp 224–234

Iketani H (1998) Classification of fruit trees – what is the problem? what is important? J Jpn Soc Hortic Sci 67(6):1193–1196

Imai R, Koizuka N, Fujimoto H, Hayakawa T, Sakai T, Imamura J (2003) Delimiting the fertility restorer locus *Rfk1* on a 43 kb contig in Kosena radish (*Raphanus sativus* L.). Mol Genet Genomics 269:388–394

Iwabuchi M, Koizuka N, Fujimoto H, Sakai T, Imamura J (1999) Identification and expression of the kosena radish (*Raphanus sativus* cv. Kosena) homologue of the ogura radish CMS-associated gene *orf138*. Plant Mol Biol 39:183–188

Jean M, Brown GG, Landry BS (1997) Genetic mapping of nuclear fertility restorer genes for the 'Polima' cytoplasmic male sterility in canola (*Brassica napus* L.) using DNA markers. Theor Appl Genet 95:321–328

Jourdren C, Barret P, Brunel D, Delourme R, Renard M (1996) Specific molecular marker of the genes controlling linolenic acid content in rapeseed. Theor Appl Genet 93:512–518

Kikuchi M, Ajisaka H, Kuginuki Y, Hirai M (1999) Conversion of RAPD markers for a clubroot resistance gene of *Brassica rapa* into sequence-tagged sites (STSs). Breed Sci 49:83–88

Kimber D, McGregor DI (eds) (1995) *Brassica* oilseeds. Production and utilization. CAB International, Cambridge, UK, pp 1–9

Klein RR, Klein PE, Chhabra AK, Dong J, Pammi S, Childs KL, Mullet JE, Rooney WL, Schertz KF (2001) Molecular mapping of the *rf1* gene for pollen fertility restoration in sorghum (*Sorghum bicolor* L.). Theor Appl Genet 102:1206–1212

Koizuka N, Imai R, Iwabuchi M, Sakai T, Imamura J (2000) Genetic analysis of fertility restoration and accumulation of ORF125 mitochondrial protein in the kosena radish (*Raphanus sativus* cv. Kosena) and a *Brassica napus* restorer line. Theor Appl Genet 100:949–955

Koizuka N, Imai R, Fujimoto H, Hayakawa T, Kimura Y, Kohno-Murase J, Sakai T, Kawasaki S, Imamura J (2003) Genetic characterization of a pentatricopeptide repeat protein gene, *orf687*, that restores fertility in the cytoplasmic male-sterile Kosena radish. Plant J 34:407–409

Kresovich S, Szewe-Mcfadden AK, Bliek SM, McFerson JR (1995) Abundance and characterization of simple-sequence repeats (SSRs) isolated from a size-fractionated genomic library of *Brassica napus* L. (rapeseed). Theor Appl Genet 91:206–211

Kuginuki Y, Yui M, Yoshikawa H, Hida K, Hirai M (1997) RAPD markers linked to a clubroot-resistance locus in *Brassica rapa* L. Euphytica 98(3):149–154

Lan TH, Paterson AH (2000) Comparative mapping of quantitative trait loci sculpting the curd of *Brassica oleracea*. Genetics 155:1927–1954

Lan TH, DelMonte TA, Reischmann KP, Hyman J, Kowalski SP, McFerson J, Kresovich S, Paterson AH (2000) An EST-enriched comparative map of *Brassica oleracea* and *Arabidopsis thaliana*. Genome Res 10(6):776–788

Landry B, Hubert N, Etoh T, Harada JJ, Lincoln SE (1991) A genetic map of *Brassica napus* based on restriction fragment length polymorphism detected with expressed DNA sequences. Genome 34:543–552

Li G, Quiros CF (2001) Sequence-related amplified polymorphism (SRAP), a new marker system based on a simple PCR reaction: its application to mapping and gene tagging in *Brassica*. Theor Appl Genet 103:455–461

Liu N, Shan Y, Wang FP, Xu CG, Peng KM, Li XH, Zhang Q (2001) Identification of an 85-kb DNA fragment containing *pms1*, a locus for photoperiod-sensitive genic male sterility in rice. Mol Genet Genomics 266:271–275

Lombard V, Delourme R (2001) A consensus linkage map for rapeseed (*Brassica napus* L.): construction and integration of three individual maps from DH populations. Theor Appl Genet 103:491–507

Ma CZ, Kimura Y, Fujimoto H, Sakai T, Imamura J, Fu TD (2000) Genetic diversity of Chinese and Japanese rapeseed (*Brassica napus* L.) varieties detected by RAPD markers. Breed Sci 50:257–265

Michelmore RW, Paran I, Kesseli RV (1991) Identification of markers linked to disease-resistance genes by bulked segregant analysis: a rapid method to detect markers in specific genomic regions by using segregating populations. Proc Natl Acad Sci USA 88:9828–9832

Mizushima U (1980) Genome analysis in *Brassica* and allied genera. In: Tsunoda S, Hinata K, Gómez-Campo C (eds) *Brassica* crop and wild allies. Japan Scientific Society Press, Tokyo, Japan, pp 89–106

Nap JP, Metz PLJ, Escaler M, Conner AJ (2003) The release of genetically modified crops into the environment. Plant J 33:1–18

Negi MS, Devic M, Delseny M, Lakshmikumaran M (2000) Identification of AFLP fragments linked to seed coat colour in *Brassica juncea* and conversion to a SCAR marker for rapid selection. Theor Appl Genet 101:146–152

Osborn TC, Kole C, Parkin IA, Sharpe AG, Kuiper M, Lydiate DJ, Trick M (1997) Comparison of flowering time genes in *Brassica rapa, B. napus* and *Arabidopsis thaliana*. Genetics 146:1123–1129

Plieske J, Struss D (2001a) Microsatellite markers for genome analysis in *Brassica*. I. Development in *Brassica napus* and abundance in *Brassicaceae* species. Theor Appl Genet 102:689–694

Plieske J, Struss D (2001b) STS markers linked to *Phoma* resistance genes of the *Brassica* B-genome revealed sequence homology between *Brassica nigra* and *Brassica napus*. Theor Appl Genet 102:483–488

Prabhu KV, Sommers DJ, Rakow G, Gugel RK (1998) Molecular markers linked to white rust resistance in mustard *Brassica juncea*. Theor Appl genet 97(5/6):865–870

Pradhan AK, Gupta V, Mukhopadhyay A, Arumugam N, Sodhi YS, Pental D (2002) A high-density linkage map in *Brassica juncea* (Indian mustard) using AFLP and RFLP markers. Theor Appl Genet 106:607–614

Saal B, Plieske J, Hu J, Quiros CF, Struss D (2001) Microsatellite markers for genome analysis in *Brassica* . II. Assignment of rapeseed microsatellites to the A and C genomes and genetic mapping in *Brassica oleracea* L. Theor Appl Genet 102:695–699

Sakai T, Imamura J (1992) Alteration of mitochondrial genomes containing *atpA* genes in the sexual progeny of cybrids between *Raphanus sativus* cms line and *Brassica napus* cv.Westar. Theor Appl Genet 84:923–929

Sakai T, Liu HJ, Iwabuchi M, Kohno-Murase J, Imamura J (1996) Introduction of a gene from fertility restored radish (*Raphanus sativus*) into *Brassica napus* by fusion of X-irradiated protoplasts from a radish restorer line and iodoacetamide-treated protoplasts from a cytoplasmic male-sterile cybrid of *B. napus*. Theor Appl Genet 93:373–379

Schmidt R (2000) Synteny: recent advances and future prospects. Curr Opin Plant Biol 3:97–102

Sebastian RL, Howell EC, King GJ, Marshall DF, Kearsey MJ (2000) An integrated AFLP and RFLP *Brassica oleracea* linkage map from two morphologically distinct doubled-haploid mapping populations. Theor Appl Genet 100:75–81

Sillito D, Parkin IA, Mayerhofer R, Lydiate DJ, Good AG (2000) *Arabidopsis thaliana* : a source of candidate disease-resistance genes for *Brassica napus*. Genome 43(3):452–460

Slocum MK, Figdore SS, Kennerd WC, Suzuki JY, Osborn TC (1990) Linkage arrangement of restriction fragment length polymorphism loci in *Brassica oleracea*. Theor Appl Genet 80:57–64

Somers DJ, Rakow G, Rimmer SR (2002) *Brassica napus* DNA markers linked to white rust resistance in *Brassica juncea*. Theor Appl Genet 104:1121–1124

Song KM, Osborn TC, Williams PH (1988) *Brassica* taxonomy based on nuclear restriction fragment length polymorphisms (RFLPs). 1. Genome evolution of diploid and amphidiploid species. Theor Appl Genet 75:784–794

Song KM, Suzuki JY, Slocum MK, Williams PH, Osborn TC (1991) A linkage map of *Brassica rapa* (syn. *campestris*) based on restriction fragment length polymorphism loci. Theor Appl Genet 82:296–304

Suwabe K, Iketani H, Nunome T, Kage T, Hirai M (2002) Isolation and characterization of microsatellites in *Brassica rapa* L. Theor Appl Genet 104:1092–1098

Tanksley SD, Ganal MW, Martin GB (1995) Chromosome landing: a paradigm for map-based gene cloning in plants with large genomes. Trends Genet 11(2):63–69

Thomas CM, Vos P, Zabeau M, Jones D, Norcott KA, Chadwick BP, Jones JDG (1995) Identification of amplified restriction fragment polymorphism (AFLP) markers tightly linked to the tomato *Cf-9* gene for resistance to *Cladsprorium fulvum*. Plant J 8(5):785–794

Tommasini L, Batley J, Arnold GM, Cooke RJ, Donini P, Lee P, Law JR, Lowe C, Moule C, Trick M, Edwards KJ (2002) The development of multiplex simple sequence repeats (SSR) markers to complement distinctness, uniformity and stability testing of rape (*Brassica napus* L.) varieties. Theor Appl Genet 106:1091–1101

Tsunoda S (1980) Eco-physiology of wild and cultivated forms in *Brassica* and allied genera. In: Tsunoda S, Hinata K, Gómez-Campo C (eds) *Brassica* crop and wild allies. Japan Scientific Society Press, Tokyo, Japan, pp 110–120

U N (1935) Genome analysis in the *Brassicae*, with special reference to the experimental formation of *Brassica napus*. Ikusyugaku Zasshi 7:389–452

Vos P, Horgers R, Bleeker M, Reijans M, Lee TVD, Hornes M, Frijters A, Pot J, Peleman J, Kuiper M, Zabau M (1995) AFLP: a new technique for DNA fingerprinting. Nucleic Acid Res 23(21):4407–4414

Voorrips RE, Jongerius MC, Kanne HJ (1997) Mapping of two genes for resistance to clubroot (*Plasmodiphora brassicae*) in a population of doubled haploid lines of *Brassica oleracea* by means of RFLP and AFLP markers. Theor Appl Genet 94:75–82

Williams JGK, Kubelik AR, Livak KJ, Rafalski JA, Tingey V (1990) DNA polymorphisms amplified by arbitrary primers are useful as genetic markers. Nucleic Acid Res 18(22):6531–6535

II.2 Genomics as Efficient Tools: Example Sunflower Breeding

A. SARRAFI and L. GENTZBITTEL[1]

1 Introduction

Sunflower (*Helianthus annuus* L.) is grown throughout the world mostly as a source of vegetable oil and protein. The main objectives of sunflower breeding programs are the development of productive F_1 hybrids with high oil and protein yield. As in all other crops, sunflower yield depends on many yield components which are controlled by several genes, the effects of which are modified by the environment. Conventional plant breeding methods are responsible for the improvement of plant yield which is provided by plant breeders in sunflower as well as in other crops.

Now, new technologies have introduced an additional means for improving sunflower yield and quality using molecular genetics. Molecular genetics and genomics are considered as tools for the genetic characterization of organisms. The aim of molecular genetics in sunflower breeding is to identify, isolate, amplify and modify genes or other sequences of DNA and to combine and express the novel or modified sequences in new genotypes. Despite some limitations, molecular genetics is now producing significant results by using the new technologies to influence basic and applied research in sunflower improvement.

Linkage map construction is usually the first step for identification of quantitative trait loci (QTLs) controlling different traits. Later, we will present some recent maps constructed for sunflower by different methods and also some QTLs identified for the traits related to yield improvement in this species. We will then present some recent results of genomics and the identification of genes of importance for sunflower breeding.

2 Linkage Mapping

The development of molecular marker techniques has provided an additional tool to determine linkage maps which are used for quantitative trait loci (QTLs) analysis. Molecular marker-based linkage maps are powerful tools for

---

[1] Department of Biotechnology and Plant Breeding, Pôle de Biotechnologie Végétale, ENSAT-BAP, 18 Chemin de Borde Rouge, BP 107, 31326 Castanet, France

breeding programs. Genetic maps in sunflower constructed by different marker systems have been described by several authors. The most important markers are: restriction fragment length polymorphism, (RFLP), amplified fragment length polymorphism (AFLP) and simple sequence repeat (SSRs).

2.1 Restriction Fragment Length Polymorphism Markers

Restriction fragment length polymorphism genetic markers have unique characteristics for crop improvement and can be used to identify crop varieties or hybrids as well as to construct genetic maps. The association between specific DNA markers and agronomic traits will facilitate marker-assisted selection in breeding programs. Early genetic maps, constructed with RFLP markers (mainly cDNAs), were reported by Berry et al. (1995) and Gentzbittel et al. (1995). These maps were incomplete and were developed independently, thus showing very few common markers such as anchor points. Later, 232 unique cDNA probes were used for the RFLP linkage analysis of F_2 population of the cross 'RHA 271 × HA 234' and revealed 271 polymorphic loci (Jan et al. 1998). This map encompasses 20 linkage groups covering 1164 cM of the sunflower genome. Twenty linkage groups of the map are more than the 17 haploid chromosome number of sunflower (Table 1).

The rapid accumulation of markers and mapping populations are challenges for the management of information and the merging of separate sets of data in order to accumulate more valuable information for further research and the better use of genetics. Combined maps provide an easy and convenient way of comparing the component maps and other important information about the reliability of markers, order and distances between markers. With this aim, Gentzbittel et al. (1999) integrated seven individual maps and constructed a near-saturated linkage map, based on RFLPs and including major phenotypic traits (Table 1). This integrated map is arranged in 17 major linkage groups containing 238 loci and covers 1534 cM.

2.2 Amplified Fragment Length Polymorphism Markers

The amplified fragment length polymorphism (AFLP) assay is a powerful technique for genome mapping and genetic variability studies in sunflower. AFLP markers are typically dominant and most fragments correspond to a unique position on the genome and, hence, can be exploited as landmarks in genetic and physical mapping (Meksem et al. 1995). A total number of 19 AFLP primer pairs selected for sunflower were used to determine the linkage map of recombinant inbred lines (RILs) of the cross 'PAC-2 × RHA-266'. Out of 333 markers, 254 were placed in 18 linkage groups; the total length of the map is 2558 cM (Flores-Berrios 2002a; Table 1). The length of this map is significantly greater than that described by Jan et al. (1998) and Gentzbittel et al.

Table 1. Sunflower linkage maps: population, number of markers, linkage groups, kind of markers and map length (cM)

Cross	Population	Number of markers	Number of six linkage groups	Marker systems	Map length (cM)	Reference
RHA271 × HA 234	F_2	271	20	RFLP	1164	Jan et al. (1998)
SD × PA4	F_2	86	17	RFLP	774	Gentzbittel et al. (1999)
SD × CP73	F_2	106	18	RFLP	1056	Gentzbittel et al. (1999)
CP73 × PAC1	F_2	111	16	RFLP	1068	Gentzbittel et al. (1999)
GH × PAC2	F_2	95	18	RFLP	1008	Gentzbittel et al. (1999)
HA89 × RHA266	F_2	76	14	RFLP	176	Gentzbittel et al. (1999)
CX × RHA 266	F_2	99	18	RFLP	582	Gentzbittel et al. (1999)
PAC2 × RHA266	F_2	144	21	RFLP	763	Gentzbittel et al. (1999)
Composite map	F_2	238	17	RFLP	1534	Gentzbittel et al. (1999)
PAC2 × RHA266	RILs (F8)	254	18	AFLP	2558	Flores-Berrios et al. (2000a)
ZENB8 × HA89	F_3 and F_4	205	17	RFLP	1380	Leon et al. (2001)
HA370 × HA372	F_2	446	17	AFLP and RFLP	1326	Gedil et al. (2001b)
HA89 × CAS3	F_2	154	17	RFLP and AFLP	1807	Pérez-Vich et al. (2002)
HAOL9 × CAS3	F_2	137	17	RFLP and AFLP	1641	Pérez-Vich et al. (2002)
LI × L2	F_3	170	20	AFLP and SSR	2539	Mokrani et al. (2002)
CmsHA89 × Wild (H. annuus Var annuus)	F_3	107	17	SSR	972	Burke et al. (2002)

(1999), constructed by RFLPs, and corresponds to an expected twofold expansion of the distances, partly because of using RILs created through selfing. Sunflower possess 17 haploid chromosomes and 18 linkage groups, as well as unlinked markers presented in the map, which indicate that further mapping is still needed to obtain a saturated one.

2.3 Simple Sequence Repeat Markers

Simple sequence repeats (SSRs), called microsatellites, are also used as molecular markers. Their polymorphism has shown high efficiency and they are used for genetic mapping, population and evolutionary studies, as well as for fingerprinting and pedigree analysis (Plaschke et al. 1995; Rongen et al. 1995; Guilford et al. 1997). SSR markers are, thus, now recognized as one of the most efficient molecular markers. SSR identification in sunflower has been reported independently by at least two groups (Paniego et al. 2002; Yu et al. 2002). Burke et al. (2002) established a genetic map of an F_3 population derived from the cross 'cms HA 89 × wild (H. *annuus var. annuus*)' using SSR markers (Table 1). This map covers 17 linkage groups with 107 SSRs and the total map distance is 972 cM. The map length is thus considerably shorter than other maps presented in Table 1. This fact is most likely due, at least in part, to incomplete genome coverage.

The most complete SSR linkage map was published by Tang et al. (2002). They describe a map of 17 linkage groups, based on 408 SSRs, genotyped on 94 RILs, producing 462 SSR marker loci segregating in the mapping population. However, until now, this map has not been associated with any key traits for sunflower breeding and is thus of less interest. Important efforts toward the integration of previously described knowledge-associated data into the frame of a dense SSR map remain to be made. Many factors such as the nature of the populations studied, the number of individuals and the number of markers might change the recombination rate and, consequently, the distance between two loci. Some of the genetic maps summarized in Table 1 are constructed using AFLP and RFLP markers (Gedil et al. 2001b; Pérez-Vich et al. 2002), or AFLPs and SSRs (Mokrani et al. 2002). The advantage of these kinds of maps is that in cases where they have common SSR or RFLP markers with some other maps, they can be combined in order to construct new maps with more markers.

2.4 Cultivated Sunflower Genetic Polymorphisms and Heterosis Modelling

The discovery of genetic structures in cultivated sunflower was one of the first aims of genetic fingerprinting using molecular markers. Initial studies using RFLPs or RAPDs (Berry et al. 1994; Gentzbittel et al. 1994; Mosges and Friedt 1994; Teulat et al. 1994; Zhang et al. 1995) revealed a clustering of inbreds between restorer and maintainer lines of the 'classical' cytoplasmic male sterility to be the most important genetic structure found in cultivated sunflower. AFLPs (Hongtrakul et al. 1997) and minisatellite (Dehmer and Friedt 1998) analyses confirmed these results and also showed that the confectionery sunflower lines are clearly outgroups of oil-producing sunflowers. Using SSRs as high throughout tools for genetic fingerprinting allowed Yu et al. (2002) to show that the results obtained by this molecular tool are consistent with the previously described structures. They also reconfirmed that polymorphism within cultivated sunflower is much more reduced than in other crops. Heterotic group modelling in sunflower using molecular tools was reported several times (Tersac et al. 1993, 1994; Cheres et al. 2000), but failed to reveal clear heterotic groups. At least AFLP-based genetic distances were poor predictors of hybrid seed yield.

3 Quantitative Trait Loci Identification

3.1 Agronomic Traits

The principal goal of sunflower breeding programs is the development of new cultivars with a high oil yield. Identification of the chromosome regions which affect grain yield, oil percentage in grain and other agronomic traits should increase our understanding of the genetic control of the characters and help us to develop marker-assisted selection programs. Days to flowering is one of these important traits because cultivars with certain ranges of cycle length provide optimum yield in specific environments. Leon et al. (2001) located six QTLs associated with growing degree days to flowering in an F_2 population presenting 76% of phenotypic variation, whereas Mokrani et al. (2002) identified two QTLs for sowing to flowering date using 118 F_3 families coming from crosses between two others genotypes with a total effect of 89.30% (Table 2).

In the same F_3 families, Mokrani et al. (2002) have also detected two QTLs for grain weight per plant, one QTL for 1000-grain weight and seven QTLs for oil percentage in grain with total phenotypic variation of: 50.70, 22.70 and 90.40%, respectively (Table 2). QTLs for percentage of oil in grain and 1000-grain weight were also detected by Mestries et al. (1998). These results complement pioneer work describing molecular markers linked to oil characteristics and quantitative genetics analyses (Leon et al. 1995, 2003). As far as oil quality is concerned, four QTLs controlling stearic acid and three for oleic acid were identified by Pérez-Vich et al. (2002). Phenotypic variation for these traits are 84.4 and 58.4%, respectively (Table 2).

3.2 Resistance to Disease

Plant improvement implies the ability to create genotype resistance to different diseases. Being originally breed in Russia under continental conditions, cultivated sunflower is susceptible to diseases under wetter conditions such as those encountered in western Europe. Downy mildew and black stem caused by *Plasmopara halstedii* and *Phoma macdonaldii* are considered important diseases in sunflower (Acimovic 1984; Mouzeyar et al. 1994; Tourvieille de Labrouhe et al. 1998).

Table 2. QTLs detected by different marker systems and their effects on some agronomic traits in sunflower

Reference	Traits	Abbreviation	Number of QTLs	Linkage groups	Total phenotypic variation (%)	Markers
Leon et al. (2001)	Growing degree days to flowering	gdd	6	A, B, F, I, J, L	76.0	RFLP
Mokrani et al. (2002)	Grain weight per plant	gwp	2	9	50.70	AFLP and SSR
	1000 grain weight	tgw	1	16	22.70	
	Oil percentage in grain	pog	7	9, 11, 12, 13	90.40	
	Sowing to flowering	stf	2	9, 10	89.30	
Pérez-Vich et al. (2002)	Stearic acid	C18:0	4	1, 3, 8, 14	84.4	AFLP and RFLP
	Oleic acid	C18:1	3	1, 8, 14	58.4	
Rachid Al-Chaarani et al. (2002)	Downy mildew	dmr	4	1, 9, 17	54.90	AFLP
	Black stem	bsr	7	3, 4, 8, 9, 11, 15, 17	93.10	
Hervé et al. (2001)	Chlorophyll concentration	chl	4	5, 8, 10, 18	53.5	AFLP
	Net photosynthesis	pho	3	9, 14	62.5	
	Stomatal conductance	sco	4	3, 8, 16, 17	61.9	
	Predawn leaf water potential	pot	3	8, 10, 14	34.1	
Flores-Berrios et al. (2000a)	Shoots per explant	ose	6	2, 4, 7, 9, 17	52.0	AFLP
Flores-Berrios et al. (2000b)	Shoots per regeneration explant	osr	7	2, 4, 6, 7, 8, 15, 17	67.0	AFLP
Flores-Berrios et al. (2000b)	Total embryogenic explants	toe	4	1, 3, 13, 15	48.0	AFLP
Flores-Berrios et al. (2000c)	Total protoplast division	ptd	12	1, 7, 8, 10, 13, 14, 15, 17	72.0	AFLP

Downy mildew is the most studied system, probably because it is considered to be a simple gene-for-gene resistance model. Several resistance clusters were described (Mouzeyar et al. 1995; Roeckel Drevet et al. 1996; Vear et al. 1997; Bert et al. 2001), and at least one of them was independently confirmed by Gedil et al. (2001a). These results are accompanied by the descriptions of molecular markers tightly linked to the downy mildew resistances, often based on candidate gene approaches, and putatively used as tools for breeding (Gentzbittel et al. 1998; Brahm et al. 2000; Gedil et al. 2001a; Bouzidi et al. 2002). Using a quantitative model, Rachid Al-Chaarani et al. (2002) identified four QTLs for resistance to downy mildew and seven QTLs for resistance to black stem using recombinant inbred lines of the cross 'PAC 2 × RHA 266' (Table 2). The four detected loci explained 54.9% of the total phenotypic variance for resistance to downy mildew, whereas the seven detected QTLs for black stem explained 92% of the variance.

Candidate genes for disease tolerance to *Sclerotinia* were described (Mouzeyar et al. 1997; Gentzbittel et al. 1998) and were evaluated in subsequent QTL analyses (Bert et al. 2002). Under semi-dry conditions (Spain, Israel, Northern Africa), *Orobanche cumana*, a parasitic weed, appears to be a potential important disease. Genetic mapping of several sources of resistance to *O. cumana* were described (Lu et al. 1999, 2000) leading the way to putative breeding for appropriate sunflower inbreds. Linkage of molecular markers with resistance genes to rust were also described (Lawson et al. 1998).

Thus, several tools are already available to breeders for the improvement of breeding processes for disease resistance.

3.3 Photosynthesis and Water Status Traits

Yield components and oil production are positively correlated with photosynthesis and water status traits in sunflower. Hervé et al. (2001), conducted an experiment to identify QTLs for photosynthesis and water status traits using RILs from the cross 'PAC 2 × RHA 266' and the results are summarized in Table 2. Four QTLs detected for chlorophyll concentration accounted for 53.5% of phenotypic variation for this trait and three QTLs for net photosynthesis accounted for 62.5% of total phenotypic variation. As far as stomatal conductance is concerned, four QTLs with 61.9% of phenotypic effect were detected, whereas predawn water potential was associated with three QTLs and only 34.1% of phenotypic variance.

3.4 In Vitro Regeneration

The ability to regenerate large numbers of plants is important for the development of biotechnology as regards genetic transformation in sunflower. In vitro regeneration was investigated in 75 RILs and their parents (PAC 2 and

RHA 266). The results summarized in Table 2 show that six putative QTLs for the number of shoots per explant and seven for shoots per regenerating explants were detected in cotyledon organogenesis culture (Flores-Berrios et al. 2000a). The same RILs were also used for somatic embryogenesis by epidermic layers; four QTLs were identified for the total number of embryogenic explants which explained 48% of the phenotypic variation for this trait (Flores-Berrios et al. 2002b). The above-mentioned RILs were also used in another experiment in which 12 QTLs were identified for protoplast division with a total phenotypic variation of 72% (Flores-Berrios et al. 2000c). Some segments of the linkage groups 1, 15 and 17 are likely to contain genes important for organogenesis, somatic embryogenesis and protoplast division. The QTLs identified in these three linkage groups should be involved in cell division in early events associated with cell differentiation.

4 Sunflower Genomics: Towards Genes and Functions

Sunflower genomics is still in its infancy, as compared to *A. thaliana*, rice or maize genomics initiatives. However, significant programs have been recently developed, with the aim of bringing to sunflower breeders the tools and knowledge to answer the major challenges of sunflower breeding. In this section, we will present results of the characterization of the oil synthesis pathways and the emergence of genomics programs for sunflower.

4.1 Oil Synthesis

Despite being biochemically well characterized, the lipid metabolism in sunflower remains undescribed at the molecular level and, as stated in previous sections, the genetics of oil (fatty acids) synthesis still remains to be discovered and understood. The 'high oleic acid' trait has mainly been studied because of its economic importance. It is unclear if other fatty acids are considered as breeding and research targets.

Sunflower possesses the unique ability to produce oleic acid at both a high percentage and high yield. This feature was obtained after mutagenesis treatment on Vnimk 8931 population (known as the Pervenets mutation). Several commercial varieties derived from Pervenets and breeding materials with a high oleic acid content have been marketed. However, the genetics of this trait are still not fully understood by breeders. To characterize the Pervenets mutation, several groups have tried to clone and discover mutations in the genes associated with oleic acid synthesis (Hongtrakul et al. 1998a, b; Lacombe and Berville 2001). Tightly linked markers were discovered, some of them displaying the characteristics of candidate genes (Lacombe and Berville 2001). However, the exact mechanism by which the Pervenets mutations provide the 'high oleic' phenotype remains unknown.

4.2 Functional Genomics

For functional genomics, the objectives are to optimize the targets for breeding by deciphering the genetics of simple or quantitative traits, for example by identifying expressed sequence tags (ESTs) putatively underlying QTLs. This method is based on the massive and parallel analysis of the expression levels of thousands of genes in key situations and rely heavily, in a first step, on massive EST sequencing and the use of so-called DNA chips. For sunflower, two major programs are emerging.

A US program for the massive sequencing of ESTs is underway. As a result of this program, the increase in GenBank entries for sunflower ESTs has been spectacular: from 191 entries in December 2000 to 49,264 entries in January 2003. These ESTs were obtained from two inbred lines (RHA801 and RHA280) and cover 11 different organs or physiological situations: callus, roots, disk and ray flowers, pre-fertilized flowers, developing kernel, chemical induction, root environmental stress, shoot environmental stress, germinating seeds, flower environmental stress and hulls. A French program of massive sunflower EST sequencing is also on the way (Caboche and Boucly 2000). More than 20 different cDNA libraries covering key traits for sunflower breeding were constructed and subjected to sequencing. The targets are plant and seed development, drought tolerance and disease resistance. The results will be deposited in public databases and are already available through a high value-added database after bioinformatics work (Samson 2003).

As a consequence of the large number of available sequences (47,000 in March 2003), high-throughput design of primers for EST-based SSRs or single nucleotide polymorphism (SNP) discovery is currently possible to significantly increase the number of expressed sequence-based molecular markers in sunflower.

The expected result of these functional genomics programs is the identification of key genes by temporal, spatial or conditional study of the gene expression level. The steps of gene validation could, however, be difficult as genetic transformation for sunflower does not reach the efficiency needed to be used as a routine tool. To overcome this problem, programs leading to the creation of large mutant collections should be underway. In any case, the functional genomics programs will led to the discovery of new targets for breeding, or to the definition of molecular covariates for the resolution of QTLs for major agronomic traits in sunflower.

4.3 Sunflower Genome Structure

As presented in previous sections, molecular linkage maps for sunflower are already available, though based on different tools and with contrasting efficiencies and knowledge-associated data. In order to facilitate positional gene cloning and to provide a public tool for the improvement of sunflower genome analyses, a BAC library of about four genome equivalents of the inbred HA821 was constructed (Gentzbittel et al. 2002).

As the sequence of the dicot model plant, *Arabidopsis thaliana*, becomes available, several groups are working towards the analysis and the fine characterization of the synteny conservation between the sunflower and *A. thaliana* genomes. Massive EST sequencing also provides a large amount of information that could be used in many different ways. For example, it could provide physical anchors to an EST-derived marker-based (SNPs, SSRs, CAPS, etc.) genetic map, which could be extended to regional or global contig building of the sunflower genome. These contigs will, thus, be of valuable use in the fast cloning of important genes. As another example, ESTs could be used to globally align the sunflower genome with that of *A. thaliana* by in silico hybridization and genetic mapping, thus using the *A. thaliana* sequence as a predictor of the sunflower genome sequence in the regions where synteny exists. For example, a set of conserved orthologous sequences (COS) between different dicots is being defined. In this respect, two websites are proposing tools in a first attempt to graphically present the results of the synteny analyses: the Composite Genome Project Database(http://cgpdb.ucdavis.edu/) and the ICCARE/HeliantSynteny information server (http://genopole.toulouse.inra.fr/~cmuller/accueil.html)

5 Conclusions

Taking into account the large size of the sunflower genome (estimated 3.10^9 bp), a global physical map and genome sequencing project are not realistic in the near future. Research programs will probably focus on a few regions of the sunflower genome governing key agronomic traits, such as oil and protein synthesis, seed development and disease resistance. Numerous molecular tools are already available to develop such programs.

References

Acimovic M (1984) Sunflower diseases in Europe, the United States and Australia, 1981–1983. Helia 7:45–54
Berry ST, Allen RJ, Barnes SR, Caligari PDS (1994) Molecular marker analysis of *Helianthus annuus* L. 1. Restriction fragment length polymorphism between inbred lines of cultivated sunflower. Theor Appl Genet 89:435–441
Berry ST, Leon AJ, Hanfrey CC, Challis P, Burkolz A, Barnes SR, Rufener GK, Lee M, Caligari PDS (1995) Molecular-marker analysis of *Helianthus annuus* L. 2. Construction of an RFLP map for cultivated sunflower. Theor Appl Genet 91:195–199
Bert PF, Tourvieille de Labrouhe D, Philippon J, Mouzeyar S, Jouan I, Nicolas P, Vear F (2001) Identification of a second linkage group carrying genes controlling resistance to downy mil-

dew (*Plasmopara halstedii*) in sunflower (*Helianthus annuus* L.). Theor Appl Genet 103:992–997

Bert PF, Jouan I, Tourvieille de Labrouhe D, Serre F, Nicolas P, Vear F (2002) Comparative genetic analysis of quantitative traits in sunflower (*Helianthus annuus* L.) 1. QTL involved in resistance to *Sclerotinia sclerotiorum* and *Diaporthe helianthi*. Theor Appl Genet 105:985–993

Bouzidi MF, Badaoui S, Cambon F, Vear F, Tourvieille de Labrouhe D, Nicolas P, Mouzeyar S (2002) Molecular analysis of a major locus for resistance to downy mildew in sunflower with specific PCR-based markers. Theor Appl Genet 4:592–600

Brahm L, Rocher T, Friedt W (2000) PCR-based markers facilitating marker assisted selection in sunflower for resistance to downy mildew. Crop Sci 40:676–682

Burke JM, Tang S, Knapp SJ, Rieseberg LH (2002) Genetic analysis of sunflower domestication. Genetics 161:1257–1267

Caboche M, Boucly M (2000) The Genoplante programme, a mobilizing programme in plant genomics. CR Acad Agric Fr 86:159–173

Cheres MT, Miller JF, Crane JM, Knapp SJ (2000) Genetic distance as a predictor of heterosis and hybrid performance within and between heterotic groups in sunflower. Theor Appl Genet 100:889–894

Dehmer KJ, Friedt W (1998) Evaluation of different microsatellite motifs for analysing genetic relationships in cultivated sunflower (*Helianthus annuus* L.). Plant Breed 117:45–48

Flores Berrios E, Gentzbittel L, Kayyal H, Alibert G, Sarrafi A (2000a) AFLP mapping of QTLs for in vitro organogenesis traits using recombinant inbred lines in sunflower (*Helianthus annuus* L.). Theor Appl Genet 101:1299–1306

Flores Berrios E, Sarrafi A, Fabre F, Alibert G, Gentzbittel L (2000b) Genotypic variation and chromosomal location of QTLs for somatic embryogenesis revealed by epidermal layers culture of recombinant inbred lines in the sunflower (*Helianthus annuus* L.). Theor Appl Genet 101:1307–1312

Flores Berrios E, Gentzbittel L, Mokrani L, Alibert G, Sarrafi A (2000c) Genetic control of early events in protoplast division and regeneration pathways in sunflower. Theor Appl Genet 101:606–612

Gedil MA, Slabaugh MB, Berry S, Johnson R, Michelmore R, Miller J, Gulya T, Knapp SJ (2001a) Candidate disease resistance genes in sunflower cloned using conserved nucleotide-binding site motifs: genetic mapping and linkage to the downy mildew resistance gene Pl1. Genome 44:205–212

Gedil MA, Wye C, Berry ST, Segers B, Peleman J (2001b) An integrated RFLP-AFLP linkage map for cultivated sunflower. Genome 44:213–221

Gentzbittel L, Zhang YX, Vear F, Griveau B, Nicolas P (1994) RFLP studies of genetic relationships among inbred lines of the cultivated sunflower, *Helianthus annuus* L.: evidence for distinct restorer and maintainer germplasm pools. Theor Appl Genet 89:419–425

Gentzbittel L, Vear F, Zhang YX, Bervillé A, Nicolas P (1995) Development of a consensus linkage RFLP map of cultivated sunflower (*Helianthus annuus* L.). Theor Appl Genet 90:1079–1086

Gentzbittel L, Mouzeyar S, Badaoui S, Mestries E, Vear F, Tourvieille de Labrouhe D, Nicolas P (1998) Cloning of molecular markers for disease resistance in sunflower (*Helianthus annuus* L.). Theor Appl Genet 96:519–525

Gentzbittel L, Mestries E, Mouzeyar S, Mazeyrat F, Badaoui S, Vear F, Tourvieille de Labrouhe D, Nicolas P (1999) A composite map of expressed sequences and phenotypic traits of the sunflower (*Helianthus annuus* L.) genome. Theor Appl Genet 99:218–234

Gentzbittel L, Abbott A, Galaud JP, Georgi L, Fabre F, Liboz T, Alibert G (2002) A bacterial artificial chromosome (BAC) library for sunflower, and identification of clones containing genes for putative transmembrane receptors. Mol Genet Genom 266:979–987

Guilford A, Prakash S, Zhu JM, Rikkerink E, Gardiner S, Bassett H, Forster R (1997) Microsatellites in *Malus* × *domestica* (apple): abundance polymorphism and cultivar identification. Theor Appl Genet 94:245–249

Hervé D, Fabre F, Flores Berrios E, Leroux N, Al Chaarani G, Planchon C, Sarrafi A, Gentzbittel L (2001) QTL analysis of photosynthesis and water status traits in sunflower (*Helianthus annuus* L.) under greenhouse conditions. J Exp Bot 52:1857–1864

Hongtrakul V, Huestis GM, Knapp SJ (1997) Amplified fragment length polymorphisms as a tool for DNA fingerprinting sunflower germplasm: genetic diversity among oilseed inbred lines. Theor Appl Genet 95:400–407

Hongtrakul V, Slabaugh MB, Knapp SJ (1998a) A seed specific DELTA-12 oleate desaturase gene is duplicated, rearranged, and weakly expressed in high oleic acid sunflower lines. Crop Sci 38:1245–1249

Hongtrakul V, Slabaugh MB, Knapp SJ (1998b) DFLP, SSCP, and SSR markers for DELTA9-stearoyl-acyl carrier protein desaturases strongly expressed in developing seeds of sunflower: intron lengths are polymorphic among elite inbred lines. Mol Breed 4:195–203

Jan CC, Vick BA, Miller JF, Kahler AL, Butler ET (1998) Construction of an RFLP linkage map for cultivated sunflower. Theor Appl Genet 96:15–22

Lacombe S, Berville A (2001) A dominant mutation for high oleic acid content in sunflower (*Helianthus annuus* L.) seed oil is genetically linked to a single oleate-desaturase RFLP locus. Mol Breed 8:129–137

Lawson WR, Goulter KC, Henry RJ, Kong GA, Kochman JK (1998) Marker-assisted selection for two rust resistance genes in sunflower. Mol Breed 4:227–234

Leon AJ, Lee M, Rufener GK, Berry ST, Mowers RP (1995) Use of RFLP markers for genetic linkage analysis of oil percentage in sunflower seed. Crop Sci 35:558–564

Leon AJ, Lee M, Andrade FH (2001) Quantitative trait loci for growing degree days to flowering and photoperiod response in sunflower (*Helianthus annuus* L.). Theor Appl Genet 102:497–503

Leon AJ, Andrade FH, Lee M (2003) Genetic analysis of seed-oil concentration across generations and environments in sunflower. Crop Sci 43:135–140

Lu Y, Gagne G, Grezes Besset B, Blanchard Pand P, Lu YH (1999) Integration of a molecular linkage group containing the broomrape resistance gene Or5 into an RFLP map in sunflower. Genome 42:453–456

Lu YH, Melero Vara JM, Garcia Tejada JA, Blanchard P (2000) Development of SCAR markers linked to the gene Or5 conferring resistance to broomrape (*Orobanche cumana* Wallr.) in sunflower. Theor Appl Genet 100:625–632

Meksem K, Leister D, Peleman J, Zabeau M, Salamini, Gebhardt C (1995) A high-resolution map of the vicinity of the R1 locus on chromosome V of potato based on RFLP and AFLP markers. Mol Gen Genet 249:74–81

Mestries E, Gentzbittel L, Tourvieille de Labrouhe D, Nicolas P, Vear F (1998) Analysis of quantitative trait loci associated with resistance to *Sclerotinia sclerotiorum* in sunflower (*Helianthus annuus* L.) using molecular markers. Mol Breed 4:215–226

Mokrani L, Gentzbittel L, Azanza F, Fitamant L, Al-Chaarani G, Sarrafi A (2002) Mapping and analysis of quantitative trait loci for grain oil and agronomic traits using AFLP and SSR in sunflower (*Helianthus annuus* L.). Theor Appl Genet 106:149–156

Mosges G, Friedt W (1994) Genetic 'fingerprinting' of sunflower lines and F1 hybrids using isozymes, simple and repetitive sequences as hybridization probes, and random primers for PCR. Plant Breed 113:114–124

Mouzeyar S, Tourvieille de Labrouhe D, Vear F (1994) Effect of host-race combination on resistance of sunflower to downy mildew. J Phytopathol 141:249–258

Mouzeyar S, Roeckel Drevet P, Gentzbittel L, Philippon J, Tourvielle de Labrouhe D, Vear F, Nicolas P (1995) RFLP and RAPD mapping of the sunflower Pl1 locus for resistance to *Plasmopara halstedii* race 1. Theor Appl Genet 91:733–737

Mouzeyar S, Gentzbittel L, Badaoui S, Bret-Mestrie E, Perrault A, De Conto V, Cock M, Dumas C, Tourvieille de Labrouhe D, Vear F, Nicolas P (1997) Molecular marker for resistance to *Sclerotinia sclerotiorum*. European patent 98917244.0

Paniego N, Echaide M, Munoz M, Fernandez L, Torales S, Faccio P, Fuxan I, Carrera M, Zandomeni R, Suarez EY, Esteban Hopp H (2002) Microsatellite isolation and characterization in sunflower (*Helianthus annuus* L.). Genome 45:34–43

Pérez-Vich B, Fernandez-Martinez JM, Grondona M, Knapp SJ, Berry ST (2002) Stearoyl-ACP and oleoyl-PC desaturase genes cosegregate with quantitative trait loci underlying high stearic and high oleic acid mutant phenotypes in sunflower. Theor Appl Genet 104:338–349

Plaschke J, Ganal MW, Röder MS (1995) Detection of genetic diversity in closely related bread wheat using microsatellites markers. Theor Appl Genet 91:1001–1007

Rachid Al-Chaarani G, Roustae L, Gentzbittel L, Mokrani L, Barrault G, Dechamp-Guillaume G, Sarrafi A (2002) A QTL analysis of sunflower partial resistance to downy mildew (*Plasmopara halstedii*) and black stem (*Phoma macdonaldii*) by the use of recombinant inbred lines (RILs). Theor Appl Genet 104:490–496

Roeckel Drevet P, Gagne G, Mouzeyar S, Gentzbittel L, Philippon J, Nicolas P, Tourvieille de Labrouhe D, Vear F (1996) Colocation of downy mildew (*Plasmopara halstedii*) resistance genes in sunflower (*Helianthus annuus* L.). Euphytica 91:225–228

Rongwen J, Akkaya MS, Bhagwat AA, Lavi U, Gregan PB (1995) The use of microsatellite DNA markers for soybean genotype identification. Theor Appl Genet 90:43–48

Samson D (2003). GenoPlante-Info (GPI): a collection of databases and bioinformatics resources for plant genomics. Nucleic Acids Res 31:179–182

Tang SX, Yu JK, Slabaugh MB, Shintani DK, Knapp SJ (2002) Simple sequence repeat map of the sunflower genome. Theor Appl Genet 105:1124–1136

Tersac M, Vares D, Vincourt P (1993) Combining groups in cultivated sunflower populations (*Helianthus annuus* L.) and their relationship with country of origin. Theor Appl Genet 87:603–608

Tersac M, Blanchard P, Brunel D, Vincourt P (1994) Relations between heterosis and enzymatic polymorphism in populations of cultivated sunflowers (*Helianthus annuus* L.). Theor Appl Genet 88:49–55

Teulat B, Zhang YX, Nicolas P (1994) Characteristics of random amplified polymorphic DNA markers discriminating *Helianthus annuus* inbred lines. Agronomie 14:497–502

Tourvieille de Labrouhe D, Champion R, Vear F, Mouzeyar S, Said J (1998) Une nouvelle race de mildiou en France. CETIOM Inform Tech 104:3–10

Vear F, Gentzbittel L, Philippon J, Mouzeyar S, Mestries E, Roeckel Drevet P, Tourvieille de Labrouhe D, Nicolas P (1997) The genetics of resistance to five races of downy mildew (*Plasmopara halstedii*) in sunflower (*Helianthus annuus* L.). Theor Appl Genet 95:584–589

Yu JK, Mangor J, Thompson L, Edwards KJ, Slabaugh MB, Knapp SJ (2002) Allelic diversity of simple sequence repeats among elite inbred lines of cultivated sunflower. Genome 45:652–660

Zhang YX, Gentzbittel L, Vear F, Nicolas P (1995) Assessment of inter- and intra-inbred line variability in sunflower (*Helianthus annuus*) by RFLPs. Genome 38:1040–1048

II.3 Genome Analysis: Mapping in Sugar Beet

C. JUNG[1]

1 Introduction: The Species of the Genus *Beta*

Sugar beet (*Beta vulgaris* ssp. *vulgaris* L. conv. *crassa* prov. *altissima*), the only sucrose-storing crop of moderate climates, is a biennial species forming a leaf rosette and a storage root in the first year. It belongs to the genus *Beta* (Chenopodiaceae) which is divided into four sections. All cultivated forms (fodder beet, leaf beet, red beet) belong to the same species *B. vulgaris* ssp. *vulgaris* which has been classified to the first section (Beta). They are closely related to the wild beet *B. vulgaris* ssp. *maritima* which is believed to be the ancestral form of cultivated beet. All species of the section Beta are readily crossable with each other. In contrast, crosses with species of the other sections are difficult due to strong crossing barriers. Crossings with species of the section 4 (Procumbentes) are extremely difficult due to the low viability of hybrids which requires embryo rescue techniques. Moreover, chromosomes of Beta and Procumbentes species lack homology, resulting in very low fertility of the hybrids. Nevertheless, alien introgression lines have been produced carrying monosomic additions and translocations of the wild beets *B. corolliflora* (Gao et al. 2001), *B. procumbens* (Savitsky 1978) and *B. webbiana* (Reamon-Ramos and Wricke 1992).

2 Resources and Techniques

2.1 Molecular Marker Maps

Sugar beet is a diploid species with $2n=2x=18$ chromosomes and a haploid genome size of 758 Mb (Arumuganathan and Earle 1991). A gametophytic self-incompatibility system with a series of S-alleles inhibits production of inbred lines. Due to its true diploid nature, neither monosomic nor nullisomic lines are available because the loss of chromosomes is not tolerated. However, two sets of primary trisomics have been established (Butterfass

---

[1] Plant Breeding Institute, Christian-Albrechts-University of Kiel, Olshausenstr. 40, 24098 Kiel, Germany

1963; Romagosa et al. 1986) and used for chromosomal assignment of genes previously mapped to linkage groups.

Genetic mapping began with morphological and isozyme markers (Wagner et al. 1992). Those maps were extended by molecular markers and later the linkage groups were assigned to the nine chromosomes of beet (Table 1). The RFLP map of Pillen et al. (1992, 1993) and Schondelmaier et al. (1995) contained the gene for red hypocotyl color R, the restorer gene X, the gene for early bolting B and the $Hs1^{pro-1}$ gene for nematode resistance together with six isozyme markers previously mapped by Wagner et al. (1992). Later, this map was extended by amplified fragment length polymorphisms (AFLPs; Schondelmaier et al. 1996). Typically for sugar beet, 12% of the RFLP markers showed distorted segregation which was explained by gametic selection of linked lethal genes, three of these genes (let-I, let-V, let-VIII) were incorporated into the map mainly at the end of the linkage groups. Duplicate regions were only found on three chromosomes, confirming the true diploid nature of sugar beet. Using different mapping populations, Barzen et al. (1992, 1995) mapped 301 restriction fragment length polymorphism (RFLP) and random amplified polymorphic DNA (RAPD) loci including the rhizomania resistance gene Rz on chromosome three and the gene for the monogerm character M to the end of chromosome 4. Schumacher et al. (1997) presented a map with 600 markers which is based on the linkage maps previously presented by Barzen et al. (1995) and Schondelmaier et al. (1996). This map comprises the highest number of markers for that species with a majority of RFLPs and full assignment to the nine chromosomes of beet.

The highest number of RFLPs incorporated in a single map was published by Halldén et al. (1996a). However, the chromosomal location of the markers is unknown since no relationship to any other linkage map could be given. With one exception, each linkage group showed clusters of markers mainly in the center which was explained by recombination hot spots close to the centromeres of sugar beet. Major clustering in the center of the linkage groups was also reported by Nilsson et al. (1997) who mapped 160 RAPD and 248 RFLP markers covering 508 cM.

Giorio et al. (1997) mapped RFLPs, RAPDs, sequence characterized amplified region (SCARs) and one sequence tagged site (STS) marker on 14 linkage groups reflecting incomplete marker density. On one linkage group the Rr1 locus for rhizomania resistance could be placed flanked by two RAPD loci. The map of Uphoff and Wricke (1995) was exclusively relying on RAPD markers linked with isozyme markers and the genes B, R, and X. The gene for nematode resistance Nema was placed at a different location to the $Hs1^{pro-1}$ gene. Both maps could not be correlated with the other linkage maps of Beta.

Comparatively less information about simple sequence repeat (SSR) markers has become public, which is mainly due to the fact that the majority of markers have been developed by private breeding companies. Mörchen et al. (1996) described four microsatellites without genetic mapping of the corre-

Table 1. Linkage maps of sugar beet. The map of Schumacher et al. (1997) combines markers from different maps previously published by Schondelmaier et al. (1996) and Barzen et al. (1995)

Size of mapping population	RFLP	RAPD	AFLP	Others	Average distance between markers (cM)	Total map distance (cM)	Genes mapped[c]	References
96	206	–	120	9	2.6[a]	557[a]	$R, X, B, Hs1^{pro-1}$	Pillen et al. (1993); Schondelmaier et al. (1995, 1996)
49	248	50	–	–	2.7	815	$Rr1, M$	Barzen et al. (1995)
355[b]	413	–	–	–	1.5	621	$Hs1^{pro-1}$	Halldén et al. (1996a, b)
161	248	160	–	–	1.2	508	–	Nilsson et al. (1997)
112[b]	20	76	–	4	8.0	688.4	$Rr1$	Giorio et al. (1997)
178[b]	32	–	–	23	8.3	458	–	Rae et al. (2000)
93[b]	1	85	–	8	7.9	738	$B, R, X, Nema$	Uphoff and Wricke (1995)
145[b]	470	–	120	10	1.1	688.4	M, R, X	Schumacher et al. (1997)
211	75	–	146	3	5.0	1119	–	Schäfer-Pregl et al. (1999)

[a] Without AFLPs
[b] >1 Population
[c] R Red hypocotyls, X restorer, $Rr1$ rhizomania resistance (Rz), M monogerm character, B early bolting, $Hs1^{pro-1}$ nematode resistance, $Nema$ nematode resistance

sponding loci, while Rae et al. (2000) reported 23 markers mapped to the nine linkage groups of beet previously established by Barnes et al. (1996).

Single nucleotide polymorphisms (SNP) are highly abundant in the genome of sugar beet. In principle, SNPs can be the reason for all kinds of molecular marker polymorphisms. However, in combination with high throughput detection technologies this marker system is most suitable for screening large populations. Schneider et al. (1999) investigated 42 genes involved in carbohydrate and nitrogen metabolism that had been assigned to the nine linkage groups of sugar beet. The genes were mainly mapped as SNPs using different techniques like single-strand conformation polymorphism (SSCP), HA heteroduplex analysis and cleaved amplified polymorphic sequences (CAPS). In a second paper, haplotypes of 37 genes were investigated in two inbred lines. The SNP rate corresponded to 1 every 130 bp (Schneider et al. 2001). Using the SSCP technique for detecting SNPs one polymorphism was found every 1470 bp corresponding to 5×10^5 SSCPs in the whole genome. The fraction of genes having different SSCP alleles in a random comparison of two lines was equal to 54%.

Assigning linkage groups to the nine chromosomes of sugar beet proved to be difficult because no nullisomic lines are available. Schondelmaier and Jung (1997) used a standard set of 24 mapped RFLPs in combination with a set of primary trisomics (Butterfass 1963). In this way, each linkage group could be assigned to one of the nine chromosomes of sugar beet. Thus, mapping data of different origins could be linked with each other using RFLPs as anchor markers. Taking the information that is publicly available, 10 Mendelian genes and 13 major quantitative trait loci (QTL) could be placed on the chromosomes together with 26 isozyme loci (Table 2).

2.2 Genomic Libraries

Three YAC libraries of sugar beet have been published (Eyers et al. 1992; Klein-Lankhorst et al. 1994; Kleine et al. 1995). Successful isolation of clones from a certain region of the genome has been described only for two libraries (Klein-Lankhorst et al. 1994; Kleine et al. 1995). Due to low genome coverage and difficulties in handling and propagating the YACs, two BAC libraries have been established and used for extracting clones with repetitive sequences clustered around the centromere (Gindullis et al. 2001a) and for creating a contig around the bolting gene *B* with the aim to clone the gene (Hohmann et al. 2003).

2.3 Expressed Sequence Tag

The availability of large expressed sequence tag (EST) collections is essential for the identification of genes. Using oligonucleotide fingerprinting, 159,936 cDNA clones from sugar beet leaf, developing root, storage root and inflores-

Table 2. Location of genes for Mendelian traits and quantitatively inherited characters (QTL) on the nine chromosomes of sugar beet. The chromosomes were designated as suggested by Butterfass (1963) and Schondelmaier and Jung (1997)

Chromosome	QTL	Genes
1		$Ak1^a$, $Let5b^b$, $PsbS^c$, $PetE^c$, 18S-5.8S-25S rRNA^d
2	Potassium content^{s,x}, Cercospora resistance^u	Y^e, $B^{e,f,a}$, C^g, $Est2^h$, $Fdp2^h$, $Got3(2)^{h,a}$, $Let1a^b$, $Nema^i$, R^b
3	Cercospora resistance^{t,u}	$Est5^h$, $X^{h,j,a,w}$, $Rr1^k$ (=Rz,= Rz-1), $Gdh2^a$, $Skdh2^l$, $Rz2^v$
4	Cercospora resistance^t, corrected sugar yield^x, beet yield^x	$Est3^h$, $Z^{h,m,w}$, $M^{k,h}$, Fas^h, Nb^n, Sc^n, $PsbO^c$, $cTRF^c$, $PsaL^c$, $PsbR^c$, $R1H^o$, $Mdh3^p$, $5SrRNA^q$
5		$Est1^h$, $Let8^b$
6	Cercospora resistance^u	$PetH^c$, $PsbP^c$
7	Nitrogen content^s, Cercospora resistance^t, ion balance^x	$Dia1^j$, $PsaE^c$
8	Sugar content^s, nitrogen content^s	$Acp1^j$, $PsbQ^c$
9	Cercospora resistance^{t,u}, potassium content^x,	$Hs1^r$, $Hs2^r$, $PsaH^c$, $PetC^c$

[a] Abe et al. (1993)
[b] Red hypocotyl color R, severely distorted segregation Let1a, Let5b, Let8 (Pillen et al. 1993; Schondelmaier et al. 1995)
[c] Chloroplast protein genes PsbS, PetE, PsbO, cTRF, PsaL, PsbR, PetH, PsbP, PsaE, PsbQ, PsaH, PetC (Pillen et al. 1996)
[d] Butterfass (1963)
[e] Keller (1936); Owen et al. (1940); Owen and Ryser (1942)
[f] Early bolting B (Boudry et al. 1994)
[g] Root color Y, curly top resistance C, cited from (Wagner et al. 1992)
[h] Stem fasciation Fas, restoration of male fertility Z and X, and more isozyme loci (Wagner et al. 1992)
[i] Nematode resistance gene Nema (Uphoff and Wricke 1995)
[j] Pillen et al. (1993)
[k] Resistance to rhizomania, Rr1, monogerm character M (Barzen et al. 1995)
[l] Skdh2 might be linked with a restorer gene, perhaps X (Abe et al. 1993)
[m] Z might be linked with M (Roundy and Theurer 1974)
[n] Late bolting, Nb, sugar content, Sc (Savitsky 1958)
[o] Laporte et al. (1998)
[p] Pgm1 is linked with Mdh3 and Pgm1 might be linked with M (Abe et al. 1993)
[q] Schondelmaier et al. (1997)
[r] Resistance against H. schachtii, $Hs1^{pro-1}$, $Hs2^{pro-7}$ (Heller et al. 1996)
[s] Weber et al. (2000)
[t] Setiawan et al. (2000)
[u] Schäfer-Pregl et al. (1999)
[v] Scholten et al. (1999)
[w] Hjerdin-Panagopoulos et al. (2002)
[x] Schneider et al. (2002)

cence were processed, resulting in the identification of 30,444 clusters (Herwig et al. 2002). After sequencing 10,961 clones, 9745 different cDNAs were identified. It was concluded that the 30,444 clusters represent up to 25,000 different genes of sugar beet.

The availability of ESTs is a prerequisite for establishing functional marker maps. Hunger et al. (2003) screened the beet EST collection for sequences with similarity to known disease resistance genes (resistance gene analogs, RGA), resulting in the identification of 29 rests. Using degenerate primers, another 47 RGA could be found. Thirty-one RGA displayed polymorphisms among the mapping population and could be placed on the combined map of sugar beet (Schumacher et al. 1997). They were found to be spread around all chromosomes of beet, except for a cluster of nine closely linked RGA on chromosome 7. Three RGA mapped to QTL for leaf spot resistance on chromosomes 5, 7, and 9 which was previously described by Schäfer-Pregl et al. (1999).

2.4 Transformation

Transgenic technology has been used to create in sugar beet new genetic variations, e.g., for rhizomania and nematode resistance and for herbicide tolerance. Compared to other dicotyledonous plants, sugar beet is a recalcitrant species for the *A. tumefaciens*-mediated gene transformation. Although sugar beet is susceptible to infection by *A. tumefaciens*, both the transformation frequency and regeneration rate of whole plants following transformed explants are very low and strongly dependent upon genotypes (Krens et al. 1988). Hall et al. (1996) reported on polyethylene glycol-mediated DNA transfer into protoplast populations enriched specifically for a single totipotent cell type derived from stomatal guard cells. Transgenic plants could be obtained within 8–9 weeks under favorable conditions. However, this technique is difficult to apply because large amounts of purified guard cells are needed and it relies on a given genotype with a high rate of regeneration. In contrast, transgenic hairy roots can be produced easily by *A. rhizogenes* transformation (Kifle et al. 1999). In this way, the activity of nematode resistance genes (Cai et al. 1997) and the activity of bacterial PHB (poly-3-hydroxybutyrate) genes could be studied (Menzel et al. 2003).

3 Repetitive DNA Classes

The genome of a higher plant is largely composed of repetitive DNA. Estimation of the proportion of repetitive DNA by reassociation kinetics revealed that the beet genome carries some 60% repetitive sequences (Flavell et al. 1974). With the advent of fluorescence in situ hybridization (FISH), sugar beet became one of the best studied plant species with respect to repetitive DNA sequences. Today, it is a model for the organization of repetitive DNA on a plant chromosome.

3.1 Tandemly Repeated DNA

The genes coding for the 5S rRNA and 18S-5.8S-25S rRNA are tandemly organized in higher plant genomes. After cytological investigations of sugar beet trisomics, the nucleolus organizer region (NOR) was assigned to chromosome 1 (Butterfass 1963; Schondelmaier and Jung 1997). The location of the 18S-5.8S-25S rRNA gene cluster at the NOR was later confirmed by FISH (Schmidt et al. 1994). The 5S rRNA gene cluster was localized on chromosome 4 by means of recombination mapping and FISH (Schondelmaier et al. 1997). A polymorphic RFLP fragment corresponding to the 5S rRNA-gene was genetically mapped to a cluster of AFLP and RFLP markers in the middle of chromosome 4. In addition, FISH revealed three signals on the trisomic line carrying three copies of that chromosome.

Many satellite DNA sequences have been cloned from sugar beet and other *Beta* species and their distribution throughout the genomes has been determined (Dechyeva et al. 2003). There is no other plant species with so many satellites investigated by FISH. Moreover, satellite DNA proved to be a valuable tool for monitoring alien gene introgression into cultivated beets. The satellites have been classified according to the restriction enzymes which had been used for cloning them. Five different satellite families have been described for the genome of *B. vulgaris*. The size of the monomers ranges between 140 and 363 bp (Schmidt et al. 1991). The chromosomal distribution of all satellites described has been determined by FISH. In general, they are present at the centromeric regions and at intercalary DAPI (4',6-diamidino-2-phenyl-indole)-positive regions of all chromosomes, however strong differences with regard to copy number become obvious. Satellites were further classified according to their genome specificity. Only the *Apa*I satellite pAV34 was found to be present in all species of the genus *Beta* while the remaining ones were specific for species of the same section (cf. 3).

3.2 Retrotransposons

Transposable DNA elements contribute to the inflation of genomes of higher plants and they are a major source of biodiversity. Retrotransposons are classified into non-LTR (long terminal repeat), SINE (short interspersed nuclear elements), LINE (long interspersed nuclear elements), and LTR retrotransposons (*Ty1-copia*, *Ty3-gypsy*). *BNR*1, the first LINE element from a dicot, was isolated from sugar beet using degenerated primers from the reverse transcriptases of *Lilium speciosum* and *Zea mays* (Schmidt et al. 1995). The element is some 5000 bp in size. It is highly repetitive and organized into clusters as revealed by FISH. It is excluded from centromeric, telomeric and nucleolus organizer regions. Kubis et al. (1998) studied the chromosomal distribution and sequence similarity of LINEs from *B. vulgaris*, *B. lomatogona* and *B. nana*. LINEs were found to be present in all species, however, variation with respect to copy number was found. As in *B. vulgaris*, LINEs were also clustered in the wild species and a high degree of sequence divergence became obvious. As representatives of the LTR retrotransposons, four *Ty1-copia*-like retrotransposons (Tbv) were isolated from sugar beet. In contrast to *BNR1*, they were found to be evenly distributed around the genome with the exception of centromeres and NORs. Sequences with high homology to *BNR1* and Tbv were also found in the other species of the genus *Beta*.

Taking together results from in situ hybridization with different repetitive sequence classes, the sugar beet chromosome was proposed to serve as a model for a plant chromosome with its typical distribution of satellites, retrotransposons and single copy sequences (Schmidt and Heslop-Harrison 1998). This information, together with a large variety of alien chromosome addition lines, provided the incentive to clone a functional centromere of a plant chromosome (Gindullis et al. 2001b). The centromere region was found to be enriched with different classes of repetitive sequences such as satellite DNA, Ty3-*gypsy*-like retrotransposons, and microsatellites. The long-range organization of centromeric DNA was studied by high-resolution multicolor-FISH on pachytene chromosomes and extended DNA fibers and by pulsed-field gel electrophoresis and a similar organization of repetitive sequences as was found in other higher eukaryotes.

4 Genetic Relationships Between Species of the Genus *Beta*

Genome evolution, genetic diversity and relatedness between Beta species was studied with molecular markers. Jung et al. (1993) investigated 41 accessions of the genus *Beta* representing wild and cultivated species of all sections. They used repetitive DNA probes yielding unique fingerprints for each of the accessions. The sugar beet cultivars examined displayed a low level of genetic diversity; they showed high similarity to *B. vulgaris* ssp. *maritima*, but low genetic similarity to the other wild species of section 1. In most cases, the present taxonomic classification of the genus *Beta* was confirmed. Species of sections 2, 3, and 4 were clearly distinguishable from those of section 1 except for *B. macrocarpa*, which showed high similarity to wild species of section 2.

None of the isolated satellites (cf. 3.10) could be found in other species of the *Chenopodiaceae* family (e.g. spinach). Procumbentes species were clearly separated from the other species with regard to repetitive sequence analysis. Two *Sau*3AI satellite families are restricted to Procumbentes species only

(Schmidt and Heslop-Harrison 1996), whereas other satellites which are highly abundant in species of the other sections do not exist in Procumbentes species. Hence, these satellites together with the *Apa*I satellite are likely to be the ancient sequences which have been established in the progenitor of *Beta* species ('Protobeta') and the Procumbentes species are regarded as phylogenetically distinct from the other *Beta* species. The *Bam*HI satellite of *B. vulgaris* was found to be present in species of the section Beta only, thus, this satellite is regarded as the youngest satellite sequence (Schmidt and Metzlaff 1991). The internal transcribed spacer region of ribosomal DNA (ITS) was used as a probe together with RAPDs for determining relatedness between 11 species or subspecies of the genus *Beta* (Shen et al. 1998). Species of section 1 formed one group, Procumbentes species formed a very distinct group, and section 2 and 3 species clustered together forming a third group.

Summarizing the experiments, cultivated species seem to be very closely related to each other and to *B. vulgaris* ssp. *maritima*. On the other hand, Procumbentes species displayed very low similarity on the nucleotide level which makes its taxonomic classification to the *Beta* species doubtful.

The low conservation of satellite DNA among *Beta* species provided the incentive to use these sequences as probes for monitoring introgression of wild species into cultivated beet. Alien addition and translocation lines carrying chromatin from *B. procumbens* (Schmidt et al. 1990; Jung and Herrmann 1991; Salentijn et al. 1992) and *B. corolliflora* (Gao et al. 2001) could be easily identified using those probes in combination with a simple squash dot technique which enabled the screening of thousands of plants in a short time. Using more sensitive genomic Southern hybridization, a correlation between the size of the introgressed alien chromatin and the signal intensity became obvious and is a prerequisite for selecting recombinant introgression lines with variable pieces of alien chromatin incorporated into the beet genome (translocation lines). In this way, the *B. procumbens* specific satellite sequence pRK643 (=pTS5) was used as a marker for positional cloning of the $Hs1^{pro-1}$ gene for nematode resistance (Cai et al. 1997).

5 Genetic Mapping of Mendelian Traits

Rhizomania is the most important disease of sugar beet. It is caused by the beet necrotic yellow vein virus (BNYVV). The only way to control this disease is the use of resistant varieties. Therefore, breeding for rhizomania resistance is a major aim all around the world. Correspondingly, major genes for rhizomania resistance have been mapped with molecular markers. The most important is a single dominant gene named *Rz* (Biancardi et al. 2002) coming from breeding material of the Holly sugar company ('Holly' resistance). The material had originally been developed from *B. vulgaris* ssp. *maritima* at Fort Collins (Colorado, USA) for resistance to *Rhizoctonia solani*. The gene was also named Rz-1 ($=Rz1$, $=Rz_1$) by Pelsy and Merdinoglu (1996) or $Rr1$ by Barzen et al. (1992). Several RFLP and RAPD markers were mapped in a short distance >5 cM from $Rr1$ on chromosome 3 (Barzen et al. 1995). Because those markers were difficult to apply for marker-assisted selection, a SCAR marker was later developed mapping at a distance of 1.3 cM from $Rr1$ (Barzen et al. 1997). A cluster of resistance genes in this region of the genome was proposed by Giorio et al. (1997) who used the same RAPDs as Barzen et al. (1995) for mapping the resistance of *B. vulgaris* ssp. *maritima*. Different RAPD markers were mapped by Pelsy and Merdinoglu (1996) to a QTL for rhizomania resistance. Although the chromosomal location of this QTL is not clear, it can be speculated that it coincides with the Rz locus. The second dominant resistance gene named $Rz2$ which was found in another *B. vulgaris* ssp. *maritima* accession WB42 was determined to be more effective than the 'Holly' $Rz1$ gene using infection studies. Both genes are located on the same chromosome at a distance of 20 cM. $Rz2$ was mapped with STS markers linked at a distance between 2 and 11 cM (Scholten et al. 1999). Two RGA were found to be linked to the $Rr1$ gene, however, recombination between the resistance and the RGAs ruled out the possibility that the RGAs could represent that gene (Hunger et al. 2003).

The **beet cyst nematode** (*Heterodera schachtii* Schm.) is a major pest in sugar beet cultivation. Resistance was only found in the three wild species of section 4. At least three different resistance genes located on different chromosomes were distinguished: $Hs1$ on the homoeologous chromosomes 1 of each species, $Hs2$ on the homoeologous chromosomes 7 of *B. procumbens* and *B. webbiana* and $Hs3$ on chromosome 8 of *B. webbiana* (Löptien 1984; Reamon-Ramos and Wricke 1992; Lange et al. 1993). Among the offspring of monosomic wild beet addition lines (2n=19), diploid translocation lines have been selected and subjected to sugar beet breeding and positional cloning of the resistance genes (Yu 1981; Jung and Wricke 1987; Heijbroek et al. 1988).

In the following, results from mapping studies with diploid nematode resistant sugar beet will be discussed. Heller et al. (1996) investigated four nonrelated translocation lines with RFLP markers. Surprisingly, the four genes mapped to the same locus in sugar beet independent of the original translocation event. Close linkage (0–4.6 cM) was found with marker loci at the very end of chromosome 9. Evidently, there are recombination hot spots in the *Beta* genome where alien chromatin is preferentially incorporated. Six RAPD markers linked to the $Hs1^{pat-1}$ gene were cloned representing repetitive DNA elements specific to the wild beet genome (Salentijn et al. 1995). Two RAPDs flanking the resistance gene were sequenced and turned into STS markers. Halldén et al. (1996b) mapped $Hs1^{pro-1}$ at a distance of 3 cM to an RFLP marker at the end of linkage group 2, thus corresponding to chromosome 9. Additional RAPD markers were placed around the $Hs1^{pro-1}$ gene ending up with a marker density of 0.3 cM in this region of the genome.

The $Hs1^{pro-1}$ gene was cloned with the help of genome-specific repetitive probes (Jung et al. 1992; Salentijn et al. 1994) and chromosomal breakpoint analysis. Expression of the corresponding cDNA in a susceptible sugar beet conferred resistance. The native $Hs1^{pro-1}$ gene, expressed in roots, encodes a 282-amino acid protein with a leucine-rich region and a putative membrane spanning segment (Cai et al. 1997). Using a PCR approach with $Hs1^{pro-1}$-derived primers, sequences with high similarity to $Hs1^{pro-1}$ were later cloned from B. webbiana and B. patellaris (Kleine et al. 1998). Sequence comparison of these genomic fragments revealed a homology of 96% between B. procumbens and B. webbiana and 93% between B. procumbens and B. patellaris. It was concluded that those sequences most likely represent the resistance genes $Hs1^{web-1}$ and $Hs1^{pat-1}$, respectively. A second nematode resistance gene on chromosome 1 of B. procumbens was proposed by Sandal et al. (1997). Recently, we have identified a RGA from the B. procumbens translocation in sugar beet which displayed high similarity to other nematode resistance genes cloned so far. The gene was named $Hs1$-1^{pro-1}.

In sugar beet, stem elongation (**bolting**) is initiated after exposure to a period of low temperatures followed by cultivation under long day conditions. A major locus B was found to cause early bolting (annuality) in wild annual beets carrying the dominant allele B. Pollen transfer from wild beets onto seed multiplication plots may introgress the B allele into cultivated biennial beets. The B locus controlling early bolting under long day conditions without requirement for vernalization was mapped with RFLPs to the center of chromosome 2 (Boudry et al. 1994). Later, tightly linked AFLPs have been described flanking the gene at a distance of 0.14 and 0.23 cM, respectively (El-Mezawy et al. 2002). The AFLPs have been cloned and the STS markers derived can be used for identifying seeds with the allele for early bolting among commercial seed lots. Moreover, these markers are being used for map-based cloning of the bolting gene (Hohmann et al. 2003). Since a number of heterozygous plants (Bb) showed the annual character, a dominant gene for long day length requirement (Lr) which should be tightly linked with B was postulated by Abe et al. (1997).

There is no need to restore male-fertility in sugar beet varieties because only vegetative parts of the plant are harvested. Nevertheless, restorer genes have been described and three of them have been mapped with molecular markers. The restorer gene X was mapped with isozyme and RFLP markers terminally on chromosome 3 (Pillen et al. 1993; Uphoff and Wricke 1995), while the restorer locus Z was mapped to chromosome 4 (Wagner et al. 1992). Accordingly, Hjerdin-Panagopoulos et al. (2002) mapped two major QTL for pollen fertility restoration linked at a distance of 15 cM on chromosome 4 while one QTL was found on chromosome 3. Laporte et al. (1998) reported a restorer locus R1H active in combination with a new cytoplasmic male-sterility. The gene was located on chromosome 4 at a distance of 1.7 cM from the next RFLP marker locus. Since Z and R1H are on the same chromosome, they may possibly be one and the same gene (Laporte et al. 1998).

Betalain pigments found in red beet have been used as natural red food colorings. Two tightly linked genes conditioning production of betalains, Y and *R*, have been located on chromosome 2. Breeding efforts to increase the betalain pigment concentration resulted in greater yields of pigments per unit area of beet. Among 161 RAPDs used in populations resulting from recurrent selection for high pigment/high solids and high pigment/low solids, molecular markers have been detected which are associated with pigment concentration and percent solids in red beet (Eagen and Goldman 1996).

6 Mapping of Quantitative Trait Loci

Cercospora **leaf spot** disease caused by the fungus *Cercospora beticola* Sacc. is one of the most serious and widely distributed foliar diseases of sugar beet in the world. Resistance breeding can help to maintain crop yield even under severe disease pressure while reducing levels of fungicide use. The development of tolerant or resistant varieties can increase sugar yield by up to 45% in the presence of *Cercospora* infection (Schäufele and Wevers 1996). However, breeding for resistance is difficult because resistance is inherited in a polygenic way and can only be assessed under field conditions. Several papers have been published reporting on molecular markers linked to resistance QTL. Schäfer-Pregl et al. (1999) analyzed F_3 families and F_2 testcrosses in the field. QTLs with high logarithm of odds (LOD) scores were localized on linkage groups 2, 3, 6, and 9. Linkage groups 4 and 5 gave a clear indication of the presence of a QTL only when F_2 data were considered. Nilsson et al. (1999) mapped five QTL for resistance to *Cercospora* leaf spot with AFLP and RFLP markers on linkage groups 1, 2, 3 and 9. The QTL were mostly additive explaining between 7–18% of the phenotypic variation.

Another study on QTL analysis of *Cercospora* resistance was carried out using a linkage map based on AFLP and RFLP markers (Setiawan et al. 2000). Two different screening methods for *Cercospora* resistance, a field test under natural infection and a leaf disk test, were used to estimate the level of resistance. The correlation between scores from the field and the leaf disc test was significant. QTL analysis was based on F_2 individuals and F_3 half-sib families. Four QTL associated with *Cercospora* resistance on chromosomes 3, 4, 7, and 9 were detected. The resistance alleles with the highest effects were from QTL located on chromosomes 4 and 9, reducing leaf spot infection by –2.03and –1.5%, respectively. The highest explained phenotypic variance was shown by the QTL located on chromosome 4 (R^2=25.1%).

Only limited data are available for the mapping of genes affecting sugar yield and quality characters with the help of molecular markers. Two approaches have been followed: mapping of QTL by analyzing quantitative variation and using candidate genes with a known function from EST libraries and other sources as molecular markers for linkage analysis. Weber et al. (2000) investigated two populations segregating for quality characters. Common QTL were found for sugar content (chromosome 8), potassium (chromosome 2) and nitrogen content (chromosome 7, 8). Following the second approach, 75 ESTs that are likely to be associated with carbohydrate and nitrogen metabolism were used for mapping QTL for sugar quality and yield parameters (Schneider et al. 2002). The genes were mainly mapped as SNPs using different techniques such as SSCP, heteroduplex analysis and CAPS. Twenty-one significant QTL were detected, all of them were flanked by ESTs, thus demonstrating the effectiveness of the candidate gene approach. The QTL with the largest effect on corrected sugar yield was mapped on chromosome 4, however, no correspondence with QTL previously published by Weber et al. (2000) was found.

7 Conclusions

The beet genome is one of the best studied plant genomes in terms of repetitive DNA sequences. Molecular markers have been mapped around the beet genome and close linkage with Mendelian traits has been established. Due to the small size of the genome, positional cloning of corresponding genes is feasible and will result in the isolation of agronomically important genes. Disease-related STS and SNP markers are routinely used as diagnostic markers for marker-assisted selection during the breeding process. On the other hand, information about QTL mapping in beet is still limited. This situation may change in the near future with the availability of large EST collections offering the possibility to select candidate genes for different traits, such as sugar yield, quality as well as resistance to *Rhizoctonia* root rot disease. This approach seems to be more promising for QTL mapping than whole genome mapping with anonymous markers.

Acknowledgements. I wish to thank Dr. M. McGrath (East Lansing) and Dr. T. Schmidt (Kiel) for helpful comments.

References

Abe J, Guan G, Shimamoto Y (1993) Linkage maps for nine isozyme and four marker loci in sugar beet (*Beta vulgaris* L.). Euphytica 66:117–126

Abe J, Guan GP, Shimamoto Y (1997) A gene complex for annual habit in sugar beet (*Beta vulgaris* L.). Euphytica 94:129–135

Arumuganathan K, Earle ED (1991) Nuclear DNA content of some important plant species. Plant Mol Biol Rep 9:208–218

Barnes S, Massaro G, Lefèbvre M, Kuiper M, Verstege E (1996) A combined RFLP and AFLP genetic map for sugar beet. Proc 59th IIRB Congress, pp 555–560

Barzen E, Mechelke W, Ritter E, Seitzer JF, Salamini F (1992) RFLP markers for sugar beet breeding – chromosomal linkage maps and location of major genes for rhizomania resistance, monogermy and hypocotyl color. Plant J 2:601–611

Barzen E, Mechelke W, Ritter E, Schulte-Kappert E, Salamini F (1995) An extended map of the sugar beet genome containing RFLP and RAPD loci. Theor Appl Genet 90:189–193

Barzen E, Stahl R, Fuchs E, Borchardt D, Salamini F (1997) Development of coupling-repulsion phase SCAR markers diagnostic for the sugar beet *Rr1* allele conferring resistance to rhizomania. Mol Breed 3:231–238

Biancardi E, Lewellen RT, de Biaggi M, Erichsen AE, Piergiorgio S (2002) The origin of rhizomania resistance in sugar beet. Euphytica 127:383–397

Boudry P, Wieber R, Saumitou-Laprade P, Pillen K, Van Dijk H, Jung C (1994) Identification of RFLP markers closely linked to the bolting gene *B* and their significance for the study of the annual habit in beets (*Beta vulgaris* L.). Theor Appl Genet 88:852–858

Butterfass T (1963) Die Chloroplastenzahlen in verschiedenartigen Zellen trisomer Zuckerrüben (*Beta vulgaris* L.). Z Bot 52:46–77

Cai D, Kleine M, Kifle S, Harloff H, Sandal NN, Marcker KA, Klein-Lankhorst RM, Salentijn EMJ, Lange W, Stiekema WJ, Wyss U, Grundler FMW, Jung C (1997) Positional cloning of a gene for nematode resistance in sugar beet. Science 275:832–834

Dechyeva D, Gindullis F, Schmidt T (2003) Divergence of satellite DNA and interspersion of dispersed repeats in the genome of the wild beet *Beta procumbens*. Chrom Res 11:3–21

Eagen KA, Goldman IL (1996) Assessment of RAPD marker frequencies over cycles of recurrent selection for pigment concentration and percent solids in red beet (*Beta vulgaris* L.). Mol Breed 2:107–115

El-Mezawy A, Dreyer F, Jacobs G, Jung C (2002) High resolution mapping of the bolting gene *B* of sugar beet. Theor Appl Genet 105:100–105

Eyers M, Edwards K, Schuch W (1992) Construction and characterisation of a yeast artificial chromosome library containing two haploid *Beta vulgaris* L. genome equivalents. Gene 121:195–201

Flavell RB, Bennet MD, Smith JB (1974) Genome size and the proportion of repeated nucleotide sequence DNA in plants. Biochem Genet 12:257–269

Gao D, Guo D, Jung C (2001) Monosomic addition lines of *Beta corolliflora* in sugar beet: cytological and molecular marker analysis. Theor Appl Genet 103:240–247

Gindullis F, Dechyeva D, Schmidt T (2001a) Construction and characterization of a BAC library for the molecular dissection of a single wild beet centromere and sugar beet (*Beta vulgaris*) genome analysis. Genome 44:846–855

Gindullis F, Desel C, Galasso I, Schmidt T (2001b) The large-scale organization of the centromeric region in *Beta* species. Genome Res 11:253–265

Giorio G, Gallitelli M, Carriero F (1997) Molecular markers linked to rhizomania resistance in sugar beet, *Beta vulgaris*, from two different sources map to the same linkage group. Plant Breed 116:401–408

Hall RD, Riksenbruinsma T, Weyens GJ, Rosquin IJ, Denys PN, Evans IJ, Lathouwers JE, Lefebvre MP, Dunwell JM, Vantunen A, Krens FA (1996) A high efficiency technique for the generation of transgenic sugar beets from stomatal guard cells. Nat Biotechnol 14:1133–1138

Halldén C, Hjerdin A, Rading IM, Säll T, Fridlundh B, Johannisdottir G, Tuvesson S, Akesson C, Nilsson NO (1996a) A high density RFLP linkage map of sugar beet. Genome 39:634–645

Halldén C, Säll T, Olsson K, Nilsson NO, Hjerdin A (1996b) The use of bulked segregant analysis to accumulate RAPD markers near a locus for beet cyst nematode resistance in *Beta vulgaris*. Plant Breed 116:18–22

Heijbroek W, Roelands AJ, de Jong JH, van Hulst C, Schoone AHL, Munning RG (1988) Sugar beets homozygous for resistance to beet cyst nematode (*Heterodera schachtii* Schm.) developed from monosomic additions of *Beta procumbens* to *B. vulgaris*. Euphytica 38:121–131

Heller R, Schondelmaier J, Steinrücken G, Jung C (1996) Genetic localization of four genes for nematode (*Heterodera schachtii* Sch.) resistance in sugar beet (*Beta vulgaris* L.). Theor Appl Genet 92:991–997

Herwig R, Schulz B, Weisshaar B, Hennig S, Steinfath M, Drungowski M, Stahl D, Wruck W, Menze A, O'Brien J, Lehrach H, Radelof U (2002) Construction of a "unigene" cDNA clone set by oligonucleotide fingerprinting allows access to 25,000 potential sugar beet genes. Plant J 32:845–857

Hjerdin-Panagopoulos A, Kraft T, Rading IM, Tuvesson S, Nilsson NO (2002) Three QTL regions for restoration of Owen CMS in sugar beet. Crop Sci 42:540–544

Hohmann U, Jacobs G, Telgmann A, Gaafar R, Alam S, Jung C (2003) A bacterial artificial chromosome (BAC) library of sugar beet and a physical map comprising the bolting gene B. Mol Gen Genom 269:126–136

Hunger S, Di Gaspero G, Möhring S, Bellin D, Schäfer-Pregl R, Borchardt D, Durel C-E, Werber M, Weisshaar B, Salamini F, Schneider K (2003) Isolation and linkage analysis of expressed disease-resistance gene analogues of sugar beet (*Beta vulgaris* L.). Genome 46:70–82

Jung C, Wricke G (1987) Selection of diploid nematode-resistant sugar beet from monosomic addition lines. Plant Breed 98:205–214

Jung C, Herrmann RG (1991) A DNA probe for rapid screening of sugar beet (*Beta vulgaris* L.) carrying extra chromosomes from wild beets of the *Procumbentes* section. Plant Breed 107:275–279

Jung C, Koch R, Fischer F, Brandes A, Wricke G, Herrmann RG (1992) DNA markers closely linked to nematode resistance genes in sugar beet (*Beta vulgaris* L.) using chromosome additions and translocations originating from wild beets of the *Procumbentes* species. Mol Gen Genet 232:271–278

Jung C, Pillen K, Frese L, Melchinger A (1993) Phylogenetic relationships between cultivated and wild species of the genus *Beta* revealed by DNA "fingerprinting". Theor Appl Genet 86:449–457

Keller W (1936) Inheritance of some major colour types in beets. J Agric Res 52:27–38

Kifle S, Shao M, Jung C, Cai D (1999) An improved transformation protocol for studying gene expression in "hairy roots" of sugar beet (*Beta vulgaris* L.). Plant Cell Rep 18:514–519

Klein-Lankhorst RM, Salentijn EMJ, Dirkse WG, Arens-de Reuver M, Stiekema WJ (1994) Construction of a YAC library from *Beta vulgaris* fragment addition and isolation of a major satellite DNA cluster linked to the beet cyst nematode resistance locus $Hs1^{pat-1}$. Theor Appl Genet 89:426–434

Kleine M, Cai D, Eibl C, Herrmann RG, Jung C (1995) Physical mapping and cloning of a translocation in sugar beet (*Beta vulgaris* L.) carrying a gene for nematode (*Heterodera schachtii*) resistance from *B. procumbens*. Theor Appl Genet 90:399–406

Kleine M, Voss H, Cai D, Jung C (1998) Evaluation of nematode resistant sugar beet (*Beta vulgaris* L.) lines by molecular analysis. Theor Appl Genet 97:896–904

Krens F, Zijstra C, van der Molen W, Huizing HJ (1988) Transformation and regeneration in sugar beet (*Beta vulgaris* L.) induced by "shooter" mutants of *Agrobacterium tumefaciens*. Euphytica S:185–194

Kubis S, Heslop-Harrison JS, Desel C, Schmidt T (1998) The genomic organization of non-LTR retrotransposons (LINEs) from three *Beta* species and five other angiosperms. Plant Mol Biol 36:821–831

Lange W, Müller J, De Bock TSM (1993) Virulence in the beet cyst nematode (*Heterodera schachtii*) versus some alien genes for resistance in beet. Fundam Appl Nematol 16:447–454

Laporte V, Saumitou-Laprade P, Butterlin G, Vernet P, Cuguen J (1998) Identification and mapping of RAPD and RFLP markers linked to a fertility restorer gene for a new source of cytoplasmic male sterility in *Beta vulgaris* ssp *maritima*. Theor Appl Genet 96:989–996

Löptien H (1984) Breeding nematode-resistant beets. I. Development of resistant alien additions by crosses between *Beta vulgaris* L. and wild species of the section *Patellares*. Z Pflanzenzücht 92:208–220

Menzel G, Harloff H, Jung C (2003) Expression of bacterial poly(3-hydroxybutyrate) synthesis genes in hairy roots of sugar beet (*Beta vulgaris* L.). Appl Microbiol Biotechnol 60:571–576

Mörchen M, Cuguen J, Michaelis G, Hänni C, Saumitou-Laprade P (1996) Abundance and length polymorphism of microsatellite repeats in *Beta vulgaris* L. Theor Appl Genet 92:326–333

Nilsson NO, Halldén C, Hansen M, Hjerdin A, Sall T (1997) Comparing the distribution of RAPD and RFLP markers in a high density linkage map of sugar beet. Genome 40:644–651

Nilsson NO, Hansen M, Panagopoulos AH, Tuvesson S, Ehlde M, Christiansson M, Rading IM, Rissler M, Kraft T (1999) QTL analysis of *Cercospora* leaf spot resistance in sugar beet. Plant Breed 118:327–334

Owen FW, Carsner E, Stout M (1940) Phototermal induction of flowering in sugar beets. J Agric Res 61:101–124

Owen FW, Ryser GK (1942) Some mendelian characters in *Beta vulgaris* L. and linkages observed in the Y-R-B group. J Agric Res 65:153–171

Pelsy F, Merdinoglu D (1996) Identification and mapping of random amplified polymorphic DNA markers linked to a rhizomania resistance gene in sugar beet (*Beta vulgaris* L.) by bulked segregant analysis. Plant Breed 115:371–377

Pillen K, Steinrücken G, Wricke G, Herrmann RG, Jung C (1992) A linkage map of sugar beet (*Beta vulgaris* L.). Theor Appl Genet 84:129–135

Pillen K, Steinrücken G, Herrmann RG, Jung C (1993) An extended linkage map of sugar beet (*Beta vulgaris* L.) including nine putative lethal genes and the restorer gene X. Plant Breed 111:265–272

Pillen K, Schondelmaier J, Jung C, Herrmann RG (1996) Genetic mapping of genes for twelve nuclear-encoded polypeptides associated with the thylakoid membranes in *Beta vulgaris* L. FEBS Lett 395:58–62

Rae SJ, Aldam C, Dominguez I, Hoebrechts M, Barnes SR, Edwards KJ (2000) Development and incorporation of microsatellite markers into the linkage map of sugar beet (*Beta vulgaris* L.). Theor Appl Genet 100:1240–1248

Reamon-Ramos SM, Wricke G (1992) A full set of monosomic addition lines in *Beta vulgaris* from *Beta webbiana*: morphology and isozyme markers. Theor Appl Genet 84:411–418

Romagosa I, Hecker RJ, Tsuchiya T, Lasa JM (1986) Primary trisomics in sugar beet. I. Isolation and morphological characterization. Crop Sci 26:243–249

Roundy TE, Theurer JC (1974) Linkage and inheritance studies involving an annual pollen restorer and other genetic characters in sugar beets. Crop Sci 14:230–232

Salentijn EMJ, Sandal NN, Lange W, de Bock TSM, Krens FA, Marcker KA, Stiekema WJ (1992) Isolation of DNA markers linked to a beet cyst nematode resistance locus in *Beta patellaris* and *Beta procumbens*. Mol Gen Genet 235:432–440

Salentijn EMJ, Sandal NN, Klein-Lankhorst R, Lange W, de Bock TSM, Marcker KA, Stiekema WJ (1994) Long-range organization of a satellite DNA family flanking the beet cyst nematode resistance locus (*Hs1*) on chromosome-1 of *B. patellaris* and *B. procumbens*. Theor Appl Genet 89:459–466

Salentijn EMJ, Arens-de Reuver M, Lange W, De Bock TSM, Stiekema WJ, Klein-Lankhorst RM (1995) Isolation and characterization of RAPD-based markers linked to the beet cyst nematode resistance locus ($Hs1^{pat1}$) on chromosome 1 of *B. patellaris*. Theor Appl Genet 90:885–891

Sandal N, Salentijn E, Kleine M, Cai D, Arens-de Reuver M, Van Druten M, De Bock TSM, Lange W, Steen P, Jung C, Marcker K, Stiekema W, Klein-Lankhorst RM (1997) Backcrossing of nematode-resistant sugar beet: a second nematode resistance gene at the locus containing $Hs1^{pro-1}$? Mol Breed 3:471–480

Savitsky VF (1958) Genetische Studien und Züchtungsmethoden bei monogermen Rüben. Z Pflanzenzücht 40:1–36

Savitsky H (1978) Nematode (*Heterodera schachtii*) resistance and meiosis in diploid plants from interspecific *Beta vulgaris* × *B. procumbens* hybrids. Can J Genet Cytol 20:177–186

Schäfer-Pregl R, Borchardt D, Barzen E, Glass C, Mechelke W, Seitzer JF, Salamini F (1999) Localization of QTLs for tolerance to *Cercospora beticola* on sugar beet linkage groups. Theor Appl Genet 99:829–836

Schäufele WR, Wevers JDA (1996) Possible contribution of tolerant and partially resistant sugar beet varieties to the control of the foliar disease *Cercospora beticola*. Proc 59th IIRB Congress, pp 19–32

Schmidt T, Metzlaff M (1991) Cloning and characterization of a *Beta vulgaris* satellite DNA family. Gene 101:247–150

Schmidt T, Heslop-Harrison JS (1996) High resolution mapping of repetitive DNA by in situ hybridization – molecular and chromosomal features of prominent dispersed and discretely localized DNA families from the wild beet species *Beta procumbens*. Plant Mol Biol 30:1099–1114

Schmidt T, Heslop-Harrison JS (1998) Genomes, genes and junk: the large-scale organization of plant chromosomes. Trends Plant Sci 3:195–199

Schmidt T, Junghans H, Metzlaff M (1990) Construction of *Beta procumbens*-specific DNA probes and their application for the screening of *B. vulgaris* × *B. procumbens* (2n=19) addition lines. Theor Appl Genet 79:177–181

Schmidt T, Jung C, Metzlaff M (1991) Distribution and evolution of two satellite DNAs in the genus *Beta*. Theor Appl Genet 82:793–799

Schmidt T, Schwarzacher T, Heslop-Harrison JS (1994) Physical mapping of rRNA genes by fluorescent in-situ hybridization and structural analysis of 5 s rRNA genes and intergenic spacer sequences in sugar beet (*Beta vulgaris*). Theor Appl Genet 88:629–636

Schmidt T, Kubis S, Heslop-Harrison JS (1995) Analysis and chromosomal localization of retrotransposons in sugar beet (*Beta vulgaris* L.): LINEs and Ty1-copia-like elements as major components of the genome. Chrom Res 3:335–345

Schneider K, Borchardt DC, Schäfer-Pregl R, Nagl N, Glass C, Jeppson A, Gebhardt C, Salamini F (1999) PCR-based cloning and segregation analysis of functional gene homologues in *Beta vulgaris*. Mol Gen Genet 262:515–524

Schneider K, Weisshaar B, Borchardt DC, Salamini F (2001) SNP frequency and allelic haplotype structure of *Beta vulgaris* expressed genes. Mol Breed 8:63–74

Schneider K, Schäfer-Pregl R, Borchardt DC, Salamini F (2002) Mapping QTLs for sucrose content, yield and quality in a sugar beet population fingerprinted by EST-related markers. Theor Appl Genet 104:1107–1113

Scholten OE, de Bock TSM, Klein-Lankhorst RM, Lange W (1999) Inheritance of resistance to beet necrotic yellow vein virus in *Beta vulgaris* conferred by a second gene for resistance. Theor Appl Genet 99:740–746

Schondelmaier J, Jung C (1997) Chromosomal assignment of the nine linkage groups of sugar beet (*Beta vulgaris* L.) using primary trisomics. Theor Appl Genet 95:590–596

Schondelmaier J, Heller R, Pillen K, Steinrücken G, Jung C (1995) A linkage map of sugar beet. Proceedings of the 58th congress of the International Institute for Beet Research, Dijon, France, pp 37–43

Schondelmaier J, Steinrücken G, Jung C (1996) Integration of AFLP markers into a linkage map of sugar beet (*Beta vulgaris* L.). Plant Breed 115:231–237

Schondelmaier J, Schmidt T, Heslop-Harrison JS, Jung C (1997) Genetic and chromosomal localization of the 5S rDNA locus in sugar beet (*Beta vulgaris* L.). Genome 40:171–175

Schumacher K, Schondelmaier J, Barzen E, Steinrücken G, Borchardt D, Weber WE, Jung C, Salamini F (1997) Combining different linkage maps in sugar beet (*Beta vulgaris* L.) to make one map. Plant Breed 116:23–38

Setiawan A, Koch G, Barnes SR, Jung C (2000) Mapping quantitative trait loci (QTLs) for resistance to *Cercospora* leaf spot disease (*Cercospora beticola* Sacc.) in sugar beet (*Beta vulgaris* L.). Theor Appl Genet 100:1176–1182

Shen Y, Ford-Lloyd BV, Newbury HJ (1998) Genetic relationships within the genus *Beta* determined using both PCR-based marker and DNA sequencing techniques. Heredity 80:624–632

Uphoff H, Wricke G (1995) A genetic map of sugar beet (*Beta vulgaris*) based on RAPD markers. Plant Breed 114:355–357

Wagner H, Weber WE, Wricke G (1992) Estimating linkage relationship of isozyme markers and morphological markers in sugar beet (*Beta vulgaris* L.) including families with distorted segregations. Plant Breed 108:89–96

Weber WE, Borchardt DC, Koch G (2000) Marker analysis for quantitative traits in sugar beet. Plant Breed 119:97–106

Yu MH (1981) Sugar beets homozygous for nematode resistance and their transmission of resistance to their progenies. Crop Sci 21:714–717

II.4 Molecular Markers in Genetics and Breeding: Improvement of Alfalfa (*Medicago sativa* L.)

I.J. Maureira and T.C. Osborn[1]

1 Introduction

Alfalfa (*Medicago sativa* L.) is one of the most important legume species used in agriculture. Its high nutritional quality and vegetative yield makes alfalfa superior to other forage crops (Sumberg et al. 1983). By using artificial selection, breeders have been able to improve many important traits of alfalfa, such as disease resistance (Barnes and Hanson 1971; Heisey and Murphy 1985; Bray and Irwin 1989; Salter et al. 1994), insect resistance (Elden and Elgin 1987), salt tolerance (Dobrenz et al. 1993), nitrogen fixation (Viands and Barnes 1979), tissue culture regenerability (Ray and Bingham 1989), ease of floral tripping (Knapp and Teuber 1994), microgametophytic vigor (Rosellini et al. 1994), tap root regeneration (Hansen and Viands 1989), large leaflets (Dobrenz et al. 1988), resistance to frequent cutting regimes (Veronesi et al. 1986), and high self-fertility (Villegas et al. 1971). The majority of these selection efforts have been based on phenotypic evaluations, and the genetic basis of most targeted traits remains unknown. Other useful traits, such as increased productivity without sacrificing winter survival, growth compatibility with other forage species, and temporal and spatial forage stability, have received little attention (Brummer 1998).

In the past few years, molecular markers, in association with more powerful statistical models, have been applied to genetic analysis and breeding of several crops. This has been also true for alfalfa, although their application has been hindered by several factors (Osborn et al. 1998). First, cultivated alfalfa is a tetraploid species (2n=4x=32) with polysomic inheritance. This complicates linkage analysis and the detection of quantitative trait loci (QTL). Second, alfalfa possesses an allogamous reproductive behavior and shows severe inbreeding depression. This limits the production of homozygous parental genotypes and recombinant inbred lines for use in genetic mapping and replicated phenotypic evaluations. Third, alfalfa cultivars are synthetic populations created by intermating several to hundreds of selected genotypes. This breeding structure makes it difficult to use marker-assisted selection for cultivar improvement.

[1] Department of Agronomy, University of Wisconsin-Madison, 1575 Linen Drive, Madison, Wisconsin 53706-1597, USA

Despite these difficulties, molecular markers have been used to generate molecular data that have helped researchers to understand the genetic nature of alfalfa. Our goal in this chapter is to review these past applications of molecular markers in alfalfa genetics and breeding, and to consider new applications that may have an impact on the generation of better alfalfa cultivars.

2 Characterization of Alfalfa Germplasms

One of the main components of plant breeding is the identification of germplasms containing genes that could improve the performance of current cultivars. Geographical origin has been commonly used as a criterion to differentiate genetic sources. However, this criterion has obvious limitations if cryptic genetic relationships are present within and among germplasms. In addition, problems such as duplication, misclassification, and/or contamination of accessions cannot be assessed based on the origin of the germplasm.

Molecular markers that detect different types of DNA sequence polymorphisms have been used to estimate genetic diversity among various alfalfa germplasm collections (e.g., Brummer et al. 1991; Kidwell and Osborn 1994; Musial et al. 2002). Using restriction fragment length polymorphism RFLP marker data, Kidwell and Osborn (1994) evaluated tetraploid *M. sativa* accessions representing the original nine alfalfa germplasm sources introduced to North America (Barnes et al. 1977). These germplasms are distributed worldwide and should include most of the variation present in current alfalfa cultivars. The authors found that molecular marker diversity was high among individuals within germplasms, and only two sources, *M. sativa* ssp. *falcata* and *M. sativa* ssp. *sativa* Peruvian, formed distinct clusters with respect to the other seven germplasms. In a study including three diploid subspecies of the *M. sativa* complex, Brummer et al. (1991) found that *M. sativa* ssp. *falcata* (2x) tended to be a separate cluster; however, some accessions were in the same clade as a few of the *M. sativa* ssp. *sativa* (2x) accessions. The third subspecies, *M. sativa* ssp. *coerulea* (2x), had no clear clustering pattern and was not differentiated from *M. sativa* ssp. *sativa* (2x) subspecies.

Conclusions about genetic relationships will depend on the genotypes analyzed, the type and number of markers surveyed, and the type of genetic diversity estimator used. For example, we used a total of 212 polymorphic DNA fragments [191 RFLPs and 21 simple sequence repeat (SSRs)] to estimate the genetic relationship among the same subspecies analyzed by Brummer et al. (1991): *M. sativa* ssp. *falcata*, *M. sativa* ssp. *sativa* and *M. sativa* ssp. *coerulea*. We found that the three subspecies clustered in three distinct clades (Fig. 1). These results do not perfectly match the findings of Brummer et al. (1991) described above. Although both studies used basically the same type of marker (RFLPs), the two studies scored a different number of fragments,

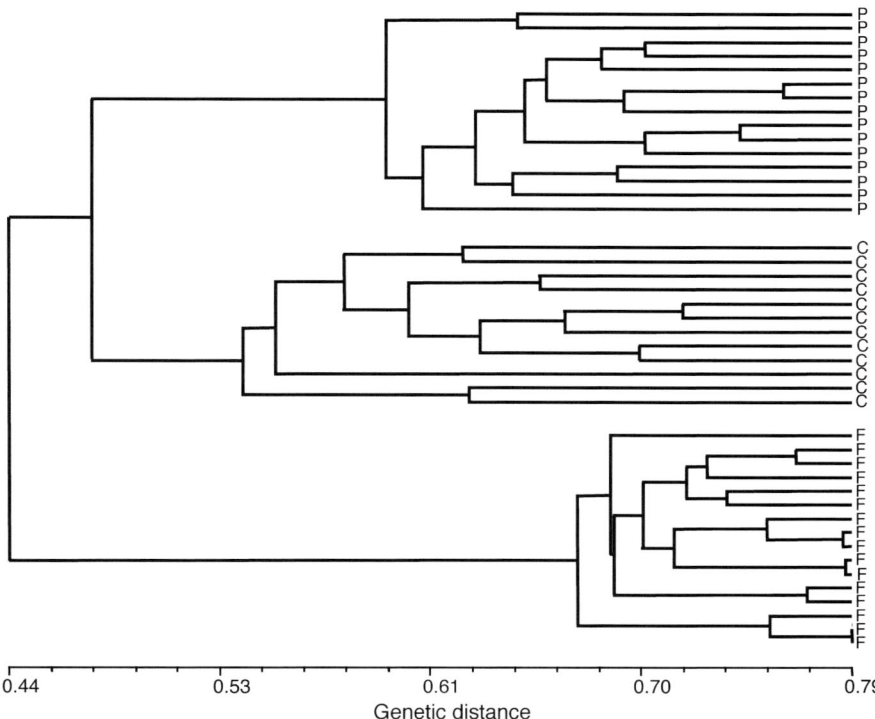

Fig. 1. Dendrogram of 43 individuals from four accessions representing three *Medicago sativa* subspecies. Cluster analysis (UPGMA) was performed on Dice coefficients calculated from 212 polymorphic fragments (detected as RFLP and SSR markers). *P* denotes individuals from *M. sativa* ssp. *sativa* Peruvian (4x, PI536535) germplasm; *F* denotes individuals from *M. sativa* ssp. *falcata* (4x, PI56033); *C* denotes individuals from *M. sativa* ssp. *coerulea* (2x, PI 15798, PI 28645)

included different accessions of the subspecies, and used different algorithms to estimate genetic relationships. Thus, discordance of results might be explained by some of these factors, or a combination of them.

Molecular markers have been used to estimate molecular variances (AMOVA) in several cultivars and ecotypes of tetraploid alfalfa (Mengoni et al. 2000; Pupilli et al. 2000). One of the main conclusions of these studies was that most of the total molecular variation was present in the within-accession component (~80–90%) and very little was in the among-accession component. This pattern of genetic variation is probably explained by the breeding methodologies used in alfalfa improvement. Most alfalfa cultivars are synthetic varieties and are created by intermating genotypes from different sources that are resistant to a wide range of diseases and insects.

It is clear that genetic variation is the propeller of plant breeding; however, the use of diverse germplasm does not ensure genetic improvement. Bonierbale et al. (1993) showed that the correlation between genetic diversity and heterosis is not consistent and depends on the genetic background of the material under evaluation. In fact, synthetic alfalfa populations developed on the basis of molecular marker diversity showed no consistent relation between diversity and forage yield (Kidwell et al. 1999). More recently, we evaluated the potential usefulness of *M. sativa* ssp. *falcata* and *M. sativa* ssp. *sativa* Peruvian accessions that were genetically distinct from alfalfa cultivars based on molecular marker data (Maureira et al. 2004). Field evaluations for forage yield suggested both accessions had favorable alleles that could be used to improve alfalfa cultivars. Thus, the potential usefulness of genetically diverse germplasms should be evaluated phenotypically before undertaking extensive efforts to incorporate the germplasm into a breeding program.

Utilization of genetic variation has not been limited to the diversity present within the *M. sativa* complex. Several attempts have been made to obtain interspecific hybrids between alfalfa and other *Medicago* species (summarized by McCoy and Bingham 1988). Although the impact of these past introgressions on cultivar development has been limited, alfalfa relatives are a potential source of favorable traits, such as resistance to extreme weather conditions and to pests. For instance, *M. marina*, *M. rhodopea* and *M. rupestris* were found to be tolerant to drought and salt stresses (Lensis and Lensis 1979; McCoy 1987), and *M. dzhawkahetica* and *M. papillosa* were resistant to Verticillimum wilt (Busch and Smith 1981) and spring blackstem (McCoy and Smith 1984). Genetic variability also has been estimated within and between *Medicago* species. Brummer et al. (1995) reconstructed a phylogeny of several accessions from six *Medicago* species using data from random amplified polymorphic DNA (RAPD) markers. Variation was present among accessions of all species and some accessions were considerably different from others within the same species.

3 Development of Genetic Maps and Identification of Regions Affecting Traits of Interest

Genetic maps based on molecular markers have proven to be very useful for understanding genome structure and for isolating important genes from many species. The comparison of maps across species has provided information about the organization and evolution of plant genomes (Livingstone et al. 1999; Perez et al. 1999; Kennard et al. 2000), and several candidate genes for disease resistance have been identified using comparative mapping (Grube et al. 2000a, b; Jahn et al. 2000). In addition, genetic maps have provided the framework for the discovery and dissection of several QTL in a number of crop species (Alpert and Tanksley 1996; Jiang et al. 1998; Lukens and Doebley 1999; Frary et al. 2000; Kole et al. 2001).

3.1 Diploid Alfalfa Maps

Statistical methodologies for mapping diploid species are well developed (Lander et al. 1987; Stam 1993), and maps for many diploid crop species have been built. The generation of diploid genotypes (2n=2x=16) of cultivated alfalfa (Bingham and McCoy 1979) and the utilization of cross-fertile diploid relatives has allowed the construction of several diploid alfalfa linkage maps (Brummer et al. 1993; Kiss et al. 1993; Echt et al. 1994; Tavoletti et al. 1996b). Recently, Kalo et al. (2000) reported an improved diploid alfalfa linkage map based on an F_2 segregating population derived from a cross between *M. sativa* ssp. *quasifalcata* × *M. sativa* ssp. *coerulea*. The map covered 754 cM and had an average marker density of 0.8/cM. By using common markers between Kalo et al.'s diploid alfalfa map and the *M. truncatula* map, Thoquet et al. (2002) were able to study chromosome homology and macrosynteny across both species. The authors concluded that the eight *M. truncatula* linkage groups were highly homologous to those of diploid *M. sativa*, implying a good level of macrosynteny between both *Medicago* genomes.

Diploid alfalfa maps have been used to map several genes including a nodulin gene (*ENOD12*, Csanadi et al. 1994), meiotic mutants producing 4n pollen and 2n eggs (Barcaccia et al. 2000; Tavolleti et al. 2000), and the unifoliate leaf-cauliflower head mutation (Brower and Osborn 1997a).

3.2 Tetraploid Alfalfa Maps

Although diploid alfalfa has been useful for mapping specific single genes, genes for some traits may have different effects in cultivated alfalfa at the tetraploid level. Clear phenotypic differences such as leaf size, vigor and forage yield plus dissimilar behavior of quantitative traits between ploidy levels (reviewed by Bingham et al. 1994) suggest that discoveries made at the diploid level may not be extrapolated to the tetraploid level. Recent information from yeast (*Saccharomyces cereviceae*) showed that gene expression is affected by ploidy levels (Galitski et al. 1999), supporting the idea that traits may be affected by ploidy level. Therefore, it is useful to have genetic maps for tetraploid alfalfa populations, although it is much more complex to develop maps for autopolyploid species than for diploid species.

The difficulty of constructing maps in an autopolyploid species is twofold: loci must be ordered along individual linkage groups and linkage groups have to be assigned to homologous sets (Ripol et al. 1999). The later problem is exclusive to autopolyploid species and creates an extra step in the mapping procedure. Wu et al. (1992) proposed a simple method to map autopolyploid species. Their approach utilized single dose restriction fragments (SDRFs) to detect and estimate linkage between markers. SDRFs segregate 1:1 (presence:absence) in the gametes and are equivalent to a dominant allele of a simplex genotype (Liu et al. 1998). Ripol et al. (1999) developed a method to include multiple dose markers into previously built SDRF linkage groups by estimating the dosage of the marker and utilizing this information later to infer the linkage relationship with the SDRFs. Although coupling and repulsion phase linkages between SDRFs can be detected and used to construct maps, repulsion phase linkages require a much larger population size to be detected (Wu et al. 1992; Ripol et al. 1999). This constraint has limited the use of repulsion phase linkages in constructing maps; and thus, most polyploid maps are based entirely on SDRFs (or higher dose markers when possible) in coupling phase linkages.

Once linkages between single dose makers have been estimated, SDRFs must be organized within linkage groups. Since SDRFs in coupling behave identically in diploids and autopolyploids, construction of single linkage groups can be achieved utilizing the same techniques and software used for mapping diploid species [i.e., MAPMAKER (Lander et al. 1987) or JoinMap (Stam 1993)].

To solve the problem of assigning single linkage groups to homologous sets, Ripol et al. (1999) suggested two alternatives: utilization of multiple dose fragments or/and highly polymorphic RFLP markers. Multiple dose markers will be linked to as many linkage groups as their dosages allow, and linkage groups containing the same multiple dose marker are assumed to be homologous. Simulations have shown that several multiple markers distributed across the genome are needed to fully connect all homologous linkage groups. Ripol et al. (1999) showed that for an autooctoploid with a monoploid number $x=8$, roughly 200 double dose fragments were needed to guarantee complete identification of all homologous sets. Homologous sets also can be determined using highly polymorphic RFLP probes or SSR primer pairs. An assumption is required that polymorphic fragments detected by a single probe or set of PCR primers represent different alleles of the same locus. Although this approach can be useful, homologous relationships among linkage groups should be confirmed, if possible, by the presence of multiple loci having polymorphic alleles or multiple dose fragments.

Currently, only one tetraploid genetic map has been reported for alfalfa (Brower and Osborn 1999). The map was built using two backcross populations of 101 individuals each and 82 segregating SDRFs in each population. Four homologous coupling-phase cosegregating linkage groups were found for seven of the eight expected homologous sets. Discordance between observed and expected number of homologous sets was explained by the lack of polymorphisms between the parents of the backcross populations. Locus positions and distances between loci were in general agreement with previous diploid maps, indicating that genome organization and recombination of tetraploid and diploid alfalfa are similar.

3.3 Detecting and Mapping Quantitative Trait Loci

Mapping of quantitative trait loci is based on the identification of simply inherited markers in close proximity with genetic factors affecting quantitative traits (Jannink and Walsh 2002). In general, two main strategies are used to identify QTLs: genome scans using anonymous DNA markers and association tests (Anderson 2001; Jannink and Walsh 2002).

3.3.1 Quantitative Trait Loci Discovery by Scanning Linkage Maps

In crop species, genomic scanning using linkage maps has been the most frequent strategy to detect genome regions associated with traits of interest. In general, one needs to have two polymorphic lines with different fixed alleles affecting a trait and a molecular marker linkage map (Mackay 2001). A large sample of individuals is generated from these two parental lines to provide observable recombinants and to allow a detailed measurement of the traits under study (Doerge 2002). The genome is scanned one marker at a time, dividing individuals into genotypic classes and making statistical tests (analysis of variance, simple regression, etc.) to determine if there are significant associations between marker genotype classes and the phenotype (Mackay 2001). This approach is known as simple marker analysis and is the simplest technique used to identify QTLs.

Detection of QTLs in polyploid species has been limited to statistical tests of single dose markers with the variance of quantitative traits (Doerge and Craig 2000). The statistical tests usually combine single factor models with multiple regression models (Sills et al. 1995; Brouwer et al. 2000). Single dose markers are tested individually to determine if the marker genotype (presence or absence) is associated with the trait variance. Single factor analyses are performed using the same techniques used in diploids, such as maximum likelihood, analysis of variance (ANOVA), or t-tests (Liu et al. 1998). A single factor model is described as follow:

$$Y_i = \mu + A_i + \varepsilon_i \tag{1}$$

where Y_i is the trait value with marker score i, μ *is the overall mean of the trait,* A_i is the effect of marker score i on the trait, and ε_i is the random variation. All single factors associated with the trait are then used in a multiple regression model:

$$Y_i = \mu + A_i + B_i + C_i + ... + Z_i + \varepsilon_i \tag{2}$$

where all terms are similar to those above, except A_i, B_i, C_i, ..., and Z_i symbolize the markers found to be significant in the single factor model (Sills et al. 1995). Multiple regression models are developed in an attempt to eliminate spurious associations between single dose markers and the traits under study. Champoux et al. (1995) showed that a highly significant QTL detected by single marker analysis of variance actually resulted from pseudolinkage to other loci. By using a multiple regression model, the authors were able to detect these false positive associations.

Although molecular markers have been used to develop diploid and tetraploid maps in alfalfa, few studies have mapped QTLs in this species (Barcaccia et al. 2000; Brouwer et al. 2000; Tavoletti et al. 2000). Two studies have focused on finding genomic regions affecting meiotic mutant phenotypes at the diploid level. Barcaccia et al. (2000) mapped at least five QTLs governing the production of 2n-eggs. Tavoletti et al. (2000) found three QTLs with highly significant effects on multinucleate-microspore formation. In the same study, the authors were able to map the jumbo pollen trait (*jp*, 4n pollen) into linkage group 6.

Thus, only one study has identified genome regions affecting traits of agronomic interest at the tetraploid level. Brouwer et al. (2000) found several QTLs affecting fall growth (FG), freezing injury (FI) and winter injury (WI) utilizing the same tetraploid mapping population described above. Although some QTLs affected all traits, some of them were specific to FG, FI, or WI. The authors suggested that these markers could be used as predictor traits in the absence of winter hardiness and that the existence of different genetic components for FG and WI could facilitate the manipulation of these traits.

Although progress has been made to promote QTL mapping in tetraploid alfalfa; several concerns need still to be addressed. To date, every model used in QTL mapping has been based on single dose markers and models utilizing double dose markers need to be developed. Furthermore, most of the QTL research has utilized single factor analysis. More sophisticated techniques such as interval, composite interval and multiple interval mapping have been developed for mapping in diploid species and these need to be adapted for use at the tetraploid level. Some attempts to use these techniques in autopolyploids have been made. Recently, construction of denser maps has allowed the utilization of interval mapping for those QTLs having close flanking markers in sugar cane (Ming et al. 2002).

3.3.2 Associative Mapping

The ultimate objective of QTL mapping is uncovering the gene(s) governing the trait of interest. Although genome scanning has been successful in the discovery of important genome regions, these regions often are large and the QTL positions cannot be precisely defined. A common strategy used to solve this problem is to fine-map the region of interest by expanding the genetic map surrounding the QTL. Backcross and advance intercross populations are frequently used for this purpose. However, a common inconvenience of fine mapping is the extra cost of developing new populations. Furthermore, recombination will still be a limiting factor when the objective is to elucidate the gene(s) responsible for the trait (Jannink and Walsh 2002).

Associative mapping has been suggested as an alternative method for mapping or fine mapping QTLs. This approach takes advantage of associations that arise relatively early in the history of populations. If several generations have passed, recombination should have removed the disequilibrium present among loosely linked markers and QTLs. Thus, associations should only remain between QTLs and tightly linked markers. This approach does not require the utilization of genetically defined bi-parental populations, allowing QTL mapping in more genetically complex populations. One of the main advantages of this approach is that it can potentially detect QTLs of small effect, especially if polymorphisms within the causative genes can be tested for association with the trait (Anderson 2001). However, screening several candidate genes might be more time-consuming than scanning the genome with anonymous markers. Furthermore, population structure, mainly caused by drift and population admixture, has been shown to be an important factor in the creation of false associations (Knowler et al. 1988; Pritchard 2001). To overcome the population structure problem, several family-based tests, such as the transmission disequilibrium test (TDT, Ewens and Spielman 1995), have been developed. In general, these tests are based on the determination of the transmission frequency of the case allele to the progeny that show the expected phenotype. The immediate problem of these techniques is the necessity of genotyping related individuals that may not exist or may not be known. Fortunately, new statistical approaches have allowed the utilization of standard association methods that are corrected for the presence of population structure (Pritchard et al. 2000a, b; Reich and Goldstein 2001). These methods assume that if there is population stratification, it will affect all loci independent of linkage to the trait loci of interest. Thus, adjustments for population structure can be made by using unlinked marker loci as controls. Early association studies were designed to test qualitative traits (Pritchard et al. 2000b); however, later studies have allowed the expansion of these techniques to quantitative traits (Thornsberry et al. 2001). Although most of the association analyses reported so far have been done in human disease research (Risch and Merikangas 1996), examples of plant studies are starting to appear in the literature (Beer et al. 1997; Obert et al. 2000; Thornesberry et al. 2001).

Skinner et al. (2000) developed a methodology for detecting marker–trait associations in alfalfa by comparing allele frequency shifts between populations selected for one trait and the original unselected population. To avoid type I error due to multiple comparisons and to estimate a more realistic significance threshold for declaring marker–trait associations, the authors proposed the use of a re-sampling method (bootstrapping). Skinner et al. (2000) also investigated the utility of the associations through conditional probability and building-model studies. The conclusions from these studies were that the presence of markers in undesired plants strongly influenced the efficiency with which the marker could be used to select desired plants. However, in some circumstances, markers could be highly efficient for selecting rare, but desirable plants from heterogeneous base populations. Obert et al. (2000) used the methodology developed by Skinner et al. (2000) to study associations between amplified fragment length polymorphism (AFLP) markers and resistance to downy mildew (*Peronospora trifoliorum* de Bary) in tetraploid alfalfa populations. By using 36 plants from two populations that differed only in selection for resistance to downy mildew, they were able to identify four AFLP fragments associated with disease susceptibility or resistance. In further F_1 and S_1 evaluations, they confirmed significant associations for two of the four AFLP markers. Although both markers were confirmed, it is important to note that the AFLP fragments were not fully linked to the disease trait. Only one maker was proven to be valuable in a breeding program after estimating their conditional probabilities. Although this type of approach could be important for identifying regions affecting specific traits, they do not account for population structure. As mentioned before, stratification can cause strong spurious associations. Using a Bayesian approach (Pritchard et al. 2000a), Thornsberry et al. (2001) removed the effect of population structure and successfully associated a deletion in the *Dwarf8* gene with flowering time variation in corn.

Although association studies have not been widely used in alfalfa breeding, they are a potential useful alternative method for identifying regions affecting traits of agronomic and economic importance. Through single plant phenotypic recurrent selection, alfalfa breeding programs have developed a large number of improved populations. Many of these populations have been selected for resistance to specific diseases or to insect damage, and they could be used for association studies. Molecular data coming from the related model plant *Medicago truncatula* will also provide a vast amount of information that could be used for these association studies. A large number of expressed sequence tag-simple sequence repeat (EST-SSR) markers developed using *M. truncatula* sequence data have been used to map traits in alfalfa populations (Sledge et al. 2003). EST-SSRs and other PCR gene-based markers should allow the targeting of candidate genes and increase the efficiency of association studies. If population structure is taken into account, populations already developed by alfalfa breeding programs could potentially be used to search for marker–trait associations.

4 Additional Uses of Molecular Markers in Alfalfa Breeding

Molecular markers have been used in genetic studies to determine patterns of inheritance among alfalfa genotypes and in interspecific hybrids between alfalfa and other *Medicago* species (Quiros 1982; McCoy et al. 1991b). By studying the segregation of three isozyme markers in progenies derived from di-, tri- or tetra-allelic plants, Quiros (1982) confirmed the pattern of tetrasomic inheritance for alfalfa, as was previously suggested using phenotypic markers (Standford 1951). Using RFLP and isozyme markers, McCoy et al. (1991) studied the polyploid nature of the *M. sativa* × *M. papillosa* hybrids. Examination of segregating loci showed that most of the markers followed a disomic inheritance pattern. The authors concluded that there was limited pairing affinity between *M. sativa* and *M. papillosa* chromosomes.

Molecular markers have also been used in studies of inbreeding and heterosis (Scotti et al. 1992, 2000; Brouwer and Osborn 1997b). Brouwer and Osborn (1997b) evaluated the reduction of heterozygosity with inbreeding by using 40 RFLP loci in diploid alfalfa. After self-pollinating several lines of *M.sativa* ssp. *sativa* and *M. sativa* ssp. *falcata* for two and four generations respectively, the authors found that the reduction of heterozygosity was lower than expected. The same phenomenon was observed by Scotti et al. (2000) in a study using tetraploid alfalfa. However, differences between expected and observed heterozygosities were smaller than at the diploid level. The authors argued that the most likely explanation for this narrower difference was that the higher buffer capacity of tetraploid alfalfa masked deleterious alleles.

The use of molecular markers has also been proposed as a method to map alfalfa centromeres and to determine the mode of $2n$ gamete formation in meiotic mutants (Tavoletti et al. 1996a, b). The model involved a maximum likelihood approach of half tetrad analysis based on multiple RFLP markers. The model was tested using the progeny of a diploid genotype characterized by high 2n egg production crossed with a tetraploid alfalfa genotype. By comparing three linked loci and one unlinked marker, the model was able to predict the most likely position of the centromeres with respect to the linked markers. The analysis also provided an estimation of the percentage of first and second division restitution 2n eggs produced by the diploid genotype.

5 Conclusions

Alfalfa (*Medicago sativa* L.) is an allogamous tetraploid species with polysomic inheritance, and not the simplest system to apply molecular maker analysis. However, new molecular techniques in association with more powerful statistical models are starting to reveal the genetic and breeding nature of this forage species. Molecular markers have facilitated the study of genetic diversity present within alfalfa, and several germplasms are being studied as potential donors of favorable alleles. Complete linkage maps of molecular markers have been developed for diploid alfalfa and several genomic regions affecting important traits have been discovered. Furthermore, utilization of common markers with the model legume species *M. truncatula* has demonstrated the homology and high level of macrosynteny between both genomes. Mapping data for tetraploid populations also have been produced to identify QTL in cultivated alfalfa. Although QTL mapping is possible at the tetraploid level, additional research will be necessary to incorporate more sophisticated mapping techniques that are used in diploid species. Association studies have been suggested as an alternative method for mapping and fine mapping QTLs in cultivated alfalfa, and improved populations produced by alfalfa breeding programs could be used for association studies without the necessity of developing extra mapping populations. However, the success of this approach will be limited by the number of gene markers available for the association tests. A large number of EST-SSR markers developed from *M. truncatula* sequence data are already available for public use. These EST-SSR markers and other PCR gene-based markers could be used to identify genomic regions of interest in association or mapping studies. This molecular data should contribute to our understanding of the genetics of cultivated alfalfa, and should improve the selection of important traits in populations used for cultivar development.

References

Alpert KB, Tanksley SD (1996) High-resolution mapping and isolation of a yeast artificial chromosome contig containing *fw2.2*: a major fruit weight quantitative trait locus in tomato. Proc Natl Acad Sci USA 93:15503–15507

Anderson L (2001) Genetic dissection of phenotypic diversity in farm animals. Nat Rev Genet 2:130–138

Barcaccia G, Albertini E, Rosellini D, Tavoletti S, Veronessi F (2000) Inheritance and mapping of 2n-egg production in diploid alfalfa. Genome 43:528–537

Barnes DK Hanson CH (1971) Recurrent selection for bacterial wilt resistance in Alfalfa. Crop Sci 11:545–546

Barnes DK, Bingham ET, Murphy RP, Hunt OJ, Beard DF, Skrdla WH, Teuber LR (1977) Alfalfa germplasm in the United States: genetic variability, use and maintenance. USDA-ARS Tech Bull 1571 USDA-ARS, Hyattsville, MD

Beer SC, Siripoonwiwat W, O'Donoughue LS, Souza E, Matthews D, Sorrells ME (1997) Associations between molecular markers and quantitative traits in an oat germplasm pool: can we infer linkages? J Agric Geno 3 [on line]. URL: http://www.cabi-publishing.org/gateways/jag/index.html

Bingham ET, McCoy TJ (1979) Cultivated alfalfa at the diploid level: origin, reproductive stability and yield of seed and forage. Crop Sci 19:97–100

Bingham ET, Groose RW, Woodfield DR, Kidwell KK (1994) Complementary gene interactions in alfalfa are greater in autotetraploids than diploids. Crop Sci 34:823–829

Bonierbale MW, Plaisted RL, Tanksley SD (1993) A test of the maximum heterozygosity hypothesis using molecular markers in tetraploid potatoes. Theor Appl Genet 86:481–491

Bray RA, Irwin JAG (1989) Recurrent selection for resistance to *Stemphylium versicarium* within the lucerne cultivars Trifecta and Sequel. Aust J Exp Agric 29:189–192

Brouwer DJ, Osborn TC (1997a) Identification of the RFLP markers linked to the unifolate leaf, cauliflower head mutation in alfalfa. J Hered 88:150–152

Brouwer DJ, Osborn TC (1997b) Molecular marker analysis of the approach to homozygosity by selfing diploid alfalfa. Crop Sci 37:1326–1330

Brouwer DJ, Osborn TC (1999) A molecular marker linkage map of tetraploid alfalfa (*Medicago sativa* L.). Theor Appl Genet 99:1194–1200

Brouwer DJ, Duke SH, Osborn TC (2000) Mapping genetic factors associated with winter hardiness, fall growth, and freezing injury in tetraploid alfalfa. Crop Sci 40:1387–1396

Brummer EC (1998) Molecular and cellular technologies in forage improvement: an overview. In: Brummer EC, Hill NS, Roberts CA (eds) Molecular and cellular technologies for forage improvement. CSSA special publication number 26, Madison, WI, pp 1–10

Brummer EC, Kochert G, Bouton JH (1991) RFLP variation in diploid and tetraploid alfalfa. Theor Appl Genet 83:89–96

Brummer EC, Bouton JH, Kochert G (1993) Development of an RFLP map in diploid alfalfa. Theor Appl Genet 86:329–332

Brummer EC, Bouton JH, Kochert G (1995) Analysis of annual *Medicago* species using RAPD markers. Genome 38:362–367

Bush LV, Smith E (1981) Susceptibility of Ontario-grown alfalfa cultivars and certain Medicago species to *Verticillium albo-atrum*. Can J Plan Path 3:169–172

Champoux MC, Wang G, Sarkarung S, Mackill DJ, O'Toole JC, Huang N, McCouch SR (1995) Locating genes associated with root morphology and drought avoidance in rice via linkage to molecular markers. Theor Appl Genet 90:969–981

Csanadi G, Szecsi J, Kalo P, Kiss P, Endre G, Kondorosi A, Kondorosi E, Kiss GB (1994) ENOD12, an early nodulin gene, is not required for nodule formation and efficient nitrogen fixation in alfalfa. Plant Cell 6:201–206

Dobrenz AK, Robinson DL, Smith SE, Stone JE (1988) Registration of AZ large leaflet nondormant alfalfa germplasm. Crop Sci 28:1034

Dobrenz AK, Smith SE, Poteet D, Miller WD (1993) Carbohydrates in alfalfa seed developed for salt tolerance during germination. Agron J 85:834–836

Doerge RW (2002) Mapping and analysis of quantitative trait loci in experimental populations. Nat Rev Genet 3:43–52

Doerge RW, Craig BA (2000) Model selection for quantitative trait locus analysis in polyploids. Proc Natl Acad Sci USA 97:7951–7956

Echt CS, Kidwell KK, Knapp SJ, Osborn TC, McCoy TJ (1994) Linkage mapping in diploid alfalfa (*Medicago sativa*). Genome 37:61–71

Elden TC, Elgin JH (1987) Recurrent seedling and individual plant selection for potato leafhopper (Homoptera: Cicadellidae) resistance in alfalfa. J Econ Entomol 80:690–695

Ewens WJ, Spielman RS (1995) The transmission/disequilibrium test: history, subdivision, and admixture. Am J Hum Genet 63:1886–1897

Frary A, Nesbitt TC, Frary A, Grandillo S, van der Knaap E, Cong B, Liu J, Meller J, Elber R, Alpert KB, Tanksley SD (2000) *fw2.2*: a quantitative trait locus key to the evolution of tomato fruit size. Science 289:85–88

Galitski T, Saldanha AJ, Styles C, Lander E, Fink GR (1999) Ploidy regulation of gene expression. Science 285:251–254

Grube RC, Blauth JR, Arnedo AMS, Caranta C, Jahn MK (2000a) Identification and comparative mapping of a dominant polyvirus resistance gene cluster in Capsicum. Theor Appl Genet 101:852–859

Grube RC, Radwanski ER, Jahn M (2000b) Comparative genetics of disease resistance within the *Solanaceae*. Genetics 155:873–887

Hansen JL, Viands DR (1989) Response from phenotypic recurrent selection for root regeneration after taproot severing in alfalfa. Crop Sci 29:1177–1181

Heisey RF, Murphy RP (1985) Phenotypic recurrent selection for resistance to *Phytophthora* root rot in two diploid alfalfa populations. Crop Sci 25:693–694

Jahn MK, Paran I, Hoffmann K, Radwanski ER, Livingstone KD, Grube RC, Aftergoot E, Lapidot M, Moyer J (2000) Genetic mapping of the *Tsw* locus for resistance to tomato spotted wilt virus in *Capsicum* and its relationship to the *Sw*-5 gene for resistance to the same pathogen in tomato. Mol Plant Microbe Int 13:673–682

Jannink JL, Walsh B (2002) Association mapping in plant populations. In: Kang MS (ed) Quantitative genetics, genomics, and plant breeding. CABI, New York, pp 59–68

Jiang CX, Wright RJ, EL-Zik KM, Patterson AH (1998) Polyploid formation created unique avenues for response to selection in Gossypium (cotton). Proc Natl Acad Sci USA 95:4419–4424

Kalo P, Endre G, Zimanyi L, Csanadi G, Kiss GB (2000) Construction of an improved linkage map of diploid alfalfa (*Medicago sativa*). Theor Appl Genet 100:641–657

Kennard WC, Phillips RL, Porter RA, Grombacher AW (2000) A comparative map of wild rice (*Zizania palustris* L. 2n=2x=30). Theor Appl Genet 101:677–684

Kidwell KK, Austin DF, Osborn TC (1994) RFLP evaluation of nine *Medicago* accessions representing the original germplasm sources for the north American alfalfa cultivars. Crop Sci 34:230–236

Kidwell KK, Hartweck LM, Yandell BS, Crump PM, Brummer JE, Moutray J, Osborn TC (1999) Forage yields of alfalfa populations derived from parents selected on the basis of molecular marker diversity. Crop Sci 39:223–227

Kiss GB, Csanadi G, Kalman K, Kalo P, Okresz L (1993) Construction of a basic genetic map for alfalfa using RFLP, RAPD, isozyme and molecular markers. Mol Gen Genet 238:129–137

Knapp EE, Teuber LR (1994) Selection progress for ease of floret tripping in alfalfa. Crop Sci 34:323–326

Knowler WC, Williams RC, Pettitt DJ, Steinberg AG (1988) Gm 3, 5, 13, 14 and type 2 *Diabetes mellitus*: an association in American Indians with genetic admixture. Am J Hum Genet 52:506–513

Kole C, Quijada P, Michaels SD, Amasino RM, Osborn TC (2001) Evidence for homology of flowering-time genes VFR2 from *Brassica rapa* and FLC from *Arabidopsis thaliana*. Theor Appl Genet 102:425–430

Lander ES, Green P, Abrahamson A, Barlow M, Daley M, Lincoln S, Newburg L (1987) MAPMAKER: an interactive computer package for constructing primary genetic linkage maps of experimental and natural populations. Genomics 1:174–181

Lensins K, Lensins I (1979) Genus Medicago (Leguminosae): a taxonomic study. Dr W Junk bv Publishers, The Hague, The Netherlands

Livingstone KD, Lackney VK, Blauth JR, Wijk RV, Jahn MK (1999) Genome mapping in *Capsicum* and the evolution of genome structure in the Solanaceae. Genetics 152:1183–1202

Liu SC, Lin YR, Irvine JE, Paterson AH (1998) Mapping QTLs in autopolyploids. In: Paterson AH (ed) Molecular dissection of complex traits. CRC Press, New York, pp 95–101

Lukens L, Doebley J (1999) Epistatic and environmental interactions for quantitative trait loci involved in maize evolution. Genet Res 74:291–302

Mackay TF (2001) Quantitative trait loci in *Drosophila*. Nat Rev Genet 2:11–21

Maureira IJ, Ortega F, Campos H, Osborn TC (2004) Population structure and combining ability of diverse Medicago sativa germplasms. Theor Appl Genet (in press)

McCoy TJ (1987) Tissue culture evaluation of NaCl tolerance in Medicago species: cellular versus whole plant response. Plant Cell Rept 6:31–34

McCoy TJ, Smith LY (1984) Uneven ploidy levels and a reproductive mutant are required for interspecific hybridization of *Medicago sativa* L. and *Medicago dzhawakhetica* Bordz. Can J Genet Cytol 26:511–518

McCoy TJ, Bingham ET (1988) Cytology and cytogenetics of alfalfa. In: Hanson AA, Barnes DK, Hill RR Jr (ed) Alfalfa and alfalfa improvement. Agron Monogr 29 ASA CSSA SSSA, Madison, WI, pp 737–776

McCoy TJ, Echt CS, Mancino LC (1991) Segregation of molecular markers supports an allotetraploid structure for *Medicago sativa*, × *Medicago papillosa*, interspecific hybrid. Genome 34:574–578

Mengoni A, Gori A, Bazzigalupo M (2000) Use of RAPD and microsatellite (SSR) to assess genetic relationships among populations of tetraploid alfalfa, *Medicago sativa*. Plant Breed 199:311–317

Ming R, Wang YW, Draye X, Moore PH, Irvine JE, Paterson AH (2002) Molecular dissection of complex traits in autopolyploids: mapping QTLs affecting sugar yield and related traits in sugarcane. Theor Appl Genet 105:332–345

Musial JM, Basford KE, Irwin JAG (2002) Analysis of genetic diversity within Australian lucerne cultivars and implications for future genetic improvement. Aust J Agric Res 53:629–633

Obert DE, Skinner DZ, Stuteville DL (2000) Association of AFLP markers with mildew resistance in autotetraploid alfalfa. Mol Breed 6:287–294

Osborn TC, Brouwer DJ, Kidwell KK, Tavoletti S, Bingham ET (1998) Molecular marker applications to genetics and breeding of alfalfa. In: Brummer EC, Hill NS, Roberts CA (eds) Molecular and cellular technologies for forage improvement. CSSA special publication number 26, Madison, WI, pp 25–31

Perez F, Menendez A, Dehal P, Quiroz CF (1999) Genomic structural differentiation in Solamun: comparative mapping of the A- and E-genomes. Theor Appl Genet 98:1183–1193

Pritchard JK (2001) Deconstructing maize populations structure. Nature Genet 28:203–204

Pritchard JK, Stevens M, Donnelly P (2000a) Inference of population structure using multilocus genotype data. Genetics 155:945–959

Pritchard JK, Stevens M, Rosenberg NA, Donnelly P (2000b) Association mapping in structured populations. Am J Hum Genet 67:170–181

Pupilli F, Lombarda P, Scotti C, Arcioni S (2000) RFLP analysis allows for the identification of alfalfa ecotypes. Plant Breed 199:271–276

Quiros CF (1982) Tetrasomic segregation for multiple alleles in alfalfa. Genetics 101:117–127

Ray IM, Bingham ET (1989) Breeding diploid alfalfa for regeneration from tissue culture. Crop Sci 29:1545–1548

Reich DE, Goldstein DB (2001) Detecting association in a case-control study while correcting for population stratification. Genet Epid 20:4–16

Ripol MI, Churchill GA, da-Silva JAG, Sorrells M (1999) Statistical aspects of genetic mapping in autopolyploids. Gene 235:31–41

Risch N, Merikangas K (1996) The future of genetic studies of complex human diseases. Science 273:1516–1517

Rosellini D, Veronesi F, Falcinelli M (1994) Recurrent selection for microgametophytic vigor in alfalfa and correlated responses at the sporophytic level. Crop Sci 34:933–936

Salter R, Miller-Garvin JE, Viands DR (1994) Breeding for resistance to alfalfa root rot caused by *Fusarium* species. Crop Sci 34:1213–1217

Scotti C, Pupilli F, Damiani F, Arcioni S (1992) Molecular marker assisted analysis of the heterozygosity in tetraploid alfalfa: inbreeding depression in S_1 families. In: Rotili P, Zannone I (eds) Proceedings of the 10th International Conference of the EUCARPIA *Medicago* spp. group. Forage Crop Institute, Lodi, Italy, pp 236–242

Scotti C, Pupilli F, Salvi S, Arcioni S (2000) Variation in vigour and in RFLP-estimated heterozygosity by selfing tetraploid alfalfa: new perspectives for the use of selfing in alfalfa breeding. Theor Appl Genet 101:120–125

Sills GR, Bridges W, Al-Janabi SM, Sobral B (1995) Genetic analysis of agronomic traits in a cross between sugarcane (*Saccharum officinarum* L.) and its presumed progenitor (*S. robustum* Brandes and Jews. ex Grassl). Mol Breed 1:355–363

Skinner DZ, Loughin T, Obert DE (2000) Segregation and conditional probability of molecular markers with traits in autotetraploid alfalfa. Mol Breed 6:295–306

Sledge M, Ray I, Rouf Mian MA (2003) EST-SSRs for genetic mapping in alfalfa. Molecular breeding of forage and turf. Abstract, Third international symposium, Dallas, TX and Ardmore, OK, USA, p 79

Sorrells ME (1992) Development and amplification of RFLPs in polyploids. Crop Sci 32:1086–1091

Stam P (1993) Construction of integrated genetic linkage maps by means of a new computer package: JoinMap. Plant J 3:739–744

Stanford EH (1951) Tetrasomic inheritance in alfalfa. Agron J 43:222–225

Sumberg JE, Murphy RP, Lowe CC (1983) Selection for fiber and protein concentration in a diverse alfalfa population. Crop Sci 23:11–14

Tavoletti S, Bingham ET, Yandell F, Veronessi F, Osborn TC (1996a) Half tetrad analysis in alfalfa using multiple RFLP markers. Proc Natl Acad Sci USA 93:10918–10922

Tavoletti S, Veronessi F, Osborn TC (1996b) RFLP linkage map of a meiotic mutant based on an F1 population. J Hered 87:167–170

Tavoletti S, Pesaresi P, Barcaccia G, Albertini E (2000) Mapping the *jp* (jumbo pollen) gene and QTLs involved in multinucleate microspore formation in diploid alfalfa. Theor Appl Genet 101:372–378

Thoquet P, Gherardi M, Journet ET, Kereszt A, Ane JM, Prosperi JM, Huguet T (2002) The molecular genetic linkage map of the model legume *Medicago truncatula*: an essential tool for comparative legume genomics and isolation of agronomically important genes. BMC Plant Biol 2:1–13

Thornsberry JM, Goodman MM, Doebley J, Kresovich S, Nielsen D, Buckler ES IV (2001) Dwarf8 polymorphisms associate with variation in flowering time. Nat Genet 28:286–289

Veronesi F, Mariani A, Falcinelli M, Arcioni S (1986) Selection for tolerance to frequent cutting regimes in alfalfa. Crop Sci 26:58–61

Viands DR, Barnes DK, Heichel GH (1981) Nitrogen fixation in alfalfa: response to bidirectional selection for associated characteristics. USDA Tech Bull 1643. US Government printing office, Washington, DC

Villegas CT, Wilsie CP, Frey KJ (1971) Recurrent selection for high self-fertility in Vernal alfalfa (*Medicago sativa* L.). Crop Sci 11:881–883

Wu KK, Burnquist W, Sorrells ME, Tew TL, Moore PH, Tanskley SD (1992) Detection and estimation of linkage in polyploids using single-dose restriction fragments. Theor Appl Genet 83:294–300

II.5 Localization of Important Traits: The Example Pea (*Pisum sativum* L.)

W.K. Swiecicki[1] and G. Timmerman-Vaughan[2]

1 Introduction

The evolution of the pea (*Pisum sativum* L.) from its wild to modern domesticated forms involved selection of plant characteristics suitable for cultivation or for improved palatability. In pea, explosive pod indehiscence and seed dormancy (hard seededness) were probably the greatest barriers to domestication (Smartt 1990) that had to be overcome. Other traits selected during domestication and development of modern cultivated forms include a number of characters that are determined by one or a few genes, such as *a* (lack of anthocyanin production) and *r* (wrinkled seed in garden types), which improved palatability, and *p* and *v* for the absence of sclerenchymatic tissue in pods. Domestication has also resulted in increased seed and pod size in pea (although not as markedly as in other crops) with a correlated increase in leaf size and stem strength. This early practice of genetics occasionally resulted in plant types that were sufficiently distinct to be given their own taxonomic classification, e.g., the old classification of the pea into garden (var. *hortense*) and field (var. *arvense*) types, largely as a result of selection for anthocyanin production.

In 1906, Bateson and Punnett demonstrated genetic linkage, based on experiments on flower color in *Lathyrus odoratus* L. Nearly a century later, genetic linkage has become a very powerful tool for plant breeding and crop improvement. This is largely due to the ability to develop molecular markers that detect DNA sequence polymorphism and characterize their association and linkage with valuable alleles. This allows the identification and tracking of genes for a wide range of simple and complex economically important characteristics. Genetic linkage maps that are well saturated with molecular markers have become an important tool for plant breeding. In the following pages, we describe the current use of genetic linkage maps in pea (using morphological, isozyme and molecular markers) to manipulate a range of economically important traits. As Tanksley et al. (1989) concluded, molecular markers are "a new tool for an old science".

---

[1] Institute of Plant Genetics, Polish Academy of Sciences, 60–479 Strzeszynska 34, Poznan, Poland
[2] Crop and Food Research, P.O. Box 4704, Christchurch, New Zealand

2 The *Pisum* Genetic Map and Loci for Agronomic Characters

The pea is a classical model for genetic studies. Key steps in the creation of the current linkage maps include:

- establishment of seven linkage groups (Lamprecht 1948),
- assigning linkage groups to particular chromosomes (Lamprecht 1961),
- construction of a chromosome map with 128 loci (Lamprecht 1974),
- updating of Lamprecht's map by Blixt (1977) with additional linkage data on 128 loci and 41 new genes,
- mapping of biochemical markers (27 isozyme loci) and relation to anchor markers (Weeden 1985),
- adding molecular markers to the pea map, structural rearrangements related to Lamprecht's version (Weeden et al. 1993),
- development of a consensus linkage map with molecular loci related to anchor markers, and linkage groups related to chromosomes (Weeden et al. 1998).

The development of a consensus pea linkage map containing molecular markers, as well as other advances (summarized in Weeden et al. 1994a; Swiecicki et al. 2000), has improved the opportunities for the use of marker-assisted selection (MAS) in pea breeding. In a relatively short time, a number of economically important genes have been mapped and tagged. Genes with a clear phenotype have played a key role in breeding different pea ideotypes, e.g., *a* and *r*, but also *le* (short stem), *af* (afila leaf), *i* (green cotyledons) and *n, p, v* (genes controlling pod characters).

Resistance to a number of important viral and fungal diseases shows monogenic inheritance. The Blixt map (1977) contained the following disease resistance loci: *er* (resistance to *Erysiphe pisi* Syd), *sbm* (resistance to pea seed-borne mosaic virus, PSbMV) and *Fw* (resistance to *Fusarium oxysporum* Schlecht. emend. f. sp. *pisi* (van Hall) Snyd. and Hans. race 1). Linked morphological markers were suggested for *er* (*Gty* – gritty seed coat) and *sbm* (*wlo* – waxless surface of leaflets and *p* – lack of sclerenchymatic pod tissue). Morphological traits used as genetic markers are often undesirable in cultivars, and if used, must be removed in the final stages of breeding. In addition, the number of morphological markers on the pea map is limited, therefore their utility for MAS is reduced in some cases due to relatively large genetic distances. Morphological markers have been used to select breeding material resistant to PSbMV, in spite of the relatively large genetic distances involved (Skarzynska 1988; Weeden et al. 1994a). In another example, *k* (reduced flower wings) is suggested as a useful tag for *mo* (resistance to bean yellows mosaic virus, BYMV; Marx and Provvidenti 1979). The linkage map of Weeden et al. (1993) summarizes the possibilities for using isozyme loci as markers of disease resistance genes. For example, *Pgmp* is linked to a group of virus resistance genes (*mo, cyv, pmv, sbm2*) on linkage

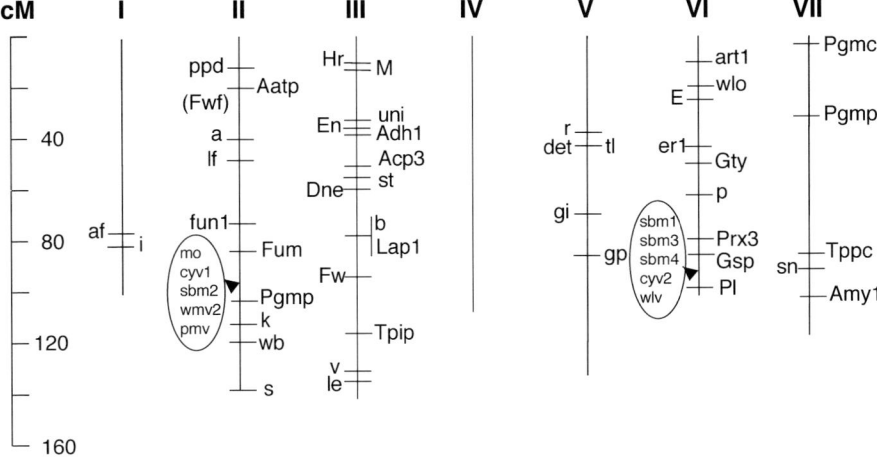

Fig. 1. *Pisum* linkage groups showing the approximate positions of agronomically important genes discussed in the text (*left*) and associated morphological and isozyme markers (*right*). Map orders and linkage relationships are based on the consensus map for *Pisum sativum*. (Weeden et al. 1998)

group II, and *Adh1* is linked to *En* (for resistance to pea enation mosaic virus, PEMV).

The identification of molecular markers associated with advantageous alleles of economically important genes provides more possibilities for the application of MAS. Since molecular linkage maps with high marker density can be constructed, the chances of finding a marker tightly linked to a given gene is increased, even when the chromosomal location of a breeding character/gene is unknown. Bulked segregant analysis (Michelmore et al. 1991) permits the identification of molecular markers associated with trait alleles without linkage mapping. This approach has been applied in pea (e.g., Timmerman et al. 1994; Timmerman-Vaughan et al. 1996; Tiwari et al. 1998; McClendon et al. 2002; Okubara et al. 2002; Schneider et al. 2002). Unlike many morphological markers, molecular markers are generally phenotypically neutral. The advantages of conducting MAS using molecular markers include the ability to analyze plants at the seeding stage, to screen multiple characters including those that would normally be epistatic with one another, to minimize linkage drag and to recover a recurrent parent's genotype rapidly when using backcross breeding strategies (Tanksley et al. 1989). The sections below summarize our current knowledge of traits and molecular markers where MAS may be useful in cultivar development in pea (Fig. 1).

3 Resistance to Powdery Mildew (*Erysiphe pisi* Syd.)

Pea powdery mildew has a worldwide distribution and a strong, negative influence on pea yield and quality. Two recessive genes controlling resistance were described, *er1* (Harland 1948) and *er2* (Heringa et al. 1969). Accessions bearing these genes (e.g., Mexique 4, Stratagem, Svp 952, Rondo) are retained in gene banks and used in breeding. Marx (1971) revealed linkage between *er1* – *Gty*. Confusion about the linkage group containing *er* and *Gty* was resolved when *Gty* was assigned to linkage group VI (Wolko and Weeden 1990). *Er1* was placed on a group VI molecular linkage map by Timmerman et al. (1994) using restriction fragment length polymorphisms (RFLPs; *Fed1*, *Gsp*, pI49, pID18), random amplified polymorphic DNAs (RAPDs; $NW04_{950}$, $PD10_{650}$) and an allozyme locus (*Prx3*), using two unrelated populations segregating for resistance (F_2 and RIL populations). The results were similar in both cases although the map distances obtained from F_2 population analysis were larger than estimates based on the RILs. The marker closest to *er1* appeared to be $PD10_{650}$, but pID18 and *Gty* were also linked. The RAPD marker $PD10_{650}$, linked to *er1* at a distance of about 2 cM, was successfully converted into a sequence characterized amplified region (SCAR). For the $PD10_{650}$ SCAR, the dominant 'band-present' allele is associated with the recessive resistance allele (therefore is in *trans*), consequently, in suitably polymorphic populations it can be used to select individuals carrying the *er1* (resistance) allele without electrophoresis (Weeden et al. 1994b). More recently, Tiwari et al. (1998) have identified three RAPD markers associated with *er1* by bulked segregant analysis, and converted two of these markers to SCARs. The two SCARs developed by Tiwari et al. (1998) will be especially useful since one is in coupling phase (*cis*) while the other is in repulsion phase (*trans*), permitting markers to be used both to select homozygous resistant progeny and to track the *er1* (resistance) allele during backcross breeding.

The introduction of both *er1* and *er2* into a cultivar could increase the durability of resistance to powdery mildew. Unfortunately, our present knowledge of *er2* is less complete than for *er1*. The *er2* chromosomal location is unknown and few lines bearing the *er2* (resistance) allele are available for breeders. Tiwari et al. (1999) used the line JI2480 as the source of *er2* resistance alleles and identified three AFLP markers linked in *trans* and two in *cis*. The *cis* phase primer combination (E+AGG/M+CTG) will be useful in MAS of heterozygous BC_nF_1 individuals for JI2480-derived resistance where the 'band-present' marker phenotype will track the resistance allele while the *trans* phase primer combinations (E+ACT/M+CGC, E+ACG/M+CCC and E+AGG/M+CTA) will be useful where the 'band-absent' marker phenotype will identify homozygous resistant individuals.

4 Resistance to Fusarium Wilt (*Fusarium oxysporum* f. sp. *pisi* (van Hall) Snyd & Hans)

Fusarium wilt caused by *Fusarium oxysporum* f. sp. *pisi* is a significant pea disease. The economically important disease is produced by races 1, 2, 5 and 6. Races 1 and 2 have worldwide distribution and dominantly inherited resistance genes *Fw* and *Fnw*, respectively, have been described (Wade 1929; Hare et al. 1949). Races 5 and 6 are found in the Pacific Northwest region of the USA and in British Columbia in Canada (Haglund 1984). Resistance to race 5 is controlled by a dominantly inherited resistance gene *Fwf* (Hagedorn 1989; Coyne et al. 2000).

Linkage studies have characterized the chromosomal locations of the *Fw* and *Fwf* genes. The gene *Fw* was originally localized on linkage group III based on a loose linkage of about 30 map units from *Le* (Wade 1929). More recently, Grajal-Martin and Muehlbauer (2002) have confirmed the placement of *Fw* on linkage group III by demonstrating linkage with *Lap1* and *b* (pink flowers in the presence of *A*) at a distance of about 13 map units, and with *Td* (leaf dentation) at a distance of about 14 map units. *Fwf* is linked to *Aatp*, a linkage group II anchor locus, at a distance of about nine map units (Coyne et al. 2000). The possibility of developing *Fusarium* wilt resistance locus-specific PCR primers for use in MAS has been demonstrated. McClendon et al. (2002) identified two AFLPs and one RAPD linked to *Fw*, and predicted a 96% probability of correctly identifying resistant lines using only the most tightly linked of these loci. Okubara et al. (2002) identified a RAPD polymorphism (U693a) linked to the *Fwf* gene at a distance of 5.6 map units. Coyne et al. (2001) have identified molecular markers linked to *Fw*, *Fnw* and *Fwf*, and confirmed that these three loci are not linked. After developing mapping populations (RILs) and using bulk segregant analysis, the following markers were identified: an AFLP marker 0.6 cM from *Fw*, a RAPD marker (primer sequence 5'-GACGAGACGG-3') 5.5 cM from *Fwf*, and a RAPD marker (primer sequence 5'-CTGCGGGTCA-3') 4 cM from *Fnw*.

5 Plant Virus Resistance

The pea is naturally infected by a number of plant viruses, which represent at least seven plant viral groups (potyviridae, enamoviridae, tobraviridae, carlaviridae, luteoviridae, cucumoviridae and the alfalfa mosaic virus group). Important rationales for characterizing the genetic basis of viral resistance include developing associated molecular markers for use in resistance breeding, and identifying the molecular basis of pathogenicity and of resistance to plant viruses.

Single locus resistance has been described for viruses in the potyvirus group and for pea enation mosaic virus (PEMV, an enamovirus). Linkage mapping has placed genetic loci for resistance to a number of potyviruses on linkage groups II and VI, and to PEMV on linkage group III. Tolerance conferred by a single genetic locus (tentatively named *Lr*) has been described for pea leaf roll virus (PLRV), a member of the luteoviruses (Hampton 1984), however, the locus that conditions this tolerance has not been characterized by linkage mapping so will not be discussed further.

Tolerance or partial resistance has been described for a further three viruses infecting pea: pea streak virus and red clover vein mosaic virus (carlaviruses; Hampton 1984), and alfalfa mosaic virus (Latham and Jones 2001). However, the genetic loci underlying these resistances have not been characterized by linkage or quantitative trait loci (QTL) mapping, therefore, the genetic basis for the possible resistance or tolerance to these viruses will not be considered further.

5.1 Potyvirus Resistance

Pea seed-borne mosaic virus (PSbMV) infection is economically significant. PSbMV is distributed throughout the world, and because it is seed-transmitted, is readily spread via infected seed lots. Pathotypes of PSbMV have been described (P-1, P-2, P-3 and P-4; Alconero et al. 1986; Hjulsager et al. 2002). Likewise, recessive resistance genes have been described (*sbm1*, *sbm2*, *sbm3* and *sbm4*). Loci *sbm1* and *sbm4* condition resistance to pathotypes P-1 and P-4, respectively, while *sbm2* and *sbm3* both condition resistance to pathotype P-2 (the L and L1 isolates). Pathotype P-3 has been described only recently (Hjulsager et al. 2002). Elegant research based on the analysis of synthetic chimeric viruses has shown that the PSbMV pathotypes can be explained by the properties of two viral cistrons, VPg and P3–6k1 (Johansen et al. 2001; Hjulsager et al. 2002).

Pea is also infected by a number of other potyviruses including bean yellow mosaic virus (BYMV), clover yellow vein virus (CYVV), pea mosaic virus (PMV), watermelon mosaic viruses I and II (WMV I, WMV II), white lupin mosaic virus (WLMV) and passionfruit woodiness virus strain K (PWV-K). Single recessive resistance genes have been identified for most of these potyviruses. Interestingly, analysis of segregation in experimental populations has shown that the loci for PSbMV resistance and other potyvirus recessive resistance loci occur in two gene clusters. Further indirect evidence for gene clustering comes from observations that multiple virus resistance phenotypes are associated in groups in diverse germplasm (e.g., Provvidenti and Hampton 1993) One gene cluster occurs on linkage group II and includes *mo* (BYMV resistance), *sbm2* (PSbMV pathotype P-2 resistance), *pmv* (PMV resistance), *wmv2* (WMV II resistance) and *cyv1* (CYVV 1 resistance; Weeden et al. 1984; Provvidenti and Alconero 1988; Provvidenti 1990; Provvidenti and Hampton

1991). The gene cluster containing *mo* was assigned to linkage group II based on demonstration of linkage to three anchor loci, *wb*, *k* and *Pgm-p* (Weeden et al. 1984). The second cluster of potyvirus resistance genes is found on linkage group VI and includes *sbm1*, *sbm3*, *sbm4*, *cyv2* and *wlv* (WLMV resistance; Provvidenti and Muehlbauer 1990; Provvidenti and Hampton 1993). The gene cluster containing *sbm1* was assigned to linkage group VI based on the demonstration of linkage with *wlo* (Gritton and Hagedorn 1975), *p* and *art1* (Skarzynska 1988) and *Prx3* (Weeden et al. 1991).

From a cultivar development point of view, the clustering of potyvirus resistance loci in two genomic regions significantly simplifies resistance breeding. Further advances have been provided by the identification of linked molecular markers and the conversion of some of these into simple, robust polymerase chain reaction (PCR) assays. Yu et al. (1996) described the development of an allele-specific associated primer (ASAP) assay based on RAPD $BC302_{1200}$, which is closely linked to *mo* (3 cM) and the linkage group II gene cluster. Three sequence tagged site (STS) assays for molecular markers associated with the *sbm1* gene cluster on linkage group VI and their suitability for use in marker-assisted selection were described by Frew et al. (2002).

5.2 Pea Enation Mosaic Virus Resistance

The gene *En*, for dominantly inherited resistance to PEMV, was first described by Schroeder and Barton (1958) and has proved to be durable since that time. *En* has been placed on linkage group III, based on association with anchor loci *st*, *uni* and *Adh1* (Gritton and Hagedorn 1980; Marx et al. 1985; Weeden and Provvidenti 1985). *Adh1* is 5 cM from *En* and, therefore, would be useful as a marker in plant breeding, except that anaerobic induction is required for expression of this isozyme. More recently, two ASAP markers linked to *En* and suitable for use in marker-assisted selection have been described (Yu et al. 1995).

6 Flowering Genetics

The genes controlling the responses of flowering time to photoperiod and temperature have an important role to play in the adaptation of pea to the broad spectrum of environments in which it is grown. Hitherto, a number of detailed analyses (reviewed by Murfet 1990; Weller et al. 1997) have characterized the genetic and physiological interactions of this quantitative character. The genes involved include *Lf* (late flowering), *Sn* (sterile nodes), *Dne* (day neutral), *E* (early), *Hr* (high response), *Ppd* (photoperiod), *Veg* (vegetative), *Dm* (diminutive), *Gi* (gigas), and *det* (determinate). Three genes affect the flowering behavior of most established cultivars, *Sn*, *Lf* and *Hr* (Weller et al. 1997) although *E*, *Dne* and *Ppd* may be important, depending on the genetic background. For practical purposes, the genotype *Lf Sn hr* has been adopted arbitrarily as the "wild-type" genotype (Murfet and Reid 1993).

The linkage relationships for most of the genes with a well-characterized role in flowering time have been determined. With regard to the consensus pea map (Weeden et al. 1998), *Ppd*, *Lf* and *Fun1* are found on linkage group II, *Dne* and *Hr* on group III, *Det* and *Gi* on group V, *E* on group VI and *Sn* on group VII. Morphological markers linked to *Dne* (*st*, 5 cM distant), *Det* (*R*, 7 cM) and *Hr* (*M*, 7 cM distant) were found before molecular linkage maps were in use (King and Murfet 1985; Marx 1986). Weeden et al. (1988) showed close linkage between *sn* and *Amy1*, and Murfet and Taylor (1999) showed that *Ppd* is linked at a 5 cM distance to isozyme locus *Aatp*. Rameau et al. (1998) used bulked segregant analysis and linkage mapping to identify and map RAPD markers linked to *Det*, *Dne* and *Sn*. Therefore, molecular tags now make it is possible to select alleles of many of the most important flowering genes in seedlings.

7 Quantitative Trait Loci

Linkage mapping of quantitative trait loci (QTLs) using molecular markers has become a basic method for characterizing the genetics of multilocus traits (Mackay 2001). QTL mapping enables the minimum number of genetic loci controlling a trait to be elucidated, the magnitude of effects to be estimated, and epistatic (genotype × genotype) interactions to be characterized. As a result of QTL mapping, associated DNA marker alleles may be identified, and these may subsequently be used in marker-assisted breeding. Ideally, molecular markers may effectively convert a quantitatively inherited trait into a series of Mendelianly inherited genetic elements that can be used to select individuals with optimal combinations of alleles. In reality, the use of QTL mapping information in plant breeding is complex and requires accurate estimates of the magnitude of effects of the QTLs in the populations being assessed, of the distances between the associated molecular markers and the QTL, and of epistatic interactions that occur between QTLs and/or background markers (Young 1999). Although there have been major improvements in QTL mapping methodology, such as composite interval mapping (Zeng et al. 1999), QTL mapping remains an imprecise science. For example, the genomic regions to which QTLs are assigned are usually relatively large (20–30 cM), the magnitudes of effects may be overestimated, and the most widely used experimental designs have limited power to detect all QTLs contributing to a phenotype (Beavis 1994; Utz and Melchinger 1994; Melchinger et al. 1998).

Quantitative trait loci mapping may also provide the starting point for cloning and characterizing the genomic sequences that control a quantitatively inherited trait. At the present time, obtaining information about the genes underlying QTLs is most successful for QTLs with major effects on phenotype. The genes underlying QTLs have been cloned in a few cases, particularly in *Arabidopsis*, rice and tomato (e.g., Fridman et al. 2000; Yano et al. 2000; El-Assal et al. 2001).

In pea, relatively few QTL mapping studies have been published. Traits characterized include resistance to economically important diseases that have been largely intractable to conventional breeding approaches – *Ascochyta* blight (Dirlewanger et al. 1994; Timmerman-Vaughan et al. 2002) and *Aphanomyces* root rot (Pilet-Nayel et al. 2002). These fungal diseases are among the most destructive diseases affecting pea crops, with impacts on many growing regions worldwide (Lawyer 1984; Pfender 1984). The development of resistant cultivars would significantly enhance our ability to control the effects of these diseases on pea yield and quality. QTL mapping studies have also characterized developmental traits (Dirlewanger et al. 1994), seed weight (Timmerman-Vaughan et al. 1996), and green seed color quality (McCallum et al. 1997).

7.1 Disease Resistance Quantitative Trait Loci

Quantitative trait loci mapping has identified genomic regions and molecular marker alleles associated with resistance to *Ascochyta pisi* race C (Dirlewanger et al. 1994), to field epidemics of Ascochyta blight most likely caused by *Mycosphaerella pinodes* and *Phoma medicaginis* var. *pinodella* (Timmerman-Vaughan et al. 2002), and to *Aphanomyces* root rot (Pilet-Nayel et al. 2002). These QTL studies represent initial efforts to understand the genetic nature of resistance in pea germplasm to these diseases, and starting points for developing resistant pea cultivars through marker-assisted selection. The QTLs detected and their association with anchor loci on the consensus linkage map of pea (Weeden et al. 1998) are summarized in Table 1.

Two QTLs for resistance to *A. pisi* race C were identified by Dirlewanger et al. (1994), based on screening $F_{2:3}$ progeny seedlings in a greenhouse test. A major QTL ($R^2=0.45$) was mapped to linkage group III, while a minor QTL was mapped to linkage group I in the vicinity of *af* and *i*. The linkage map developed by these authors only covered a portion of the pea genome, therefore, other QTLs may have not been detected. For most linkage groups, the maps presented in this study are anchored to the "consensus pea map" with one or a few loci. The major QTL identified in this study maps to the same linkage group as *Fw* (but is not associated with *Fw*), a linkage group III anchor (Weeden et al. 1998).

Many QTLs for field resistance to *Ascochyta* blight were identified (Timmerman-Vaughan et al. 2002), based on assessment of $F_{2:3}$ and $F_{2:4}$ progeny for disease symptoms on stems, leaves and pods in naturally occurring field epidemics in Western Australia over 3 years. The eight QTLs listed in

Table 1. Summary of pea disease resistance QTLs

Resistance trait	LG	Locus	Linked anchor loci[a]	Reference
A. pisi race C	I		af, i	Dirlewanger et al. (1994)
	III			
Ascochyta blight field epidemics	I	Asc1.1[b]	c206	Timmerman-Vaughan et al. (2002)
	II	Asc2.1[b]	TPP	
	III	Asc3.1[b]	M27	
	IV	Asc4.1[b]	P628	
	IV	Asc4.2[b]	P357	
	IV	Asc4.3[b]		
	V	Asc5.1[b]	gp, Pgd-c	
	VII	Asc7.1[b]	PyrB2, Cab	
Aphanomyces root rot	I	Aph3[b]	af	Pilet-Nayel et al. (2002)
	Ib	Aph4		
	Ib	Aph5		
	IV	Aph1[b]		
	V	Aph2[b]	r, P108	
	VII	Aph6	Pgd-p	
	B	Aph7		

[a] Anchor loci placed on the consensus linkage map for *Pisum sativum* (Weeden et al. 1998)
[b] QTL detected using two or more trait measures

Table 1 were detected using at least two trait measures, and another five QTLs were detected using only a single trait measure. All the QTLs explained relatively small fractions of the trait variation, therefore, none can be characterized as being "major" QTLs. The linkage maps used to map *Ascochyta* blight resistance QTLs were also anchored to the consensus pea map.

Likewise, a number of QTLs were discovered for resistance to *Aphanomyces* root rot (Pilet-Nayel et al. 2002). QTL detection in this study was based on assessment of recombinant inbred lines in disease nurseries at two sites in the USA over a 2-year period. The traits measured included disease progression on the roots, above-ground disease symptoms and dried weight losses in diseased versus nondiseased plants. Three QTLs (*Aph1*, *Aph2*, and *Aph3*) were detected using two or more trait measures or in two or more environments, while the remaining four were identified using a single trait measure (Table 1). *Aph1* has been classified by the authors as a "major" QTL because it was detected by most of the trait measures and because it explained up to 45% of the trait variation.

Of these disease resistance QTLs, only two are clearly associated with the same genomic region. *Aph3* and the minor *A. pisi* race C QTL are associated with the region of linkage group I containing *af*. Interestingly, this genomic region also contains genes for chalcone synthase (Laucou et al. 1998), an enzyme involved in the phenylpropanoid biosynthesis pathway and a candidate defense gene (summarized in Trognitz et al. 2002).

7.2 Quantitative Trait Loci for Seed Characteristics and Developmental Traits

Quantitative trait loci studies have characterized the genetics of seed weight (Timmerman-Vaughan et al. 1996). Seed weight QTLs were mapped in two experimental populations ('Primo' × OSU442-15 and JI1794 × 'Slow'). With reference to the current consensus map of the pea genome (Weeden et al. 1998), seed weight QTLs were identified in the 'Primo' × OSU442-15 population on linkage groups I, III and IV (two QTLs), while in the JI1794 × 'Slow' population QTLs were mapped to linkage groups II, III (two QTLs) and VII. Only one genomic region, on linkage group III (associated with RAPD marker B08-1250), contained seed weight QTLs in both populations. QTL mapping in other legume species, particularly *Vigna* spp. (Fatokun et al. 1992) and soybean (Maughan et al. 1996) has shown that seed weight QTLs map to orthologous genomic regions. Additional support for the observations that seed weight QTLs in legumes occur in orthologous genomic regions was obtained by mapping *Vigna* probes associated with seed weight QTLs to regions of pea linkage groups III and IV containing seed weight QTLs (Timmerman-Vaughan et al. 1996).

Quantitative trait loci mapping has also characterized the genetics of green seed color quality in field pea genotypes (McCallum et al. 1997). Using the 'Primo' × OSU442-15 population, seed color quality was measured objectively by video image analysis to provide quantitative measures of the color space components Y (color density), U and V. A 'major' QTL affecting Y, and showing dominant inheritance of the pale color phenotype, was mapped to linkage group V in the genomic region containing *r* locus. Three other QTLs affecting the U or V chrominance components of color space were mapped (with reference to the consensus pea linkage map) to linkage groups II, III and VII.

Quantitative trait loci for three developmental traits, plant height, earliness, and number of nodes, were mapped by Dirlewanger et al. (1994) to four genomic regions. Three QTLs were discovered affecting node number, two QTLs affecting earliness, and one QTL affecting plant height. There were two genomic regions where QTLs for different traits were colocalized, which either indicates that the traits involved are controlled by the same genetic locus, or that the genes involved are tightly linked. The resolution of QTL mapping studies is usually not sufficient to distinguish these two possibilities.

8 Conclusions

The pea entered the genomic era with a well-developed chromosome map having numerous, easily recognizable markers. Some of them are linked to agricultural characters used in breeding. This resulted in the so-called common map which combined classical as well as molecular markers and as a result gave the possibility of using MAS for many different characters of a cultivar ideotype. QTLs clearly showed potential as molecular markers. However, in applied pea breeding, the same problems still exist and must be overcome, e.g., increasing protein content and decreasing oligosaccharide content in seeds as well as improving resistance to lodging and to ascochytosis. Are breeders' expectations too high or are they not able to use the available tools?

References

Alconero R, Provvidenti R, Gonsalves D (1986) Three pea seedborne mosaic virus pathotypes from pea and lentil germplasm. Plant Dis 70:783–786

Beavis WD (1994) The power and deceit of QTL experiments: Lessons from comparative QTL studies. 49th annual corn and sorghum industry research conference. American Seed Trade Association, Washington, DC, pp 250–266

Blixt S. (1977) The gene symbols of *Pisum*. Pisum Newslett 9:1–59

Coyne CJ, Inglis DA, Whitehead SJ, Mc Clendon MT, Muehlbauer FJ (2000) Chromosomal location of *Fwf*, the Fusarium wilt race 5 resistance gene in *Pisum sativum*. Pisum Genet 32:20–22

Coyne CJ, Meksem K, Mc Phee KE, Inglis DA, Lightfoot D, Mc Clendon MT, Shultz J, Muehlbauer FJ (2001) Positional cloning of *Fusarium* wilt resistance genes in pea. In: Towards the sustainable production of healthy food, feed and novel products. European Association for Grain Legume Research, Paris, pp 16–17

Dirlewanger E, Isaac P, Ranade S, Belajouza M, Cousin R, Devienne D (1994) Restriction fragment length polymorphism analysis of loci associated with disease resistance genes and developmental traits in *Pisum sativum* L. Theor Appl Genet 88:17–27

El-Assal S, Alonso-Blanco C, Peeters AJ, Raz V, Koornneef M (2001) A QTL for flowering time in *Arabidopsis* reveals a novel allele of CRY2. Nat Genet 29:435–440

Fatokun CA, Menancio-Hautea DI, Danesh D, Young ND (1992) Evidence for orthologous seed weight genes in cowpea and mung bean based on RFLP mapping. Genetics 132:841–846

Frew TJ, Russell AC, Timmerman-Vaughan GM (2002) Sequence-tagged site markers linked to the *sbm*1 gene for resistance to pea seed-borne mosaic virus in pea. Plant Breed 121:512–516

Fridman E, Pleban T, Zamir D (2000) A recombination hotspot delimits a wild-species quantitative trait locus for tomato sugar content to 484 bp within an invertase gene. Proc Natl Acad Sci USA 97:4718–4723

Grajal-Martin MJ, Muehlbauer FJ (2002) Genomic location of the *Fw* gene for resistance to fusarium wilt race 1 in peas. J Hered 93:291–293

Gritton ET, Hagedorn DJ (1975) Linkage of the genes *sbm* and *wlo* in peas. Crop Sci 11:945–946

Gritton ET, Hagedorn DJ (1980) Linkage of *En* and *St* genes in pea. Pisum Newslett 12:26–27

Hagedorn WA (1989) Compendium of pea diseases. American Phytopathological Society, St Paul, Minnesota

Haglund WA (1984) Fusarium wilts. In: Hagedorn DJ (ed) Compendium of pea diseases. American Phytopathological Society Press, St Paul, Minnesota, pp 22–25

Hampton RO (1984) Diseases caused by viruses. In: Hagedorn DJ (ed) Compendium of pea diseases. American Phytopathological Society Press, St Paul, Minnesota, pp 31–37

Hare WW, Walker JC, Delwiche EH (1949) Inheritance of a gene for near-wilt resistance in the garden pea. J Agric Res 78:239–250

Harland SC (1948) Inheritance of immunity to mildew in Peruvian forms of *Pisum sativum*. Heredity 2:263–269

Heringa RJ, van Norel A, Tazelaar MF (1969) Resistance to powdery mildew (*Erysiphe polygoni* DC) in peas (*Pisum sativum* L.) Euphytica 18:163–169

Hjulsager CK, Lund OS, Johansen E (2002) A new pathotype of pea seedborne mosaic virus explained by the properties of the p3-6k1 and viral genome-linked protein (VPg)-coding regions. Mol Plant Microbe Interact 15:169–171

Johansen IE, Lund OS, Hjulsager CK, Laursen J (2001) Recessive resistance in *Pisum sativum* and potyvirus pathotype resolved in a gene-for-cistron correspondence between host and virus. J Virol 75:6609–6614

King WM, Murfet IC (1985) Flowering in *Pisum*: a sixth locus, *Dne*. Ann Bot 56:835–846

Lamprecht H (1948) The variation of linkage and the course of crossingover. Agri Hort Genet 6:10–48

Lamprecht H (1961) Die Genenkarte von *Pisum* bei normaler Struktur der Chromosomen. Agri Hort Genet 19:360–401

Lamprecht H (1974) Monographie der Gattung *Pisum*. Steiermarkische Landesdruckerei, Graz

Latham LJ, Jones RAC (2001) Alfalfa mosaic and pea seed-borne mosaic viruses in cool season crop, annual pasture, and forage legumes: susceptibility, sensitivity and seed transmission. Aust J Agric Res 52:771–790

Laucou V, Haurogné, Ellis N, Rameau C (1998) Genetic mapping in pea. 1. RAPD-based genetic linkage map of *Pisum sativum*. Theor Appl Genet 97:905–915

Lawyer AS (1984) Diseases caused by *Ascochyta* spp. In: Hagedorn DJ (ed) Compendium of pea diseases. American Phytopathological Society Press, St Paul, Minnesota, pp 11–15

Mackay TFC (2001) The genetic architecture of quantitative traits. Annu Rev Genet 35:303–339

Marx GA (1971) New linkage relations for chromosome III of *Pisum*. Pisum Newslett 3:18–19

Marx GA (1986) Linkage relationships of *Curl*, *Orc* and "*Det*" with markers on chromosome 7. Pisum Newslett 19:31–32

Marx GA, Provvidenti R (1979) Linkage relations of *mo*. Pisum Newslett 11:28–29

Marx GA, Weeden NF, Provvidenti R (1985) Linkage relationships among markers in chromosome III and *En*, a gene conferring virus resistance. Pisum Newslett 17:57–60

Maughan PJ, Maroof MAS, Buss GR (1996) Molecular marker analysis of seed-weight: genomic locations, gene action and evidence for orthologous evolution among three legume species. Theor Appl Genet 93:574–579

McCallum JA, Timmerman-Vaughan G, Frew T, Russell A (1997) Biochemical and genetic linkage analysis of green seed color in field pea. J Am Soc Hortic Sci 122:218–225

McClendon MT, Inglis DA, McPhee KE, Coyne CJ (2002) DNA markers linked to Fusarium wilt race 1 resistance in pea. J Am Soc Hortic Sci 127:602–607

Melchinger AE, Utz HF, Schon CC (1998) Quantitative trait locus (QTL) mapping using different testers and independent population samples in maize reveals low power of QTL detection and large bias in estimates of QTL effects. Genetics 149:383–403

Michelmore RW, Paran I, Kesseli RV (1991) Identification of markers linked to disease resistance genes by bulked segregant analysis: a rapid method to detect markers in specific genomic regions by using bulked segregant populations. Proc Natl Acad Sci USA 88:9828–9832

Murfet IC (1990) Flowering genes in pea and their use in breeding. Pisum Newslett 22:78–86

Murfet IC, Reid JB (1993) Developmental mutants. In: Casey R, Davies DR (eds) Peas – genetics, molecular biology and biotechnology. CAB International, Wallingford, pp 165–216

Murfet IC, Taylor SA (1999) Flowering gene *Ppd* in pea: map position and disturbed segregation of allele *ppd-2*. J Hered 90:548–550

Okubara PA, Inglis DA, Muehlbauer FJ, Coyne CJ (2002) A novel RAPD marker linked to the Fusarium wilt race 5 resistance gene (*Fwf*) in *Pisum sativum*. Pisum Genet 34:6–8

Pfender WF (1984) Aphanomyces root rot. In: Hagedorn DJ (ed) Compendium of pea diseases. American Phytopathological Society Press, St Paul, Minnesota, pp 25–28

Pilet-Nayel ML, Muehlbauer FJ, McGee RJ, Kraft JM, Baranger A, Coyne CJ (2002) Quantitative trait loci for partial resistance to Aphanomyces root rot in pea. Theor Appl Genet 106:28–39

Provvidenti R (1990) Inheritance of resistance to pea mosaic virus in *Pisum sativum*. J Hered 81:143–145

Provvidenti R, Alconero R (1988) Inheritance of resistance to a lentil strain of pea seed-borne mosaic virus in *Pisum sativum*. J Hered 79:45–47

Provvidenti R, Muehlbauer FJ (1990) Evidence of a cluster of linked genes for resistance to pea seedborne mosaic virus and clover yellow vein virus on chromosome 6. Pisum Newslett 22:43–45

Provvidenti R, Hampton RO (1991) Chromosomal distribution of genes for resistance to seven potyviruses in *Pisum sativum*. *Pisum* Genetics 23:26–28

Provvidenti R, Hampton RO (1993) Inheritance of resistance to white lupin mosaic virus in common pea. HortScience 28:836–837

Rameau C, Denoue D, Fraval F, Haurogne K, Josserand J, Laucou V, Batge S, Murfet IC (1998) Genetic mapping in pea. 2. Identification of RAPD and SCAR markers linked to genes affecting plant architecture. Theor Appl Genet 97:916–928

Schneider A, Walker SA, Sagan M, Duc G, Ellis THN, Downie JA (2002) Mapping of the nodulation loci *sym9* and *sym10* of pea (*Pisum sativum* L.). Theor Appl Genet 104:1312–1316

Schroeder WT, Barton DW (1958) The nature and inheritance of resistance to pea enation mosaic virus in garden pea, *Pisum sativum* L. Phytopathology 48:628–632

Skarzynska A (1988) Supplemental mapping data for chromosome 6. Pisum Newslett 20:34–36

Smartt J (1990) Grain legumes: evolution and genetic resources, Chap. 6. Cambridge Univ Press, Cambridge

Swiecicki WK, Wolko B, Weeden NF (2000) Mendel's genetics, the *Pisum* genome and pea breeding. Vortr Pflanzenzuchtg 48:65–76

Tanksley SD, Young ND, Paterson AH, Bonierbale MW (1989) RFLP mapping in plant breeding: new tools for an old science. Biotechnology 7:257–264

Timmerman GM, Frew TJ, Weeden NF, Miller AL, Goulden DS (1994) Linkage analysis of *er-1*, a recessive *Pisum sativum* gene for resistance to powdery mildew fungus (*Erysiphe pisi* DC). Theor Appl Genet 88:1050–1055

Timmerman-Vaughan GM, McCallum JA, Frew TJ, Weeden NF, Russell AC (1996) Linkage mapping of quantitative trait loci controlling seed weight in pea (*Pisum sativum* L.). Theor Appl Genet 93:431–439

Timmerman-Vaughan GM, Frew TJ, Russell AC, Khan T, Butler R, Gilpin M, Murray S, Falloon K (2002) QTL mapping of partial resistance to field epidemics of ascochyta blight of peas. Crop Sci 42:2100–2111

Tiwari KR, Penner GA, Warkentin TD (1998) Identification of coupling and repulsion phase RAPD markers for powdery mildew resistance gene *er-1* in pea. Genome 41:440–444

Tiwari KR, Penner GA, Warkentin TD (1999) Identification of AFLP markers for the powdery mildew resistance gene *er-2* in pea. Pisum Genet 31:27–29

Trognitz F, Manosalva P, Gysin R, Ninio-Lui D, Simor R, del Herrera MR, Trognitz B, Ghislain M, Nelson R (2002) Plant defense genes associated with quantitative resistance to potato late blight in *Solanum phureja* × dihaploid *S. tuberosum* hybrids. Mol Plant Microbe Interact 15:587–597

Utz HF, Melchinger AE (1994) Comparison of different approaches to interval mapping of quantitative trait loci. In: Van Ooijen JW, Jansen J (eds) Biometrics in plant breeding: applications of molecular markers. Proceedings of the 9th Meeting of the EUCARPIA section biometrics in plant breeding. CPRO-DLO, Wageningen, Netherlands, pp 195–204

Wade BL (1929) The inheritance of Fusarium wilt resistance in canning peas. Wis Agric Exp Sta Res Bull 97:1–32

Weeden NF (1985) An isozyme linkage map for *Pisum sativum*. In: Habblethwaite PD, Heath MC, Dawkins TCK (eds) The pea crop. Butterworths, London, pp 55–66

Weeden NF, Provvidenti R (1985) A marker locus, *Adh-1*, for resistance to pea enation mosaic virus. J Hered 79:128–131

Weeden NF, Provvidenti R, Marx GA (1984) An isozyme marker for resistance to bean yellow mosaic virus in *Pisum sativum*. J Hered 75:411–412

Weeden NF, Kneed BE, Murfet IC (1988) Mapping of the *Sn* locus to chromosome 2. Pisum Newslett 20:49–51

Weeden NF, Provvidenti R, Wolko B (1991) *Prx-3* is linked to *sbm*, the gene conferring resistance to pea seedborne mosaic virus. Pisum Genet 23:42–43

Weeden NF, Swiecicki WK, Ambrose M, Timmerman GM (1993) Linkage groups of pea. Pisum Genet 25:4

Weeden NF, Timmerman GM, Lu J (1994a) Identifying and mapping genes of economic significance. Euphytica 73:191–198

Weeden NF, Wu WY, Gu WK, Cargnoni TL, Lu J, Timmerman GM, Wolko B, Zhu Z (1994b) Applications of DNA amplification technology to vegetable breeding. Proc 7th Int Cong Soc Adv Breed Res Asia Oceania. Taipei, ROC, pp 437–445

Weeden NF, Ellis THN, Timmerman-Vaughan GM, Swiecicki WK, Rozov SM, Bernikov VA (1998) A consensus linkage map for *Pisum sativum*. Pisum Genet 30:1–4

Weller JL, Reid JB, Taylor SA, Murfet IC (1997) The genetic control of flowering in pea. Trends Plant Sci 2:412–418

Wolko B, Weeden NF (1990) Additional markers for chromosome 6. Pisum Newslett 22:71–74

Yano M, Katayose Y, Ashikari M, Yamanouchi U, Monna L, Fuse T, Baba T, Yamamoto K, Umehara Y, Ngamura Y, Sasaki T (2000) *Hd1*, a major photoperiod sensitivity quantitative trait locus in rice, is closely related to the *Arabidopsis* flowering time gene *CONSTANS*. Plant Cell 12:2473–2484

Young ND (1999) A cautiously optimistic vision for marker-assisted breeding. Mol Breed 5:505–510

Yu J, Gu WK, Provvidenti R, Weeden NF (1995) Identifying and mapping two DNA markers linked to the gene conferring resistance to pea enation mosaic virus. J Am Soc Hortic Sci 120:730–733

Yu J, Gu WK, Weeden NF (1996) Development of an ASAP marker for resistance to bean yellow mosaic virus in *Pisum sativum*. Pisum Genet 28:31–32

Zeng Z-B, Kao C-H, Basten CJ (1999) Estimating the genetic architecture of quantitative traits. Genet Res 74:279–289

II.6 Molecular Markers in *Vigna* Improvement: Understanding and Using Gene Pools

A. KAGA, D.A. VAUGHAN, and N. TOMOOKA[1]

1 Introduction

The genus *Vigna* is a large genus with about 90 species distributed worldwide in warm and tropical regions. The genus includes 13 cultigens (Table 1), of these cowpea, mungbean and azuki bean are the most important (Table 2). Perhaps due to *Vigna* cultigens being mainly crops in the developing world, molecular marker and genomic studies have lagged behind those of other major crops. For example, genome designations are known for species in many crop genera, such as *Glycine*, but these have not been established for *Vigna* species. There has recently been considerable progress in using molecular markers to understand *Vigna* genetic resources and in developing genome maps and associating molecular markers to agronomically important traits in the *Vigna* cultigens. However, the actual use of molecular markers in *Vigna* breeding at the end of 2002 is still being planned.

In this chapter we review recent progress in the use of molecular markers in understanding *Vigna* genetic resources, *Vigna* linkage map development, particularly in relation to the location of genes for agronomic traits, and transformation systems.

2 Application of Molecular Markers to Understand the *Vigna* Crop Gene Pools

2.1 Genus *Vigna*

Current understanding of the taxonomy of the genus *Vigna* rests largely on the work of Maréchal and coworkers (1978). Several insights into *Vigna* taxonomy have resulted from molecular analyses of *Vigna*. Restriction fragment length polymorphism (RFLP) analysis supports the taxonomic opinion that *Phaseolus* and *Vigna* belong to a common complex of species particularly when the poorly studied New World *Vigna* species are considered (Fatokun et

---

[1] National Institute of Agrobiological Sciences, Kannondai 2-1-2, Tsukuba, Ibaraki 305-8602, Japan

al. 1993). Chloroplast DNA, RFLP and isozyme analyses suggest that both section *Catiang* (the cowpea section) of the subgenus *Vigna* and subgenus *Ceratotropis* are well-defined groups (Jaaska and Jaaska et al. 1988, 1990; Fatokun et al. 1993; Vaillancourt and Weeden 1993)

Table 1. The cultivated and domesticated *Vigna* species

Subgenus *Section*	Cultigen species name (common name)	Presumed progenitor
Vigna		
Catiang	*V. unguiculata* (L.) Walpers subsp. *unguiculata* var. *unguiculata* (cowpea)	ssp. *unguiculata* var. *spontanea* (Schweinf.) Pasquet
Vigna	*V. subterranea* L. (bambara groundnut)[a]	*V. subterranea* L.
	V. luteola (Jacq.) Benth.[a]	*V. luteola* (Jacq.) Benth.
	V. marina (Burm.) Merrill[a]	*V. marina* (Burm.) Merrill
Plectotropis	*V. vexillata* (L.)A. Richard[a]	*V. vexillata* (L.) A. Richard
Ceratotropis		
Angulares	*V. angularis* (Willd.) Ohwi and Ohashi var. *angularis* (azuki bean)	var. *nipponensis* (Ohwi) Ohwi and Ohashi
	V. reflexo-pilosa Hayata var. *glabra* (Roxb.) N. Tomooka and Maxted	var. *reflexo-pilosa*
	V. trinervia (Heyne ex Wight and Arnott) Tateishi and Maxted[a]	*V. trinervia* (Heyne ex Wight and Arnott) Tateishi and Maxted
	V. umbellata (Thunb.) Ohwi and Ohashi (rice bean)[a]	*V. umbellata* (Thunb.) Ohwi and Ohashi
Ceratotropis	*V. mungo* (L.) Hepper var. *mungo* (black gram)	var. *silvestris* Lukaki, Maréchal and Otoul
	V. radiata (L.) Wilczek var. *radiata* (mungbean)	var. *sublobata* (Roxb.) Verdcourt
Aconitifoliae	*V. aconitifolia* (Jacquin)Maréchal (moth bean)	Unknown
	V. trilobata (L.) Verdcourt (jungli bean)[a]	*V. trilobata* (L.) Verdcourt

[a] Species cultivated, but not fully domesticated

Table 2. The production, production area and main areas of production of the main *Vigna* cultigens. (Sources: Poehlman 1991; Lumpkin and McClary 1994; Singh et al. 1997)

Species	Production (10^3 tonnes)	Area (10^3 ha)	Main areas of production
Cowpea	3000	12,500	64% Western and central Africa
Mungbean	2500–3000	5000	45% India
Azuki bean	600	1000	60% China

2.2 Subgenus *Vigna*

Cowpea was domesticated in Africa, though where is unclear (Pasquet 1999), from the wild annual form (*V. unguiculata* ssp. *unguiculata* var. *spontanea*). Amplified fragment length polymorphism (AFLP), chloroplast DNA and isozyme data reveal that domestication resulted in a major reduction in genetic diversity, suggesting a single domestication event followed by genetic differentiation under domestication (Weeden et al. 1996; Pasquet 1999; Coulibaly et al. 2002). Despite much effort to collect cowpea genetic resources from throughout Africa (Ng and Monti 1990), large gaps in the collection remain (Pasquet 1999). Cultivated cowpea (var. *unguiculata*) can intercross with wild annual (var. *spontanea*) and some wild perennial (various subspecies of *V. unguiculata*) forms (Ng 1995; Pasquet 1996). Gene flow between wild and domesticated cowpea, revealed by AFLP analysis, has resulted in a large crop weed complex and thus broadened the genetic diversity of *V. unguiculata* (Coulibaly et al. 2002).

Chloroplast DNA analysis suggests the cowpea section is phylogenetically closer to subgenus *Plectotropis* than other species of subgenus *Vigna* and this may have importance in relation to introducing useful characters into cowpea (Vaillancourt and Weeden 1993). Studies of diversity within *V. unguiculata* have shown the usefulness of random amplified polymorphic DNA (RAPD) over isozymes for revealing variation (Vaillancourt and Weeden 1993; Mignouna et al. 1998). Microsatellite markers developed by Li et al. (2001) will be powerful tools to understand the genetic diversity of the cowpea gene pool.

RAPD markers have been used to analyse diversity in *V. subterranea* (Amadou et al. 2001), *V. luteola* and *V. marina* (Sonnante et al. 1997). *V. subterranea* accessions were shown to be differentiated based on geographic origin, western and southern Africa. *V. marina* ssp. *oblonga* was found to be more closely related to *V. luteola* than *V. marina* ssp. *marina*, suggesting that taxonomic revision of *V. marina* may be in order.

2.3 Subgenus *Ceratotropis*

The subgenus *Ceratotropis*, which includes eight cultigens (Table 1), is a difficult group of species to distinguish based on morphological characters (Baudoin and Maréchal 1988). However, the application of various molecular marker techniques has greatly improved understanding of the subgenus. Molecular analyses based on AFLP, chloroplast and rDNA variation have supported the division of the subgenus *Ceratotropis* into three sections (Doi et al. 2002; Tomooka et al. 2002a, b). These three sections, section *Angulares* (azuki and rice bean group), section *Ceratotropis* (mungbean and black gram group) and section *Aconitifoliae* (moth bean group), can be considered separate gene pools for breeding purposes. Of the three sections, section *Angulares* is the most complex and speciation appears to be recent as interspecific genetic divergence is not great, although between some species interspecific hybridization barriers exist (Tomooka et al. 2002b, c). There remains much to be understood about barriers to hybridization and speciation processes in the genus *Vigna*.

Studies of *V. radiata* suggest that Afghanistan-Iran retains more genetic diversity than other regions (Tomooka et al. 1992). However, the presence of wild and weedy races of mungbean, archaeological remains of *Vigna* and landrace diversity suggest that India is the most likely area of domestication (Tomooka et al. 2003). Studies of the diversity of *V. radiata* var. *sublobata*, using RAPD and AFLP methods, suggest considerable geographic variation in this presumed progenitor of mungbean (Savaranakumar et al. 2004), but comprehensive studies of this taxa from throughout its range from Africa to Australia are lacking.

RAPD and AFLP markers have also been used to understand the domestication process in azuki bean, *V. angularis* (Yee et al. 1999; Mimura et al. 2000; Xu et al. 2000a, b; Isemura et al. 2002; Zong et al. 2003). The most comprehensive of these studies suggests that there are four different groups of germplasm related to geographic origin, China, Korea (and some Japanese germplasm), Japan and the Himalayan region. Among these, Himalayan germplasm is well differentiated from the other groups (Zong et al. 2003). The results suggest that azuki bean was probably domesticated independently in the Himalayan region and East Asia.

3 Linkage Maps

To date there have been ten different linkage maps based on eight different crosses developed for *Vigna* species. Four have been developed for *V. radiata* (Menancio-Hautea et al. 1992; Lambrides et al. 2000), four have been developed for *V. unguiculata* (Fatokun et al. 1992b; Menendez et al. 1997; Ubi et al. 2000; Ouédraogo et al. 2002a) and two have been developed for *V. angularis* (Kaga et al. 1996, 2000). A comparison among these maps is shown in Table 3. Only one of the maps developed so far has resolved the 11 linkage groups (Ouédraogo et al. 2002a), equivalent to the haploid chromosome number of these three *Vigna* species (Sinha and Roy 1979). To overcome the limited number of marker clones available for *Vigna* species, many DNA clones from related species such as *Glycine max* and *Phaseolus vulgaris* have been used to increase the saturation of the *Vigna* genome maps (Menacio-Hautea et al. 1992; Boutin et al. 1995; Kaga et al. 2000; Lambrides et al. 2000; Chaitieng et al. 2002). Recently, microsatellites have been identified in *Vigna* species based on database searches (Yu et al. 1999) and microsatellite libraries have been specifically developed from cowpea (Li et al. 2001), mungbean (Kumar et al. 2002) and azuki bean (Wang et al. 2004).

Table 3. Genome linkage maps for *Vigna* species

Cross combination	Population (plants/lines) analyzed	Markers used	Linkage groups resolved	Map distance (cM)	Level of distortion (%)	Reference
V. angularis (cv. Erimoshouzu) × *V. nakashimae*	F$_2$ population (80)	19 RFLP, 108 RAPD and 5 morphological markers	14	1250	19.7	Kaga et al. (1996)
V. angularis (cv. Erimoshouzu) × *V. umbellata* (cv. Kagoshima)	F$_2$ population (86)	114 RFLP, 74 RAPD, 1 morphological marker	14	1702	29.8	Kaga et al. (2000)
V. radiata var. *radiata* × *V. radiata* var. *sublobata* (from Madagascar)	F$_2$ population (58)	151 RFLP, 20 cDNA and 1 pest locus	14	1570	12	Menacio-Hautea et al. (1992)
V. radiata var. *radiata* × *V. radiata* var. *sublobtata* (from Australia)	F$_2$ population (67)	52 RFLP, 56 RAPD, 2 morphological markers	12	758.3	14.5	Lambrides et al. (2000)
V. radiata var. *radiata* × *V. radiata* var. *sublobata* (from Australia)	Recombinant inbreed (67)	113 RAPD, 2 morphological markers	12	691.7	24	Lambrides et al. (2000)
V. radiata (cv. Berken) × *V. radiata* ssp. *sublobata* (ACC41)	Recombinant inbreed (80)	255 RFLP markers	13	737.9	30.8	Humphrey et al. (2002)
V. unguiculata (IT84S-2246-4 improved line) × *V. unguiculata* ssp. *dekindtiana* var. *pubescens* (TVNu-110-3A)	Recombinant inbreed (94)	77 RAPD, 3 morphological markers	12	669.8	21.7	Ubi et al. (2000)
V. unguiculata (IT2246-4) × *V. unguiculata* ssp. *dekindtiana* (TVN1963)	F$_2$ population (58)	79 genomic, 4 cDNA, 6 RAPD, 2 aphid, 1 seed coat texture markers	10	>800	22	Menacio-Hautea et al. (1993); Fatokun et al. (1997)
V. unguiculata (IT84S-2049) × *V. unguiculata* (524B)	F$_2$ population (94)	133 RAPD, 19 RFLP, 25 AFLP, 3 morphological, 1 biochemical markers	12	972	18	Menéndez et al. (1997)
V. unguiculata (IT84S-2049) × *V. unguiculata* (524B)	Recombinant inbreed (94)	133 RAPD, 36 RFLP, 267 AFLP, 3 morphological, 1 biochemical markers	11	2670	19.7	Ouédraogo et al. (2002a)

The linkage maps for *V. radiata* have been based on crosses between *V. radiata* var. *radiata* and *V. radiata* var. *sublobata*. In one linkage map the *V. radiata* var. *sublobata* accession came from Madagascar, in the other the accession came from Australia. In the resulting genome maps the order of markers was similar. However, the level of distortion was higher in the cross that involved the Australian accession and regions of distortion did not coincide with those produced using the Madagascar accession. This suggests that *V. radiata* var. *sublobata* has considerable intraspecific genetic diversity and that the Australian form of var. *sublobata* is more distantly related to cultivated *V. radiata* than the Madagascar form (Lambrides et al. 2000).

Of the genome maps reported that between *V. angularis* and *V. umbellata* had the highest level of segregation distortion (29.8%; Table 3). High levels of segregation distortion may not enable the trait(s) of interest to be found in segregating populations. When using distantly related species in crosses, it is necessary to generate very large segregating populations, since the likelihood of recombination drops and thus, the likelihood of getting the traits needed in the desired background is low. However, large populations may not be easy to obtain as weak or sterile F_2 plants can limit seed production (Kaga et al. 1996). Ways to overcome this problem in *Vigna* section *Angulares* may include using bridging species to facilitate gene transfer (Tomooka et al. 2002c).

The latest and most detailed genetic linkage map of cowpea spans a total of 2670 cM with an average of 6.43 cM between markers (Ouédraogo et al. 2002a). This cowpea linkage map revealed how important using a variety of molecular markers is to obtain a saturated linkage map. Ouédraogo et al. (2002a) added AFLP markers to the cowpea linkage map and this revealed a large segment of 580 cM on linkage group 1 that was undetected when only RFLP and RAPD markers were used to create the linkage map. Thus, further major improvement to the genetic linkage map for cowpea and other *Vigna* species may be expected when microsatellite markers are used.

4 Synteny

4.1 *Vigna* and Other Genera

There have been efforts to understand the comparative genome organization across *Vigna* and related cultigens (Menacio-Hautea et al. 1993; Boutin et al. 1995). Comparisons of genome maps of *V. radiata* with *V. unguiculata* and *Phaseolus vulgaris* have revealed conserved blocks of considerable size some containing loci for important traits. The comparison with *P. vulgaris* showed that average size of conserved blocks is about 36.6 cM with the longest being 103.5 cM (Table 4). Therefore, there is considerable scope for understanding genome organization in cultigens of genus *Vigna* by using probes from and comparison with better developed genome maps in other related species.

Table 4. Comparison of lengths of genome blocks conserved between mungbean and soybean, and common bean and soybean. (Adapted from Boutin et al. 1995)

Species compared	Average length of conserved block (cM)	Standard deviation	Length of the longest conserved block (cM)
Mungbean and soybean	12.2	9.4	37.8
Common bean and soybean	13.9	9.5	34.8

One of the best-developed genome maps among legumes is that of soybean. Comparison of *V. radiata* and *Glycine max* revealed a different type of genome organization than the comparison of *Glycine max* and *Phaseolus vulgaris*. Conserved linkage blocks are smaller and are highly scattered in the *V. radiata* comparison compared to *P. vulgaris* (Table 4). However, specific analysis of a genomic region influencing seed weight in soybean showed co-linearity of RFLP markers with mungbean (Maughan et al. 1996).

Comparative mapping of *V. radiata* and *Lablab purpureus* (hyacinth pea), both belonging to subtribe Phaseolinae, revealed that the order of markers is highly conserved and enabled suggestions of which linkage group belonged on the same chromosome in lablab and mungbean (Humphrey et al. 2002). Surprisingly, the results suggest that mungbean shares a higher level of genome organization with lablab than taxonomically more closely related species in the subgenus *Ceratotropis* (*V. angularis* and *V. umbellata*). However, while mungbean and lablab maintain the same marker order, they have accumulated a large number of deletions/duplications after divergence (Humphrey et al. 2002).

Despite the incompleteness of the genetic map data, comparisons between *Phaseolus vulgaris* and *V. radiata* and *Arabidopsis* have enabled a reconstruction of a proposed ancestral DNA segment in the present genome of soybean (Lee et al. 2001).

4.2 Within the Genus *Vigna*

Early comparison of cowpea and mungbean linkage maps revealed that 90% (48 out of 53) RFLP probes hybridized with both species. While marker order was often similar, distances between markers varied. Ten regions of the linkage maps of these two cultigens showed syntenic association (Menacio-Hautea et al. 1993). A specific study of the genetics of seed weight resulted in finding quantitative trait loci (QTLs) that accounted for 52.7 and 49.7% of the variation for this trait in cowpea and mungbean, respectively (Fatokun et al. 1992a). The genomic region with the greatest effect on seed weight spanned the same RFLP markers in the same order (Fatokun et al. 1992a)

4.3 Within *Vigna* Subgenus *Ceratotropis*

Comparison of two interspecific linkage maps (rice bean × azuki bean and azuki bean × *V. nakashimae*) revealed seven conserved linkage blocks (size range 7–115 cM; Kaga et al. 2000). Comparison of the rice bean × azuki bean linkage map with the mungbean linkage map of Menacio-Hautea et al. (1993) revealed 16 conserved segments without regions of inversion and translocation (size range 2–95 cM; Kaga et al. 2000). This study enabled orthologous linkage groups in the different maps to be proposed.

5 Gene Mapping

A list of the loci for major agronomic traits that have been associated with linkage groups in *Vigna* is provided (Table 5). Among these, the progress in cowpea in mapping for resistance to the parasitic plant *Striga* and in mungbean mapping resistance to bruchid beetles and powdery mildew provide the best examples of the state of gene mapping in *Vigna*. These examples are discussed here.

The parasitic plant *Striga gesnerioides* can result in 100% yield loss in cowpea. AFLP markers were used to finely map *Striga* resistance. Five races of *Striga* are known to affect cowpea. Of these five races, AFLP markers have been found linked to genes for resistance to races 1 and 3 (Ouédraogo et al. 2001, 2002b). The AFLP marker studies revealed that the genes (or alleles) for resistance are clustered on at least one linkage group (linkage group 1 of the genome map of Ouédraogo et al. 2002a; Fig. 1). The identification of molecular markers associated with clustered *Striga* resistance genes are now leading to plans to use molecular marker selection in cowpea breeding (B.B. Singh 2002, pers. comm.).

RFLP analysis was used to map a bruchid resistance gene in wild mungbean. The gene was a single major locus on linkage group 8 (subsequently revised to linkage group 9; Young et al. 1992; Fig. 2a, b). In a mapping population between azuki bean and rice bean the main QTL for bruchid resistance in rice bean was linked to one of the same RFLP probes (pR26) linked to bruchid resistance in mungbean (Kaga 1996). The nearest RFLP marker to the bruchid resistance gene in the mungbean map was 3.6 cM distant (Fig. 2a). Since this resistance gene, from *V. radiata* var. *sublobata* (TC1966), also has an inhibitory activity against bean bug (*Riptortus clavatus* Thunberg) and it was associated with novel cyclopeptide alkaloids, further efforts were made to map this gene (Kaga and Ishimoto 1998). The resulting genetic map enabled the resistant dominant gene to be located to within 0.2 cM of the nearest RFLP markers (Fig. 2c). This map distance may enable the gene to be cloned within a large genomic library for eventual introduction into susceptible mungbean lines or other crops.

Table 5. Molecular mapping of agronomically important traits in *Vigna*

Species	Molecular markers used	Trait of interest	Linkage Group (LG)	Reference
V. unguiculata	RFLP	Aphid resistance	LG.1 (Fatokun et al. 1992b)	Myers et al. (1996)
V. unguiculata	AFLP	*Striga* resistance	Resistance to race 1 and 3 on LG.1 and 6 (Ouédraogo et al. 2002a)	Ouédraogo et al. (2001, 2002a, b)
V. unguiculata	AFLP	Cowpea mosaic virus	LG.3 (Ouédraogo et al. 2002a)	Ouédraogo et al. (2002a)
V. unguiculata	AFLP	Cowpea severe mosaic virus	LG.3 (Ouédraogo et al. 2002a)	Ouédraogo et al. (2002a)
V. unguiculata	AFLP	Blackeye cowpea mosaic virus (BlCMV)	LG.8 (Ouédraogo et al. 2002a)	Ouédraogo et al. (2002a)
V. unguiculata	AFLP	Southern bean mosaic virus (SBMV)	LG.6 (Ouédraogo et al. 2002a)	Ouédraogo et al. (2002a)
V. unguiculata	AFLP	Fusarium wilt	LG.3 (Ouédraogo et al. 2002a)	Ouédraogo et al. (2002a)
V. unguiculata	AFLP	Root-knot nematode	LG.1 (Ouédraogo et al. 2002a)	Ouédraogo et al. (2002a)
V. radiata	RFLP	Bruchid resistance	LG.8 (Menancio-Hautea et al. 1992)	Young et al. (1992)
V. radiata	RFLP, RAPD	Bruchid resistance	LG.8 (Menancio-Hautea et al. 1992)	Kaga and Ishimoto (1998)
V. radiata	RFLP, AFLP	Powdery mildew resistance	QTL	Chaitieng et al. (2002)
V. radiata	RFLP	Powdery mildew resistance	QTL	Young et al. (1993)
V. unguiculata	RFLP	Seed weight	QTL	Fatokun et al. (1992a)
V. radiata	RFLP	Seed weight	QTL	Fatokun et al. (1992a)
V. unguiculata	RFLP, RAPD	Multiple quantitative traits	QTL	Menéndez et al. (1997)

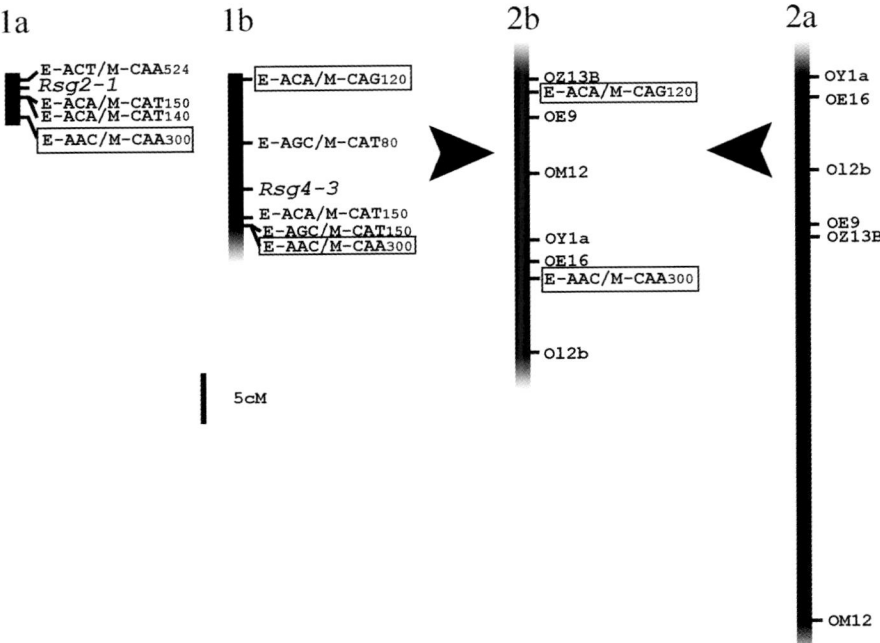

Fig. 1. The use of DNA markers (*boxes*) to finely map the location of genes (alleles) for resistance to *Striga* (based on Ouédraogo et al. 2001, 2002a, b). **1a** AFLP markers found linked to resistance (*Rsg2-1*) for *Striga* race 1 from Burkino Faso in cross Tvx3236 (sus) × IT82D-849 (res). **1b** AFLP markers found to be linked to resistance (*Rsg4-3*) for *Striga* race 3 from Niger in cross IT84S-2246-4(sus)×Tvu14676 (res). **2a** Location of RFLP markers on linkage group 1 of cowpea linkage map. **2b** Location of RFLP and AFLP markers on linkage group 1 of the cowpea linkage map

Powdery mildew resistance is a multi-genic trait. The first attempt to map resistance in a breeding line of mungbean identified three QTLs on three different linkage groups accounting for 58% of the total variation (Young et al. 1993). Using a different powdery mildew-resistant line to develop a mapping population, 96 RFLP probes failed to identify any QTL associated with resistance (Chaitieng et al. 2002). Subsequently, 100 AFLP primer pair combinations were tested and 4 out of more than 5000 AFLP bands were found to be associated with resistance. The main QTL associated with resistance was found on a new linkage group and accounted for 68% of the total variation.

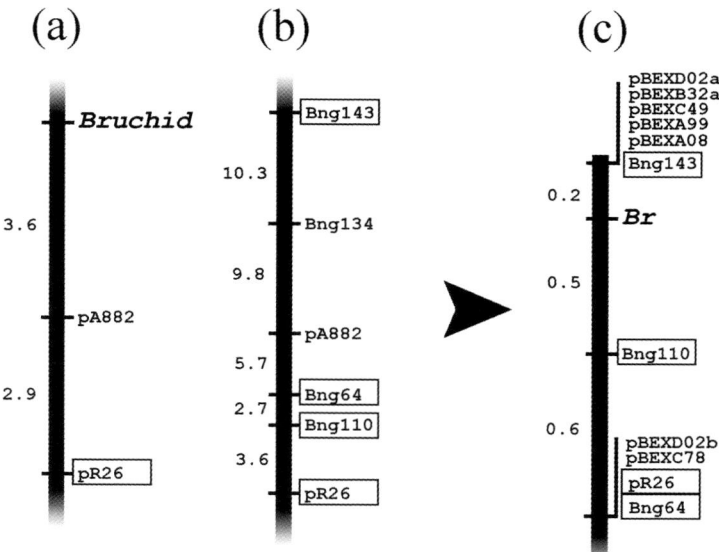

Fig. 2. Use of RFLP markers (*boxes*) to finely map the location of bruchid resistance on the mungbean linkage map. Distances in cM. **a** Initial location of bruchid resistance in the cross *V. radiata* (VC3890) × *V. radiata* var. *sublobata* (TC1966) (Young et al. 1992). **b** Revised linkage map of part of linkage group 9 with information from Bng probes added (Mungbean: U–Minnesota-9 at http://beangenes.cws.ndsu.nodak.edu/). **c** Fine mapping of the location of bruchid resistance in the cross *V. radiata* (Osaka-ryokuto) × *V. radiata* var. *sublobata* (TC1966) based on combination of previously used RFLP probes and newly developed probes from RAPD markers (Kaga and Ishimoto 1998). Distance differences between these figures reflect different parents in the crosses and mapping population size

6 Transformation Systems

Only in azuki bean (*V. angularis*) has genetic transformation been reported to result in improved breeding lines. Ishimoto et al. (1996) developed a bruchid-resistant azuki bean line that had an α-amylase inhibitor gene driven by a seed-specific promoter from common bean (*Phaseolus vulgaris*). This gene was introduced using *Agrobacterium*-mediated gene transfer. The transgenic azuki bean could completely block larvae development of three bruchid species that are the major storage pests of mungbean and cowpea. The method was refined by Yamada et al. (2001) and was found to be reproducible with high transformation efficiency. Using the same method, 15 independent transgenic lines with sGFPs65T (modified green fluorescent protein gene) and 15 independent lines with mt-sHSP (heat shock protein gene) from tomato have been produced (Kaga et al. 2003). Thus, the transformation system of azuki bean can be used for routine transformation.

Another well-established transformation system is in moth bean (*V. aconitifolia*) using protoplasts. Although it has been shown that transformation success is cultivar-dependent, the efficiency is the highest among *Vigna* species. Plant transformation systems have been reported for six *Vigna* cultigens (Table 6).

Since the success of *Agrobacterium*-mediated gene transfer, on which most methods rely, is largely influenced by a combination of culture condition, tissue type or genotype of cultivar, the direct gene transfer is an attractive alternative approach to overcoming these complicating factors. Using particle bombardment, Bhargava and Smigocki (1994) obtained transformants of mungbean, black gram and moth bean, but molecular evidence of gene integration was not shown. In planta electroporation-mediated gene transfer has been demonstrated in cowpea and several other grain legumes (Chowrira et al. 1996) and may be a more practical and rapid method to produce plants in *Vigna* species. Plans have been made to transform cowpea using gene constructs encoding *Bt* toxin, α-amylase inhibitor and cysteine proteinase inhibitor for resistance to *Maruca* pod borers and cowpea bruchids. A separate project has been proposed to transfer gene constructs for resistance to cowpea aphid-borne mosaic virus (de Vries and Toenniessen 2001).

Further information on transformation and regeneration systems in *Vigna* cultigens can be found in Nagl et al. (1997).

Table 6. Status of plant transformation in the main *Vigna* crops

Species	Transformed plants produced	Method for gene transfer
V. unguiculata	*hpt* (hygromycin phosphotransferase) gene[a]	*Agrobacterium*
	uidA (β-glucuronidase, GUS) gene[b]	In planta electroporation
	uidA and *nptII* gene[c]	*Agrobacterium*
V. radiata	*uidA* and *nptII* genes[d]	*Agrobacterium*
	uidA and *nptII* (neomycin phosphotransferase) genes[e]	Particle bombardment
	uidA, *nptII* and *hpt* genes[f]	*Agrobacterium*
V. angularis	α AI (α-amylase inhibitor) and *nptII* genes[g]	*Agrobacterium*
	sGFP(S65T) (modified green fluorescent protein), *uidA* and *nptII* genes[h]	*Agrobacterium*
V. aconitifolia	*nptII* gene[i]	Electroporation
	nptII gene[j]	*Agrobacterium*
	uidA and *nptII* genes[e]	Particle bombardment
V. mungo	*uidA* and *nptII* genes[e]	Particle bombardment

[a] Muthukumar et al. (1996); [b] Ignacimuthu (2000); [c] Chowrira et al. (1996); [d] Pal et al. (1991); [e]Bhargava and Smigocki (1994); [f]Jaiwal et al. (2001); [g]Ishimoto et al. (1996); [h]Yamada et al. (2001); [i]Köhler et al. (1987); [j]Eapen et al. (1987)

7 Conclusions

The genome size of cowpea and mungbean are small, ranging from 470–613 Mb, about half that of soybean (Murray et al. 1979; Arumuganathan and Earle 1991). Therefore, *Vigna* species are good candidates for more in-depth genome analysis in the future. The results of such research would likely have an impact beyond *Vigna* as the discussion above on comparative genome mapping suggests.

Published information on the actual use of molecular markers in *Vigna* improvement is lacking. It is only in the last year that a *Vigna* linkage map has resolved the 11 linkage groups that correspond to the haploid chromosome number of diploid *Vigna*. The latest cowpea linkage map has associated many agronomically important traits to molecular markers, thus, we may also be "cautiously optimistic" (Young 1999) that molecular markers will play a role in the future improvement of the *Vigna* cultigens.

This is a field where information rapidly becomes out of date. Recent information related to this topic may be found at:

- http://beangenes.cws.ndsu.nodak.edu
- http://www.ncbi.nlm.nih.gov/Taxonomy/Browser/wwwtax.cgi?id=3913

Acknowledgements. This review was completed in January 2003 and all authors contributed equally to this chapter. The authors acknowledge support from the Global Environment Research Fund (TY2003-F7) of the Japanese Ministry of the Environment to author AK.

References

Amadou HI, Bebeli PJ, Kaltsikes PJ (2001) Genetic diversity in Bambara groundnut (*Vigna subterranea* L.) germplasm revealed by RAPD markers. Genome 44:995–999

Arumuganathan K, Earle ED (1991) Nuclear DNA content of some important plant species. Plant Mol Biol Rep 9:208–218

Baudoin JP, Maréchal R (1988) Taxonomy and evolution in the genus *Vigna*. In: Shanmugasundaram S, McLean BT (eds) Mungbean. Proc 2nd Int Symp AVRDC, Shanhua, Tainan, Taiwan, pp 2–12

Bhargava SC, Smigocki AC (1994) Transformation of tropical grain legumes using particle bombardment. Curr Sci 66:439–442

Boutin SR, Young ND, Olson TC, Yu ZH, Shoemaker RC, Vallejos CE (1995) Genome conservation among three legume genera detected with DNA markers. Genome 38:928–937

Chaitieng B, Kaga A, Han OK, Wang XW, Wongkaew S, Laosuwan P, Tomooka N, Vaughan DA (2002) Mapping a new source of resistance to powdery mildew (*Erisiphe polygoni* DC.) in mungbean [*Vigna radiata* (L.) Wilczek]. Plant Breed 121:521–525

Chowrira GM, Akella V, Fuerst PE, Lurquin PF (1996) Transgenic grain legumes obtained by in planta electroporation-mediated gene transfer. Mol Biotech 5:85–96

Coulibaly S, Pasquet RS, Papa R, Gepts P (2002) AFLP analysis of the phenetic organization and genetic diversity of *Vigna unguiculata* L. Walp. reveals extensive gene flow between wild and domesticated types. Theor Appl Genet 104:358–366

DeVries J, Toenniessen G (2001) Securing the harvest: Biotechnology and seed systems for African crops. CABI Publishing, Wallingford, 208 pp

Doi K, Kaga A, Tomooka N, Vaughan DA (2002) Molecular phylogeny of genus *Vigna* subgenus *Ceratotropis* based on rDNA ITS and *atpB-rbcL* intergenic spacer region of cpDNA sequences. Genetica 114:129–145

Eapen S, Kohler F, Gerdemann M, Schieder O (1987) Cultivar dependence of transformation rates in moth bean after co-cultivation of protoplasts with *Agrobacterium tumefaciens*. Theor Appl Genet 75:207–210

Fatokun CA, Menacio-Hautea DI, Danesh D, Young ND (1992a) Evidence of orthologous seed weight genes in cowpea and mungbean based on RFLP mapping. Genetics 132:841–846

Fatokun CA, Danesh D, Menancio-Hautea DI, Young ND (1992b) A linkage map for cowpea [*Vigna unguiculata* (L.)Walp.] based on DNA markers (2N=22). In: O'Brien JS (ed) Genetic maps 1992. A compilation of linkage and restriction maps of genetically studied organisms. Cold Spring Harbor, New York, pp 6.257–6.258

Fatokun CA, Danesh D, Young ND, Stewart EL (1993) Molecular taxonomic relationships in the genus *Vigna* based on RFLP analysis. Theor Appl Genet 86:97–104

Fatokun CA, Young ND, Myers GO (1997) Molecular markers and genome mapping in cowpea. In: Singh BB, Mohan Raj DR, Dashiell KE, Jackai LEN (eds) Advances in cowpea research. International Institute of tropical Agriculture (IITA) and Japan International Research Center for Agricultural Sciences (JIRCAS). IITA, Ibadan, Nigeria

Humphry ME, Konduri V, Lambrides CJ, Magner T, McIntyre CL, Aiten EAB, Liu CJ (2002) Development of a mungbean (*Vigna radiata*) RFLP linkage map and its comparison with lablab (*Lablab purpureus*) reveals a high level of colinearity between the two genomes. Theor Appl Genet 105:160–166

Ignacimuthu S (2000) *Agrobacterium* mediated transformation of *Vigna sesquipedalis* Koern (Asparagus pea). Indian J Exp Biol 38:493–498

Isemura T, Ishii T, Saito H, Noda C, Misoo S, Kamijima O (2002) Genetic diversity in azuki bean landraces as revealed by RAPD analysis. Breed Res 4:125–135 (in Japanese with English summary)

Ishimoto M, Sato T, Chrispeels MJ, Kitamura K (1996) Bruchid resistance of transgenic azuki bean expressing seed α-amylase inhibitor of common bean. Entomol Exp Appl 79:309–315

Jaaska V, Jaaska V (1988) Isozyme variation in the genera *Phaseolus* and *Vigna* (*Fabaceae*) in relation to their systematics: aspartate aminotransferase and superoxide dismutase. Plant Syst Evol 159:145–159

Jaaska V, Jaaska V (1990) Isozyme variation in Asian beans. Bot Acta 103:281–290

Jaiwal PK, Kumari R, Ignacimuthu S, Potrykus I, Sautter C (2001) *Agrobacterium tumefaciens*-mediated genetic transformation of mungbean (*Vigna radiata* L. Wilczek) – a recalcitrant grain legume. Plant Sci 161:239–247

Kaga A (1996) Construction and application of linkage maps for azuki bean (*Vigna angularis*). PhD Thesis, Kobe University, 210 pp

Kaga A, Ishimoto M (1998) Genetic localization of a bruchid resistance gene and its relationship to insecticidal cyclopeptide alkaloids, the vignatic acids, in mungbean (*Vigna radiata* L. Wilcek). Mol Gen Genet 258:378–384

Kaga A, Ohnishi M, Ishii T, Kamijima O (1996) A genetic linkage map of azuki bean constructed with molecular and morphological markers using an interspecific population (*Vigna angularis* × *V. nakashimae*). Theor Appl Genet 93:658–663

Kaga A, Ishii T, Tsukimoto K, Tokoro E, Kamijima O (2000) Comparative molecular mapping in *Ceratotropis* species using an interspecific cross between azuki bean (*Vigna angularis*) and rice bean (*V. umbellata*). Theor Appl Genet 100:207–213

Kaga A, Han OK, Wang XX, Egawa Y, Tomooka N, Vaughan DA (2003) *Vigna angularis* as a model for legume research. In: Jayasuriya AHM, Vaughan DA (eds) Conservation and use of wild relatives of crops. Proceedings of the joint Department of Agriculture, Sri Lanka and National Institute of Agrobiological Sciences, Japan workshop. Department of Agriculture, Peradeniya, Sri Lanka

Köhler F, Golz C, Eapen S, Kohn H, Schieder O (1987) Stable transformation of moth bean *Vigna aconitifolia* via direct gene transfer. Plant Cell Rep 6:313–317

Kumar SV, Tan SG, Quah SC, Yusoff K (2002) Isolation of microsatellite markers in mungbean, *Vigna radiata*. Mol Ecol Notes 2:96–98

Lambrides CJ, Lawn RJ, Godwin ID, Manners J, Imrie BC (2000) Two genetic linkage maps of mungbean using RFLP and RAPD markers. Aust J Agric Res 51:415–425

Lee JM, Grant D, Vallejos CE, Shoemaker RC (2001) Genome organization in dicots. II. *Arabidopsis* as a 'bridging species' to resolve genome evolution events among legumes. Theor Appl Genet 103:765–773

Li C-D, Fatokun CA, Ubi B, Singh BB, Scoles GJ (2001) Determining genetic similarities and relationships among cowpea breeding lines and cultivars by microsatellite markers. Crop Sci 41:189–197

Lumpkin TA, McClary DC (1994) Azuki bean: botany, production and uses. CAB International, Wallingford, UK, 268 pp

Maréchal R, Mascherpa JM, Stainier F (1978) Etude taxonomique d'un groupe complexe d'espèces des genres *Phaseolus* et *Vigna* (Papilionaceae) sur la base de données morphologiques et polliniques, traitées par l'analyse informatique. Boissiera 28:1–273

Maughan PJ, Saghai Maroof MA, Buss GR (1996) Molecular-marker analysis of seed weight: genome locations, gene action, and evidence for orthologous evolution among three legume species. Theor Appl Genet 93:574–579

Menacio-Hautea D, Kumar L, Danesh D, Young ND (1992) A genome map for mungbean [*Vigna radiata* (L.) Wilczek] based on DNA genetic markers (2n=2x=22). In: O'Brien JS (ed) Genetic maps 1992. A compilation of linkage and restriction maps of genetically studied organisms. Cold Spring Harbor, New York, pp 6.259–6.261

Menacio-Hautea D, Fatokun CA, Kumar L, Danesh D, Young ND (1993) Comparative genome analysis of mungbean (*Vigna radiata* L. Wilczek) and cowpea (*V. unguiculata* L. Walpers) using RFLP mapping data. Theor Appl Genet 86:797–810

Menéndez CM, Hall AE, Gepts P (1997) A genetic linkage map of cowpea (*Vigna unguiculata*) developed from a cross between two inbred, domesticated lines. Theor Appl Genet 95:1210–1217

Mignouna HD, Quat NQ, Ikea J, Thottapilly G (1998) Genetic diversity in cowpea as revealed by random amplified polymorphic DNA. J Genet Breed 53:151–159

Mimura M, Yasuda K, Yamaguchi H (2000) RAPD variation in wild, weedy and cultivated azuki beans in Asia. Genet Res Crop Evol 47:603–610

Murray MG, Palmer JD, Cuellar RE, Thompson WF (1979) Deoxyribonucleic acid sequence organization in the mungbean genome. Biochemistry 18:5259–5264

Muthukumar B, Mariamma M, Veluthambi K, Gnanam A (1996) Genetic transformation of cotyledon explants of cowpea (*Vigna unguiculata* L. Walp.) using *Agrobacterium tumefaciens*. Plant Cell Rep 15:980–985

Myers GO, CA Fatokun, Young ND (1996) RFLP mapping of an aphid resistance gene in cowpea (*Vigna unguiculata* L. Walp.). Euphytica 91:181–187

Nagl W, Ignacimuthu S, Becker J (1997) Genetic engineering and regeneration of *Phaseolus* and *Vigna*. State of the art and new attempts. J Plant Physiol 150:625–644

Ng NQ (1995) Cowpea. In: Smartt J, Simmonds NW (eds) Evolution of crop plants. Longman, London, pp 326–332

Ng NQ, Monti LM (1990) Cowpea genetic resources. Int Inst Trop Agric, Ibadan, Nigeria, 200 pp

Ouédraogo JT, Maheshwari V, Berner DK, St.-Pierre CA, Belzile F, Timko MP (2001) Identification of AFLP markers linked to resistance of cowpea (*Vigna unguiculata* L.) to parasitism by *Striga gesnerioides*. Theor Appl Genet 102:1029–1036

Ouédraogo JT, Gowda BS, Jean M, Close TJ, Ehlers JD, Hall AE, Gillaspie, Roberts PA, Ismail AM, Bruening G, Gepts P, Timko MP, Belzile FJ (2002a) An improved genetic linkage maps for cowpea (*Vigna unguiculata* L.) combining AFLP, RFLP, RAPD, biochemical markers and resistance traits. Genome 45:175–188

Ouédraogo JT, Tignegre JB, Timko MP, Belzile FJ (2002b) AFLP markers linked to resistance against *Striga gesnerioides* race 1 in cowpea (*Vigna unguiculata*). Genome 45:787–793

Pal M, Ghosh U, Chandra M, Pal A, Biswas BB (1991) Transformation and regeneration of mungbean (*Vigna radiata*). Indian J Biochem Biophys 28:449–455

Pasquet RS (1996) Wild cowpea (*Vigna unguiculata*) evolution. In: Pickersgill B, Lock JM (eds) Advances in legume systematics 8: legumes of economic importance. Royal Botanical Gardens, Kew, pp 95–100

Pasquet RS (1999) Genetic relationships among subspecies of *Vigna unguiculata* (L.) Walp. based on allozyme variation. Theor Appl Genet 98:104–119

Poehlman JM (1991) The mungbean. IBH Publ, New Delhi, 375 pp

Saravanakumar P, Kaga A, Tomooka N, Vaughan DA (2004) AFLP and RAPD analyses of intra- and interspecific variation in some *Vigna* subgenus *Ceratotropis* (Leguminosae) species. Aust J Bot (in press)

Singh BB, Chambliss OL, Sharma B (1997) Recent advances in cowpea breeding. In: Singh BB, Mohan Raj DR, Dashiell KE, Jackai LEN (eds) Advances in cowpea research. JIRCAS and IITA, Ibadan, Nigeria, pp 30–49

Sinha SSN, Roy H (1979) Cytological studies in the genus *Phaseolus*. I. Mitotic analysis in fourteen species. Cytologia 44:191–199

Sonnante G, Spinosa A, Marangi A, Pignone D (1997) Isozyme and RAPD analysis of the genetic diversity within and between *Vigna luteola* and *V. marina*. Ann Bot 80:741–746

Tomooka N, Lairungreang C, Nakeeraks P, Egawa Y, Thavarasook C (1992) Center of genetic diversity and dissemination pathways in mung bean deduced from seed protein electrophoresis. Theor Appl Genet 83:289–293

Tomooka N, Maxted N, Thavarasook C, Jayasuriya AHM (2002a) Two new species, sectional designations and new combinations in *Vigna* subgenus *Ceratotropis* (Piper) Verdc. (*Leguminosae, Phaseoleae*). Kew Bull 57:613–624

Tomooka N, Yoon MS, Doi K, Kaga A, Vaughan DA (2002b) AFLP analysis of diploid species in the genus *Vigna* subgenus *Ceratotropis*. Genet Res Crop Evol 49:521–530

Tomooka N, Vaughan DA, Moss H, Maxted N (2002c) The Asian *Vigna*: the genus *Vigna* subgenus *Ceratotropis* genetic resources. Kluwer, Dordrecht, 270 pp

Tomooka N, Vaughan DA, Kaga A (2004) Mungbean. In: Singh RJ, Jauhar PP (eds) Genetic resources, chromosome engineering and crop improvement. II. Grain legumes. CRC Press, Boca Raton (in press)

Ubi BE, Mignouna H, Thottappilly G (2000) Construction of a genetic linkage map and QTL analysis using a recombinant inbred population derived from an intersubspecific cross of cowpea (*Vigna unguiculata* (L.) Walp.). Breed Sci 50:161–172

Vaillancourt RE, Weeden NF (1993) Chloroplast DNA phylogeny of Old World *Vigna* (Leguminosae). Syst Bot 18:642–651

Wang XW, Kaga A, Tomooka N, Vaughan DA (2004) The development of SSR markers by a new method in plants and their application to gene flow studies in azuki bean [*Vigna angularis* (Willd.) Ohwi and Ohashi] Theor Appl Genet (on line)

Weeden NF, Wolko B, Vaillancourt R (1996) Contrasting patterns of partitioning genetic diversity in cultivated legumes. In: Pickergill B, Lock JM (eds) Advances in legume systematics 8 legumes of economic importance. Royal Botanic Gardens, Kew, pp 1–9

Xu RQ, Tomooka N, Vaughan DA (2000a) AFLP markers for characterizing the azuki bean complex. Crop Sci 40:808–815

Xu RQ, Tomooka N, Vaughan DA, Doi K (2000b) The *Vigna angularis* complex: genetic variation and relationships revealed by RAPD analysis, and their implications for in situ conservation and domestication. Genet Res Crop Evol 47:123–134

Yamada T, Teraishi M, Hattori K, Ishimoto M (2001) Transformation of azuki bean by *Agrobacterium tumefaciens*. Plant Cell Tissue Org Cult 64:47–54

Yee E, Kidwell KK, Sills GR, Lumpkin TA (1999) Diversity among selected *Vigna angularis* (Azuki) accessions on the basis of RAPD and AFLP markers. Crop Sci 39:268–275

Young ND (1999) A cautiously optimistic vision for molecular marker breeding. Mol Breed 5:505–510

Young ND, Kumar L, Menacio-Hautea D, Danesh D, Talekar NS, Shanmugasundarum S, Kim DH (1992) RFLP mapping of a major bruchid resistance gene in mungbean (*Vigna radiata*, L. Wilczek). Theor Appl Genet 84:839–844

Young ND, Danesh D, Menancio-Hautea D, Kumar L (1993) Mapping oligogenic resistance to powdery mildew in mungbean with RFLP's. Theor Appl Genet 87:243–249

Yu K, Park SJ, Poysa V (1999) Abundance and variation of microsatellite DNA sequences in bean (*Phaseolus* and *Vigna*). Genome 42:27–34

Zong XX, Kaga A, Tomooka N, Wang XW, Han OK, Vaughan DA (2003) The genetic diversity of the *Vigna angularis* complex in Asia. Genome 46:647–658

II.7 Molecular Markers for Genetics and Breeding: Development and Use in Pepper (*Capsicum* spp.)

V. LEFEBVRE[1]

1 Introduction

The genus *Capsicum* is a member of the Solanaceae family that includes tomato, potato, eggplant, tobacco, petunia and others. Pepper is grown worldwide. It is the world's second most important Solanaceae vegetable after tomato. Because of its very large variability and its great geographical distribution, pepper has multiple uses. It can be consumed fresh, cooked or dried. It is also used as an alimentary colorant or by the chemical industries for the composition of drugs. Finally, it is being used more and more as an ornamental plant (Palloix et al. 2003). In 2002, world production was more than 22 million tons (FAO 2002; http://www.fao.org/) on about 1.6 million ha, with China, Mexico, and Turkey as the main growers. Spain, Italy and now the Netherlands are the main growers in Europe. The seed industries are very active in Europe and particularly in France, with an increasing number of inscriptions of F_1 hybrids. The breeding objectives are essentially yield improvement, environment adaptation, fruit traits, and disease resistance. Because of its widespread geographical distribution all over the world from the intertropical belt to northern Europe latitudes, pepper is vulnerable to a great number of pathogens. Complexes of viruses transmitted by aphids and the Oomycete *Phytophthora capsici* cause major damage all over the world (Yoon et al. 1991). The great intra- and interspecific variability is of great help for breeding programs. In Europe, the germplasm numbers about 12,000 accessions, located mainly in seven European research centres (Daunay et al. 2001). In addition, numerous tools of cellular biology are available, such as the immature embryo rescue, in order to assist interspecific crosses, and the production of doubled haploid lines, to accelerate the fixation of improved material. The stable transformation to produce transgenic plants is really difficult in pepper. Some research groups published positive results (Lim et al. 1999; Shin et al. 2002; Li et al. 2003). Finally, the advent of DNA-based genetic markers offers useful tools for genome-wide studies and for assisting selection.

This chapter gives an overview of the development and use of molecular markers in plant breeding and crop improvement related to pepper. It is sub-

---

[1] Institut National de la Recherche Agronomique, Unité de Génétique et Amélioration des Fruits et Légumes, Domaine Saint Maurice, BP94, 84143 Montfavet cedex, France

divided into seven subsections dealing with a survey of molecular markers developed on pepper, diversity examination, variety identification, genetic maps, trait locus mapping, available genomic resources, and I conclude with the exploration of these different resources for marker-assisted selection in pepper and its perspectives.

2 Molecular Markers

Different kinds of molecular genetic markers were developed on *Capsicum*. Isozymes revealed low polymorphism. A very few isozyme polymorph markers were detected inside *C. annuum* (Conicella et al. 1990). In general, restriction fragment length polymorphism (RFLP) markers enable the detection of more polymorphism among *C. annuum* accessions than isozyme markers (Lefebvre et al. 1993). A few pepper libraries of cDNA clones and selected single-copy genomic clones (prefixed PC and PG clones) were used as a source of probes for RFLP markers. However, tomato-derived probes (designated CD, CT, TG) were often preferred to identify orthologous genes and to gain a better understanding of genome organisation within the Solanaceae. They particularly allowed the comparative mapping of genomes of sexually incompatible species. Previously, most molecular maps were constructed by using RFLPs. However, RFLPs require a great deal of time and effort, limiting their use in plant breeding. The random amplified polymorphic DNA (RAPD) and amplified fragment length polymorphism (AFLP) markers have largely been explored in pepper by different research groups. Presently, AFLPs are becoming more and more important. A few simple sequence repeat (SSR) markers, characterised from *Capsicum* sequences available in databases, have been published (Huang et al. 2001). Finally, several locus-specific PCR-based markers or STS (sequence tagged sites) have been described for use in molecular-assisted selection (see below). In the next decade, many single nucleotide polymorphisms (SNPs) are expected (Cheng et al. 2002), and should be mapped. Studying sequence variation of genes across a range of cultivars, primitive cultivated and wild species, will provide unlimited molecular genetic markers with diverse genetic backgrounds which will be useful for breeding.

3 Genetic Diversity

The genus *Capsicum* includes 22 wild species and five domesticated species. The five domesticated species are *C. annuum* L., *C. frutescens* L., *C. chinense* Jacq., *C. baccatum* L., and *C. pubescens* R. & P. (Eshbaugh 1980; IBPGR 1983). *C. annuum* is the most dispersed and cultivated species through the world.

The *Capsicum* genus originates from the New World. It spread rapidly throughout the world after the voyage of C. Columbus who first introduced it to Europe at the end of fifteenth century. It was further dispersed, first to Mediterranean countries, second, to Africa, India, China, and at last, returned to oriental Europe through Asia.

The assessment of genetic variability is of interest for cultivar protection and also for practical applications such as the establishment of core collections for the conservation of genetic resources and the choice of parental lines to be used for crossing in order to broaden the genetic basis of the cultivars. Molecular markers are very useful in complementing the morphological characterisation to describe the genetic diversity available in the centres of diversification, and to allow the detection of additional sources of genetic variation useful for pepper improvement as has been recently experienced on a Nepalese collection of *C. annuum* landraces (Baral and Bosland 2002). The genetic diversity among and within the *Capsicum* species has been investigated in several studies using DNA markers (Loaiza-Figueroa et al. 1989; Livneh et al. 1990; Prince et al. 1992, 1995; Lefebvre et al. 1993, 2001; Rodriguez et al. 1999). A molecular classification of the *Capsicum* genus was proposed based on a phylogenetic study using DNA sequences from one non-coding chloroplast region and one non-coding nuclear region. The phenetic tree based on molecular data was consistent with the previous morphological and cytological botany classification. This study demonstrated that species of *Capsicum* (excluding *C. ciliatum*) form a monophyletic group, and supplied sufficient data to be useful in species delimitation (Walsh and Hoot 2001). Multidimensional scaling analysis of the genetic distances among 134 accessions analysed with 110 RAPD markers resulted in clustering corresponding to a previous species assignment, except for six accessions. Diagnostic RAPDs were identified which discriminate the *Capsicum* species (Rodriguez et al. 1999). A larger molecular diversity was demonstrated among and within the wild species compared to those measured within the cultivated species, although the cultivated species display a larger phenotypic diversity for numerous traits. It was easy to distinguish exotic lines from large-fruited improved pepper lines that are based on a narrow genetic basis (Lefebvre et al. 1993, 2001; Paran et al. 1998).

4 Variety Identification and Hybrid Purity

Most of the present-day commercial cultivars are F_1 hybrids of *C. annuum* L. species. The species *C. annuum* is preferentially autogamous, which is supposed to largely result from the domestication process. Geno-cytoplasmic or genic male sterility has been used for F_1 seed production in a few cultivars, but the unstable behaviour of the male sterility as well as its unfavourable effect on fruit quality did not permit hand pollination being avoided in most of the cases. Finally, the hermaphrodite type of the flowers requires great attention to castration before anthesis for F_1 production.

Molecular markers are now largely used by seed companies for hybrid purity control and cultivar identification in pepper. Paran et al. (1998) and Ballester and de Vicente (1998) evaluated the effectiveness of PCR-based markers to detect polymorphism among closely related sweet pepper cultivars. Other authors developed cultivar-specific markers suitable for large-scale quality-control assessment (Livneh et al. 1992; Ilbi 2003). Despite frequent dominant inheritance, PCR-based markers are efficient for routine assessment of F_1 hybrid seed purity.

DNA profiling may be a useful tool for cultivar identification as well as for variety protection. The cultivated *C. annuum* species were classified according to the fruit shape (Pochard 1966) as blocky, triangular, long or half-long types. RAPD and AFLP markers allow us to easily split pepper large-fruited lines into the blocky types and the long types; the half-long types being located at an intermediate position resulting from a cross between the two former types. Within the blocky types, the US, the Dutch, and the Italian types are currently distinguished. They correspond to different markets and resulted from distinct genetic pools, as was confirmed at the molecular level (Lefebvre et al. 2001).

5 Genetic Maps

The *Capsicum* genus is diploid, with rare exceptions mentioned, with a total number of 24 chromosomes ($2n=2x=24$). It shares the same basic chromosome number as the other members of the family ($x=12$). The haploid DNA content of *Capsicum annuum* was estimated to be 2.76 pg and its genome size to be 2702–3420 Mbp per haploid genome (Galbraith et al. 1983; Arumuganathan and Earle 1991). Its haploid genome is three- to four-fold larger than that of tomato, potato and eggplant, and about 20 times larger than that of *Arabidopsis thaliana*.

The basic *Capsicum* karyotype consists of one pair of acrocentric chromosomes (the XI) and 11 pairs of metacentric or submetacentric chromosomes that are very difficult to distinguish from each other. However, variations between the species exist with two pairs of acrocentric chromosomes (the XI and the XII) in most of the domesticated *C. annuum*, and the number of chromosomes bearing satellites (nucleolar organiser) that vary from one in *C. annuum* and *C. pubescens* to two in the other species (Pickersgill 1991). The domesticated *C. annuum* karyotype is thus characterised by three easily cytogenetically recognisable chromosomes: chromosome I is the largest, chromosomes XI and XII are both acrocentric, and chromosome XII brings a satellite with the nucleolar organiser (Pochard 1970). Chromosome pairing at meiosis of interspecific hybrids also indicates that reciprocal translocations occurred during speciation involving at least five chromosome pairs (Greenleaf 1986; Pickersgill 1991).

The first studies on chromosome mapping in pepper were performed thanks to a collection of 12 primary trisomics obtained by Pochard (1970). The 12 chromosomes were named by French colour names (*violet, indigo, bleu, vert, jaune, orange, rouge, pourpre, noir, brun, bistre, gris*). The primary trisomics permitted the identification, by cytogenetic observations, of the three recognisable chromosomes (the I corresponds to the *Violet*, the XI to the *Jaune*, and the XII to the *Pourpre*) and the localisation of ten monogenic traits (Pochard 1977; Pochard and Dumas de Vaulx 1982). Tanksley published isozyme markers in interspecific crosses in 1984, and developed a linkage map of pepper based on nine isozymes arranged in four linkage groups. However, he revealed a very low level of polymorphism even in an interspecific cross. Advances in marker technologies allowed the construction of more complete genetic maps. The two first molecular linkage maps of pepper were constructed with tomato-derived RFLP markers from interspecific crosses *C. annuum* × *C. chinense* in order to maximize the polymorphism level (Tanksley et al. 1988; Prince et al. 1993). These studies demonstrated that the gene repertoire was conserved in pepper and tomato, but the gene order was different. Several other molecular linkage maps have been constructed for pepper, based mainly on RFLP, RAPD, and AFLP markers (Table 1). Most interspecific maps were constructed using *C. annuum* × *C. chinense* crosses, probably because these crosses present relatively good fertility and a high level of polymorphism. A linkage map from a cross of *C. annuum* × *C. frutescens* was also reported by Rao et al. (2003). A few linkage maps were constructed from intraspecific crosses within *C. annuum* by crossing bell-pepper lines with small-fruited exotic lines; Lefebvre et al. (1993) demonstrated that polymorphism could reach up to 40% in these kinds of crosses whatever kind of molecular marker was observed. Intraspecific crosses are advantageous for mapping because they help to prevent the low fertility, low recombination, distorted segregations and chromosomal aberrations often encountered with interspecific crosses. Moreover, mapping loci determining agronomic traits in intraspecific crosses facilitate the exploitation of results for marker-assisted selection since breeding schemes in pepper abundantly use the intraspecific variability, particularly for polygenic traits (Lefebvre et al. 1995). Three reference intraspecific genetic linkage maps were constructed. RFLP-based anchoring of pepper genetic maps demonstrated that comigrating RAPD (Lefebvre et al. 1997) and AFLP (Lefebvre et al. 2002) markers correspond to homologous loci across crosses. The alignment of the three individual maps permitted the arrangement of 12 consensus major linkage groups that contain 100 known function gene markers and on which nine phenotypic loci were positioned (*L, pvr2, Pvr4, Tsw, Me3, Bs3, C, up, y*; Lefebvre et al. 1995, 1997, 2002). Finally, another *C. annuum* intraspecific map was constructed by Ben Chaim et al. (2001b).

Table 1. List of published molecular genetic linkage maps of the pepper genome

Reference	Mapping population	Number of markers mapped (predominant type of markers)	Number of linkage groups (LG)	Total size (H: Haldane, K: Kosambi) (cM)
Tanksley (1984)	80–153 Interspecific BC [(C. annuum cv. NM6-4 × C. chinense CA4) × NM6-4] and 85–295 F₂ progenies (C. annuum cv. NM6-4 × C. chinense CA4)	9 Isozymes	4	~52
Tanksley et al. (1988)	46 Interspecific BC progeny [(C. annuum Doux des Landes CA50 × C. chinense CA4) × CA50]	80 Markers (RFLP + isozymes)	14	634
Prince et al. (1993)	46 Interspecific F₂ progeny (C. annuum CA133 × C. chinense CA4)	192 Markers (RFLP + isozymes)	19	720
Lefebvre et al. (1995)	A consensus map from 3 intraspecific C. annuum populations: DH200=PY (94 doubled haploid progeny Perennial × Yolo Wonder) + DH591 (44 doubled haploid progeny Vat × CM334) + DH702 (31 doubled haploid progeny Yolo × CM334)	85 Markers (RFLP + RAPD)	14	823 (K)
Lefebvre et al. (1997)	2 Intraspecific C. annuum populations: DH200=PY (94 doubled haploid progeny Perennial × Yolo Wonder) + HV (98 doubled haploid progeny H3 × Vania)	83–189 Markers (RFLP + RAPD)	15–16	417–1515 (K)
Kim et al. (1997)	86 Interspecific F₂ progeny (C. annuum 2002 × C. chinense 1679)	174 Markers (AFLP)	12	n.a.
Livingstone et al. (1999)	75 Interspecific F₂ progeny (C. annuum NuMex Rnaky × C. chinense PI159234)	677 Markers (RFLP + isozymes + RAPD + AFLP)	13	1246 (K)
Kang et al. (2001)	107 Interspecific F₂ progeny (C. annuum TF68 × C. chinense Habanero)	580 Markers (RFLP + AFLP)	16	1320 (H)
Ben Chaim et al. (2001b)	180 Intraspecific C. annuum F₂ progeny (Maor × Perennial)	177 Markers (RFLP + RAPD + AFLP + morphological)	12	1740 (K)

Table 1. (Continue)

Reference	Mapping population	Number of markers mapped (predominant type of markers)	Number of linkage groups (LG)	Total size (H: Haldane, K: Kosambi) (cM)
Lefebvre et al. (2002)	3 Intraspecific *C. annuum* populations: HV (101 doubled haploid progeny H3 × Vania) + PY (114 doubled haploid progeny Perennial × Yolo Wonder) + YC (151 F$_2$ progeny Yolo Wonder × CM334)	208–630 Markers (RFLP + RAPD + AFLP + morphological)	16–20 per map arranged in 12 consensus LG	685–1668 (H)
Rao et al. (2003)	248 Interspecific BC$_2$ progeny [((*C. annuum* cv. Maor × *C. frutescens* BG2816) × BG2816) × BG2816]	92 RFLP markers	12	1100 (K)
Paran et al. (2004)	75 Interspecific F$_2$ progeny (*C. annuum* NuMex Rnaky × *C. chinense* PI159234) 83 Interspecific BC$_1$ progeny [(*C. annuum* 100/63 × *C. chinense* PI152225) × 100/63] 180 Intraspecific F$_2$ progeny (*C. annuum* Maor × *C. annuum* Perennial) 101 Intraspecific doubled haploid progeny (H3 × Vania) = HV 114 Intraspecific doubled haploid progeny (Perennial × Yolo Wonder) = PY 151 Intraspecific F$_2$ progeny (Yolo Wonder × CM334) = YC$_2$	2262 Markers on the integrated map	13	1832

The integration of the French, Israeli, and American maps was performed (Paran et al. 2004) using 320 common RFLP and AFLP markers. The integrated map consisted of 2262 markers and gave an average marker density in the genome of one marker per 0.8 cM. This allows one to precisely compare the location of loci and genes between the different crosses used in genetic analyses. The integration of the different maps is very useful for the breeders: for each map position they can choose a large set of markers that will be useful with different recurrent germplasms. Scientists could also use the integrated map to develop core maps in new crosses or to add markers in a particular region in order to clone a gene.

To date, there is not yet a saturated linkage map even if the integration of several maps allow one to design the 12 consensus major linkage groups (Lefebvre et al. 2002; Paran et al., online 2003). A discrepancy between all the published maps concerned the P1 and P8 chromosomes. Indeed, the reciprocal translocations between *C. annuum* and *C. chinense* would cause pseudo-linkage between markers near the interchange breaks on the chromosomes involved. For example, markers from the P1 linkage group published by Livingstone et al. (1999) in *C. annuum* × *C. chinense* map split into the P1 and the P8 chromosomes of the *C. annuum* × *C. annuum* map of Lefebvre et al. (2002); this is in agreement with the chromosomal interchanges reported by Pickersgill (1991).

A great variability in marker density along the linkage groups was observed on pepper maps, similar to that already observed in tomato (Tanksley et al. 1992), where cytogenetic data enable one to show that the marker clusters probably figure the centromeres. The majority of the AFLP markers clustered in each linkage group, although *PstI/MseI* markers were more evenly distributed than *EcoRI/MseI* markers within the linkage groups (Kang et al. 2001). No more than one marker clustering region was observed per linkage group. Moreover, marker clusters observed in the pepper maps contain markers located in tomato centromeres. The clusters of markers are thus presumed to correspond to the centromeric regions of pepper chromosomes (Lefebvre et al. 1997; Livingstone et al. 1999).

A total of seven linkage groups were assigned to pepper chromosomes (Fig. 1; Table 2). Tanksley (1984) and Tanksley et al. (1988) assigned two linkage groups to the chromosomes *Pourpre* and *Noir* by allele dosage of isozyme markers and by in situ hybridisation assays in primary trisomics. Other linkage groups were assigned through mapping major genes (*L*, *C*, *up*, *pvr2*, *y*) previously assigned to a chromosome by segregation analyses of the primary trisomics. It concerns respectively the chromosomes *Brun*, *Jaune*, *Noir*, *Orange* and *Indigo* (Lefebvre et al. 1995, 1997, 1998, 2002). The linkage group 10 was recently assigned to the chromosome *Rouge* thanks to the anchor phenotypic marker *A* controlling the presence of anthocyanin pigments in the plant tissues (Chaim et al. 2003).

Although most of the tomato and pepper clones reciprocally hybridised to pepper and tomato genomic DNA, the linear order of clones along the chro-

Table 2. The different nomenclatures of the pepper linkage groups, their anchor markers and correspondence with the tomato chromosomes

Pepper chromosome on molecular linkage maps	Anchor markers	Corresponding primary trisomic	Cytogenetic pepper chromosome number (Pochard 1977)	References	Corresponding tomato chromosomes (in Lefebvre et al. 2002)
P1	–	–	–	–	T1, T8
P2	C	Jaune	XI	Lefebvre et al. (1995, 2002)	T2
P3	–	–	–	–	T3, T9
P4	pvr2	Orange	–	Lefebvre et al. (1997, 2002)	T3, T4
P5	–	–	–	–	T2, T4, T5
P6	y (CCS)	Indigo	–	Lefebvre et al. (1998, 2002)	T6
P7	–	–	–	–	T7
P8	R45s, Idh-1	Pourpre	XII	Tanksley (1984); Lefebvre et al. (2002)	T1, T8
P9	–	–	–	–	T1, T9, T12
P10	A	Rouge	–	Chaim et al. (2003)	T4, T10
P11	L	Brun	–	Lefebvre et al. (1995, 2002)	T5, T11, T12
P12	up, 6Pgdh-1	Noir	–	Tanksley (1984); Lefebvre et al. (1995, 2002)	T4, T11, T12

mosomes was not conserved between the two species. No important gene losses were observed between the two maps, but many rearrangements were observed between the two species with many pepper chromosomes containing several distinct tomato segments (Table 2). At a microsynteny level, rearrangements appeared more complex (Lefebvre et al. 2002). Comparison of the pepper and tomato genetic maps showed that 18 homeologous linkage blocks cover 98.1% of the tomato genome and 95.0% of the pepper genome. By comparing the saturated tomato map (Tanksley et al. 1992) with the pepper map, a minimum of 32 chromosome breaks are necessary to transform the order and position of orthologous genes in the tomato map to that observed in pepper (Livingstone et al. 1999). The comparison of the number of loci revealed by tomato-derived RFLP probes demonstrated that duplicated loci are not confined to the pepper genome, thus ruling out gene duplication as an explanation for the three- to four-fold higher DNA content in pepper (Tanksley et al. 1988). Livingstone et al. (1999) suggested that the difference in nuclear DNA content between pepper and tomato could be accounted for by retrotransposons and by constitutive heterochromatin (An et al. 1996).

6 Map Position and Markers of Loci Governing Traits of Interest

Many genes have been characterised in pepper, particularly for fruit characteristics and disease resistance (Lippert et al. 1965). Some were previously assigned to pepper chromosomes by using the 12 primary trisomics, but much progress was recently made from molecular mapping of major genes and quantitative trait loci (QTLs). Narrowly linked genetic markers are now available for marker-assisted selection (Tables 3, 4). Development of QTL-linked PCR-based markers is underway in several laboratories.

For the past few years, a large effort has been concentrated on disease resistance loci. Among all the resistance reactions characterised in pepper, numerous are under a polygenic control compared to other crops such as tomato. Regarding certain pathogens, total resistance under a monogenic control (R gene) has been described. It is the case in resistance to *tobamoviruses, potyviruses*, to tomato spotted wilt virus (*TSWV*), to *Xanthomonas campestris* and to *Meloidogyne* spp. Partial resistance with a polygenic control was also found for certain pathogens. For other pathogens, only partial and polygenic resistance was described, such as the resistance to cucumber mosaic virus (*CMV*), *Phytophthora capsici, Verticillium dahliae*, and *Leveillula taurica*. Figure 1 illustrates the organisation of pathogen resistance loci on the pepper genome, and demonstrates that resistance loci were clustered on several resistance hot spots dispersed on several chromosomes. Resistance hot spots suggest that the same resistance gene could have a common action regarding several pathogens or that linked genes could evolve from a common ancestral gene.

DNA markers make the identification of complex loci having several R genes possible. At least three dominant R genes (*Tsw-Pvr4-Pvr7*) for resistance to potyviruses and to the tospovirus *TSWV* are organised in a cluster linked to the marker loci CD72, CT124 and GC082 on the distal portion of chromosome P10 (Grube et al. 2000a; Lefebvre et al. 2002). The loci *Me3* and *Me4* conferring resistance to *Meloidogyne* species are separated by 10 cM near the marker locus CT135 on the chromosome P9 (Djian-Caporalino et al. 2001). Allelism tests could also indicate or confirm the observation of clusters. At least three recessive loci (*pvr1, pvr2* and *pvr5*) for resistance to potyviruses in different *Capsicum* species clustered on chromosome P4 (Caranta

Table 3. Major gene mapping in pepper

Locus name	Effect (gene)	Useful marker	Carrier-chromosome	Reference
L	Resistance to TMV	RFLP (tg036) at 6 cM	P11	Lefebvre et al. (1995)
$pvr2^a$	Resistance to PVY(0), PVY(1) (gene encoding eIF4E)	Cloned gene: SCAR	P4	Caranta et al. (1997a); Ruffel et al. (2002)
Pvr4	Resistance to PVY(0), PVY(1), PVY(1–2)	CAPS at 2.1 cM + SCAR	P10	Caranta et al. (1999); Arnedo-Andres et al. (2002)
pvr6	Resistance to PVMV when associated to $pvr2^2$ gene	RFLP marker (tg057)	P3	Caranta et al. (1996)
Pvr7	Resistance to PVY(0), PVY(1), PVY(1–2)	CAPS linked to Pvr4	P10	Grube et al. (2000a)
Tsw	Resistance to TSWV	CAPS at 0.9 cM	P10	Moury et al. (2000)
$Bs2^a$	Resistance to *Xanthomonas campestris* race 1 (gene encoding a NBS-LRR)	AFLP at 0 cM Cloned gene + SCAR at 4.9 and 5.3 cM	Unknown	Tai et al. (1999a, b); Kim et al. (2001)
Bs3	Resistance to *Xanthomonas campestris* race 2	SCAR at 2.1 cM	P2	Pierre et al. (2000)
Me3, Me4	Heat-stable resistance to root-knot nematodes (*Meloidogyne* spp.)	AFLP at 0.5 cM and at 10.0 cM	P9	Djian-Caporalino et al. (2001)
up	Erected fruit	AFLP at 5 cM	P12	Lefebvre et al. (1995, 2002)
C	Fruit pungency	AFLP at 5 cM RFLP (TG205) at 0 cM CAPS at 0.4 cM	P2	Lefebvre et al. (1995, 2002); Blum et al. (2002)
y^a	Red versus yellow fruit colour (gene encoding capsantin-capsorubin synthase)	Cloned gene: SCARs	P6	Lefebvre et al. (1998); Popovsky and Paran (2000)
$C2^a$	Orange fruit colour (gene encoding phytoene synthase)	Cloned gene: RFLP	P4	Thorup et al. (2000); Huh et al. (2001)
A	Anthocyanin pigments in the tissues	RFLP (TG63)	P10	Chaim et al. (2003)
Rf	Fertility restorer	RAPD at 0.37 cM	Unknown	Zhang et al. (2000)
S^a	Soft flesh and deciduous fruit (gene encoding a polygalacturonase)	RFLP at 0 cM (PG)	P10	Rao and Paran (2003)
(Without name)	Stunted growth when associated with the cytoplasm of *C. chinense*	RAPD at 6 cM	Unknown	Inai et al. (1993)

[a] Cloned gene

Table 4. Identification of QTLs associated with agronomic interest traits in pepper

Trait	Experimental design	Number of mapped markers	Number of additive QTLs detected	Effect of the QTLs	Reference
Number of flowers per node	46 F_2 progeny	192 Markers	2 QTLs on the same LG→T2)	28.8+39.9%	Prince et al. (1993)
Fruit-related traits	180 F3 family progeny C. annuum cv. Maor × C. annuum Perennial	177 Markers	55 QTLs	6–67% according to the QTL	Ben Chaim et al. (2001b)
Yield and fruit-related traits	248 interspecific BC_2 progeny [((C. annuum cv. Maor × C. frutescens BG2816) × BG2816]	92 Markers	58 QTLs	1–25% according to the QTL	Rao et al. (2003)
Fruit shape	Several C. annuum intraspecific crosses and interspecific crosses with C. frutescens and C. chinense	12–21 Markers of the chromosome P3	Effect of one QTL named fs3.1	2.9–66.7% according to the trait and to the cross	Ben Chaim et al. (2003)
Capsaicinoid content	242 Interspecific F_2 progeny [((C. annuum cv. Maor × C. frutescens BG2816]	10 Markers of the chromosome P7	1 Major QTL on P7	34–38% according to the experimental year	Blum et al. (2003)
Resistance to Phytophthora capsici	94 Doubled haploid progeny Perennial × Yolo Wonder	119 Markers	5 QTLs + 7 digenic interactions	21–90% according to the resistance components	Lefebvre and Palloix (1996)

Table 4. (Continue)

Trait	Experimental design	Number of mapped markers	Number of additive QTLs detected	Effect of the QTLs	Reference
Resistance to *Phytophthora capsici*	101 Doubled haploid progeny H3 × Vania (HV)	HV: 135 markers	HV: 7 QTLs + digenic interactions	HV: 43–69%	Thabuis et al. (2003)
	114 Doubled haploid progeny Perennial × Yolo Wonder (PY)	PY: 154 markers	PY: 5 QTLs + digenic interactions	PY: 52–65%	
	151 F_2 progeny Yolo Wonder × Criollo de Morelos 334 (YC_2)	YC_2: 64 markers	YC_2: 9 QTLs (a conserved major effect QTL on P5)	YC_2: 22–81% according to the resistance components	
Resistance to potyviruses	94 Doubled haploid progeny Perennial × Yolo Wonder	172 Markers	11 QTLs + 1 digenic interaction	66–76% according to the potyvirus strain	Caranta et al. (1997a)
Restriction of cucumber mosaic virus installation in host-cells	94 Doubled haploid progeny Perennial × Yolo Wonder	138 Markers	2 QTLs + 1 digenic interaction	Together explaining 57% of the phenotypic variation	Caranta et al. (1997b)
Resistance to cucumber mosaic virus	180 F_3 family progeny *C. annuum* Maor × *C. annuum* Perennial	177 Markers	4 QTLs + 2 digenic interactions	7–33% according to the QTL	Ben Chaim et al. (2001a)
Restriction of cucumber mosaic virus long-distance movement	101 Doubled haploid progeny H3 × Vania	184 Markers	4 QTLs (1 major on P12) + 2 digenic interactions	4.0–63.6% according to the QTL	Caranta et al. (2002)
Resistance to *Leveillula taurica*	101 Doubled haploid progeny H3 × Vania	134 markers	5 QTLs + 2 digenic interactions	Together explaining more than 50% of the phenotypic variation	Lefebvre et al. (2003)

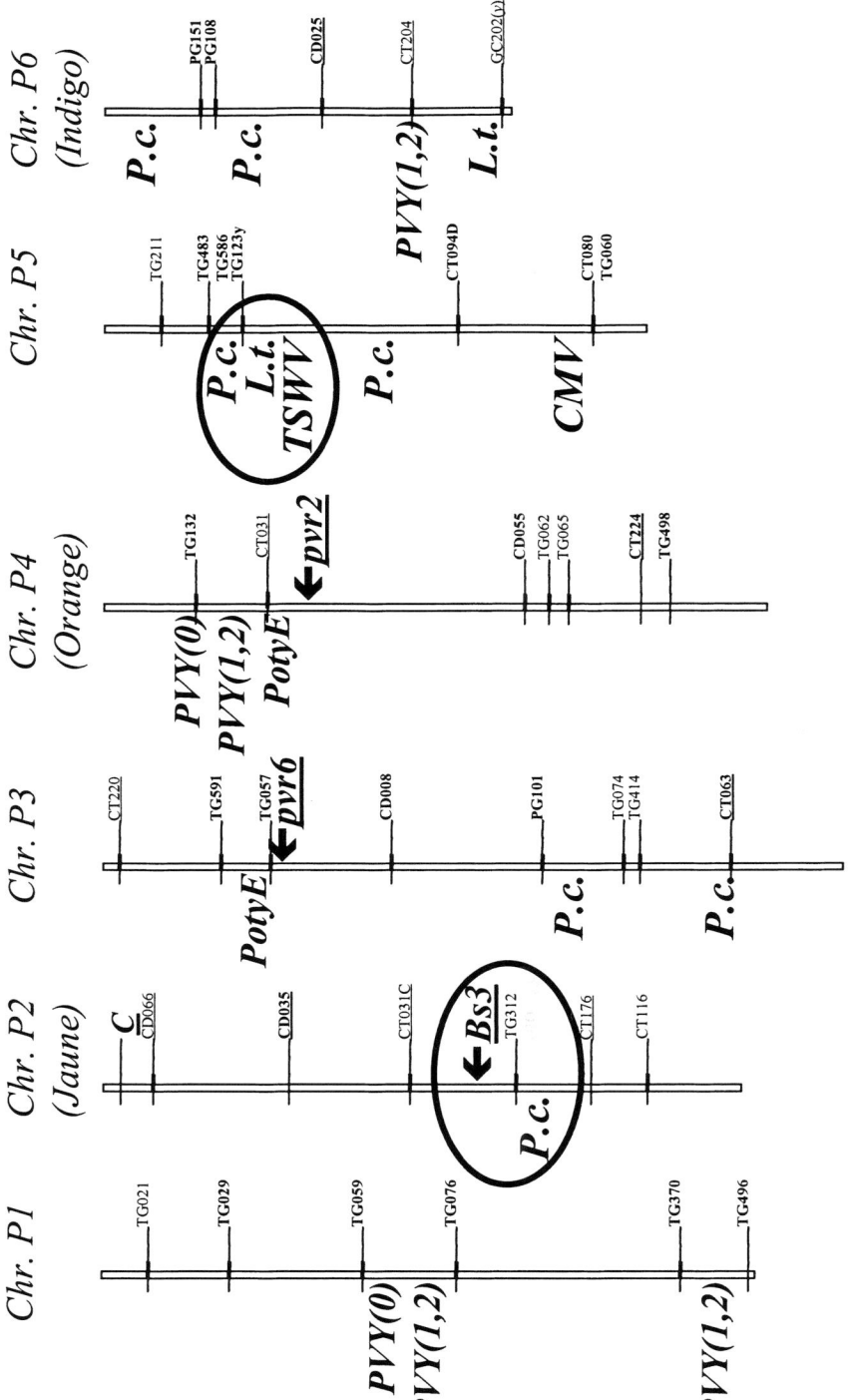

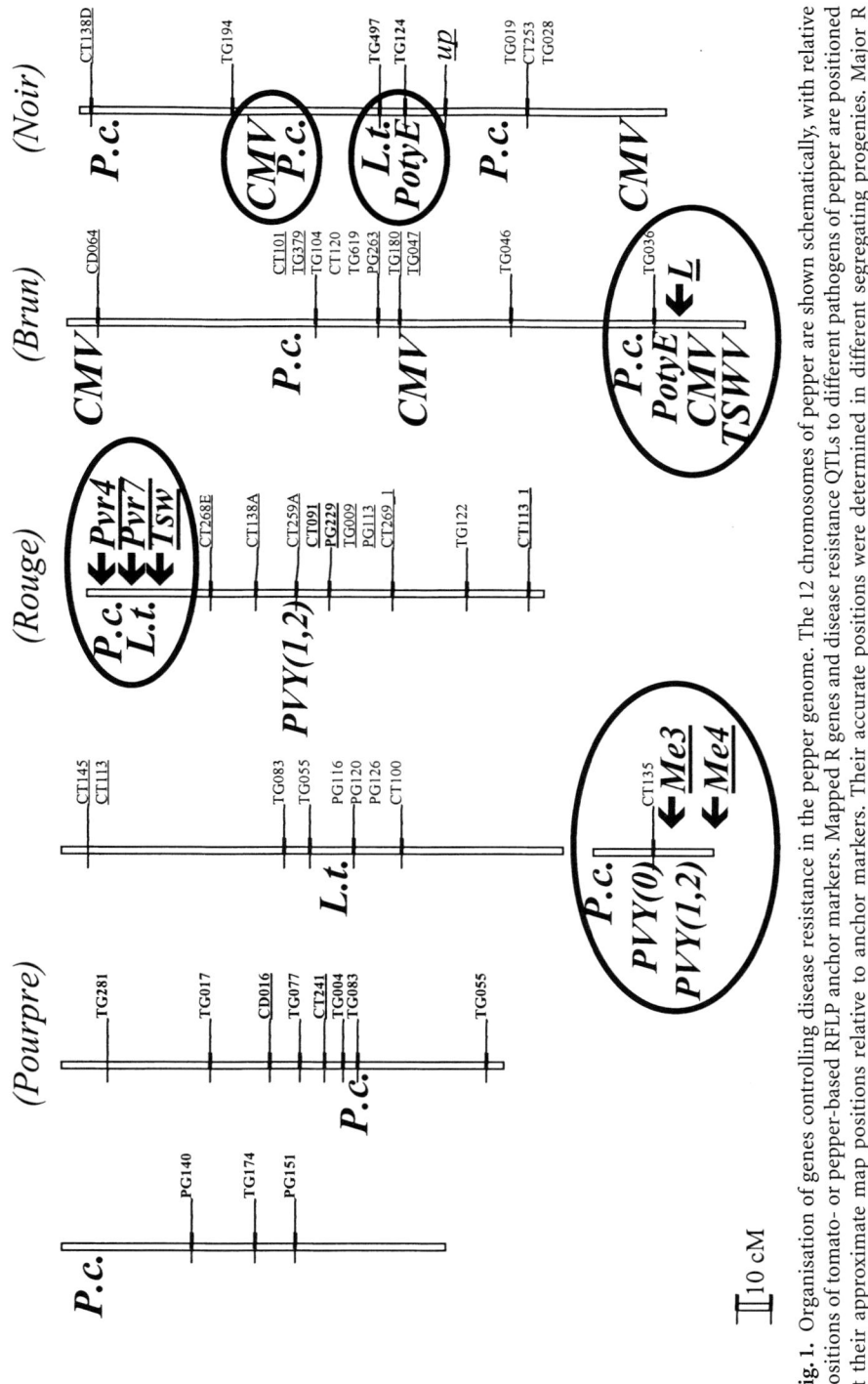

Fig. 1. Organisation of genes controlling disease resistance in the pepper genome. The 12 chromosomes of pepper are shown schematically, with relative positions of tomato- or pepper-based RFLP anchor markers. Mapped R genes and disease resistance QTLs to different pathogens of pepper are positioned at their approximate map positions relative to anchor markers. Their accurate positions were determined in different segregating progenies. Major R genes are represented by an *arrow*. QTLs are represented by abbreviations of the pathogen against which they act. Regions involved in epistatic relationships were omitted for simplification. *Pc. Phytophthora capsici* (Lefebvre and Palloix 1996; Pflieger et al. 2001b; Thabuis et al. 2003), *L.t. Leveillula taurica* (Lefebvre et al. 2003), *CMV* cucumber mosaic virus (Caranta et al. 1997b, 2002; Pflieger et al. 1999), *PVY* potato virus Y, strain (0) or (1,2), *PotyE* potyvirus E (Caranta et al. 1997a), *TSWV* (Moury 1997)

et al., pers. comm.). Clusters can consist of a series of allelic genes such as in the *L* locus located on the chromosome P11 (Lefebvre et al. 1995) where five alleles were described (Boukema 1980), or in the *pvr2* locus where at least three alleles are known ($pvr2^1$, $pvr2^2$, $pvr2^+$, Kyle and Palloix 1997). Regarding the potyviruses, Kyle and Palloix (1997) described seven major genes that display different specificity actions against strains of potato virus Y (*PVY*), tobacco etch virus (*TEV*), pepper mottle virus (*PepMov*), and pepper veinal mottle virus (*PVMV*). Mapping results showed that these genes were mainly organised in two clusters: one cluster of recessive genes on the chromosome P4 and one cluster of dominant resistant genes on the chromosome P10. So far, no R gene cluster has been mapped close to the *pvr6* locus on the chromosome P3 (Caranta et al. 1996) and the *pvr3* locus has not yet been mapped.

Independent QTL maps can also be compared using anchor markers. Integration of QTLs for resistance to *P. capsici*, *L. taurica*, *CMV* and potyviruses, and R genes on a synthetic representation of the pepper genome reveals linkages between QTLs and between QTLs and R genes. Regarding several genitors, QTLs were detected in the vicinity of all the R genes presently mapped. Clusters of QTLs were also mapped at other loci, for instance on the chromosomes P5 and P12. The most evident clusters of resistance loci to different pathogens are located on the chromosomes P2, P5, P9, P10, P11 and P12. For instance, QTLs for resistance to *P. capsici* and to *L. taurica* located in the vicinity of the R gene cluster *Tsw-Pvr4-Pvr7* on the chromosome P10. QTLs for resistance to potyvirus, to *CMV*, to *P. capsici*, and to *TSWV* (Moury 1997) were mapped in the vicinity of the locus *L* on the chromosome P11. One cluster of R genes and QTLs for resistance to potyviruses, located on P4, is under characterisation since the *pvr2* gene was recently cloned. This gene belongs to a new class of resistance genes. It codes for eIF4E, an eukaryotic initiation factor involved in RNA translation (Ruffel et al. 2002).

Epistasis interactions were often evoked as probably involved in the determination of disease resistance in pepper (e.g. Bartual et al. 1994; Daubèze et al. 1995). QTL dissection with molecular markers confirmed this hypothesis. One obvious example concerns the complementation between *pvr2* and *pvr6* that confers a complete resistance to PVMV though none of these two single genes confer any resistance to this virus (Caranta et al. 1996). Important digenic effects were reported for the partial resistance to *P. capsici* (Lefebvre and Palloix 1996; Thabuis et al. 2003), to *CMV* (Caranta et al. 1997b, 2002; Ben Chaim et al. 2001a), to potyviruses (Caranta et al. 1997a) and to *L. taurica* (Lefebvre et al. 2003). One major digenic interaction associated with the resistance to *P. capsici* was validated between the chromosomes P5 and P10 by isolating one plant having both the concerned QTLs and studying its selfing progeny (Thabuis et al. 2004). Whereas the loci involved in epistasis were not reported in the Fig. 1, it was observed that QTLs involved in digenic interactions mapped frequently in the vicinity of QTLs having an additive effect to the same pathogen or to another.

More and more time is being invested in horticultural and fruit traits. Several genes determining fruit colour were mapped and characterised by a candidate gene approach, thanks to the well-known biosynthesis pathway of pigments. The molecular characterisation of the *y* locus controlling the red vs. yellow fruit colours and the *C2* locus involved in the red vs. orange colours supplied powerful markers for selection since they correspond to the functional gene itself and therefore cancel the recombination risk between the gene and the marker (Lefebvre et al. 1998; Huh et al. 2001). The dominant *S* gene determining the soft flesh and deciduous fruit of pepper was mapped on the chromosome P10 and was demonstrated to cosegregate with a tomato polygalacturonase gene marker (Rao and Paran 2003). Male sterility is a very important trait for hybrid production. Both nuclear recessive and genocytoplasmic sterilities were characterised (Shifriss 1997). A major fertility restorer was flanked by PCR-based markers (Zhang et al. 2000), and a quantitative restoration system has been mapped (Wang et al. 2004). The level of pungency is also a breeding objective for most of the commonly grown varieties of hot peppers. The major gene *C* for fruit pungency was mapped on the chromosome P2 (Lefebvre et al. 1995), and a cleaved amplified polymorphic sequence (CAPS) marker located 0.4 cM from *C* was developed (Blum et al. 2002) using the sequence of the *Capsicum* fibrillin gene that encodes a protein strongly regulated during fruit ripening (Deruère et al. 1994). The competition is now open between several research groups for the isolation of the *C* locus controlling the capsaicin synthesis, a characteristic of the *Capsicum* genus. Pungency is also a very complex trait since it is determined by the quantitative control of several capsaicinoids. QTLs for the capsaicin and dihydrocapsaicin contents were recently mapped (Blum et al. 2003). Capsaicinoids, amino acids, carotenoids, and vitamins are major components of the pepper fruit that determine pigment contents, taste and nutritive quality. The integration of these components is now being considered more and more. Chemical fingerprints of these secondary metabolites will be integrated into the molecular maps in the near future.

Concerning the fruit architecture, QTLs influencing horticultural characteristics were reported in an intraspecific *C. annuum* cross (Ben Chaim et al. 2001b) and in an interspecific *C. annuum* × *C. frutescens* cross (Rao et al. 2003). Some QTLs for fruit traits, including fruit diameter and fruit weight, are closely linked, reflecting the high genetic correlations observed among these traits. A major effect QTL influencing the fruit shape located on the chromosome P3 is conserved in intraspecific *C. annuum* crosses and in crosses of *C. annuum* with *C. chinense* and *C. frutescens* (Ben Chaim et al. 2003).

The genomic positions of resistance loci mapped in pepper were compared to those of tomato and potato (Grube et al. 2000b). Resistance to tobacco mosaic virus (TMV) is controlled by the genes *Tm-1* and *Tm-2* in tomato and *L* in pepper. These three genes do not map to syntenic regions in the two species (Young et al. 1988; Levesques et al. 1990; Lefebvre et al. 1995). The tomato

Sw-5 locus and the pepper *Tsw* locus, controlling the resistance to TSWV, do not map to corresponding genomic regions either (Jahn et al. 2000). Despite these two examples, several cases of functional colinearities were described: a corresponding position in at least two genera of Solanaceae controls a similar trait. The tomato *pot-1* gene conferring resistance to PVY and TEV was identified in a colinear genomic region to the *pvr2* pepper locus (Parrella et al. 2002) and was demonstrated to correspond to the same coding gene (S. Ruffel et al., pers. comm.). The *Me3-Me4* cluster was supposed to be in a colinear region of the tomato *Mi-3* mapped on chromosome T12 and potato *Gpa2* on chromosome XII (Djian-Caporalino et al. 2001). These loci are all involved in the resistance to nematodes. Colinearities were also suspected for QTLs. The major-effect resistance QTL to *P. capsici* on the chromosome P5 in pepper is in a syntenic position to a resistance QTL to *P. infestans* on the potato chromosome IV (Pflieger et al. 2001b; Thabuis et al. 2003). Two resistance QTLs to powdery mildew due to *L. taurica*, which have been mapped on the pepper chromosomes P6 and P9, show colinear positions with R genes and QTLs involved in the resistance to powdery mildew on the tomato chromosomes T6 and T12 (Lefebvre et al. 2003). Several pepper fruit-related QTLs also appeared to correspond to positions in tomato for loci controlling the same traits, suggesting the hypothesis that these genes may be orthologous in the two species (Ben Chaim et al. 2001b). More recently, comparison of the genomic locations of the eggplant fruit weight, shape and colour QTLs with the positions of similar loci in tomato, potato and pepper revealed that 40% of the different loci have putative orthologous counterparts in at least one of these crop species (Doganlar et al. 2002). Thorup et al. (2000) demonstrated two cases of orthologous loci involved in the organ pigmentation: in pepper and potato for the *CrtZ-2* marker locus encoding the β-carotene hydroxylase, and in pepper and tomato for the *CCS* marker locus encoding the capsanthin capsorubin synthase. These results suggest that the colinear genes originated from common ancestral genes. These observations support the hypothesis that orthologous genetic networks could control related complex phenotypes.

7 Genomic Resources

Genomic resources constitute useful tools for exploring the genome of an organism on a large scale, for cataloguing all its genes and understanding their functions. They include high-through-put sequencing, analysis of gene expression, integration of high density genetic, physical and functional maps. In pepper, genomic resources are still weakly developed compared to model plants or certain important crops, such as maize or tomato. However, several research programs are underway for increasing them.

A number of known genes were mapped in pepper (Lefebvre et al. 1998, 2002; Pflieger et al. 1999, 2001b; Thorup et al. 2000; Huh et al. 2001; Cheng et al. 2002; Blum et al. 2003; Rao and Paran 2003). Functional genetic maps facilitate the identification of positional candidate genes to characterise loci of interest (Pflieger et al. 2001a). Several examples of the candidate gene approach were described in pepper. The first one concerned the molecular characterisation of the *y* locus controlling the yellow vs. red fruit colour that was attributed to the capsanthin-capsorubin synthase gene (Lefebvre et al. 1998). Another candidate gene, phytoene synthase, was proposed to be responsible for the *C2* gene involved in the orange fruit colour (Thorup et al. 2000; Huh et al. 2001). Resistance gene homologues and defence response gene homologues were proposed as candidate genes for quantitative resistance gene loci (Pflieger et al. 1999, 2001b; Grube et al. 2000b). Pepper homologues of cloned tomato R genes were found in syntenous positions in other Solanaceous genomes, but can also be mapped to additional positions (Pflieger et al. 1999; Grube et al. 2000b). Colocalisations between nucleotide-binding-site (NBS)-containing sequences and resistance QTLs suggest that mechanisms of qualitative and quantitative resistance may be similar (Pflieger et al. 1999). More recently, Ruffel et al. (2002) demonstrated by a candidate gene approach that the *pvr2* locus corresponds to an eukaryotic initiation factor 4E gene. A number of pepper sequences were isolated by differential expression analysis from pepper cDNA libraries constructed with the mRNA from pepper plants infected with various pathogens. They correspond to genes induced by biotic or abiotic stresses, and could further serve the candidate gene approach.

In June 2003, more than 22,500 expressed sequence tags (ESTs) from *C. annuum* and *C. chinense* were available in genomic databases. This number is still low compared to the genus *Lycopersicon* that counts more than 160,000 ESTs and the genus *Solanum* that counts more than 94,000 ESTs (mostly potato). However, some of these clones were spotted on glass slides to study the expression profiling (Lee and Choi 2002). More than 22,000 other pepper nucleotide sequences were deposited in databases that have also gathered more than 175,000 tomato sequences and 95,000 potato sequences (http://www.ncbi.nlm.nih.gov/).

A yeast artificial chromosome library of 19,000 clones with an average insert size of 500 Kb and representing approximately three haploid genomes was constructed from high-molecular weight DNA isolated from *Capsicum annuum* leaf protoplasts (Tai and Staskawicz 2000). This YAC library was used to isolate the *Bs2* gene (Tai et al. 1999a). A bacterial artificial chromosome library consisting of 235,000 clones with an average insert size of 130 Kb was constructed from the Mexican pungent *Capsicum* annuum 'CM334' line. It was estimated to contain approximately 12 genome equivalents and to represent at least 99% of the pepper genome (Yoo et al. 2003). Another BAC library was obtained from a doubled haploid line issued from the cross Perennial × Yolo Wonder, segregating for several resistance and horticultural traits (Ruffel et al. 2004). It comprised more than 239,000 BAC clones with an average insert size of 125 Kb and was estimated to represent

10.8 genome equivalents. It was used for isolating the genomic sequence of the *pvr2* locus (Ruffel et al. 2002). BAC libraries constitute a useful tool for studying microsynteny and for aligning contigs from different species.

By considering the pepper genome size (2702–3420 Mbp) and the average length of the different linkage maps of pepper (about 1500 cM), 1 cM should correspond to an average of 1800–2200 Kb. However, this correspondence between physical and genetic size may vary greatly along the genome as shown by the physical and genetic data around the region of the *Bs2* gene, where 1 cM is about 250 Kb (Tai et al. 1999b).

8 Marker-Assisted Selection

Selecting simultaneously for multigenic resistance and for polygenic fruit quality traits may be eased by the development of molecular markers and molecular linkage maps. Several molecular markers are already available in *Capsicum* for marker-assisted selection (Table 3). Some of them are supposed to correspond to the gene itself, such as for the loci *y* (Lefebvre et al. 1998) and *C2* (Huh et al. 2001). Easy-to-use specific markers for QTLs are in development. Until now, breeders have mainly exploited monogenic trait markers (*y*, *Tsw*, *Pvr4*, *pvr2*, *Me3* and *L*) for breeding commercial cultivars with resistance and fruit colour traits. The marker-assisted selection was particularly interesting for the lately expressed traits like colour of matured fruit or for breeding for multi-resistance.

Different marker-assisted selection programs in pepper are now in use for QTL exploitation. For instance, two introgression populations were bred for improving the resistance to *P. capsici* that was previously dissected in QTLs (Thabuis 2002; Thabuis et al. 2003). One aimed at transferring five QTL-resistant alleles from Criollo de Morelos into a partially resistant bell pepper genetic background. One out of 450 plants obtained from a recurrent phenotypic selection population using Criollo de Morelos 334 as the resistant parent was selected as having the five QTL-resistant alleles, thanks to 12 linked-markers covering the whole confidence interval of the five QTLs selected (Thabuis et al. 2001). Another experimental study aiming at transferring four QTL-resistant alleles from the small fruited genitor Perennial into a bell pepper recipient line was performed by marker-assisted backcrosses during three cycles (Thabuis et al. 2004). A population size of 350 plants enabled both efficient control of the QTLs and efficient background selection. This experiment permitted the effect of the genomic regions detected in the mapping population to be validated, while a decrease of effect was noted for low-effect QTLs and epistatic interactions. The phenotypic evaluations of each backcross generation demonstrated an increased level of resistance and an efficient return to the recipient phenotype for the fruit weight. Taking into account the confidence intervals of the QTL positions by using several markers, contributes to the optimisation for a successful QTL transfer. Using markers of the genetic background accelerates the recovery of the recipient parent genome and limits the number of selection cycles (Hospital and Charcosset 1997). QTL-linked markers also enabled the a posteriori analysis of a phenotypic selection program for *P. capsici* resistance. Thabuis et al. (2004) demonstrated that low-effect resistance QTLs were not retained through the phenotypic selection cycles and that selection for *P. capsici* had no significant impact on the horticultural advance. These different experiences demonstrated the feasibility of marker-assisted selection for polygenic traits in pepper.

Map integration of QTL information from various analyses, such as horticultural and resistance genetic dissection, will help breeders to challenge the unlikely genetic linkage between favourable and unfavourable traits. Marker-assisted selection should help to pyramid QTLs and to control the genetic background by using a limited population size. More and more specific easy-to-use markers are expected to be described in the literature. Finally, fine-mapping experiments of particular genomic regions and creation of near-isogenic lines except for one QTL (so-called NIL-QTLs) will contribute to a better assessment of a single QTL effect and to construct elite genotypes having a chosen set of loci.

9 Conclusion

Saturated maps and global genome knowledge still have much progress to make in the *Capsicum* genus although some original traits, such as capsaicin synthesis, fruit colour control and certain resistance to diseases, brought this species into the spotlight. Its large genome size and lower research investment, compared to major important crops, hinder the genomic advancements in the *Capsicum* genus. More genetic and genomic tools have to be produced to complement those already available. Several introgression line populations of related *Capsicum* species in a *C. annuum* genetic background are under production in several laboratories. This genetic resource will facilitate the identification and fine-mapping of agronomically interesting loci, particularly QTLs. Moreover, the association studies aiming at investigating the natural allele diversity are very promising in *Capsicum* since this genus possesses a particularly rich germplasm collection and shows a very large phenotypic variability. Finally, extensive comparative genomic approaches will benefit pepper genome analysis for exploring gene functions and regulations. Indeed, the *Capsicum* genus constitutes a good model to transfer and validate tools and knowledge generated in model genomes, and especially from the related genus such as tomato and potato, given that the synteny relationships with pepper have been well described.

Acknowledgements. I am grateful to A Palloix for his helpful comments of the review. I also thank all my collaborators for sharing up-to-date information.

References

An CS, Kim SC, Go SL (1996) Analysis of red pepper (*Capsicum annuum*) genome. J Plant Biol 39:57–61
Arnedo-Andres MS, Gil-Ortega R, Luis-Arteaga M, Hormaza JI (2002) Development of RAPD and SCAR markers linked to the *Pvr4* locus for resistance to PVY in pepper (*Capsicum annuum* L.). Theor Appl Genet 105:1067–1074
Arumuganathan K, Earle ED (1991) Nuclear DNA content of some important plant species. Plant Mol Biol Rep 9:208–219
Ballester J, de Vicente MC (1998) Determination of F_1 hybrid seed purity in pepper using PCR-based markers. Euphytica 103:223–226
Baral J, Bosland PW (2002) Genetic diversity of a *Capsicum* germplasm collection from Nepal as determined by randomly amplified polymorphic DNA markers. J Am Soc Hortic Sci 127:318–324
Bartual R, Lacasa A, Marsal JI, Tello JC (1994) Epistasis in the resistance of pepper to *Phytophthora* stem blight (*Phytophthora capsici* L.) and its significance in the prediction of double cross performances. Euphytica 72:149–152
Ben Chaim A, Grube RC, Lapidot M, Jahn M, Paran I (2001a) Identification of quantitative trait loci associated with resistance to cucumber mosaic virus in *Capsicum annuum*. Theor Appl Genet 102:1213–1220
Ben Chaim A, Paran I, Grube RC, Jahn M, van Wijk R, Peleman J (2001b) QTL mapping of fruit-related traits in pepper (*Capsicum annuum*). Theor Appl Genet 102:1016–1028
Ben Chaim A, Borovsky Y, Rao GU, Tanyolac B, Paran I (2003) *fs3.1*: a major fruit shape QTL conserved in *Capsicum*. Genome 46:1–9
Blum E, Liu K, Mazourek M, Yoo EY, Jahn M, Paran I (2002) Molecular mapping of the *C* locus for presence of pungency in *Capsicum*. Genome 45:702–705
Blum E, Mazourek M, O'Connell M, Curry J, Thorup T, Liu K, Jahn M, Paran I (2003) Molecular mapping of capsaicinoid biosynthesis genes and quantitative trait loci analysis for capsaicinoid content in *Capsicum*. Theor Appl Genet: online
Boukema IW (1980) Allelism of genes controlling resistance to TMV in *Capsicum* L. Euphytica 29:433–439
Caranta C, Palloix A, Gebre-Selassie K, Lefebvre V, Moury B, Daubèze AM (1996) A complementation of two genes originating from susceptible *Capsicum annuum* lines confers a new and complete resistance to pepper veinal mottle virus. Phytopathology 86:739–743
Caranta C, Lefebvre V, Palloix A (1997a) Polygenic resistance of pepper to potyviruses consists of a combination of isolate-specific and broad-spectrum quantitative trait loci. Mol Plant-Microbe Interact 10:872–878
Caranta C, Palloix A, Lefebvre V, Daubèze AM (1997b) QTLs for a component of partial resistance to cucumber mosaic virus in pepper: restriction of virus installation in host-cells. Theor Appl Genet 94:431–438
Caranta C, Thabuis A, Palloix A (1999) Development of a CAPS marker for the *Pvr4* locus: a tool for pyramiding potyvirus resistance genes in pepper. Genome 42:1111–1116
Caranta C, Pflieger S, Lefebvre V, Daubèze AM, Thabuis A, Palloix A (2002) QTLs involved in the restriction of cucumber mosaic virus (CMV) long-distance movement in pepper. Theor Appl Genet 104:586–591
Chaim AB, Borovsky Y, de Jong W, Paran I (2003) Linkage of the *A* locus for the presence of anthocyanin and *fs10.1*, a major fruit-shape QTL in pepper. Theor Appl Genet 106:889–894

Cheng CM, Palloix A, Lefebvre V (2002) Isolation, mapping and characterization of allelic polymorphism of *Chi3-P1*, a class III chitinase of *Capsicum annuum* L. Plant Sci 163:481–489

Conicella C, Errico A, Saccardo F (1990) Cytogenetic and isozyme studies of wild and cultivated *Capsicum annuum*. Genome 33:279–282

Daubèze AM, Hennart JW, Palloix A (1995) Resistance to *Leveillula taurica* in pepper (*Capsicum annuum*) is oligogenically controlled and stable in Mediterranean regions. Plant Breed 114:327–332

Daunay MC, Jullian E, Dauphin F (2001) Management of eggplant and pepper genetic resources in Europe: networks are emerging. Proceedings of XIth EUCARPIA meeting on genetics and breeding of *capsicum* and eggplant, 9–13 April 2001, Antalya, Turkey, pp 1–5

Deruère J, Romer S, d'Harlingue A, Backhaus RA, Kuntz M, Camara B (1994) Fibril assembly and carotenoid overaccumulation in chromoplasts: a model for supramolecular lipoprotein structures. Plant Cell 6:119–133

Djian-Caporalino C, Pijarowski L, Fazari A, Samson M, Gaveau L, O'Byrne C, Lefebvre V, Caranta C, Palloix A, Abad P (2001) High-resolution genetic mapping of the pepper (*Capsicum annuum* L.) resistance loci *Me3* and *Me4* conferring heat-stable resistance to root-knot nematodes (*Meloidogyne* spp.). Theor Appl Genet 103:592–600

Doganlar S, Frary A, Daunay MC, Lester RN, Tanksley SD (2002) Conservation of gene function in the Solanaceae as revealed by comparative mapping of domestication traits in eggplant. Genetics 161:1713–1726

Eshbaugh WH (1980) The taxonomy of the genus *Capsicum* (Solanaceae). Phytologia 47:153–165

Galbraith DW, Harkins KR, Maddox JM, Ayres NM, Sharma DP, Firoozabady E (1983) Rapid flow cytogenetic analysis of the cell cycle in intact plant tissues. Science 220:1049–1051

Greenleaf WH (1986) Pepper breeding. In: Bassett MJ (ed) Breeding vegetable crops. AVI Publ, Westport, Connecticut, 134 pp

Grube RC, Blauth JR, Arnedo AMS, Caranta C, Jahn MK (2000a) Identification and comparative mapping of a dominant potyvirus resistance gene cluster in *Capsicum*. Theor Appl Genet 101:852–859

Grube RC, Radwanski ER, Jahn M (2000b) Comparative genetics of disease resistance within the Solanaceae. Genetics 155:873–887

Hospital F, Charcosset A (1997) Marker-assisted introgression of quantitative trait loci. Genetics 147:1469–1485

Huang SW, Zhang BX, Milbourne D, Cardle L, Yang GM, Guo JZ (2001) Development of pepper SSR markers from sequence databases. Euphytica 117:163–167

Huh JH, Kang BC, Nahm SH, Kim S, Ha KS, Lee MH, Kim BD (2001) A candidate gene approach identified phytoene synthase as the locus for mature fruit color in red pepper (*Capsicum* spp.). Theor Appl Genet 102:524–530

IBPGR (1983) Genetic resources of *Capsicum*. International Board for Plant Genetic Resources. AGPG/IBPGR/82/12, Rome, Italy, 49 pp

Ilbi H (2003) RAPD markers assisted varietal identification and genetic purity test in pepper, *Capsicum annuum*. Sci Hortic 97:211–218

Inai S, Ishikawa K, Nunomura O, Ikehashi H (1993) Genetic analysis of stunted growth by nuclear-cytoplasmic interaction in interspecific hybrids of *Capsicum* by using RAPD markers. Theor Appl Genet 87:416–422

Jahn M, Paran I, Hoffmann K, Radwanski ER, Livingstone KD, Grube RC, Aftergoot E, Lapidot M, Moyer J (2000) Genetic mapping of the *Tsw* locus for resistance to the Tospovirus tomato spotted wilt virus in *Capsicum* spp. and its relationship to the *Sw-5* gene for resistance to the same pathogen in tomato. Mol Plant Microbe Interact 13:673–682

Kang BC, Nahm SH, Huh JH, Yoo HS, Yu JW, Lee MH, Kim BD (2001) An interspecific (*Capsicum annuum* × *C. chinense*) F_2 linkage map in pepper using RFLP and AFLP markers. Theor Appl Genet 102:531–539

Kim BD, Kang BC, Nam SH, Kim BS, Kim NS, Lee MH, Ha KS (1997) Construction of a molecular map and development of a molecular breeding technique. J Plant Biol 40:156–163

Kim KT, Choi HS, Kim HJ, Pae DH, Yoon JY, Kim BD (2001) Development of DNA markers linked to bacterial leaf spot resistance of chilli. Acta Hortic 546:597–601

Kyle MM, Palloix A (1997) Proposed revision of nomenclature for potyvirus resistance genes in *Capsicum*. Euphytica 97:183–188

Lee S, Choi D (2002) Toward functional genomics of plant–pathogen interactions: isolation and analysis of defense-related genes of hot pepper expressed during resistance against pathogen. Plant Pathol J 18:63–67

Lefebvre V, Palloix A (1996) Both epistatic and additive effects of QTLs are involved in polygenic-induced resistance to disease: a case study, the interaction pepper–*Phytophthora capsici* Leonian. Theor Appl Genet 93:503–511

Lefebvre V, Palloix A, Rives M (1993) Nuclear RFLP between pepper cultivars (*Capsicum annuum* L.). Euphytica 71:189–199

Lefebvre V, Palloix A, Caranta C, Pochard E (1995) Construction of an intraspecific integrated linkage map of pepper using molecular markers and doubled-haploid progenies. Genome 38:112–121

Lefebvre V, Caranta C, Pflieger S, Moury B, Daubèze AM, Blattes A, Ferrière C, Phaly T, Nemouchi G, Ruffinatto A, Palloix A (1997) Updated intraspecific maps of pepper. Capsicum Eggplant Newslett 16:35–41

Lefebvre V, Kuntz M, Camara B, Palloix A (1998) The capsanthin-capsorubin synthase gene: a candidate gene for the *y* locus controlling the red fruit colour in pepper. Plant Mol Biol 36:785–789

Lefebvre V, Goffinet B, Chauvet JC, Caromel B, Signoret P, Brand R, Palloix A (2001) Evaluation of genetic distances between pepper inbred lines for cultivar protection purposes: comparison of AFLP, RAPD and phenotypic data. Theor Appl Genet 102:741–750

Lefebvre V, Pflieger S, Thabuis A, Caranta C, Blattes A, Chauvet JC, Daubèze AM, Palloix A (2002) Towards the saturation of the pepper linkage map by alignment of three intraspecific maps including known-function genes. Genome 45:839–854

Lefebvre V, Daubèze AM, Rouppe van der Voort J, Peleman J, Bardin M, Palloix A (2003) QTLs for resistance to powdery mildew in pepper under natural and artificial infections. Theor Appl Genet 107:661–666

Levesque H, Vedel F, Mathieu C, de Courcel AGL (1990) Identification of a short rDNA spacer sequence highly specific of a tomato line containing *Tm-1* gene introgressed from *Lycopersicon hirsutum*. Theor Appl Genet 80:602–608

Li D, Zhao K, Xie B, Zhang B, Luo K (2003) Establishment of a highly efficient transformation system for pepper (*Capsicum annuum* L.). Plant Cell Rep 21:785–788

Lim HT, Lee GY, You YS, Park EJ, Song YN, Yang DC, Choi KH (1999) Regeneration and genetic transformation of hot pepper plants. Acta Hortic 483:387–396

Lippert LF, Bergh DO, Smith PG (1965) Gene list for the pepper. J Hered 56:30–34

Livingstone KD, Lackney VK, Blauth JR, van Wijk R, Jahn MK (1999) Genome mapping in *Capsicum* and the evolution of genome structure in the Solanaceae. Genetics 152:1183–1202

Livneh O, Nagler Y, Tal Y, Harush SB, Gafni Y, Beckmann JS, Sela I (1990) RFLP analysis of a hybrid cultivar of pepper (*Capsicum annuum*) and its use in distinguishing between parental lines and in hybrid identification. Seed Sci Technol 18:209–214

Livneh O, Vardi E, Stram Y, Edelbaum O, Sela I (1992) The conversion of a RFLP assay into PCR for the determination of purity in a hybrid pepper cultivar. Euphytica 62:97–102

Loaiza-Figueroa F, Ritland K, Cancino JAL, Tanksley SD (1989) Patterns of genetic variation of the genus *Capsicum* (Solanaceae) in Mexico. Plant Syst Evol 165:159–188

Moury B (1997) Evaluation de sources de résistance au Tomato Spotted Wilt Virus chez le piment et création d'outils d'aide à la sélection. Thèse de doctorat, Ecole Nationale Supérieure Agronomique de Rennes, France

Moury B, Pflieger S, Blattes A, Lefebvre V, Palloix A (2000) A CAPS marker to assist selection of tomato spotted wilt virus (TSWV) resistance in pepper. Genome 43:137–142

Palloix A, Daubèze AM, Pochard E (2003) Le Piment. In: Pitrat M, Foury C (eds) Histoire de légumes: Des origines à l'orée du XXIe siècle. Edition INRA, Paris, France, pp 278–290

Paran I, Aftergoot E, Shifriss C (1998) Variation in *Capsicum annuum* revealed by RAPD and AFLP markers. Euphytica 99:167–173

Paran I, Rouppe van der Voort J, Lefebvre V, Jahn M, Landry L, van Schriek M, Tanyolac B, Caranta C, Ben Chaim A, Livingstone K, Palloix A, Peleman J (2004) An integrated genetic linkage map of pepper (*Capsicum* spp.) Mol Breed 13(3):251–261

Parrella G, Ruffel S, Moretti A, Morel C, Palloix A, Caranta C (2002) Recessive resistance genes against potyviruses are localized in colinear genomic regions of the tomato (*Lycopersicon* spp.) and pepper (*Capsicum* spp.) genomes. Theor Appl Genet 105:855–861

Pflieger S, Lefebvre V, Caranta C, Blattes A, Goffinet B, Palloix A (1999) Disease resistance gene analogs as candidates for QTLs involved in pepper–pathogen interactions. Genome 42:1100–1110

Pflieger S, Lefebvre V, Causse M (2001a) The candidate gene approach in plant genetics: a review. Mol Breed 7:275–291

Pflieger S, Palloix A, Caranta C, Blattes A, Lefebvre V (2001b) Defense response genes co-localize with quantitative disease resistance loci in pepper. Theor Appl Genet 103:920–929

Pickersgill B (1991) Cytogenetics and evolution of *Capsicum*. In: Tsuchiya T, Gupta PK (eds) Chromosome engineering in plants: genetics, breeding, evolution. Elsevier, Amsterdam, pp 139–160

Pierre M, Noel L, Lahaye T, Ballvora A, Veuskens J, Ganal M, Bonas U (2000) High-resolution genetic mapping of the pepper resistance locus *Bs3* governing recognition of the *Xanthomonas campestris* pv *vesicatoria* AvrBs3 protein. Theor Appl Genet 101:255–263

Pochard E (1966) Données expérimentales sur la sélection du piment (*Capsicum annuum* L.). Ann Amélior Plant 16:185–197

Pochard E (1970) Description des trisomiques de piment (*Capsicum annuum* L.) obtenus dans la descendance d'une plante haploïde. Ann Amélior Plant 20:233–256

Pochard E (1977) Localization of genes in *Capsicum annuum* L. by trisomic analysis. Ann Amélior Plantes 27:255–266

Pochard E, Dumas de Vaulx R (1982) Localization of *vy2* and *fa* genes by trisomic analysis. Capsicum Newslett 1:18–19

Popovsky S, Paran I (2000) Molecular genetics of the *y* locus in pepper: its relation to capsanthin-capsorubin synthase and to fruit color. Theor Appl Genet 101:86–89

Prince JP, Loaiza-Figueroa F, Tanksley SD (1992) Restriction fragment length polymorphism and genetic distance among Mexican accessions of *Capsicum*. Genome 35:726–732

Prince JP, Pochard E, Tanksley SD (1993) Construction of a molecular linkage map of pepper and a comparison of synteny with tomato. Genome 36:404–417

Prince JP, Lackney VK, Angeles C, Blauth JR, Kyle MM (1995) A survey of DNA polymorphism within the genus *Capsicum* and the fingerprinting of pepper cultivars. Genome 38:224–231

Rao GU, Paran I (2003) Polygalacturonase: a candidate gene for the soft flesh and deciduous fruit mutation in *Capsicum*. Plant Mol Biol 51:135–141

Rao GU, Ben Chaim A, Borovsky Y, Paran I (2003) Mapping of yield-related QTLs in pepper in an interspecific cross of *Capsicum annuum* and *C. frutescens*. Theor Appl Genet 106:1457–1466

Rodriguez JM, Berke T, Engle L, Nienhuis J (1999) Variation among and within *Capsicum* species revealed by RAPD markers. Theor Appl Genet 99:147–156

Ruffel S, Dussault MH, Palloix A, Moury B, Bendahmane A, Robaglia C, Caranta C (2002) A natural recessive resistance gene against potato virus Y in pepper corresponds to the eukaryotic initiation factor 4E (eIF4E). Plant J 32:1067–1075

Ruffel S, Carante C, Palloix A, Lefebvre V, Cabsche M, Bendahmane A (2004) Structural analysis of the eukaryotic initiation factor 4E (eIF4E) gene controlling potyvirus resistance in pepper: exploitation of a BAC library. Gene, in press

Shifriss C (1997) Male sterility in pepper (*Capsicum annuum* L.). Euphytica 93:83–88

Shin R, Han JH, Lee GJ, Peak KH (2002) The potential use of a viral coat protein gene as a transgene screening marker and multiple virus resistance of pepper plants coexpressing coat proteins of cucumber mosaic virus and tomato mosaic virus. Transgenic Res 11:215–219

Tai T, Staskawicz BJ (2000) Construction of a yeast artificial chromosome library of pepper (*Capsicum annuum* L.) and identification of clones from the *Bs2* resistance locus. Theor Appl Genet 100:112–117

Tai TH, Dahlbeck D, Clark ET, Gajiwala P, Pasion R, Whalen MC, Stall RE, Staskawicz BJ (1999a) Expression of the *Bs2* pepper gene confers resistance to bacterial spot disease in tomato. Proc Natl Acad Sci USA 96:14153–14158

Tai T, Dahlbeck D, Stall RE, Peleman J, Staskawicz BJ (1999b) High-resolution genetic and physical mapping of the region containing the *Bs2* resistance gene of pepper. Theor Appl Genet 99:1201–1206

Tanksley SD (1984) Linkage relationships and chromosomal locations of enzyme coding genes in pepper, *Capsicum annuum*. Chromosoma 89:352–360

Tanksley SD, Bernatzky R, Lapitan NL, Prince JP (1988) Conservation of gene repertoire but not gene order in pepper and tomato. Proc Natl Acad Sci USA 85:6419–6423

Tanksley SD, Ganal MW, Prince JP, de Vicente MC, Bonierbale MW, Broun P, Fulton TM, Giovannoni JJ, Grandillo S, Martin GB (1992) High density molecular linkage maps of the tomato and potato genomes. Genetics 132:1141–1420

Thabuis A (2002) Construction de résistance polygénique assistée par marqueurs: application à la résistance quantitative du piment (*Capsicum annuum* L.) à *Phytophthora capsici*. Thèse de Docteur en Sciences, Institut National Agronomique Paris-Grignon, France

Thabuis A, Lefebvre V, Daubèze AM, Signoret P, Phaly T, Nemouchi G, Blattes A, Palloix A (2001) Introgression of a partial resistance to *Phytophthora capsici* Leon. into a pepper elite line by marker assisted backcrosses. Acta Hortic 546:645–650

Thabuis A, Palloix A, Pflieger S, Daubèze AM, Caranta C, Lefebvre V (2003) Comparative mapping of *Phytophthora resistance* loci in pepper germplasm: evidence for conserved resistance loci across Solanaceae and for a large genetic diversity. Theor Appl Genet 106:1473–1485

Thabuis A, Palloix A, Servin B, Daubèze AM, Signoret P, Hospital F, Lefebvre V (2004) Marker-assisted introgression of 4 *Phytophthora capsici* resistance QTL alleles into a bell pepper line: validation of additive and epistatic effects. Mol Breed (online 2003)

Thabuis A, Lefebvre V, Bernard G, Daubèze AM, Pochard E, Palloix A (2004) Phenotypic and molecular evaluation of a recurrent selection program for a polygenic resistance to *Phytophthora capsici* in pepper. Theor Appl Genet (in press)

Thorup TA, Tanyolac B, Livingstone KD, Popovsky S, Paran I, Jahn M (2000) Candidate gene analysis of organ pigmentation loci in the Solanaceae. Proc Natl Acad Sci USA 97:11192–11197

Walsh BM, Hoot SB (2001) Phylogenetic relationships of *Capsicum* (Solanaceae) using DNA sequences from two noncoding regions: the chloroplast atpB-rbcL spacer region and nuclear waxy introns. Int J Plant Sci 162:1409–1418

Wang LH, Zhang BX, Lefebvre V, Huang SW, Daubèze AM, Palloix A (2004) QTL analysis of fertility restoration in cytoplasmic male sterile pepper. Theor Appl Genet, in press

Yoo EY, Kim S, Kim YH, Lee CJ, Kim BD (2003) Construction of a deep coverage BAC library from *Capsicum annuum*, 'CM334'. Theor Appl Genet 107:540–543

Yoon JY, Green SK, Talekar NS, Chen JT (1991) Pepper improvement in the tropics: problems and the AVRDC approach. Tomato and pepper production in the tropics. Proceedings of the international symposium on integrated management practices, Tainan, Taiwan, 21–26 March 1988, AVRDC, pp 86–98

Young ND, Zamir D, Ganal MW, Tanksley SD (1988) Use of isogenic lines and simultaneous probing to identify DNA markers tightly linked to the *Tm-2a* gene in tomato. Genetics 120:579–585

Zhang BX, Huang SW, Yang GM, Guo JZ (2000) Two RAPD markers linked to a major fertility restorer gene in pepper. Euphytica 113:155–161

II.8 Potato Genetics: Molecular Maps and More

C. Gebhardt[1]

1 Introduction: The Potato as an Object of Genetic Analysis

With 300,000 million tons produced per year, the potato is the fourth most important crop worldwide, after maize, rice and wheat (FAO statistics). The largest producers in terms of area planted and million tons harvested are China, Russia and India, whereas the most efficient producers with 37–45 t/ha are the western European countries and the United States (Graf 2002). Potatoes are grown for direct consumption, for processed food products such as chips and french fries, for animal feed and for industrial production of starch and starch derivatives. Potatoes may also be used in the future as bioreactors for producing specific organic molecules (Börnke et al. 2002).

After the rediscovery of Mendel's laws, potato was among the first plant species investigated for the mode of inheritance of an agronomic character. This was resistance to wart disease caused by the fungal pathogen *Synchytrium endobioticum* (Salaman and Lesley 1923). During the following 60 years, genetic analysis of potato was restricted mainly to single dominant characters (e.g., Cockerham 1970). The construction of classical genetic linkage maps based on morphological characters as done, for example, for the closely related tomato (Rick 1975), was not practical in potato, due to the tetraploidy of this crop species combined with tetrasomic inheritance, which aggravates the detection of linkage and largely prevents the recovery of recessive phenotypes.

Two technical developments made the construction of detailed genetic maps for the potato over the last 15 years possible: the ploidy reduction from the tetraploid to the diploid level, and the advent of DNA-based markers.

Diploid fertile plants can be generated from 2n gametes either by pollination of tetraploid plants with certain genotypes of *Solanum phureja*, which induces the parthenogenetic development of 2n female gametes into plants (Hougas et al. 1964; Hermsen and Verdenius 1973) or by regenerating plants from in vitro cultures of 2n microspores of tetraploid parents (Dunwell and Sunderland 1973; Powell and Uhrig 1987). Diploid potato plants are, however, largely self-incompatible and, therefore, highly heterozygous. Linkage analy-

---

[1] Max-Planck Institute for Plant Breeding Research, Carl-von-Linne-Weg 10, 50829 Köln, Germany

sis in progeny of diploid potato parents follows the same principles as human genetics: Partially heterozygous parents generate segregating F_1 offspring. The heterozygosity of the parents allows the construction of two linkage maps in a single F_1 mapping population, based on meiotic recombination in the female parent and in the male parent (Ritter et al. 1990).

DNA-based markers originate from the natural DNA variation present in a population of individuals of the same species. The molecular basis of the variation are point mutations (SNP, single nucleotide polymorphism) and insertions, deletions (InDels) or inversions of DNA fragments in one allele versus another. In contrast to induced mutations that often have a severe phenotypic effect, these DNA polymorphisms have survived evolutionary times because they did not compromise the viability and competitiveness of the individuals that carry them. Over the last 23 years, molecular genetics has developed an array of tools to detect natural DNA variation, beginning with hybridization-based analysis of restriction fragment length polymorphisms (RFLP; Botstein et al. 1980), progressing to PCR-based marker systems and currently ending with the detection of SNPs by comparative sequencing of alleles (reviewed by Reiter 2001).

2 Reference Molecular Maps of Potato

Using various types of DNA-based markers, detailed molecular linkage maps have been constructed for the 12 potato chromosomes in several diploid mapping populations, as summarized in Table 1. In most cases, the parents crossed for generating the mapping population included, in addition to *Solanum tuberosum*, other closely related tuber-bearing *Solanum* species that can easily be hybridized with *S. tuberosum*. One linkage map covering parts of the 48 linkage groups of tetraploid potato was also constructed (Meyer et al. 1998), based on AFLP markers (amplified fragment length polymorphism; Vos et al. 1995). The cytogenetic map of potato was anchored to the molecular maps by FISH (fluorescence in situ hybridization) markers derived from 12 chromosome-specific RFLP markers of known position on the molecular maps (Dong et al. 2000). These molecular maps and the DNA markers constituting them are the current backbone of genome analysis in cultivated potato and related tuber-bearing *Solanum* species.

3 Synteny of Potato with Other Plant Genomes

The RFLP assay is based on nucleic acid hybridization between a labeled marker probe and a membrane-bound genomic target sequence. Hence, depending on the experimental conditions used, cross-hybridization is

Table 1. Reference molecular maps constructed for potato chromosomes

Parental species[a]	Population name	Size and type of progeny	Marker type	No. of marker loci	Reference
S. phu × [S. tbr × S. chc]	?	65 F$_1$	Isozyme, RFLP	134	Bonierbale et al. (1988)
[S. tbr × S. tbr] × S. tbr[b]	BC916[2]	67 BC1	RFLP, SSR	~450	Gebhardt et al. (1989, 1991, 2001); Milbourne et al. (1998)
[S. tbr × S. ber] × S. ber	?	155 BC1	RFLP	~180	Tanksley et al. (1992)
[S. phu × S. tbr] × ([S. phu × S. tbr] × S. tbr)	CxE	67 BC1	Morphological, isozyme, RFLP, AFLP, cDNA-AFLP	~1200	Jacobs et al. (1995); van Eck et al. (1995); Brugmans et al. (2002)
S. tbr × [S. tbr × S. spg]	F1840	92 F$_1$	RFLP	445	Gebhardt et al. (1991, 2003); Leister et al. (1996)
S. tbr × S. tbr[b]	SHxRH	136 F$_1$	AFLP	~10,000	Rouppe van der Voort et al. (1997); http://www.dpw.wageningen-ur.nl/uhd/

[a] tbr, tuberosum; ber, berthaultii; phu, phureja; chc, chacoense; spg, spegazzinii
[b] S. tuberosum clones with genetic material from several wild species in their pedigree

detected between DNA sequences that are not identical, but similar, and RFLP markers originating from one species can be used for the construction of linkage maps in related species. Conserved genetic linkage between loci that share sequence similarity in different species indicates structural similarity between the different genomes. RFLP markers made it possible to compare the genome structure of potato, tomato and pepper, the three most important crop species of the Solanaceae family (Bonierbale et al. 1988; Gebhardt et al. 1991; Tanksley et al. 1992; Livingston et al. 1999). Comparative mapping revealed that the genomes of potato and tomato are co-linear except for paracentric inversions of five chromosome arms (Bonierbale et al. 1988; Tanksley et al. 1992). The larger genome of pepper is also syntenic with tomato/potato, but shows numerous rearrangements of chromosome fragments (Livingston et al. 1999). Synteny with potato/tomato, although with rearrangements, has also been demonstrated for genomes of nontuberbearing *Solanum* species (Perez et al. 1999). Structural similarity between genomes makes it possible "to look over the fence", i.e., to make functional comparisons between sexually incompatible species based on positional information. For example, the positions of genes controlling quantitative and qualitative resistance to pathogens can now be compared between potato, tomato and pepper (Leister et al. 1996; Grube et al. 2000; Gebhardt and Valkonen 2001). Recently, the comparison of a potato genetic map with the physical map of the sequenced *Arabidopsis* genome (Arabidopsis Genome Initiative 2000) revealed syntenic relationships between circa 40% of the potato genetic map and circa 50% of the physical map of this very distantly related plant species (Gebhardt et al. 2003).

4 Potato Function Map for Pathogen Resistance

The prevention of crop losses in yield and quality by disease is of primary importance in potato cultivation. Due to the changing populations of pests and pathogens, the selection of cultivars with improved genetic resistance continues, therefore, to be high on the agenda in potato breeding. Knowledge of the genetic basis of pathogen resistance in potato has broadened tremendously over the last 15 years, thanks to DNA-marker technologies. DNA-based markers have been used to localize on the potato molecular maps genes for qualitative (*R* genes) and quantitative (QTL, quantitative trait locus) resistance to various pests and pathogens: *Potato Viruses X, Y, N* and *A, Potato Leaf Roll Virus* (PLRV), the oomycete *Phytophthora infestans* (late blight), the fungus *Synchytrium endobioticum* (wart disease), the root nematodes *Globodera rostochiensis, Globodera pallida* and *Meloidogyne chitwoodi* and the soil bacterium *Erwinia carotovora* ssp. *atroseptica* (reviewed in Geb-

Fig. 1. Potato function map for pathogen resistance, modified after Gebhardt and Valkonen ▶ (2001, with permission from the Annual Review of Phytopathology, Vol. 39 2001 by Annual Reviews). The 12 schematic linkage groups correspond to the 12 potato chromosomes, with RFLP anchor markers shown to the *left*. Candidate gene loci are shown to the *right*. Candidate gene loci that were detected by resistance gene-like (RGL) markers are *underlined* (Leister et al. 1996; Hehl et al. 1999; Zimnoch-Guzowska et al. 2000). Other candidate gene loci were detected by pathogenesis-related (PR) markers (Leonards-Schippers et al. 1994; Gebhardt et al. 2001, 2003; Trognitz et al. 2002). *R*-genes and major resistance QTL of potato shown (*bold*) on the map are: virus resistance genes *Rx1, Rx2, Na, Nb, Ns, Nx, Ry* and *PLRV.1*; nematode resistance genes *Gro*, Gpa*, H1* and R_{mc1}; resistance genes to oomycetes or fungi *R1, R2, R3, R6, R7, RB, R_{ber}* and *Sen1*. Some resistance loci of tomato, tobacco and pepper that can be anchored to potato linkage groups are shown (*outlined*) at syntenic positions: Virus resistance genes: *Tm-1, Tm-2a, TY-1, Sw-5* (all from tomato), *L* (from pepper) and *N* (from tobacco); tomato nematode resistance genes *Hero, Mi* and *Mi-3*; tomato resistance genes to fungi *Cf*, I*, Asc, Ol-1, Lv* and *Sm*; tomato resistance genes to bacteria *Pto* and *Ve*; tomato gene *Meu1* for resistance to aphids. Potato QTL for resistance to *Phytophthora infestans* (late blight) and *Erwinia carotovora* ssp. *atroseptica* are indicated as *black* and *gray bars*, respectively, on the linkage groups. For more detailed information and further references, see Leister et al. (1996), Grube et al. (2000) and Gebhardt and Valkonen (2001)

Potato Genetics: Molecular Maps and More

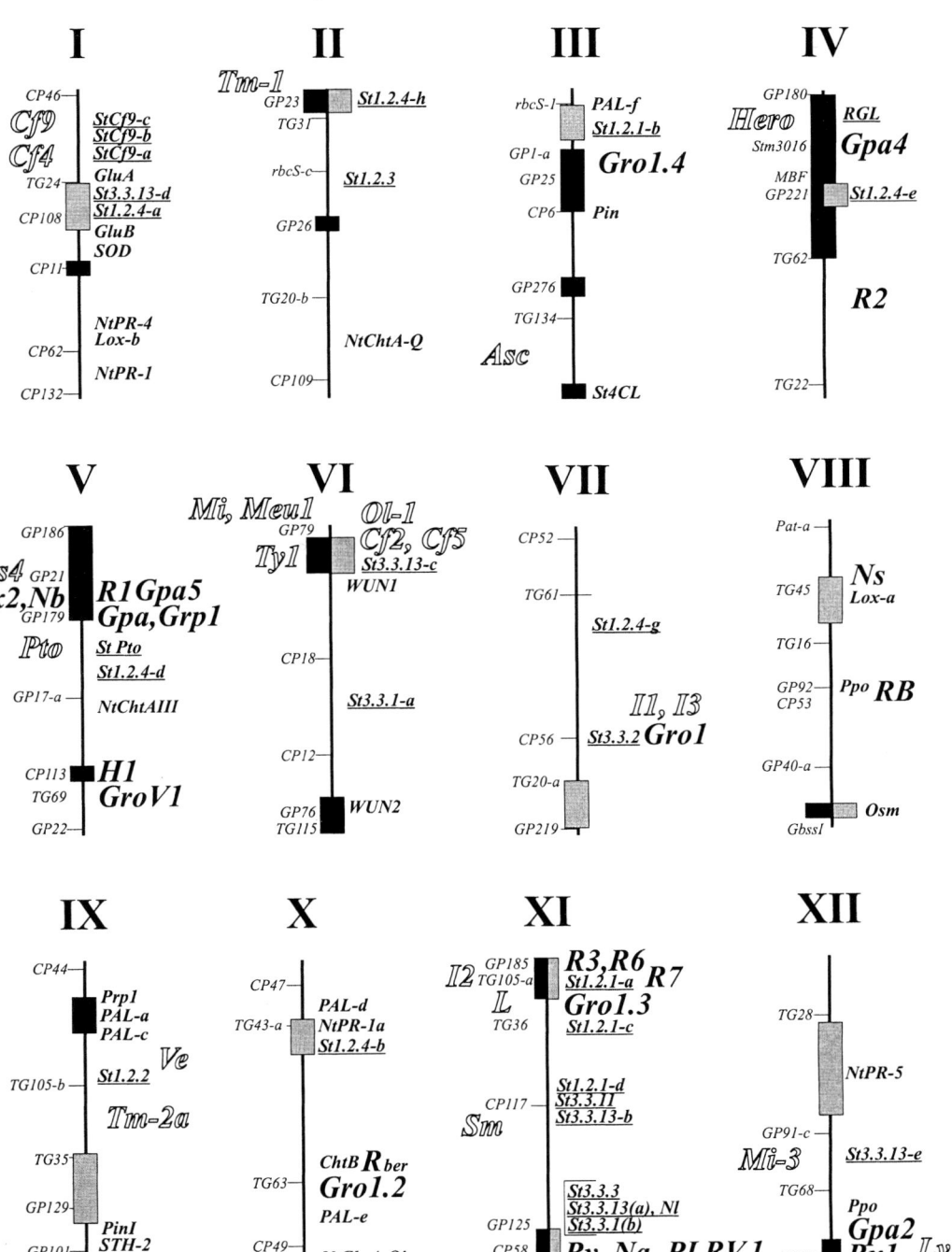

hardt and Valkonen 2001; see also Marczewski et al. 2001, 2002). Most genes for resistance have been introgressed into potato breeding lines over the last 80 years from other tuber-bearing *Solanum* species by sexual hybridization and back-crossing (Ross 1986). More recently, somatic hybridization has also been used for introgression of resistance factors (Brown et al. 1996; Helgeson et al. 1998). Mapping experiments performed in potato are very heterogeneous. They are carried out in different genetic materials, either based on whole genome mapping or bulked segregant analysis and using different sets of RFLP, RAPD (random amplified polymorphic DNA), AFLP and SSR (simple sequence repeat) markers. In many cases, however, at least some RFLP or SSR markers from the potato/tomato reference maps (Table 1) were included in the mapping experiments. Such anchor markers make it possible to compare the positions of resistance factors across otherwise independent mapping experiments, even across species borders, and to integrate this information into a potato function map for pathogen resistance (Fig. 1).

From this map the following observation is evident: a number of *R* genes and QTL for resistance to different types of pathogens map to similar positions, so-called hot spots for resistance in the potato genome. For example, there is such a hot spot for resistance on potato chromosome V, which includes major genes *Rx2* and *Nb* both for resistance to *Potato Virus X*, the *R1* gene for race-specific resistance to late blight, QTL for resistance to late blight and QTL for resistance to the nematode *G. pallida* (Fig. 1). At least three further resistance hot spots are located in both distal regions of chromosome XI and on chromosome XII. From classical genetic studies in plants, it is known that single genes for resistance to different races of a pathogen may be tightly linked, either because they are multiple alleles of one gene, or because several related genes are located next to each other in the same narrow genome segment (Pryor and Ellis 1993). Such clustered gene families evolve from common ancestors by local gene duplications followed by structural and functional diversification. The same model may be used to explain the observed clustering of genes for qualitative and quantitative resistance to different pathogens.

Molecular characterization of the first four resistance genes from potato showed that they all belong to the same superfamily of plant genes for pathogen resistance, which share a putative nucleotide-binding and a leucine-rich repeat domain (NB-LRR type genes; Bendahmane et al. 1999, 2000; van der Vossen et al. 2000; Ballvora et al. 2002). Moreover, RFLP mapping of DNA fragments with sequence similarity to NB-LRR type genes revealed close linkage of RGLs (resistance-gene-like sequences) to resistance loci, particularly in resistance hot spots (Fig. 1; Leister et al. 1996). This suggests that NB-LRR type genes are candidates for being the molecular basis of a considerable proportion of qualitative and quantitative resistance factors in potato. Based on this candidate gene hypothesis, the *Gro1* gene for resistance to the root cyst nematode *G. rostochiensis* has been cloned (Paal et al. 2004). Candidates for participating in the control of quantitative resistance may be, in addition to NB-LRR type genes, genes functional in pathogenesis (pathogenesis-related, PR genes). A number of such genes have also been mapped (Leonards-Schippers et al. 1994; Gebhardt et al. 2001, 2003; Trognitz et al. 2002) and are included in the function map for resistance (Fig. 1).

5 Potato Function Map for Tuber Traits

Most characters relevant for the tuber crop show continuous phenotypic variation because they are controlled by multiple genes and by environmental factors. Tuber flesh color, tuber skin color and tuber shape segregated and were mapped, however, as Mendelian factors (Bonierbale et al. 1988; Gebhardt et al. 1991; van Eck et al. 1994). DNA-based markers made it possible to dissect the genetic components of quantitative tuber traits for the first time. QTL analysis has been done for tuberization (van den Berg et al. 1996a), tuber dormancy (Freyre et al. 1994; van den Berg et al. 1996b), tuber shape (van Eck et al. 1994), tuber starch content (Freyre and Douches 1994; Schäfer-Pregl et al. 1998), tuber yield (Schäfer-Pregl et al. 1998), chip color (Douches and Freyre 1994) and cold sweetening (Menendez et al. 2002). Similar to resistance factors, it is possible to compile, based on anchor markers, the results of mapping experiments in different genetic materials into a potato function map for tuber traits (Fig. 2). This function map also contains the approximate chromosomal positions of candidate gene loci for controlling starch and sugar QTL (Chen et al. 2001; Gebhardt et al. 2001; Menendez et al. 2002). Candidate genes for controlling natural variation of tuber starch and sugar content are particularly genes that function in carbohydrate metabolism and transport.

It may be observed on this map that some QTL for tuber starch content map to similar positions as QTL for sugar content and even yield QTL. Particularly intriguing is a segment of chromosome V, where QTL for tuber starch content, sugar content and yield are linked. This same genomic region also harbors QTL for plant maturity and vigor (Collins et al. 1999; Oberhagemann et al. 1999; Visker et al. 2003). The resistance hot spot mentioned above is also located in this same region. Overlapping of QTL for different traits is expected if the observed phenotypes result either from pleiotropic effects of a single gene or from the effects of closely linked, but otherwise unrelated genes. Which of these possibilities applies, is subject to detailed molecular and functional analysis in such genomic regions. Some genes functional in carbohydrate metabolism or transport are closely linked to QTL for tuber traits (Fig. 2; Menendez et al. 2002). Allelic variants of such genes may, indeed, be responsible for the observed QTL effects.

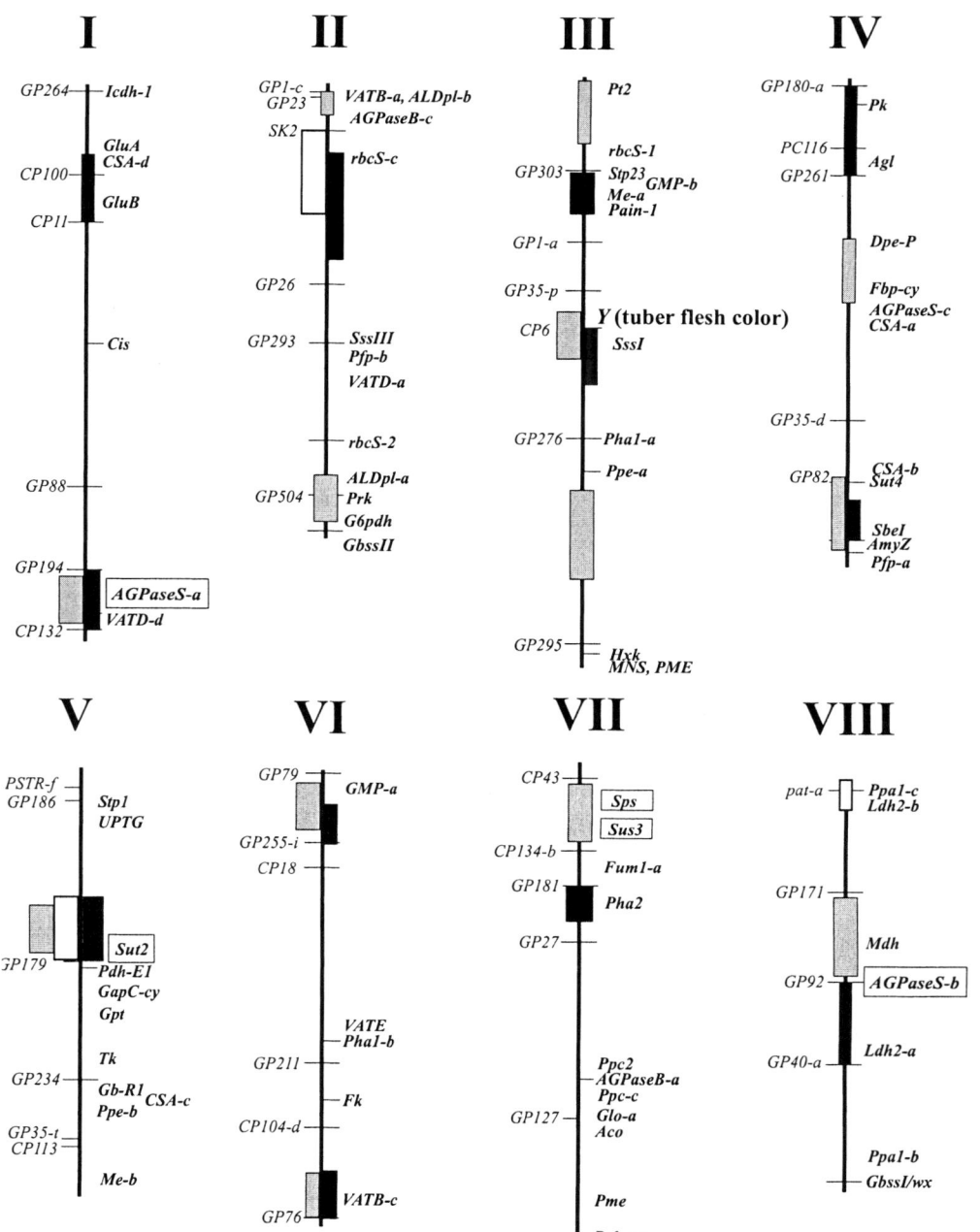

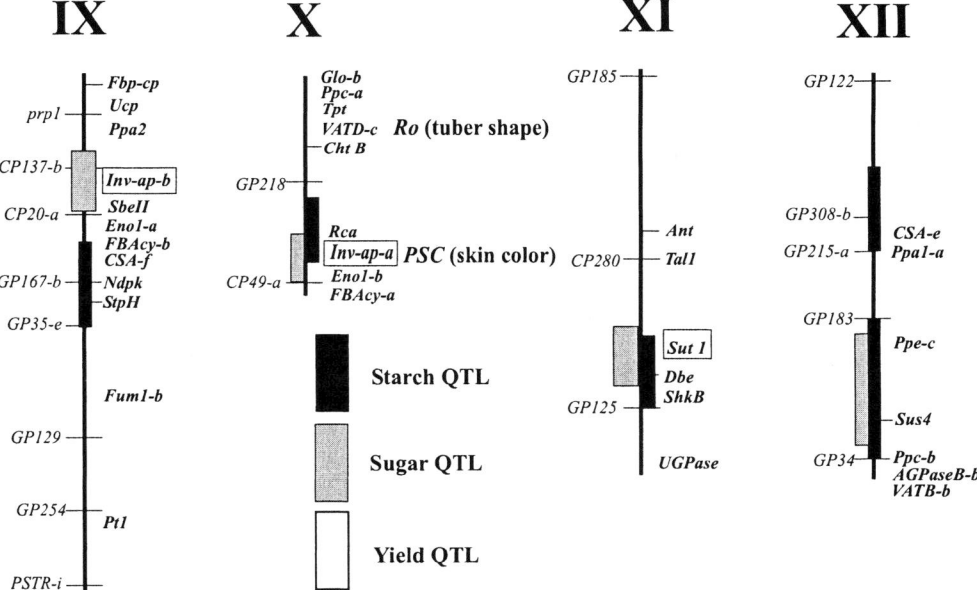

Fig. 2. Potato function map for tuber traits. The 12 linkage groups are based on the RFLP map constructed for population K31 described in Schäfer-Pregl et al. (1998). Anchor RFLP markers are shown to the *left*, and candidate gene loci (Chen et al. 2001; Gebhardt et al. 2003) are shown to the *right* of the linkage groups. QTL for tuber starch, tuber yield (Schäfer-Pregl et al. 1998) and cold sweetening (Menendez et al. 2002) are shown as *black*, *white* and *gray bars*, respectively. Some promising candidate gene loci for QTL for tuber traits (Menendez et al. 2002) are framed

6 Potato Function Maps as Basis for Innovative Approaches to Breeding

Integrating positional information of genetic factors controlling agronomic characters across mapping experiments and even across related species gives access to numerous, potentially useful DNA markers beyond those employed within an individual mapping experiment. Based on positional information, crossing strategies can be designed for combining genes from different sources. Linked markers can be selected for the development of PCR-based, allele-specific, diagnostic assays (examples in: Niewöhner et al. 1995; Marczewski et al. 2001; Bryan et al. 2002). Allele-specific markers can then be used to select in crosses plants that have or do not have a particular trait allele. The reliability of the marker test depends, however, on the rate of recombination between the marker and the gene of interest. The predictive value of a marker decreases with increasing genetic distance and is often restricted to specific crosses where the marker allele is known to be linked in coupling phase with the trait allele (known marker phase). Breeding potato cultivars, however, is based on intercrossing hundreds of genotypes of various pedigrees. In such a wide gene pool, prediction of phenotype based on marker genotype is not straightforward because, due to an unknown number of meiotic generations separating the individuals in the population, linkage equilibrium may have been reached between a specific trait allele and a linked marker allele (unknown marker phase). Only when the marker resides within or is physically close to the gene that controls the phenotype, recombination between marker allele and trait allele is absent or rare even after many generations of meiotic recombination. In this case, linkage disequilibrium between marker allele and trait allele persists, even in wide gene pools (Collins et al. 1997). Markers based on DNA polymorphism in those genes that control agronomic characters are, therefore, the ultimate diagnostic tools for marker-assisted breeding.

7 Conclusions

The first potato markers have been identified that have more general diagnostic value. Markers based on a candidate resistance-gene-like sequence mapping to the resistance hot spot on potato chromosome XI, which includes the *Ry* gene for resistance to *Potato Virus Y* (Fig. 1; Leister et al. 1996) were diagnostic for the presence of the Ry_{adg} gene in different potato germplasm (Sorri et al. 1999; Kasai et al. 2000). Thus, candidate gene markers are excellent tools when searching for "universal" markers for use in marker-assisted breeding. A marker diagnostic for a DNA segment introgressed from *Solanum vernei*, which carries genes for resistance to the root cyst nematode *Globodera pallida* on chromosome V has also been found (Bryan et al. 2002). Markers located in the resistance hot spot on potato chromosome V that includes the *R1* gene for resistance to late blight (Fig. 1) were associated with QTL for late blight resistance and plant maturity when scored in more than 400 potato cultivars (Gebhardt et al. 2004). Thus, function maps, which integrate positions of candidate genes with positions of factors controlling agronomic traits can provide a knowledge base for selecting the most appropriate markers to develop marker-assisted breeding.

References

Arabidopsis Genome Initiative (2000) Analysis of the genome sequence of the flowering plant *Arabidopsis thaliana*. Nature 408:796–815
Ballvora A, Ercolano MR, Weiß J, Meksem K, Bormann C, Oberhagemann P, Salamini F, Gebhardt C (2002) The *R1* gene for potato resistance to late blight (*Phytophthora infestans*) belongs to the leucine zipper/NBS/LRR class of plant resistance genes. Plant J 30:361–371

Bendahmane A, Kanyuka K, Baulcombe DC (1999) The *Rx* gene from potato controls separate virus resistance and cell death responses. Plant Cell 11:781–791

Bendahmane A, Querci M, Kanyuka K, Baulcombe DC (2000) *Agrobacterium* transient expression system as a tool for the isolation of disease resistance genes: application to the *Rx2* locus in potato. Plant J 21:73–81

Bonierbale M, Plaisted RL, Tanksley SD (1988) RFLP maps based on a common set of clones reveal modes of chromosomal evolution in potato and tomato. Genetics 120:1095–1103

Börnke FM, Hajirezaei M, Sonnewald U (2002) Potato tubers as bioreactors for palatinose production. J Biotechnol 96:119–124

Botstein D, White RL, Skolnick M, Davis RW (1980) Construction of a genetic linkage map in man using restriction fragment length polymorphisms. Am J Hum Genet 32:314–331

Brown CR, Yang C-P, Mojtahedi H, Santo GS (1996) RFLP analysis of resistance to Columbia root-knot nematode derived from *Solanum bulbocastanum* in a BC$_2$ population. Theor Appl Genet 92:572–576

Brugmans B, Fernandez del Carmen A, Bachem CWB, van Os H, van Eck H, Visser RGF (2002) A novel method for the construction of genome wide transcriptome maps. Plant J 31:211–222

Bryan GJ, McLean K, Bradshaw JE, de Jong WS, Phillips M, Castelli L, Waugh R (2002) Mapping QTLs for resistance to the cyst nematode *Globodera pallida* derived from the wild potato species *Solanum vernei*. Theor Appl Genet 105:68–77

Chen X, Salamini F, Gebhardt C (2001) A potato molecular function map for carbohydrate metabolism and transport. Theor Appl Genet 102:284–295

Cockerham G (1970) Genetical studies on resistance to potato viruses X and Y. Heredity 25:309–348

Collins A, Milbourne D, Ramsay L, Meyer R, Chatot-Balandras C, Oberhagemann P, de Jong W, Gebhardt C, Bonnel E, Waugh R (1999) QTL for field resistance to late blight in potato are strongly correlated with earliness and vigour. Mol Breed 5:387–398

Collins FS, Guyer MS, Chakravarti A (1997) Variations on a theme: cataloguing human DNA sequence variation. Science 278:1580–1581

Dong F, Song J, Naess SK, Helgeson JP, Gebhardt C, Jiang J (2000) Development and applications of a set of chromosome-specific cytogenetic DNA markers in potato. Theor Appl Genet 101:1001–1007

Douches DS, Freyre R (1994) Identification of genetic factors influencing chip color in diploid potato (*Solanum* spp.). Am Potato J 71:581–590

Dunwell JM, Sunderland N (1973) Anther culture of *Solanum tuberosum* L. Euphytica 22:317–323

Freyre R, Douches DS (1994) Development of a model for marker-assisted selection of specific gravity in diploid potato across environments. Crop Sci 34:1361–1368

Freyre R, Warnke S, Sosinski B, Douches DS (1994) Quantitative trait locus analysis of tuber dormancy in diploid potato (*Solanum* spp.). Theor Appl Genet 89:474–480

Gebhardt C, Valkonen JPT (2001) Organization of genes controlling disease resistance in the potato genome. Annu Rev Phytopathol 39:79–102

Gebhardt C, Ritter E, Barone A, Debener T, Walkemeier B, Schachtschabel U, Kaufmann H, Thompson RD, Bonierbale MW, Ganal MW, Tanksley SD, Salamini F (1991) RFLP maps of potato and their alignment with the homeologous tomato genome. Theor Appl Genet 83:49–57

Gebhardt C, Ritter E, Debener T, Schachtschabel U, Walkemeier B, Uhrig H, Salamini F (1989) RFLP analysis and linkage mapping in *Solanum tuberosum*. Theor Appl Genet 78:65–75

Gebhardt C, Ritter E, Salamini F (2001) RFLP map of the potato. In: Phillips RL, Vasil IK (eds) DNA-based markers in plants, 2nd edn. Advances in cellular and molecular biology of plants, vol 6. Kluwer, Dordrecht, pp 319–336

Gebhardt C, Walkemeier B, Henselewski H, Barakat A, Delseny M, Stüber K (2003) Comparative mapping between potato (*Solanum tuberosum*) and *Arabidopsis thaliana* reveals structurally conserved domains and ancient duplications in the potato genome. Plant J 34:529–541

Gebhardt C, Ballvora A, Walkemeier B, Oberhagemann P, Schüler K (2004) Genetic potential assessment in germ plasm collections of crop plants by marker-trait association: a case study for potatoes with quantitative variation of resistance to late blight and maturity type. Mol Breed 13:93–102

Graf G (2002) Die Weltkartoffelernte 2001. Kartoffelbau 11:440–442

Grube RC, Radwanski ER, Jahn M (2000) Comparative genetics of disease resistance within the Solanaceae. Genetics 155:873–887

Hehl R, Faurie E, Hesselbach J, Salamini F, Whitham S, Baker B, Gebhardt C (1999) TMV resistance gene *N* homologues are linked to *Synchytrium endobioticum* resistance in potato. Theor Appl Genet 98:379–386

Helgeson JP, Pohlman JD, Austin S, Haberlach GT, Wielgus SM, Ronis D, Zambolim L, Tooley P, McGrath JM, James RV, Stevenson WR (1998) Somatic hybrids between *Solanum bulbocastanum* and potato: a new source of resistance to late blight. Theor Appl Genet 96:738–742

Hermsen JGTH, Verdenius J (1973) Selection from *Solanum tuberosum* group phureja of genotypes combining high-frequency haploid induction with homozygosity for embryo spot. Euphytica 22:244–259

Hougas RW, Peloquin SJ, Gabert AC (1964) Effect of seed parent and pollinator on the frequency of haploids in *Solanum tuberosum*. Crop Sci 4:593–595

Jacobs JME, van Eck HJ, Arens P, Verkerk-Bakker B, te Lintel Hekkert B, Bastiaanssen HJM, El Kharbotly A, Pereira A, Jacobsen E, Stiekema WJ (1995) A genetic map of potato (*Solanum tuberosum*) integrating molecular markers, including transposons, and classical markers. Theor Appl Genet 91:289–300

Kasai K, Morikawa Y, Sorri VA, Valkonen JPT, Gebhardt C, Watanabe KN (2000) Development of SCAR markers to the PVY resistance gene Ry_{adg} based on a common feature of plant disease resistance genes. Genome 43:1–8

Leister D, Ballvora A, Salamini F, Gebhardt C (1996) A PCR based approach for isolating pathogen resistance genes from potato with potential for wide application in plants. Nature Genet 14:421–429

Leonards-Schippers C, Gieffers W, Schäfer-Pregl R, Ritter E, Knapp SJ, Salamini F, Gebhardt C (1994) Quantitative resistance to *Phytophthora infestans* in potato: a case study for QTL mapping in an allogamous plant species. Genetics 137:67–77

Livingstone KD, Lackney VK, Blauth JR, van Wijk R, Kyle-Jahn M (1999) Genome mapping in *Capsicum* and the evolution of genome structure in the Solanaceae. Genetics 152:1173–1202

Marczewski W, Flis, B, Syller J, Schäfer-Pregl R, Gebhardt C (2001) A major QTL for resistance to *Potato leafroll virus* (PLRV) is located in a resistance hotspot on potato chromosome XI and is tightly linked to *N*-gene-like markers. Mol Plant Microbe Interact 14:1420–1425

Marczewski W, Hennig J, Gebhardt C (2002) The potato virus S gene *Ns* maps to potato chromosome VIII. Theor Appl Genet 105:564–567

Menendez CM, Ritter E, Schäfer-Pregl R, Walkemeier B, Kalde A, Salamini F, Gebhardt C (2002) Cold-sweetening in diploid potato. Mapping QTL and candidate genes. Genetics 162:1423–1434

Meyer RC, Milbourne D, Hackett CA, Bradshaw JE, McNichol JW, Waugh R (1998) Linkage analysis in tetraploid potato and association of markers with quantitative resistance to late blight (*Phytophthora infestans*). Mol Gen Genet 259:150–160

Milbourne D, Meyer RC, Collins AJ, Ramsay LD, Gebhardt C, Waugh R (1998) Isolation, characterisation and mapping of simple sequence repeat loci in potato. Mol Gen Genet 259:233–245

Niewöhner J, Salamini F, Gebhardt C (1995) Development of PCR assays diagnostic for RFLP markers closely linked to alleles *Gro1* and *H1*, conferring resistance to the root cyst nematode *Globodera rostochiensis* in potato. Mol Breed 1:65–78

Oberhagemann P, Chatot-Balandras C, Bonnel E, Schäfer-Pregl R, Wegener D, Palomino C, Salamini F, Gebhardt C (1999) A genetic analysis of quantitative resistance to late blight in potato: towards marker assisted selection. Mol Breed 5:399–415

Paal J, Henselewski H, Muth J, Meksem K, Menéndez CM, Salamini F, Ballvora A, Gebhardt C (2004) Molecular cloning of the potato *Gro1* gene for resistance to the root cyst nematode *Globodera rostochiensis* based on a candidate gene approach. Plant J 38:285–297

Perez F, Menendez A, Dehal P, Quiros CF (1999) Genomic structural differentiation in Solanum: comparative mapping of the A- and E-genomes. Theor Appl Genet 98:1183–1193

Powell W, Uhrig H (1987) Anther culture of *Solanum* genotypes. Plant Cell Tissue Org Cult 11:13–24

Pryor T, Ellis J (1993) The genetic complexity of fungal resistance genes in plants. Adv Plant Pathol 10:281–305

Reiter R (2001) PCR-based marker systems. In: Phillips RL, Vasil IK (eds) DNA-based markers in plants, 2nd edn. Advances in cellular and molecular biology of plants, vol 6. Kluwer, Dordrecht, pp 9–29

Rick CM (1975) The tomato. In: King RC (ed) Handbook of genetics, vol 2. Plenum Press, New York, pp 247–280

Ritter E, Gebhardt C, Salamini F (1990) Estimation of recombination frequencies and construction of RFLP linkage maps in plants from crosses between heterozygous parents. Genetics 125:645–654

Ross H (1986) Potato breeding. Problems and perspectives. Adv Plant Breed Suppl 13

Rouppe van der Voort J, Wolters P, Folkertsma R, Hutten R, van Zandvoort P, Vinke H, Kanyuka K, Bendahmane A, Jacobsen E, Janssen R, Bakker J (1997) Mapping of the cyst nematode resistance locus *Gpa2* in potato using a strategy based on comigrating AFLP markers. Theor Appl Genet 95:874–880

Salaman RN, Lesley JW (1923) Genetic studies in potatoes: the inheritance of immunity to wart disease. J Genet 13:177–186

Schäfer-Pregl R, Ritter E, Concilio L, Hesselbach J, Lovatti L, Walkemeier B, Thelen H, Salamini F, Gebhardt C (1998) Analysis of quantitative trait loci (QTL) and quantitative trait alleles (QTA) for potato tuber yield and starch content. Theor Appl Genet 97:834–846

Sorri VA, Watanabe KN, Valkonen JPT (1999) Predicted kinase 3a motif of a resistance gene-like fragment as a unique marker for PVY resistance. Theor Appl Genet 99:164–170

Tanksley SD, Ganal MW, Prince JP, de Vicente MC, Bonierbale MW, Broun P, Fulton TM, Giovannoni JJ, Grandillo S, Martin GB, Messeguer R, Miller JC, Miller L, Paterson AH, Pineda O, Röder MS, Wing RA, Wu W, Young ND (1992) High density molecular linkage maps of the tomato and potato genomes. Genetics 132:1141–1160

Trognitz FC, Manosalva PM, Nino-Liu DO, Herrera MR, Ghislain M, Trognitz BR, Nelson RJ (2002) Plant defense genes associated with quantitative resistance to potato late blight in *Solanum phureja* × *S. tuberosum* hybrids. Mol Plant Microbe Interact 15:587–597

Van den Berg JH, Ewing EE, Plaisted RL, McMurry S, Bonierbale MW (1996a) QTL analysis of potato tuberization. Theor Appl Genet 93:307–316

Van den Berg JH, Ewing EE, Plaisted RL, McMurry S, Bonierbale MW (1996b) QTL analysis of potato tuber dormancy. Theor Appl Genet 93:317–324

Van der Vossen EAG, Rouppe van der Voort JNAM, Kanyuka K, Bendahmane A, Sandbrink H, Baulcombe DC, Bakker J, Stiekema WJ, Klein-Lankhorst RM (2000) Homologues of a single resistance-gene cluster in potato confer resistance to distinct pathogens: a virus and a nematode. Plant J 23:567–576

Van Eck HJ, Jacobs JME, Stam P, Ton J, Stiekema WJ, Jacobsen E (1994) Multiple alleles for tuber shape in diploid potato detected by qualitative and quantitative genetic analysis using RFLPs. Genetics 137:303–309

Van Eck HJ, Rouppe van der Voort J, Draaistra J, van Zandvoort P, van Enckevort E, Segers B, Peleman J, Jacobsen E, Helder J, Bakker J (1995) The inheritance and chromosomal localization of AFLP markers in a non-inbred potato offspring. Mol Breed 1:397–410

Visker MHPW, Keizer LCP, Van Eck HJ, Jacobsen E, Colon LT, Struik PC (2003) Can the QTL for late blight resistance on potato chromosome 5 be attributed to foliage maturity type? Theor Appl Genet 106:317–325

Vos P, Hogers R, Bleeker M, Reijans M, van de Lee T, Hornes M, Frijters A, Pot J, Peleman J, Kuiper M, Zabeau M (1995) AFLP: a new technique for DNA fingerprinting. Nucleic Acids Res 23:4407–4414

Zimnoch-Guzowska E, Marczewski W, Lebecka R, Flis B, Schäfer-Pregl R, Salamini F, Gebhardt C (2000) QTL analysis of new sources of resistance to *Erwinia carotovora* ssp. *atroseptica* in potato done by AFLP, RFLP and resistance-gene-like markers. Crop Sci 40:1156–1167

II.9 Molecular Marker Maps of Barley: A Resource for Intra- and Interspecific Genomics

R.K. Varshney, M. Prasad, and A. Graner[1]

1 Introduction

Barley (*Hordeum vulgare* L.) is an important staple crop ranking fourth in the world food production area. It is grown mainly for animal feed and as raw material for beer production in a wide range of temperate and semi-arid environments with major areas of production in the European Union, Russia, and North America.

It belongs to the Triticeae tribe of the Poaceae, the largest family within the monocotyledonous plants. The genus *Hordeum* comprises 32 species and altogether 45 taxa of which *H. vulgare* ssp. *vulgare* is the only species that underwent domestication (von Bothmer et al. 1995). Indeed, barley is considered as one of the founder species of modern agriculture as it was domesticated about 10,000 years ago from the wild progenitor, *Hordeum vulgare* ssp. *spontaneum* most probably in the western part of the Fertile Crescent (Zohary and Hopf 2001; Salamini et al. 2002). Barley is a self-pollinating diploid with 2n=2x=14 chromosomes. Its genome has been estimated to comprise approximately 5.3×10^9 bp (Bennett and Smith 1976). Like other cereals, its genome consists of a complex mixture of unique and repeated nucleotide sequences (Flavell 1980). While repeated sequences frequently may present an obstacle to genome analysis, several other features make barley the organism of choice for genome analysis within the Triticeae tribe. These include the possibility to develop doubled haploid (DH) lines, the availability of numerous mutants and cytogenetic stocks like wheat-barley addition lines, and the availability of large DNA-insert libraries (for details see Kleinhofs and Graner 2001). Moreover, after more than a decade of intense research, efficient transformation protocols in barley have been developed for the stable transformation of both immature embryos and microspores (for review, see Lemaux et al. 1999). In this chapter, we will review recent progress related to the generation of genetic and physical maps in barley. Moreover, we will summarize the available data on comparative mapping between rice and barley and discuss strategies of how to use that knowledge for various aspects of basic and applied genome research.

---

[1] Institute of Plant Genetics and Crop Plant Research (IPK), Corrensstrasse 3, 06466 Gatersleben, Germany

2 Molecular Markers

During the last two decades, a variety of molecular markers have been developed. Restriction fragment length polymorphism (RFLP) markers became available as early as 1980 (Botstein et al. 1980), which opened an era of molecular plant genetics. A second marker type, randomly amplified polymorphic DNA (RAPDs) markers became available in 1990 (Williams et al. 1990). Although these two markers systems were used in many plant species, they did not prove to be the markers of choice because RFLPs involve much time and labor while RAPDs suffer from reproducibility. Consequently, new molecular markers like SSRs (simple sequence repeats; Tautz 1989), AFLPs (amplified fragment length polymorphisms; Vos et al. 1995) and their modified forms were developed (for review, see Gupta et al. 2002). More recently, SNP markers (single nucleotide polymorphisms; Wang et al. 1998) have been added to this list. In addition to these generic marker systems, DNA markers that exploit the specificities of the barley genome have been developed. They are based on the presence of highly abundant, dispersed repetitive retroelements. In particular, sequence features of the long terminal repeats have been exploited for the development of marker assays such as sequence-specific amplification polymorphism (S-SAP; Waugh et al. 1997), inter-retrotransposon amplified polymorphism (IRAP; Kalendar et al. 1999), retrotransposon-microsatellite amplified polymorphism (REMAP, Kalendar et al. 1999), and *copia*-SSR (Provan et al. 1999). The association of miniature inverted repeat elements (MITEs) with transcribed regions as has been described for the maize genome (Bureau and Wessler 1994; Bhattarammaki et al. 2002), has been exploited in barley for the detection of inter-miniature repeat polymorphisms (IMP; Chang et al. 2001). Each marker system mentioned above has some advantages as well as disadvantages (discussed in Gupta et al. 2002) and the user has to make his choice based on the intended objective, convenience and costs.

3 Construction of Molecular Maps

In the past, genetic maps of higher plants were based almost entirely on morphological and biochemical markers. However, these markers did not prove useful in the preparation of maps with good resolution as their numbers are limited and allelic variants are frequently restricted to exotic germplasm, which precludes their application in breeding programs. The availability of DNA-based markers facilitated the construction of maps with a high marker density in almost all the crop species (Phillips and Vasil 2001). These genetic maps proved very useful for trait mapping, QTL identification, and the characterization of germplasm collections (see Jain et al. 2002). Moreover, the correlation of genetic and physical maps provides insight in the organization of the barley genome in terms of recombination frequency and the distribution of genes along the individual chromosomes.

3.1 Genetic Maps

In barley the first report of a molecular marker map for chromosome 6H appeared in 1988 and a partial map of the whole genome incorporating RFLPs, morphological, isozyme and PCR markers was published in 1990 (Kleinhofs et al. 1988; Shin et al. 1990). The first comprehensive marker maps were developed using three different F1-derived doubled haploid populations, namely Igri × Franka (Graner et al. 1991), Proctor × Nudinka (Heun et al. 1991), and Steptoe × Morex (Kleinhofs et al. 1993). Since then the number of maps has further increased. Due to the more or less unlimited availability of DNA from the DH-populations, the initial RFLP maps could be further extended as new marker types emerged, thus forming a dynamic resource that can be continuously improved. The majority of mapping efforts were confined to germplasm of the primary gene pool, which comprises only cultivated barley (*Hordeum vulgare* ssp. *vulgare*) and its wild progenitor (*H. vulgare* ssp. *spontaneum*). More recently, however, a map of *H. bulbosum* has been published (Jaffe et al. 2000; Salvo-Garrido et al. 2001). This is the only species representing the secondary genepool of barley. The availability of a marker map for *H. bulbosum* may represent the first step in systematically exploring the genetics of resistance to biotic and abiotic stress that is characteristic for this species. The same will apply to the map of *H. chilense*, being a member of the tertiary gene pool (Hernandez et al. 2001). Although there is a sterility barrier preventing the meiotic transfer of genes from the corresponding species into *H. vulgare*, transgenic strategies may provide a means to access the genetic diversity of this gene pool. A list of all major maps is shown in Table 1. Detailed and updated information on most of these maps is available at the GrainGenes website (http://wheat.pw.usda.gov/ggpages/maps.shtml).

So far, more than 3000 genetic markers have been mapped in barley. However, this large number may not belie the difficult integration of the individual maps. Only a limited number of maps contain sufficient common markers to merge their information in consensus maps as they were constructed by several groups (Langridge et al. 1995; Sherman et al. 1995; Qi et al. 1996). These consensus maps display higher marker densities than their individual components, which makes them a highly useful resource, if markers are required for specific chromosomal regions. On the other hand, the accuracy of consensus maps may decrease at higher resolution, in particular in those regions where marker densities are high and the number of markers in common to the maps that were merged is low.

In an attempt to develop a more generic solution for the integration of information from the different genetic maps at medium genetic resolution,

Table 1. List of some comprehensive genetic maps[a] available for barley

Map type	Population used for mapping	Number of mapped loci	Genome coverage (cM)	References
RFLP maps	Doubled haploids (DH; Proctor × Nudinka)	154	1091	Heun et al. (1991)
	2 DH populations (Igri × Franka, H. vulgare ssp. Vada × H. vulgare ssp. spontaneum line 1B-87)	226	1453	Graner et al. (1991)
	DH (Steptoe × Morex)	295	1250	Kleinhofs et al. (1993)
	DH (Harrington × TR306)	216	1060	Kasha et al. (1994)
	F_2 population (Ko A × Mokusekko)	222	1389	Miyazaki et al. (2000)
	F_1 full-sib families (H. bulbosum; PB1 × PB11)	136	621	Salvo-Garrido et al. (2001)
SSR maps	DH (H. vulgare var Lina × H. spontaneum Canada Park)	242	1173	Ramsay et al. (2000)
AFLP maps	DH (Proctor × Nudinka)	118	1096	Becker et al. (1995)
	Recombinant inbred lines (RILs; L94 × Vada)	561	1062	Qi et al. (1998)
	DH (Proctor × Nudinka)	511	2673	Castiglioni et al. (1998)
Composite maps[b]	Consensus map from 7 maps	587	1087	Langridge et al. (1995)
	Consensus map from 4 maps	898	1060	Qi et al. (1996)
	F_2 population (H. chilense)	123	694	Hernandez et al. (2001)
	DH (OWB_{Dom} × OWB_{Rec})	713	1387	Costa et al. (2001)
	RILs (Azumamugi × Kato Nakate Gold)	272	925.6	Mano et al. (2001)
	RILs (Russia 6 × H.E.S. 4)	1172	1596	Hori et al. (2003)

[a] Only maps comprising more than 100 loci are listed
[b] Composite maps include more than one type of marker types. An updated version of these maps can be found at GrainGenes website (http://wheat.pw.usda.gov/)

Kleinhofs and Graner (2001) have divided the barley genome into approximately 10-cM intervals ("BINs"). Each BIN is defined by its two flanking markers, which have been anchored in the Steptoe/Morex and the Igri/Franka maps. The BIN map readily allows the placement of markers mapped in different mapping populations. Although its genetic resolution is only limited, it accommodates the information from a large number of maps. Thus, it is relatively easy to identify a large number of markers for a given chromosomal region, which is sufficient for most issues that are addressed in mapping projects. An updated version of the barley BIN map can be found at (http://barleygenomics.wsu.edu).

Based on the data of the most comprehensive maps, the genetic length of the barley genome can be estimated to be somewhere between 1050 and 1400 cM. Several factors may account for the differences in map size. Säll

(1991) observed genotypic effects that influence map distances. In addition, some maps may not cover the complete genome, since the parents may share the portion of their genomes that are identical by descent. An example for this is chromosome 2HL of the Igri/Franka map, where a large portion of this chromosome arm is devoid of markers (Graner et al. 1994). Moreover, double crossovers increase the map length of chromosomes. Apart from their real existence, e.g., as a result of (rare) gene conversion events, the occurrence of double crossovers in the range of a few centiMorgans is very unlikely. Hence, closer examination in most cases reveals that double crossovers are the results of scoring errors. In conjunction with AFLP marker technology, it was observed that partial cytosine methylation of CpG or CpXpG motifs within restriction sites may result in the occurrence of "double crossovers" (unpubl. data). The commonly used mapping software treats double crossovers in different ways. As a result, one and the same data set may yield maps of different lengths, depending on the software used. Despite these inaccuracies and varying chromosome numbers and genome sizes, there is striking conservation of the genetic length of individual chromosomes between the various grass genomes, suggesting that the crossover frequency during meiosis underlies a relatively stringent, evolutionary conserved control mechanism.

3.2 Physical Maps

While genetic maps are based on statistical frequencies, physical maps rely on direct size estimates as they can be obtained using microscopic analysis or, at the highest possible resolution, via DNA sequencing.

By means of in situ hybridization, the localization of a given gene may be directly visualized on a mitotic metaphase chromosome. Examples are the B-hordeins (Lehfer et al. 1993), or 5S rDNA (Leitch and Heslop-Harrison 1993; Fukui et al. 1994). Using a fluorescent in situ hybridization technique, the relationship between physical and genetic distances at *Hor1* and *Hor2* loci was investigated (Pedersen and Linde-Laursen 1995). However, in situ hybridization techniques are laborious and not routinely practicable on a genome-wide scale. Its strength lies in the visualization of repeated DNA elements, whose distribution provides clues regarding genome structure and genome evolution (Vershinin et al. 2002). In wheat, physical mapping of single and low copy sequences was started using terminal chromosomal deletion stocks (Endo and Gill 1996). By using defined deletions, the distribution of markers and recombination events can be studied for all wheat chromosomes (for review, see Gupta et al. 1999). Although there is a report on deletion-based physical mapping in barley by using chromosome 7H-specific AFLP and STS markers (Serizawa et al. 2001), the diploid nature of the barley genome so far prevented the generation of such stocks for all barley linkage groups. To achieve direct physical mapping of single copy sequences, Sorokin et al. (1994) devised a PCR-mediated technique to integrate translocation breakpoints (TBs) into the genetic map of barley. By using this approach, 240 translocation breakpoints were integrated as physical landmarks into the linkage maps of all the seven chromosomes of barley (Künzel et al. 2000). The comparison of the genetic and the physical TB-map provided experimental evidence of the uneven distribution of recombination along the barley chromosomes. These could be partitioned into regions of high recombination (<1 Mb/cM), medium recombination (1.1–4.4 Mb/cM) and low recombination (7.9≥200 Mb/cM). Regions of high or medium recombination are mainly confined to a few relatively small areas located in the distal parts of the chromosomes while in the proximal regions recombination is highly suppressed. Regions of high recombination amount to only 4.9% of the physical size of the barley genome, but they harbor 47.3% of the 429 markers from the Igri × Franka map. This lends strength to the hypothesis that there is a "gene space", which is characterized by high recombination frequencies, whereas the gene-poor regions are characterized by low recombination frequencies. Since similar findings were observed for wheat, it may be speculated that the above-mentioned observations represent general features of the *Triticeae* genomes (Sandhu and Gill 2002). The availability of genome-wide BAC-contigs has been a prerequisite for sequencing the model genomes of *Arabidopsis* and rice (Sasaki and Burr 2000; TAGI 2000). Similar efforts are currently underway to prepare contig maps of the genomes of maize (http://www.maizemap.org/iMapDB/iMap.html) and sorghum (Klein et al. 2000). As a resource for contig construction, several large insert DNA libraries have been constructed for barley. These include two YAC libraries having 2× and 4× genome coverage (Kleine et al. 1993; Simons et al. 1997) and two BAC libraries having 6.3× and 1× genome coverage, respectively (Yu et al. 2000; Saisho et al. 2002). However, the large size of its genome and the limited financial resources presently available oppose the development of a full genome contig map for barley. Therefore, attempts for the construction of physical maps remain restricted to distinct regions that are targeted in the context of map-based cloning projects (Lahaye et al. 1998; Druka et al. 2000).

As an alternative to the resource-intense development of contig maps, efforts are underway to establish subgenomic physical maps from radiation hybrid (RH) populations (Cox et al. 1990) or by the so-called HAPPY (<u>hap</u>loid genome <u>p</u>olymerase chain reaction) mapping procedure (Dear and Cook 1989). Both methods do not require BAC contigs or cloned DNA fragments and may be suitable for the high throughput mapping of PCR-based markers independent of the presence of polymorphism once their successful application in barley has been demonstrated (Waugh et al. 2002; Thangavelu et al. 2003).

3.3 Functional Maps

In the course of systematic attempts of gene isolation, EST sequencing projects are underway in several laboratories (see Close et al. 2001; Michalek et al. 2002). As of June 2003 the total number of barley ESTs listed in dbEST amounts to 370,258 (http://www.ncbi.nlm.nih.gov/dbEST/dbEST_summary.html). After cluster analysis, this set has been estimated to contain about 40,000 distinct genes, which may represent some 85% of the gene complement of barley (unpubl. data). Ideally, all of these genes should be placed on a genetic map, to facilitate their integration with agronomic traits. Since this represents an insurmountable task, present activities aim at construction of a high-density transcript map comprising about 1000 mapped ESTs. These will form anchor points for the construction of subgenomic physical maps (see above) and for the analysis of evolutionary relationships between the individual grass genomes. Moreover, EST-derived markers complement existing marker resources regarding diversity analysis and trait mapping. Apart from being mapped as RFLPs, the available sequence information makes ESTs ideally suitable for the development of SSR and SNP markers (Kota et al. 2001a, b). Software tools have been developed that allow the rapid screening of any EST collection for the presence of SSR motifs and the development of the corresponding primers for PCR detection in genomic DNA (Varshney et al. 2002; Thiel et al. 2003). Similarly, the availability of EST sequences derived from different cultivars has been exploited for the computational identification of SNP-markers (Kota et al. 2003). Using this strategy, genes that display threefold increase in sequence diversity can be selected as can be seen from the sequence diversity (π)-values of 0.0028 for unselected and 0.009 for the computationally selected ESTs. Based on RFLPs, SSRs, and SNPs a nonredundant set of >900 EST-derived markers has been recently placed on a barley consensus map (Table 2). Along with those cDNA-markers that were previously mapped, the number of genetically mapped genes exceeds 1100.

Table 2. Progress in the preparation of the 'transcript map' of barley

Markers	Barley-chromosome							Total
	1H	2H	3H	4H	5H	6H	7H	
RFLP	73	91	82	48	82	51	83	510
SNP	27	36	42	25	41	26	37	234
SSR	25	31	35	26	22	25	21	185
Total	125	158	159	99	145	102	141	**929**

4 Comparative Mapping and Synteny

Despite more than 60 million years of evolution within the subfamily of the *Poaceae*, the individual grass genomes are characterized by large segments of conserved linkage blocks that display colinear marker orders between different species (in the context of this paper, the term synteny refers to a colinear marker order). Similar to a LEGO-model, grass genomes are considered to be made up from conserved segments (Moore 1995). As an extension of this model, the individual grass genomes can be arranged in concentric circles in which orthologous genes, which are derived from a common ancestor locus, are located on a radial line (Devos and Gale 2000). Our present knowledge of synteny is mainly based on comparative mapping of cross-hybridizing RFLP markers requiring extensive experimental efforts. Because of their conserved nature, gene-derived sequences such as cDNA probes proved superior to genomic RFLP probes because the latter detect less conserved sequences and, therefore, show less interspecific cross hybridization. A summary of studies on comparative mapping in barley and other cereals is presented in Table 3.

In addition to revealing evolutionary patterns within the Poaceae subfamily, comparative mapping provides access to the model genome of rice. An obvious strategy emerging from the concept of syntenous relationships is the transfer to the barley genome of the vast amount of genomic information and resources available for rice. In this context, ESTs from the barley transcript map provide entry gates to identify markers or genes in the rice genome that can be used for the saturation of a target region of interest in barley.

Together with the barley ESTs, the availability of (1) a nearly complete physical contig map of the rice genome (Chen et al. 2002), (2) a large number of rice ESTs (6591) that have been anchored to this map (Wu et al. 2002) and (3) the availability of the complete draft sequence of rice chromosomes (Goff et al. 2002; Yu et al. 2002; http://rgp.dna.affrc.go.jp/IRGSP/) facilitate computational approaches to identify syntenic regions between rice and barley at high resolution. Based on BlastN analysis of 894 mapped barley EST against

Table 3. Syntenic relationship between barley and other cereals, studied on basis of molecular markers

Cereal species	References
Barley, wheat	Hohmann et al. (1995); Dubcovsky et al. (1996); Hernandez et al. (2001); Salvo-Garrido et al. (2001)
Barley, rye	Wang et al. (1992)
Barley, rice	Kilian et al. (1995, 1997); Saghai-Maroof et al. (1996); Han et al. (1998, 1999); Smilde et al. (2001)
Barley, wheat, rye	Devos and Gale (1993); Devos et al. (1993); Börner et al. (1998)
Barley, oat, maize	Yu et al. (1996)
Barley, wheat, rice	Dunford et al. (1995); Kato et al. (2001)

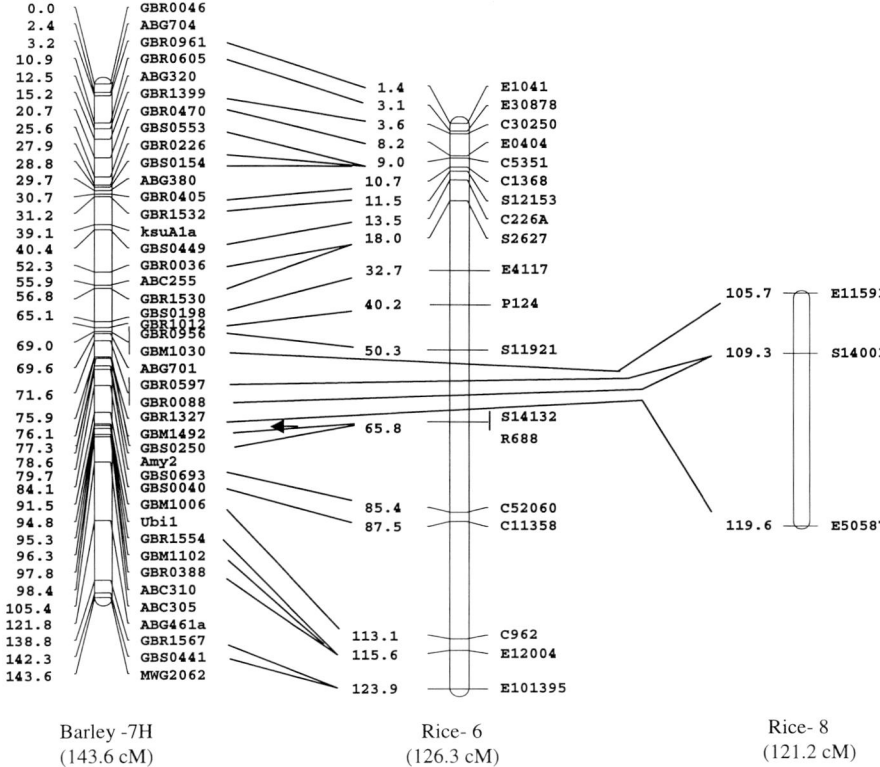

Fig. 1. Comparative maps of barley chromosome 7H and rice chromosomes 6 and 8. Barley chromosome 7H is represented as a consensus map of three different crosses (for details, see text). For clarity, only anchor markers and markers colinear with rice are shown. Cumulative map distances are given on the *left* side and marker designations on the *right* side of the maps. Colinear loci are connected by *lines* and were identified in silico by comparing the mapped barley EST sequences to the rice BAC/PAC database (RGP). The position of the centromere in barley is indicated by an *arrow*. The complete length of the respective chromosomes is indicated at the bottom of each map

the rice BAC/PAC database, a total number of 650 hits with an expected value less than 1E-5.0 (sequence identity is more than 80%) were observed. Of these rice entries, 414 (63%) displayed a syntenic relationship to the corresponding barley entries at the chromosomal level of which 242 showed colinearity at marker order level. Using this computational approach, we identified >30 syntenic genes (ESTs) on every chromosome of barley (Fig. 1). As a result, the 'transcript map' will enable barley geneticists to rapidly identify a target region in rice, which may be already available as a final sequence, or will be before long. In the case of map-based cloning, the rice sequence can then be searched for the presence of candidate genes. Although in several cases the conservation of synteny between barley and rice extends to the sequence level (Dubcovsky et al. 2001), there is ample evidence for the occurrence of rearrangements at the level of microsynteny, i.e., at the physical level (see Bennetzen and Ma 2003). For instance, no rice ortholog of the barley stem rust resistance gene *Rpg1* exists in rice although the gene order surrounding the locus was highly conserved between the two species (Kilian et al. 1997; Han et al. 1999). Similarly, in a study on comparative mapping of the barley *ppd-h1* (photoperiod response) on barley chromosome 2HS and its orthologous region *Hd2* (heading date) on rice chromosome 7L, disruption of colinearity was observed (Dunford et al. 2002; Griffiths et al. 2003). However, even if no candidate gene is present, the rice sequence information can be used to identify homologous barley ESTs for marker saturation of the target interval region.

5 Conclusions

During the past decade, molecular marker maps have been the basis for successful tagging a number of agronomically important traits and an even larger number of quantitative trait loci (e.g., Graner et al. 2000; Hayes and Jones 2000). However, further progress in trait mapping will critically depend on the availability of appropriate plant material. In this context, the generation and phenotypic analysis of experimental populations (F_2, DH, RIL, etc.) is time consuming. Moreover, only two alleles will segregate in a conventional progeny, thus limiting the scope of genetic analysis. The development of novel approaches of association genetics based on the exploitation of linkage disequilibrium (LD) may lead to the verification of candidate genes in natural populations or collections of various genotypes (Rafalski and Morgante 2004). As to barley, the successful application of this approach will require further knowledge of population structures and the decay of LD with physical distance.

The available resource of barley ESTs already includes the vast majority of the genes. Their systematic correlation with phenotypes not least suffers from the lack of a comprehensive physical map. As a result, the available genetic maps need to be further refined and complemented by physical maps for selected chromosomal regions. In the future, intergenomic approaches which are based on the exploitation of the genomic information and the resources that are available for rice will blaze the trail for the establishment of high-density EST-based maps. Together with procedures such as HAPPY mapping, these could provide improved access to candidate genes underlying the traits of interest.

While on the one hand, the isolation of a given gene is a prerequisite to understanding its cellular function, the identification and subsequent introgression of superior alleles will be of seminal importance for breeding improved cultivars. With about 485,000 barley accessions stored in ex situ collections worldwide, the approaches outlined above may help to tap into the vast reservoir of genetic diversity waiting in the gene banks to unlock the future.

References

Becker J, Vos P, Kuiper M, Salamini F, Heun M (1995) Combined mapping of AFLP and RFLP markers in barley. Mol Gen Genet 249:65–73

Bennett MD, Smith LB (1976) Nuclear DNA amounts in angiosperms. Philos Trans R Soc Lond B Biol Sci 274:227–274

Bennetzen JL, Ma J (2003) The genetic colinearity of rice and other cereals on the basis of genomic sequence analysis. Curr Opin Plant Biol 6:128–133

Bhattramakki D, Dolan M, Hanafey M, Wineland R, Vaske D, Register JC, Tingey SV, Rafalski A (2002) Insertion-deletion polymorphisms in 3' regions of maize genes occur frequently and can be used as highly informative genetic markers. Plant Mol Biol 48:539–547

Börner A, Korzun V, Worland AJ (1998) Comparative genetic mapping of loci affecting plant height and development in cereals. Euphytica 100:245–248

Botstein D, White RL, Skolnick M, Davis RW (1980) Construction of a genetic linkage map in man using restriction fragment length polymorphism. Am Jour Hum Genet 32:314–331

Bureau TE, Wessler SR (1994) Mobile inverted-repeat elements of the tourist family are associated with the genes of many cereal grasses. Proc Natl Acad Sci USA 91:1411–1415

Castiglioni P, Pozzi C, Heun M, Terzi V, Muller KJ, Rohde W, Salamini F (1998) An AFLP-based procedure for the efficient mapping of mutations and DNA probes in barley. Genetics 149:2039–2056

Chang RY, O'Donoughue LS, Bureau TE (2001) Inter-MITE polymorphism (IMP):a high throughput transposon-based genome mapping and fingerprinting approach. Theor Appl Genet 102:773–781

Chen M, Presting G, Barbazuk WB, Goicoechea JL, Blackmon B, Fang G, Kim H, Frisch D, Yu Y, Sun S et al. (2002) An integrated physical and genetic map of the rice genome. Plant Cell 14:537–545

Close TJ, Wing R, Kleinhofs A, Wise R (2001) Genetically and physically anchored EST resources for barley genomics. Barley Genet Newslett 31:29–30

Costa JM, Corey A, Hayes PM, Jobet C, Kleinhofs A, Kopisch-Obusch A, Kramer SF, Kudrna D, Li M, Riera-Lizarazu O, Sato K et al. (2001) Molecular mapping of the Oregon Wolfe Barleys: a phenotypically polymorphic doubled-haploid population. Theor Appl Genet 103:415–424

Cox DR, Burmeister M, Price ER, Kim S, Mayers RM (1990) Radiation hybrid mapping – a somatic-cell genetic method for constructing high-resolution maps of mammalian chromosomes. Science 250:245–250

Dear PH, Cook RR (1989) HAPPY mapping – a proposal for linkage mapping the human genome. Nucleic Acids Res 17:6795–6807

Devos KM, Gale MD (1993) Extended genetic maps of the homoeologous group-3 chromosomes of wheat, rye and barley. Theor Appl Genet 85:649–652

Devos KM, Gale MD (2000) Genome relationships: the grass model in current research. Plant Cell 12:637–646

Devos KM, Millan T, Gale MD (1993) Comparative RFLP maps of homoeologous group-2 chromosomes of wheat, rye, and barley. Theor Appl Genet 85:784–792

Druka A, Kudrna D, Han F, Kilian A, Steffenson B, Frisch D, Tomkins J, Wing R, Kleinhofs A (2000) Physical mapping of the barley stem rust resistance gene *rpg4*. Mol Gen Genet 264:283–290

Dubcovsky J, Luo M-C, Zhong G-Y, Bransteitter R, Desai A, Kilian A, Kleinhofs A, Dvorak J (1996) Genetic map of diploid wheat, *Triticum monococcum* L, and its comparison with maps of *Hordeum vulgare* L. Genetics 143:983–999

Dubcovsky J, Ramakrishna W, San Miguel PJ, Busso CS, Yan LL, Shiloff BA, Bennetzen JL (2001) Comparative sequence analysis of colinear barley and rice bacterial artificial chromosomes. Plant Physiol 125:1342–1353

Dunford RP, Kurata N, Laurie DA, Money TA, Minobe Y, Moore G (1995) Conservation of fine-scale DNA marker order in the genomes of rice and the Triticeae. Nucleic Acids Res 23:2724–2728

Dunford RP, Yano M, Kurata N, Sasaki T, Huestis G, Rocheford T, Laurie DA (2002) Comparative mapping of the barley Ppd-H1 photoperiod response gene region, which lies close to a junction between two rice linkage segments. Genetics 161:825–834

Endo TR, Gill BS (1996) The deletion stocks of common wheat. J Hered 87:295–307

Flavell R (1980) The molecular characterization and organization of plant chromosomal DNA sequences. Annu Rev Plant Physiol 31:569–596

Fukui K, Kamisugi Y, Sakai F (1994) Physical mapping of 5S rDNA loci by direct cloned biotinylated probes in barley chromosomes. Genome 37:105–111

Goff SA, Ricke D, Lan T-H, Presting G, Wang R, Dunn M, Glazebrook J, Sessions A, Oeller P, Varma H, Hadley D et al. (2002) A draft sequence of the rice genome (*Oryza sativa* L. ssp. *japonica*). Science 296:92–100

Graner A, Jahoor A, Schondelmaier J, Siedler H, Pillen K, Fischbeck G, Wenzel G, Herrmann RG (1991) Construction of an RFLP map of barley. Theor Appl Genet 83:250–256

Graner A, Bauer E, Kellermann A, Kirchner S, Muraya JK, Jahoor A, Wenzel G (1994) Progress of RFLP-map construction in winter barley. Barley Genet Newslett 23:53–59

Graner A, Michalek W, Streng S (2000) Molecular mapping of genes conferring resistance to viral and fungal pathogens. In: Logue S (ed) Proc 8th Barley Gene Symp, vol I. Adelaide University, Australia, pp 45–52

Griffiths S, Dunford RP, Coupland G, Laurie DA (2003) The evolution of CONSTANS-like gene families in barley, rice and *Arabidopsis*. Plant Physiol 13:1855–1867

Gupta PK, Varshney RK, Sharma PC, Ramesh B (1999) Molecular markers and their application in wheat breeding. Plant Breed 118:369–390

Gupta PK, Varshney RK, Prasad M (2002) Molecular markers: principles and methodology. In: Jain SM, Brar DS, Ahloowalia BS (eds) Molecular techniques in crop improvement. Kluwer, Dordrecht, , pp 9–54

Han F, Kleinhofs A, Ullrich SE, Kilian A, Yano M, Sasaki T (1998) Synteny with rice: analysis of barley malting quality QTLs and *rpg4* chromosome regions. Genome 41:373–380

Han F, Kilian A, Chen JP, Kudrna D, Steffenson B, Yamamoto K, Matsumoto T, Sasaki T, Kleinhofs A (1999) Sequence analysis of a rice BAC covering the syntenous barley *Rpg1* region. Genome 42:1071–1076

Hayes PM, Jones BL (2000) Malting quality from a QTL perspective. In: Logue S (ed) Proc 8th Barley Gene Symp, vol I. Adelaide University, Australia, pp 99–106

Hernandez P, Dorado G, Prieto P, Gimenez MJ, Ramirez MC, Laurie DA, Snape JW, Martin A (2001) A core genetic map of *Hordeum chilense* and comparisons with maps of barley (*Hordeum vulgare*) and wheat (*Triticum aestivum*). Theor Appl Genet 102:1259–1264

Heun M, Kennedy AE, Anderson JA, Lapitan NLV, Sorrells ME, Tanksley SD (1991) Construction of a restriction fragment length polymorphism map for barley (*Hordeum vulgare*). Genome 34:437–447

Hohmann U, Graner A, Endo TR, Gill BS, Herrmann RG (1995) Comparison of wheat physical maps with barley linkage maps for group 7 chromosomes. Theor Appl Genet 91:618–626

Hori K, Kobayashi T, Shimizu A, Sato K, Takeda K, Kawasaki S (2003) Efficient construction of high-density linkage map and its application to QTL analysis in barley. Theor Appl Genet 107:806–813

Jaffe B, Caligari PDS, Snape JW (2000) A skeletal linkage map of *Hordeum bulbosum* L. and comparative mapping with barley (*H. vulgare* L.). Euphytica 115:115–120

Jain SM, Brar DS, Ahloowalia BS (2002) Molecular techniques in crop improvement. Kluwer, Dordrecht

Kalendar R, Grob T, Regina M, Suoniemi A, Schulman AH (1999) IRAP and REMAP: two new retrotransposon-based DNA fingerprinting techniques. Theor Appl Genet 98:704–711

Kasha KJ, Kleinhofs A, the North American Barley Genome Mapping Project (1994) Mapping of the barley cross Harrington × TR306. Barley Genet Newslett 23:65–69

Kato K, Nakamura W, Tabiki T, Miura H, Sawada S (2001) Detection of loci controlling seed dormancy on group 4 chromosomes of wheat and comparative mapping with rice and barley genomes. Theor Appl Genet 102:980–985

Kilian A, Kudrna DA, Kleinhofs A, Yano M, Kurata N, Steffenson B, Sasaki T (1995) Rice-barley synteny and its application to saturation mapping of the barley *Rpg1* region. Nucleic Acids Res 23:2729–2733

Kilian A, Chen J, Han F, Steffenson B, Kleinhofs A (1997) Towards map-based cloning of the barley stem rust resistance genes *Rpg1* and *rpg4* using rice as an intergenomic cloning vehicle. Plant Mol Biol 35:187–195

Klein PE, Klein RR, Cartinhour SW, Ulanch PE, Dong J, Obert JA, Morishge DT, Schlueter SD, Childs KL, Ale M et al. (2000) A high throughput AFLP based method for constructing integrated genetic and physical maps: progress toward a sorghum genome map. Genome Res 10:789–807

Kleine M, Michalek W, Diefenthal H, Dargatz H, Jung C (1993) Construction of a barley (*Hordeum vulagre* L.) YAC library and isolation of a Hor1-specific clone. Mol Gen Genet 240:265–272

Kleinhofs A, Graner A (2001) An integrated map of the barley genome. In: Phillips RL, Vasil IK (eds) DNA-based markers in plants, 2nd edn. Kluwer Academic, Dordrecht, The Netherlands, pp 187–200

Kleinhofs A, Chao S, Sharp PJ (1988) Mapping of nitrate reductase genes in barley and wheat. In: Miller TE, Koebner RMD (eds) Proc 7th Int Wheat Genet Symp, vol I. Institute of Plant Science Research, Cambridge, pp 541–546

Kleinhofs A, Kilian A, Saghai-Maroof MA, Biyashev RM, Hayes PM, Chen FQ, Lapitan N, Fenwick A, Blake T, Kanazin V et al. (1993) A molecular, isozyme and morphological map of barley (*Hordeum vulgare*) genome. Theor Appl Genet 86:705–712

Kota R, Varshney RK, Thiel T, Dehmer KJ, Graner A (2001a) Generation and comparison of EST-derived SSRs and SNPs in barley (*Hordeum vulgare* L.). Hereditas 135:145–151

Kota R, Wolf M, Michalek W, Graner A (2001b) Application of DHPLC for mapping of single nucleotide polymorphisms (SNPs) in barley (*Hordeum vulgare* L.). Genome 44:523–528

Kota R, Rudd S, Facius A, Kolesov G, Thiel T, Zhang H, Stein N, Mayer K, Graner A (2003) Snipping polymorphisms from large EST collections in barley (*Hordeum vulgare* L.). Mol Gen Genom 270:224–233

Künzel G, Korzun L, Meister A (2000) Cytologically integrated physical restriction fragment length polymorphism maps for the barley genome based on translocation breakpoints. Genetics 154:397–412

Lahaye T, Shirasu K, Schulze-Lefert P (1998) Chromosome landing at the barley *Rar1* locus. Mol Gen Genet 260:92–101

Langridge P, Karakousis A, Collins N, Kretchmer J, Manning S (1995) A consensus linkage map of barley. Mol Breed 1:389–395

Lehfer H, Busch W, Martin R, Hermann RG (1993) Localization of the B-hordein locus on barley chromosomes using fluorescence in situ hybridisation. Chromosoma 102:428–432

Leitch IJ, Heslop-Harrison JS (1993) Physical mapping of four sites of 5S rDNA sequences and one site of the α-amylase 2 gene in barley (*Hordeum vulgare* L.). Genome 36:517–523

Lemaux PG, Cho MJ, Zhang S, Bregitzer P (1999) Transgenic cereals: *Hordeum vulgare* L. (barley). In: Vasil IK (ed) Molecular improvement of cereal crops. Elsevier, Amsterdam, pp 255–316

Mano Y, Kawasaki S, Takaiwa F, Komatsuda T (2001) Construction of a genetic map of barley (*Hordeum vulgare* L.) cross 'Azumamugi' × 'Kanto Nakate Gold' using a simple and efficient amplified fragment-length polymorphism system. Genome 44:284–292

Michalek W, Weschke W, Pleissner KP, Graner A (2002) EST analysis in barley defines a unique set comprising 4000 genes. Theor Appl Genet 104:97–103

Miyazaki C, Osanai E, Saeki K, Hirota N, Ito K, Ukai Y, Konishi T, Saito A (2000) Construction of a barley RFLP linkage map using an F_2 population derived from a cross between Ko A and Mokusekko 3. Barley Genet Newslett 30:41–43

Moore G (1995) Cereal genome evolution – pastoral pursuits with Lego genomes. Curr Opin Genet Dev 5:717–724

Pedersen C, Linde-Laursen I (1995) The relationship between physical and genetic distances at the *Hor1* and *Hor2* loci of barley estimated by two-colour fluorescent in situ hybridization. Theor Appl Genet 91:941–946

Phillips RL, Vasil IK (2001) DNA-based markers in plants, 2nd edn. Kluwer Academic, Dordrecht, The Netherlands

Provan J, Thomas WTB, Forster BP, Powell W (1999) *Copia*-SSR: a simple marker technique which can be used on total genomic DNA. Genome 42:363–366

Qi X, Stam P, Lindhout P (1996) Comparison and integration of four barley genetic maps. Genome 39:379–394

Qi X, Stam P, Lindhout P (1998) Use of locus-specific AFLP markers to construct a high-density molecular map in barley. Theor Appl Genet 96:376–384

Rafalski A, Morgante (2004) Corn and humans: recombination and linkage disequilibrium in two genomes of similar size. Trends Genet 20:103–111

Ramsay L, Macaulay M, Ivanissevich DS, MacLean K, Cardle L, Fuller J, Edwards KJ, Tuvesson S, Morgante M, Massari A et al. (2000) A simple sequence repeat-based linkage map of barley. Genetics 156:1997–2005

Saghai Maroof MA, Tang GP, Biyashev RM, Maughan PJ, Zhang Q (1996) Analysis of the barley and rice genomes by comparative RFLP linkage mapping. Theor Appl Genet 92:541–551

Saisho D, Kawasaki S, Sato K, Takeda K (2002) Construction of a BAC library from Japanese malting barley Harunanijo. In: Plant, animal and microbe genomes X Conference, 12–16 Jan 2002, San Diego, p 393 (http://www.intl-pag.org/pag/10/abstracts/PAGX-P393.html)

Salamini F, Özkan H, Brandolini A, Schäfer-Pegl R, Martin W (2002) Genetics and geography of wild cereal domestication in the near East. Nat Rev Genet 3:429–441

Säll T (1991) Genetic-control of recombination in barley. 3. recombination between the Hordein loci in 3 different genotypes. Hereditas 115:13–16

Salvo-Garrido H, Laurie DA, Jaffe B, Snape JW (2001) An RFLP map of diploid *Hordeum bulbosum* L. and comparison with maps of barley (*H. vulgare* L.) and wheat (*Triticum aestivum* L.). Theor Appl Genet 103:869–880

Sandhu D, Gill KS (2002) Gene-containing regions of wheat and the other grass genomes. Plant Physiol 128:803–811

Sasaki T, Burr B (2000) International rice genome sequencing project: the effort to completely sequence the rice genome. Curr Opin Plant Biol 3:138–141

Serizawa N, Nasuda S, Shi F, Endo TR, Prodanovic S, Schubert I, Kuenzel G (2001) Deletion-based physical mapping of barley chromosome 7H. Theor Appl Genet 103:827–834

Sherman JD, Fenwick AL, Namuth DM, Lapitan NLV (1995) A barley RFLP map-alignment of three barley maps and comparisons to Gramineae species. Theor Appl Genet 91:681–690

Shin JS, Corpuz L, Chao S, Blake TK (1990) A partial map of the barley genome. Genome 33:803–808

Simons G, van der Lee T, Diergaarde P, Daelen RV, Groenendijk J, Frijters A, Büschges R, Hollricher K, Töpsch S, Schulze-Lefert P et al. (1997) AFLP-based fine mapping of the *Mlo* gene to a 30-kb DNA segment of the barley genome. Genomics 44:61–70

Smilde DW, Haluskova J, Sasaki T, Graner A (2001) New evidence for the synteny of rice chromosome 1 and barley chromosome 3H from rice expressed sequence tags. Genome 44:361–367

Sorokin A, Marthe F, Houben A, Pich U, Graner A, Künzel G (1994) Polymerase chain-reaction mediated localization of RFLP clones to microisolated translocation chromosomes of barley. Genome 37:550–555

Tautz D (1989) Hypervariability of simple sequences as a general source for polymorphic DNA markers. Nucleic Acids Res 17:6443–6471

TAGI, The Arabidopsis Genome Initiative (2000) Analysis of the genome sequence of the flowering plant *Arabidopsis thaliana*. Nature 408:796–815

Thangavelu M, James AB, Bankier A, Bryan GJ, Dear PH, Waugh R (2003) HAPPY mapping in plant genome: reconstruction and analysis of a high-resolution physical map of 1.9 Mpp region of *Arabidopsis thaliana* chromosome 4. Plant Biotech Jour 1:23-31

Thiel T, Michalek W, Varshney RK, Graner A (2003) Exploiting EST databases for the development of cDNA derived microsatellite markers in barley (*Hordeum vulgare* L.). Theor Appl Genet 106:411-422

Varshney RK, Thiel T, Stein N, Langridge P, Graner A (2002) *In silico* analysis on frequency and distribution of microsatellites in ESTs of some cereal species. Cell Mol Biol Lett 7:537-546

Vershinin AV, Druka A, Alkhimova AG, Kleinhofs A, Heslop-Harrison JS (2002) LINEs and gypsy-like retrotransposons in *Hordeum* species. Plant Mol Biol 49:1-14

Von Bothmer R, Jacobsen N, Baden C, Jørgensen RB, Linde-Laursen I (1995) An ecogeographical study of the genus *Hordeum*, 2nd edn. Systematic and ecogeographic studies on crop genepools 7. IPGRI, Rome, 129 pp

Vos P, Hogers R, Bleeker R, Reijans M, van de Lee T, Hornes M, Frijters A, Pot J, Peleman J, Kupier M et al. (1995) AFLP: a new technique for DNA fingerprinting. Nucleic Acids Res 23:4407-4414

Wang ML, Atkinson MD, Chinoy CN, Devos KM, Gale MD (1992) Comparative RFLP-based genetic maps of barley chromosome-5 (1H) and rye chromosome-1 R. Theor Appl Genet 84:339-344

Wang DG, Fan JB, Siao CJ, Berno A, Young P, Sapolsky R, Ghandour G, Perkins N, Winchester E, Spencer J et al. (1998) A large-scale identification, mapping, and genotyping of single nucleotide polymorphisms in the human genome. Science 280:1077-1082

Waugh R, Dear PH, Powell W, Machray GC (2002) Physical education – new technologies for mapping plant genomes. Trends Plant Sci 7:521-523

Waugh R, McLean K, Flavell AJ, Pearce SR, Kumar A, Thomas BBT, Powell W (1997) Genetic distribution of *BARE-1*-like retrotransposable elements in the barley genome revealed by sequence-specific amplification polymorphisms (S-SAP). Mol Gen Genet 255:687-694

Williams JGK, Kubelik AR, Livak KJ, Rafalski JA, Tingey SV (1990) DNA polymorphisms amplified by arbitrary primers are useful as genetic markers. Nucleic Acids Res 18:6531-6535

Wu J, Maehara T, Shimokawa T, Yamamoto S, Harada C, Takazaki Y, Ono N, Mukai Y, Koike K, Yazaki J et al. (2002) A comprehensive rice transcript map containing 6591 expressed sequence tag sites. Plant Cell 14:525-535

Yu GX, Bush AL, Wise RP (1996) Comparative mapping of homoeologous group 1 regions and genes for resistance to obligate biotrophs in *Avena*, *Hordeum*, and *Zea mays*. Genome 39:155-164

Yu Y, Tomkins JP, Waugh R, Frisch DA, Kudrna D, Kleinhofs A, Brueggeman RS, Muehlbauer GJ, Wise RP, Wing RA (2000) A bacterial artificial chromosome library for barley (*Hordeum vulgare* L.) and the identification of clones containing putative resistance genes. Theor Appl Genet 101:1093-1099

Yu J, Hu S, Wang J, Wong GKS, Li S, Liu B, Deng Y, Dai L, Zhou Y, Zhang X, Cao M, Liu J, Sun J, Tang J, Chen Y et al. (2002) A draft sequence of the rice genome (*Oryza sativa* L. ssp. indica). Science 296:79-92

Zohary D, Hopf M (2001) Domestication of plants in the old world – the origin and spread of cultivated plants in West Asia, Europe, and the Nile Valley, 3rd edn. Oxford Univ Press, Oxford, UK

II.10 Genomics of Rice: Markers as a Tool for Breeding

Y. Kishima, K. Onishi, and Y. Sano[1]

1 Introduction

Rice is unique among plants because of its dual utility as food and genetic material, i.e., it is the principal food of nearly half of the world's people, and it possesses a small and thus tractable genome size (430 Mb pairs) relative to those of other monocotyledonous plants. Due to its being the staple diet in a number of countries, rice improvement programs have aimed at diverse breeding objectives. To satisfy these various demands, rapid, efficient and reliable breeding programs have been required. Consequently, rice has become one of the crops with the most advanced application of molecular marker techniques for marker-assisted selection in breeding. Furthermore, early in 2002, draft data of the entire genomic sequences of Japonica and Indica were published (Goff et al. 2002; Yu et al. 2002), and at the end of the year, the International Rice Genome Sequencing Project announced the high-accuracy sequence of every chromosome (http://rgp.dna.affrc.go.jp/rgp/Dec18–NEWS.html). The release of these high-quality data to the public provides great advantages for rice breeding and will open a new era of rice breeding strategies associated with sophisticated DNA marker technology. The development of DNA markers has made it possible to identify quantitative trait loci (QTL) through high-density molecular linkage maps (McCouch et al. 1988; Saito et al. 1991; Causse et al. 1994; Kurata et al. 1994; Harushima et al. 1998). Many traits dealt with by plant breeders are the genetic variation of quantitative traits in nature. Most traits of agricultural significance vary continuously in natural populations due to the segregation of multiple QTLs with small and conditional effects. One of the most important applications of DNA markers is thus to dissect the naturally occurring allelic variations underlying complex traits at QTLs. In this review, we would like to summarize the status of the development of molecular markers for use in a genome-wide survey of rice. We will also discuss the use of DNA marker technology in rice breeding in the post-genome era.

---

[1] Laboratory of Plant Breeding, Graduate School of Agriculture, Hokkaido University, Kita-9, Nishi-9, Kita-ku, Sapporo 060-8589, Japan

2 Conventional Markers

Classical markers are genetically determined morphological characteristics that can be monitored visually. Isozymes that can detect variant forms of the same enzyme are also another type of classical marker (Ishikawa et al. 1992). In rice, a total of 463 classical markers have been identified and assigned since Nagao and Takahashi (1963) proposed the first rice map consisting of 12 linkage groups, and 174 of these markers have been mapped. Compared to other crops, rice has a greater number of conventional markers that have been characterized and located on the chromosomes. However, the usefulness of these markers is limited by factors such as the number of distinct traits to be detected and the reliability of assessment. All the classical markers correspond to major gene traits and have now become attractive targets for gene isolation. Several of these genes have already been isolated using modern molecular marker techniques.

3 The Beginnings of Molecular Marker Analysis

The advent of DNA marker technology commenced with restriction fragment length polymorphism (RFLP) analyses based on the Southern hybridization technique (Bostein et al. 1980). RFLP analysis requires a specific probe to hybridize with a single-copy region in genomic DNA. Based on RFLP analysis, linkage maps of the rice genome were constructed by several groups. In the beginning, using mainly genomic DNA segments as probes, McCouch et al. (1988) and Saito et al. (1991) located 135 and 347 loci, respectively. Subsequently, Causse et al. (1994), Kurata et al. (1994) and Harushima et al. (1998) developed high-density maps with more than 2000 loci, and in the latest report (as of the end of 2002), 3267 loci were mapped, mostly using cDNA probes. The number of RFLP markers located on the rice genetic map increased tenfold in just a decade. RFLP markers are generally thought to be suitable for testing F_2 progeny in Japonica-Indica crosses via the detection of polymorphism of restriction fragments (Zhang et al. 1992). However, this technique is laborious owing to its need for a large quantity of DNA and time for various procedures (Kumar 1999). Therefore, researchers are not currently employing this technique much for mapping in rice.

Because of the simplicity and convenience of PCR-based techniques employing random primers, random amplified polymorphic DNA (RAPD) used to be popular for exploring markers linked to specific traits starting in the early 1990s (Williams et al. 1990). However, RAPD often generates artifacts or uncertain results and also has a reputation of being highly unreliable (Mackill and Ni 2001). Therefore, RAPD is being replaced by other PCR techniques that require specific primers designed based on sequences in the database.

Although RFLP and RAPD still have advantages in providing markers for some specific traits or acting as a means of surveying linkage with particular phenotypes, the demand for their use as a source of DNA markers in mapping is decreasing since the sequence of the entire rice genome has been published.

4 Molecular Markers Currently Used in Rice Breeding

Following *Arabidopsis*, rice has entered the post-genomic era (The Arabidopsis Genome Initiative 2000). The rice genomic sequence is available from the database, so we can design specific primers at desired sites. The public availability of rice genomic sequences has dramatically altered the strategies used for marker development and has largely removed technical and economic limitations. Specific PCR markers have been applied to mapping and marker-assisted selection at a number of loci. Sequence tagged sites (STSs) are available for any single-copy region, and include expressed sequence tag (EST) sites. Even if a PCR product does not show length polymorphism, it often becomes a marker that generates a polymorphism after cutting with a suitable restriction enzyme (CAPS: cleaved amplified polymorphic sequence). The rice genome website (http://rgp.dna.affrc.go.jp/Publicdata.html) provides information about specific locations, primers and resultant band patterns of STSs and CAPSs. Simple sequence repeats (SSRs, also referred to as microsatellites) are abundant and distributed throughout the rice genome (Wu and Tanksley 1993; Panaud et al. 1995, 1996; Akagi et al. 1996; Chen et al. 1997; Temnykh et al. 2000, 2001). This kind of repeats consists of di- to tetra-nucleotide sequence motifs flanked by unique sequences (Temnykh et al. 2001). The length of an SSR locus generally shows a high level of allelic diversity, marking SSRs as valuable genetic markers (Weber and May 1989; McCouch et al. 1997). Specific primers in the flanking sequences are used for PCR amplification. More than 500 microsatellite markers have been located on the rice genetic map (McCouch et al. 2001). Furthermore, more than 7000 uncharacterized SSR sequences have been recently released from the International Rice Microsatellite Initiative (IRMI; http://www.gramene.org/microsat/). IRMI also distributes information about primer sequences, PCR conditions, polymorphism, and map positions. STS, CAPS and SSR markers are expressed in a co-dominant manner, and are thus more useful for analyzing genetic differences than markers expressed in a dominant manner such as by RAPD markers. They are increasingly available at a number of mapped loci in the rice genome due to their technical advantages and efficiency in obtaining polymorphisms (Kumar 1999).

5 Amplified Fragment Length Polymorphism Analysis in Rice

The objectives of rice breeding are remarkably diverse because of the rich genetic diversity and local differences in consumer preferences in rice. Therefore, it will be important to involve thousands of landraces and the wild relatives of rice in the breeding programs. It will be necessary to prepare molecular markers specific to these materials, in addition to the popular markers provided by international projects. For certain purposes among the breeding objectives, amplified fragment length polymorphism (AFLP) markers are useful in order to identify hundreds of fragments (Vos et al. 1995). The AFLP technique has the capacity to detect thousands of independent loci with minimal cost and time. AFLP analysis has performed well at producing polymorphic bands in *Oryza* species (Maheswaran et al. 1997; Zhu et al. 1998; Aggarwal et al. 1999). For instance, Maheswaran et al. (1997) reported that AFLP detected an average of 47.3 bands using a single primer pair and 10.4 bands (21.8% of the total number of bands) as polymorphism between Indica (IR64) and Azucena (Japonica). The use of AFLP for finding linkage markers of specific traits or for the evaluation of genetic relationships is rapidly increasing.

6 MITE-Transposon Display in Rice

Abundant miniature inverted repeat transposable elements (MITEs) have been found in the spacer regions of the rice genome (Bureau and Wessler 1994a, b; Lisch et al. 2001; Nagano et al. 2002). MITE is a collective name for small transposon-like elements with terminal inverted repeats (TIRs) at the ends (Wessler et al. 1995). MITE families have been found in plants and animals, and most seem to be inactive (Feschotte et al. 2002). In contrast to active inverted repeat elements such as Ac/Ds and En/Spm, MITEs are characterized by their uniformly small size (<500 bp) and by target site preferences (Feschotte et al. 2002). MITEs also display a very high copy number and an even distribution throughout the chromosomes (Casa et al. 2000). They are known to be major elements contributing to the repetition in the rice genome architecture (Bureau et al. 1996; Turcotte et al. 2001; Nagano et al. 2002). High-copy number, conserved structure and unique insertion sites are the essential features of MITEs as efficient DNA markers used at the whole-genome level. Via a computer-assisted search of 30 Mb of a bacterial artificial chromosome (BAC)-end sequence, Mao et al. (2000) estimated the copy number of six MITEs; namely, Tourist: 4000, Stowaway: 3000, Gaijin: 2200, Ditto: 2000, Olo24: 2000 and Castaway: 1500. MITE-transposon display (MITE-TD) is an AFLP-related technique that is based on MITE sequences and used to detect polymorphisms (Fig. 1). MITE-TD provides the most effi-

Genomics of Rice: Markers as a Tool for Breeding 249

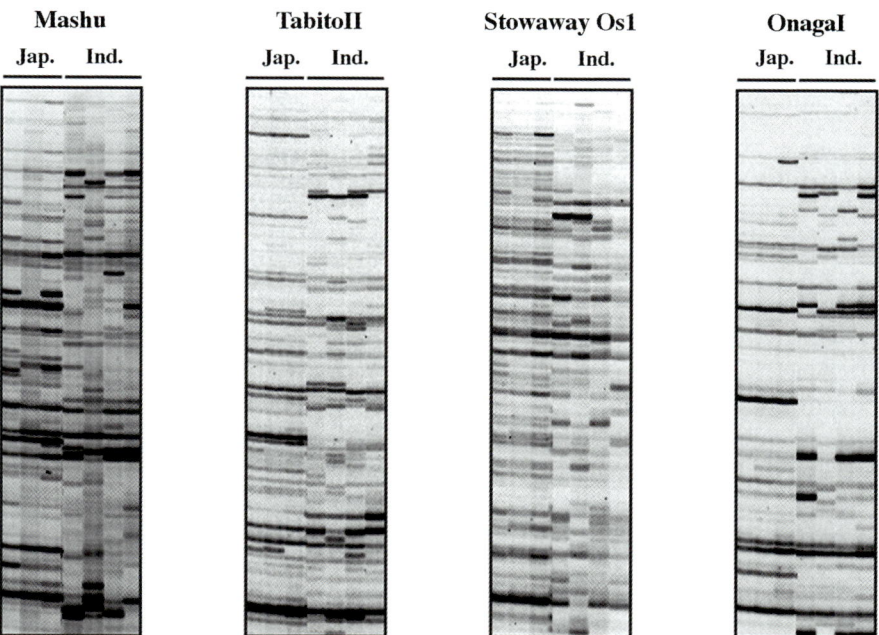

Fig. 1. A new tool for the whole-genomic marker system, MITE-Transposon Display, using four MITE elements (Stowaway, Tabito II, Onaga I and Mashu) in rice. Three different lines from each of Japonica and Indica were subjected to the MITE-TD. (Takagi et al. 2003)

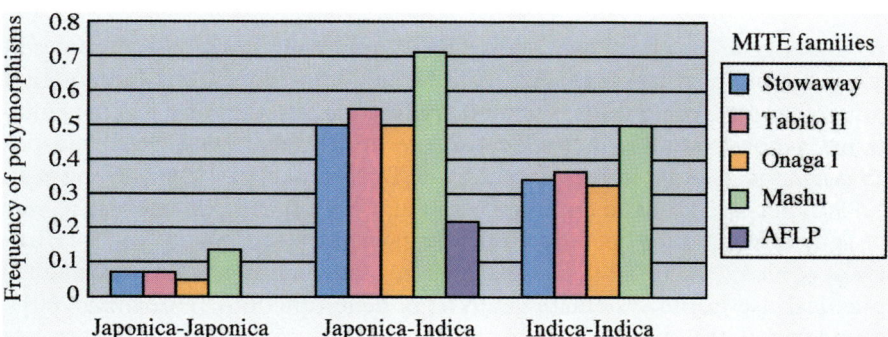

Fig. 2. The frequencies of occurrence of the polymorphisms in Japonica–Japonica, Japonica–Indica and Indica–Indica detected by transposon displays of the four MITE elements (Stowaway, Tabito II, Onaga I and Mashu) and by AFLP. The results of the MITE-TD analyses performed here revealed that each of the four trials was superior to AFLP analysis in terms of the frequencies of polymorphisms between Indica and Japonica (Nagano et al. 2002; Takagi et al. 2003). The level of polymorphisms between a pair of two lines was estimated as frequency of polymorphisms=$N/(N_i+N_j-N)$ where N is the number of shared fragments and N_i and N_j are the total numbers of fragments for entries i and j

cient detection of polymorphisms of all the molecular marker techniques applied to date in *Oryza* species. In fact, MITE-TD analysis was superior to AFLP analyses in terms of the frequency of occurrence of the polymorphisms between Indica and Japonica (Fig. 2).

Nagano et al. (2002) characterized repetitive sequences in a 200-kb region around the rice *waxy* locus, and identified 55 transposable element (TE)-like sequences therein. Fifty out of the 55 TE-like sequences consisted of MITEs or MITE-like sequences, most of which were categorized into three representative MITE families, the Stowaway, Tourist and Mu-like families (the Mu-like family is usually distinguishable from MITEs, since it normally consists of copies larger than 500 bp). Each of these three families comprises several subfamilies (Nagano et al. 2002). In addition, five new MITE families were identified around the rice *waxy* locus. Of the new MITE families, four subfamilies showed a high copy number in a BLAST search (version 2.0, http://www.ncbi.nlm.nih.gov/blast/) against the rice genome sequence. Tabito II (Tourist family), Stowaway OS-1 (Stowaway/Tnr1 family), Onaga I (Mu-like family) and Mashu (a new MITE family) were selected, and Mashu was found to have a markedly conserved copy organization, and to create polymorphisms at high frequency (Figs. 1, 2). So far, the MITE-TD system using Mashu is the best system for the detection of the polymorphisms, and resulted in detection at frequencies 1.5–3 times higher than the other rice MITE elements. Mashu-mediated TD analysis is useful for the identification of Japonica varieties (Fujino, Kishima and Sano, unpubl. data).

7 Transposable Elements as Markers for Major Genes

A "major gene" has been defined as a locus conferring discrete phenotypes in a segregating population (Mackill and Ni 2001). Identification of a mutation in a major gene often leads to understanding the function of the gene. Functional genomic studies in *Arabidopsis* are undertaken using DNA transposons like those found in maize, such as Ac/Ds (Smith et al. 1996; Parinov et al. 1999; Ito et al. 2002) and Spm (Tissier et al. 1999). In rice, an alternative way of gene targeting has been developed using a retrotransposon, Tos17 (Hirochika et al. 1996). Tos17 is highly active in tissue-cultures and is suitable for practical use in the systematic analysis of gene functions (Hirochika 2001). Compared with other plant retrotransposons, its copy number is very low (1–5 copies in the rice genome), and 5–30 copies were observed in regenerated plants after tissue-culture (Hirochika et al. 1996; Yamazaki et al. 2001). Most of the mutagenized lines were derived from 5-month-old cultures and the mean number of newly transposed copies was eight per line. In order to provide a 99% chance of finding a mutant of any one gene, a total of 50,000 mutant lines are expected to be required. According to Hirochika (2001), a rice mutagenized system using Tos17 has been developed with a collection of

32,000 regenerated lines carrying about 256,000 insertion sites, and several important genes have been cloned in these lines. It is also possible to map the insertion sites on the chromosomes by searching the flanking sequences against the rice database.

To construct a saturated mutant line, it will be necessary to develop several other systems, because each element seems to have different target-site preference. Very recently (January 2003), the discovery of an active endogenous transposon in the rice genome has been reported. This transposon, named Ping, is the first movable MITE element family and carries 15-bp TIRs that are flanked by a 3-bp AT-rich target site duplicated sequence (Jiang et al. 2003; Kikuchi et al. 2003; Nakazaki et al. 2003). The putative autonomous elements (Pong) of Ping contain two putative ORFs, one of which has partial homology with putative transposases in the PIF family. The copy number of the Ping family in Japonica is about 60–80, which is much lower than that of other MITEs. Moreover, in Indica and *O. rufipogon* strains, no Ping sequence has been identified so far. These facts indicate that the Ping family has been activated relatively recently. It is notable that Ping is active in cells derived from anther cultures of Japonica rice. Like Tos17, Ping will be useful for rice functional genomics and identification of specific molecular markers of rice genes.

8 Conclusions: Quantitative Trait Loci and Future Prospects

QTLs are involved in many agronomical traits such as those for yield, adaptation and stress tolerance, which are complex traits in the naturally occurring genetic diversity (Tanksley 1993; Paterson 1995). Fine mapping of a QTL is required if it should be feasible to delimit a candidate genomic region. To reduce the genetic factors involved in a QTL, recombinant inbred lines (RILs) or nearly isogenic lines (NILs) have been effectively utilized in rice (Yano and Sasaki 1997). In fact, a number of the QTLs of rice have been extracted by using RILs and NILs (Yano 2001). A high-density map is also required to narrow down the candidate genomic region to within a span about the size of a single BAC clone (150 kb). In rice, several high-density molecular linkage maps have been constructed (McCouch et al. 1988; Saito et al. 1991; Causse et al. 1994; Kurata et al. 1994; Harushima et al. 1998). These maps integrated with SSR and CAPS markers provide a framework for the detection of individual factors controlling complex traits. *Heading-date1* (*Hd1*), which controls the heading date of rice, was the first QTL identified by a map-based strategy using advanced backcross progenies (Yano et al. 2000). This gene was found to be a homologue of CONSTANS in *Arabidopsis* (Putterill et al. 1995). This is an example of a QTL that was successfully extracted as a major gene by using sophisticated materials and a high-density molecular map.

Use of the advanced molecular mapping method will promote identification of more of the genes at QTLs for traits of interest. Integration of the information from various genetic linkage maps will be necessary to facilitate comparison between detected QTLs and known major genes. A comprehensively integrated genetic map including QTL information would enhance our understanding and facilitate positional cloning of genes underlying quantitative traits in rice. Consequently, molecular markers are required for the process of marker-assisted selection in the advanced rice breeding program. Since efficient molecular markers are now available in rice, as described above, elucidating the genetic architecture associated with quantitative traits is a challenge that can be tackled by the advanced rice breeding program.

Acknowledgements. We thank Kyoko Takagi and Hironori Nagano for providing their unpublished data described here.

References

Aggarwal RK, Brar DS, Nandi S, Huang N, Khush GS (1999) Phylogenetic relationships among *Oryza* species revealed by AFLP markers. Theor Appl Genet 98:1320–1328

Akagi H, Yokozeki Y, Inagaki A, Fujimura T (1996) Microsatellite DNA markers for rice chromosomes. Theor Appl Genet 93:1071–1077

Bostein D, White RL, Skolnick M, Davis RW (1980) Construction of a genetic linkage map in man using restriction fragment length polymorphism. Am J Hum Genet 32:314–331

Bureau TE, Wessler SR (1994a) Stowaway: a new family of inverted repeat elements associated with the genes of both monocotyledonous and dicotyledonous plants. Plant Cell 6:907–916

Bureau TE, Wessler SR (1994b) Mobile inverted-repeat elements of the Tourist family are associated with the genes of many cereal grasses. Proc Natl Acad Sci USA 91:1411–1145

Bureau TE, Ronald PC, Wessler SR (1996) A computer-based systematic survey reveals the predominance of small inverted-repeat elements in wild-type rice genes. Proc Natl Acad Sci USA 93:8524–8529

Casa AM, Brouwer C, Nagel A, Wang L, Zhang Q, Kresovich S, Wessler SR (2000) The MITE family heartbreaker (Hbr): molecular markers in maize. Proc Natl Acad Sci USA 97:10083–10089

Causse MA, Fulton TM, Cho YG, Ahn SN, Chunwongse J, Wu KS, Xiao JH, Yu ZH, Ronald PC, Harrinton SE, Second G, McCouch SR, Tanksley SD (1994) Saturated molecular map of the rice genome based on an interspecific backcross population. Genetics 138:1251–1274

Chen X, Temnykh S, Xu Y, Cho YG, McCouch SR (1997) Development of a microsatellite framework map providing genome-wide coverage in rice (*Oryza sativa* L.). Theor Appl Genet 95:553–567

Feschotte C, Jiang N, Wessler SR (2002) Plant transposable elements: where genetics meets genomics. Nat Rev Genet 3:329–341

Goff SA, Ricke D, Lan TH et al. (2002) A draft sequence of the rice genome (*Oryza sativa* L. ssp. japonica). Science 296:92–100

Harushima Y, Yano M, Shomura P, Sato M, Shimano T, Kuboki Y, Yamamoto T, Lin SY, Antonio BA, Parco A, Kajiya H, Huang N, Yamamoto K, Nagamura Y, Kurata N, Khush GS, Sasaki T (1998) A high-density rice genetic linkage map with 2275 markers using a single F-2 population. Genetics 148:479–494

Hirochika H (2001) Contribution of the Tos17 retrotransposon to rice functional genomics. Curr Opin Plant Biol 4:118–122

Hirochika H, Sugimoto K, Otsuki Y, Tsugawa H, Kanda M (1996) Retrotransposons of rice involved in mutations induced by tissue culture. Proc Natl Acad USA 93:7783–7788

Ishikawa R, Harada T, Niizeki M, Saito K (1992) Reconstruction of linkage map with isozyme, morphological and physiological markers in rice chromosome-12. Jpn J Breeding 42:235–244

Ito T, Motohashi R, Kuromori T, Mizukado S, Sakurai T, Kanahara H, Seki M, Shinozaki K (2002) A new resource of locally transposed dissociation elements for screening gene-knockout lines in silico on the *Arabidopsis* genome(1)[(w)]. Plant Physiol 129:1695–1699

Jiang N, Bao Z, Zhang X, Hirochika H, Eddy SR, McCouch S, Wessler SR (2003) An active DNA transposon family in rice. Nature 421:163–167

Kikuchi K, Terauchi K, Wada M, Hirano H-Y (2003) The plant MITE *mPing* is mobilized in anther culture. Nature 421:167–170

Kumar LS (1999) DNA markers in plant improvement: An overview. Biotechnology Advances 17:143–182

Kurata N, Nagamura Y, Yamamoto K et al. (1994) A 300 kilobase interval genetic-map of rice including 883 expressed sequences. Nat Genet 8:365–372

Lisch DR, Freeling M, Langham RJ, Choy MY (2001) Mutator transposase is widespread in the grasses. Plant Physiol 125:1293–1303

Mackill DJ, Ni J (2001) Molecular mapping and marker-assisted selection for major-gene traits in rice. In: Khush GS, Brar DS, Hardy B (eds) Rice genetics IV. International Rice Research Institute, Science Publ, Philippines, pp 137–151

Maheswaran M, Subudhi PK, Nandi S, Xu JC, Parco A, Yang DC, Huang N (1997) Polymorphism, distribution, and segregation of AFLP markers in a doubled haploid rice population. Theor Appl Genet 94:39–45

Mao L, Wood TC, Yu YS, Budiman MA, Tomkins J, Woo SS, Sasinowski M, Presting G, Frisch D, Goff S, Dean RA, Wing RA (2000) Rice transposable elements: a survey of 73,000 sequence-tagged-connectors. Genome Res 10:982–990

McCouch SR, Kochert G, Yu ZH, Wang ZY, Khush GS, Coffman WR, Tanksley SD (1988) Molecular mapping of rice chromosome. Theor Appl Genet 76:815–829

McCouch SR, Chen XL, Panaud O, Temnykh S, Xu YB, Cho YG, Huang N, Ishii T, Blair M (1997) Microsatellite marker development, mapping and applications in rice genetics and breeding. Plant Mol Biol 35:89–99

McCouch SR, Temnykh S, Lukashova A, Coburn J, DeClerck G, Cartinhour S, Harrington S, Thomson M, Septiningsih E, Semon M, Moncada P (2001) Microsatellite markers in rice: abundance, diversity, and applications. In: Khush GS, Brar DS, Hardy B (eds) Rice genetics IV. International Rice Research Institute, Science Publ, Philippines, pp 117–135

Nagano H, Kunii M, Azuma T, Kishima Y, Sano Y (2002) Characterization of the repetitive sequences in a 200-kb region around the rice waxy locus: diversity of transposable elements and presence of veiled repetitive sequences. Genes Genet Syst 77:69–79

Nagao S, Takahashi M-E (1963) Trial construction of twelve linkage groups in Japanese rice. J Fac Agric Hokkaido Univ 53:72–130

Nakazaki T, Okumoto Y, Horibata A, Yamahira S, Teraishi M, Nishida H, Inoue H, Tanisaka T (2003) Mobilization of a transposon in the rice genome. Nature 421:170–172

Panaud O, Chen X, McCouch SR (1995) Frequency of microsatellite sequences in rice (*Oryza sativa* L.). Genome 38:1170–1176

Panaud O, Chen X, McCouch SR (1996) Development of microsatellite markers and characterization of simple sequence length polymorphism (SSLP) in rice (*Oryza sativa* L.). Mol Gen Genet 252:597–607

Parinov S, Sevugan M, Ye D, Yang WC, Kumaran M, Sundaresan V (1999) Analysis of flanking sequences from Dissociation insertion lines: a database for reverse genetics in Arabidopsis. Plant Cell 11:2263–2270

Paterson AH (1995) Molecular dissection of quantitative traits –progress and prospects. Genome Res 5:321–333

Putterill J, Robson F, Lee K, Simon R, Coupland G (1995) The CONSTANS gene of *Arabidopsis* promotes flowering and encodes a protein showing similarities to zinc finger transcription factors. Cell 80:847–857

Saito A, Yano M, Kishimoto N et al. (1991) Linkage map of restriction-fragment-length-polymorphism loci in rice. Jpn J Breed 41:665–670

Smith D, Yanai Y, Liu YG, Ishiguro S, Okada K, Shibata D, Whittier RF, Fedoroff NV (1996) Characterization and mapping of Ds-GUS-T-DNA lines for targeted insertional mutagenesis. Plant J 10:721–732

Takagi K, Nagano H, Kishima Y, Sano Y (2003) MITE-transposon display efficiently detects polymorphisms among the *Oryza* AA-genome species. Breed Sci 53:125–132

Tanksley SD (1993) Mapping polygenes. Annu Rev Genet 27:205–233

Temnykh S, Park WD, Ayres N, Cartinhour S, Hauck N, Lipovich L, Cho YG, Ishii T, McCouch SR (2000) Mapping and genome organization of microsatellite sequences in rice (*Oryza sativa* L.). Theor Appl Genet 100:697–712

Temnykh S, DeClerck G, Lukashova A, Lipovich L, Cartinhour S, McCouch S (2001) Computational and experimental analysis of microsatellites in rice (*Oryza sativa* L.): frequency, length variation, transposon associations, and genetic marker potential. Genome Res 11:1441–1452

The Arabidopsis Genome Initiative (2000) Analysis of genome sequence of flowering plant *Arabidopsis thaliana*. Nature 408:796–815

Tissier AF, Marillonnet S, Klimyuk V, Patel K, Torres MA, Murphy G, Jones JDG (1999) Multiple independent defective Suppressor-mutator transposon insertions in Arabidopsis: a tool for functional genomics. Plant Cell 11:1841–1852

Turcotte K, Srinivasan S, Bureau T (2001) Survey of transposable elements from rice genomic sequences. Plant J 25:169–179

Vos P, Hogers R, Bleeker M et al. (1995) AFLP: a new technique for DNA fingerprinting. Nucleic Acids Res 23:4407–4414

Weber JL, May PE (1989) Abundant class of human DNA polymorphisms which can be typed using the polymerase chain-reaction. Am J Hum Genet 44:388–396

Wessler SR, Bureau TE, White SE (1995) LTR-retrotransposons and MITEs: important players in the evolution of plant genomes. Curr Opin Genet Dev 5:814–821

Williams JG, Kubelik AR, Livak KJ, Rafalski JA, Tingey SV (1990) DNA polymorphisms amplified by arbitrary primers are useful as genetic markers. Nucleic Acids Res 18:6531–6535

Wu KS, Tanksley SD (1993) Abundance, polymorphism and genetic mapping of microsatellites in rice. Mol Gen Genet 241:225–235

Yamazaki M, Tsugawa H, Miyao A, Yano M, Wu J, Yamamoto S, Matsumoto T, Sasaki T, Hirochika H (2001) The rice retrotransposon Tos17 prefers low-copy-number sequences as integration targets. Mol Genet Genom 265:336–344

Yano M (2001) Genetic and molecular dissection of naturally occurring variation. Curr Opin Plant Biol 4:130–135

Yano M, Sasaki T (1997) Genetic and molecular dissection of quantitative traits in rice. Plant Mol Biol 35:145–153

Yano M, Katayose Y, Ashikari M, Yamanouchi U, Monna L, Fuse T, Baba T, Yamamoto K, Umehara Y, Nagamura Y, Sasaki T (2000) *Hd1*, a major photoperiod sensitivity quantitative trait locus in rice, is closely related to the Arabidopsis flowering time gene *CONSTANS*. Plant Cell 12:2473–2483

Yu J, Hu SN, Wang J et al. (2002) A draft sequence of the rice genome (*Oryza sativa* L. ssp. *indica*). Science 296:79–92

Zhang QF, Maroof MAS, Lu TY, Shen BZ (1992) Genetic diversity and differentiation of Indica and Japonica rice detected by RFLP. Theor Appl Genet 83:495–499

Zhu J, Gale MD, Quarrie S, Jackson MT, Bryan GJ (1998) AFLP markers for the study of rice biodiversity. Theor Appl Genet 96:602–611

II.11 Wheat Microsatellites: Potential and Implications

M.S. Röder[1], X.-Q. Huang[1], and M.W. Ganal[2]

1 Introduction

Microsatellites, also called simple sequence repeats (SSRs) are a PCR-based marker system which exploits the high variability in the repeat number of simple tandemly repeated DNA motifs. In most cases, microsatellites containing dinucleotide or trinucleotide motifs are used for marker development. Microsatellite markers consist of a defined primer pair flanking a specific microsatellite site in the genome, which can be used for PCR amplification. The variation in repeat number of the microsatellites results in PCR products of varying lengths, which are stably inherited and thus can serve as genetic markers. Microsatellite markers combine a number of advantages for practical applications: they are codominant and multiallelic; they are amenable for automation and high throughput analysis; they are highly variable and in many plant species they detect a higher level of polymorphism per locus than other marker systems such as restriction fragment length polymorphism (RFLPs) or amplified fragment length polymorphisms (AFLPs; Röder et al. 1995). A special advantage in wheat is the genome specificity of most microsatellite markers, which allows the analysis of the three homoeologous genomes of this allohexaploid species individually. Microsatellite markers can be easily transferred between wheat mapping populations, since they identify a specific locus in various genetic backgrounds. This property makes wheat microsatellite markers a valuable tool for determining the chromosomal identity of anonymous linkage groups created by other marker systems such as AFLPs.

2 Development of Microsatellite Markers

The development of good functional wheat microsatellite markers, i.e., high amplifying primer pairs yielding one specific amplification product is a cumbersome task because of the complex nature of the genome which contains an

---

[1] Institut für Pflanzengenetik und Kulturpflanzenforschung (IPK), Corrensstr. 3, 06466 Gatersleben, Germany
[2] TraitGenetics GmbH, Am Schwabeplan 1b, 06466 Gatersleben, Germany

Table 1. The present status of development of microsatellite markers in wheat

Designation of SSR	Number of SSR published	Remarks	Reference
taglgap, taglut	2	Isolated from gene sequences of wheat storage proteins; chromosomal location via NT lines	Devos et al. (1995)
gwm	230	Isolated from genomic libraries of A, B, and D-genomes; mapped in the ITMI population	Röder et al. (1998b)
psp	53	Isolated from genomic libraries of A, B, and D-genomes; mapped in CS × synthetic population	Stephenson et al. (1998)
gdm	65	Isolated from genomic libraries of D-genome; 46 SSR mapped in the ITMI population; chromosomal location of 19 SSR via NT lines	Pestsova et al. (2000)
barc	168	Isolated from genomic libraries of A, B, and D-genomes; mapped in the ITMI population; and in CS deletion lines	Song et al. (2002) http://www.scabusa.org/research_bio.html
DupW	22	Isolated from ESTs; mapped in the ITMI population	Eujayl et al. (2002)
cfd	84	Isolated from genomic libraries of D-genome; mapped in Courtot × CS population	Guyomarc'h et al. (2002)
wmc	225	Isolated from genomic libraries of A, B, and D-genomes; mapped in ITMI population	Gupta et al. (2002) http://res2.agr.ca/winnipeg/mg_wmcinfo-e.htm
No symbol	56	Isolated from ESTs; chromosomal location via NT lines	Holton et al. (2002)
gwm	638	Isolated from genomic libraries of A, B, and D-genomes; mapped in ITMI population; or located via NT lines	Röder and Trait Genetics (unpubl.)
Total	1543		

Abbreviations: *NT lines* nulli-tetrasomic lines, *ITMI* International Triticeae mapping initiative, *CS* Chinese Spring

excess of repetitive sequences. Nevertheless, a number of sources have published wheat microsatellite markers available for practical application (Table 1). The primer pairs were derived from genomic sequences either enriched for microsatellites (Gupta et al. 2002), or for low-copy DNA (Röder et al. 1998b). A few microsatellites derived from expressed sequences (ESTs) have recently become available (Eujayl et al. 2002; Holton et al. 2002). For chromosomal location, the international wheat community is relying mainly on one mapping population, the so-called ITMI-population (International Triticeae Mapping Initiative). This recombinant inbred population was derived from a cross of the Mexican cultivar 'Opata' and a synthetic wheat generated at CIMMYT, Mexico. An RFLP map was created for the ITMI population (Nelson et al. 1995a–c; van Deynze et al. 1995; Marino et al. 1996), which was also used as a framework to map microsatellite markers. Besides segregating populations for genetic mapping, cytogenetic stocks such as nulli-tetrasomic lines or defined deletion stocks (Endo and Gill 1996), were employed for the location of microsatellite markers onto defined chromosomes or chromosomal regions (Plaschke et al. 1996; Bryan et al. 1997; Röder et al. 1998a).

3 The Bridge to Practical Applications

Microsatellite markers have been exploited for three main types of investigations: these are the genetic mapping of single genes, the dissection of quantitative traits (QTLs) and the analysis of genetic diversity. Here, we will discuss the application of wheat microsatellite markers for practical use in plant breeding.

4 Diagnostic Markers for Traits of Interest

Microsatellite markers are ideally suited for genetic mapping of genes of agronomic interest in segregating populations and is documented by a wealth of publications (for review: Gupta et al. 1999; Table 2). However, for the practical utility of a marker, not only is the tight linkage of the marker to a gene of interest required, but the marker has to be diagnostic in various genetic backgrounds. This goal is most efficiently achieved when the gene of interest has been introduced into the breeding material from only one well-defined source. The ideal marker is tightly linked and exhibits a unique allele which is not observed in the plant material serving as genetic background. These aspects are illustrated in the following examples:

- The gibberellin-sensitive dwarfing gene *Rht8* was introduced in the 1930s from the Japanese variety 'Akakomugi' into Italian bread wheat varieties. It was possible to map *Rht8* using single chromosome recombinant lines and to identify a tightly linked microsatellite marker *Xgwm261* on the short arm of wheat chromosome 2D (Korzun et al. 1998). Three main alleles of *Xgwm261* were diagnostic for three phenotypes of *Rht8* and allowed to investigate the distribution of allelic variants of *Rht8* in over 100 wheat varieties (Worland et al. 1998). The results indicated a predominance of the *Rht8* allele for reduced plant height derived from 'Akakomugi' in varieties of southern and southeastern Europe, while in central and northern European varieties the allele for a neutral phenotype of *Rht8* predominated

Table 2. Molecular mapping of major genes/QTL in wheat using SSR markers

Trait	Gene/QTL	Chromosomal location	Population type/ strategy	Reference
Powdery mildew resistance	Pm5e	7BL	$F_{2:3}$ lines, BSA	Huang et al. (2003a)
	Pm24	1DS	$F_{2:3}$ lines, BSA	Huang et al. (2000)
	Pm27	6B-6G	F_2 lines	Järve et al. (2000)
	Pm30	5BS	BC_2F_2 lines, BSA	Liu et al. (2002b)
	MlG	6AL	BC_2F_2 lines, BSA	Xie et al. (2003)
	MlRE	6AL	F_3 lines, BSA	Chantret et al. (2000)
Adult plant resistance to powdery mildew	QTL	5D	F_3 lines, BSA	Chantret et al. (2000)
	QTL	1B, 2A, 2B	$F_{2:3}$ lines, BSA	Liu et al. (2001a)
	QTL	2B, 5D, 6A	DH	Mingeot et al. (2002)
Leaf rust resistance	Lr13	2B	F_2 lines	Seyfarth et al. (2000)
	Lr39	2DS	F_2 lines	Raupp et al. (2001)
	LrTr	4BS	F_2 lines	Sarbarzeh et al. (2001)
Yellow rust resistance	Yr10	1B	F_3 lines	Bariana et al. (2002)
	Yr10	1B	F_2 lines	Wang et al. (2002)
	Yr15	1B	F_2 lines, BSA	Chagué et al. (1999)
	Yr26	1BS	F_2 lines, BSA	Ma et al. (2001)
	YrH52	1B	F_2 lines	Peng et al. (1999)
Adult plant resistance to yellow rust	Yrns-B1	3B	$F_{2:3}$ lines	Börner et al. (2000)
Septoria tritici blotch resistance	Stb5	7D	Chromosome-recombinant lines	Arraiano et al. (2001)
	Stb6	3AS	F_2 lines	Brading et al. (2002)
Fusarium head blight resistance	QTL	5A, 3B	DH lines	Buerstmayr et al. (2002)
	QTL	3BS, 6BS	RILs	Anderson et al. (2001)
	QTL	3B	Advanced lines	del Blanco et al. (2003)
	QTL	2AS, 2BL, 3BS	RILs	Zhou et al. (2002)
Russian wheat aphid resistance	Dn1, Dn2, Dn5, Dn8, Dnx	7DS	F_2 lines	Liu et al. (2001b)
	Dn2	7DS	F_2 lines	Miller et al. (2001)
	Dn4	1D	F_2 lines	Liu et al. (2002a)
	Dn6	7D	F_2 lines	Liu et al. (2002a)
	Dn9	1DL	F_2 lines	Liu et al. (2001b)
Greenbug resistance	Gb3	7D	$F_{2:3}$ lines	Weng and Lazar (2002)
Cyst nematode resistance	Cre5	2AS	NILs	Jahier et al. (2001)
Pseudocercosporella herpotrichoides resistance	Pch1	7A	F_3 lines	Huguet-Robert et al. (2001)
Barley yellow dwarf virus	BYDV	7DL	Recombination lines	Ayala et al. (2001)
Preharvest sprouting tolerance	Major gene	6BS	RILs, BSA	Roy et al. (1999)
	QTL	6A, 3B, 7B	RILs	Zanetti et al. (2000)

Table 2. (Continue)

Trait	Gene/QTL	Chromosomal location	Population type/ strategy	Reference
Vernalization response	*Vrn1*	5AL	F$_2$ lines	Korzun et al. (1997)
Dwarfing genes	*Rht8*	2DS	Single chromosome substitution lines	Korzun et al. (1998)
	Rht12	5AL	F$_2$ lines	Korzun et al. (1997)
Induced sphaerococcoid mutation genes	*S1,S2,S3*	3D, 3B, 3A	F$_2$ lines	Salina et al. (2000)
Grain protein content	QTL	2DL	RILs, BSA	Prasad et al. (1999)
	QTL	6BS	RSL	Khan et al. (2000)
	QTL	5A	RILs, BSA	Singh et al. (2001)
	QTL	2A, 2B, 2D, 3D, 4A, 6B, 7A, 7D	RILs	Prasad et al. (2003)
Waxy	*Wx-D1*	7D	–	Shariflou and Sharp (1999)
Milling yield	QTL	3A, 7D	SSD lines, BSA	Parker et al. (1999)
Milling and baking quality	QTL	2B, 5D	RILs	Campbell et al. (2001)
Grain weight	QTL	1AS	RILs, BSA	Varshney et al. (2000)
Yield and yield components	QTL	See reference	Advanced BC$_2$F$_2$ lines	Huang et al. (2003b)

Abbreviations: *BSA* bulked segregant analysis, *DH* doubled haploids, *RIL* recombinant inbred line, *NIL* near-isogenic line, *RSL* recombinant substitution line, *SSD* single seed descent

(Worland et al. 1998; Röder et al. 2002). This observation is possibly explained by the close linkage of the *Ppd1* gene rendering photoperiodic insensitivity in short day conditions to *Rht8*, so that both genes are often transferred as a linkage block and have a selective advantage at lower latitudes. A pedigree analysis in varieties of breeding programs using microsatellite markers of chromosome 2D illustrated the linkage drag around the *Rht8/Ppd1* genes through the generations resulting in a slow diminution of the chromosomal segment originally introduced from 'Akakomugi' (Pestsova and Röder 2002).

- The dominant powdery mildew resistance gene *Pm24* from the Chinese landrace 'Chiyacao' was mapped on wheat chromosome 1D. The fragment size of the closely linked wheat microsatellite *Xgwm337* was specific for 'Chiyacao' in comparison to 35 wheat cultivars carrying known powdery mildew resistance genes from other sources (Huang et al. 2000). Therefore, *Xgwm337* can be used to monitor and control the introduction of the novel resistance gene *Pm24* into the European breeding material.

In a similar manner, the wheat powdery mildew resistance gene *Pm5e* originating from the Chinese wheat variety 'Fuzhuang 30' was mapped in the distal region of chromosome 7BL. It was suggested that the allele of microsatellite *Xgwm1267* linked to *Pm5e* in 'Fuzhuang 30' can be used for marker-assisted selection in the background of European breeding material (Huang et al. 2003a).

- Several research groups mapped QTLs for Fusarium head blight resistance originating from the Chinese cultivar 'Sumai 3' or its descendants. A main QTL was found on chromosome arm 3BS in all studies associated with the microsatellite markers *Xgwm533*, *Xgwm493*, *Xgwm389* and *Xbarc147* (Anderson et al. 2001; Buerstmayr et al. 2002; Zhou et al. 2002). Further QTLs were detected on chromosomes 3AL, 6AS and 6BS (Anderson et al. 2001), on chromosome 5A associated with the markers *Xgwm293*, *Xgwm304* and *Xgwm156* and on chromosome 1B associated with the high-molecular-weight glutenin locus *XgluB1* (Buerstmayr et al. 2002) and on chromosome arms 2BL and 2AS (Zhou et al. 2002). The diagnostic value of the linked markers remains to be established. However, since the resistance was introduced from one defined source, it is likely that combinations of marker alleles originating from 'Sumai 3' will be of diagnostic value in other genetic backgrounds. This assumption is confirmed by the observation that the QTL on 3BS was detected in various crosses.

4.1 Complex Agronomic Traits: Marker-Based Quantitative Trait Loci Detection in Advanced Backcross Populations

Most characters of economic interest such as quality and yield are defined by multiple genes. While in the past, QTL mapping was often accomplished in recombinant inbred populations or doubled haploids (Keller et al. 1999a, b; Perretant et al. 2000; Börner et al. 2002), the so-called advanced backcross breeding has been suggested to combine marker-based QTL detection with the introduction of novel germplasm into breeding material (Tanksley and Nelson 1996). An exotic donor line, which may be a wild species or unadapted germplasm is introduced into the background of an elite cultivar and reduced to few genomic introgressions by several backcrosses. The phenotypic variation caused by the introgressions is measured in the BC_2 or BC_3 generation and a marker analysis is performed in order to locate the respective QTLs. The first advanced backcross analysis described in wheat was performed in a BC_2-population derived from a cross between the winter wheat variety 'Prinz' and a synthetic wheat. A total of 40 putative QTLs were detected of which 11 were for yield, 16 for yield components, 8 for ear emergence time and 5 for plant height (Huang et al. 2003b). For 24 of them, alleles from the synthetic wheat were associated with a positive effect on agronomic traits, despite the fact that the synthetic wheat was overall inferior with respect to agronomic appearance and performance.

Using the advanced backcross strategy markers linked to QTLs, i.e., genes from a defined source can be identified. Such markers have a high potential of being diagnostic in the genetic background of adapted breeding material and, therefore can serve as tools for marker-assisted breeding and the pyramiding of QTLs.

5 Analysis of Genetic Diversity

Since microsatellite markers detect a high level of variability, they are ideal markers for the identification of varieties, the analysis of germplasm in germplasm collections and the analysis of genetic relationships (Plaschke et al. 1995; Donini et al. 1998; Prasad et al. 2000; Stachel et al. 2000). For example, a limited set of 19 wheat microsatellite markers was able to discriminate nearly all European wheat varieties in a study of 500 varieties (Röder et al. 2002). Thus, with a few microsatellite markers, varieties can be identified in a fairly cheap and routine manner. This could have long-term implications during the variety registration process, where related varieties are currently compared based on morphological characters and can now be discriminated by microsatellite fingerprinting and a comparison to a marker database.

For the management of germplasm collections, microsatellite fingerprinting can provide information concerning potential duplicates and the genetic relationships of accessions collected in certain habitats and provide information for an eventual need of further collections. An example of a large-scale analysis of wheat gene bank material has been published by Huang et al. (2002). A comparison of data from gene bank material and currently grown varieties will, in the long term, provide information which alleles have not yet been introduced into varieties and, thus, could increase the genetic base for breeding. The hypothesis that during modern plant breeding there is a narrowing of the genetic base on which new varieties are developed was investigated in several studies comparing new and old varieties. In three studies no apparent loss of genetic diversity was observed (Donini et al. 2000; Manifesto et al. 2001; Christiansen et al. 2002).

Analysis of genetic relationships by microsatellite markers is also important for the prediction of variance in progenies and the definition of heterotic groups during hybrid breeding (Bohn et al. 1999).

6 Conclusions

The use of microsatellite markers during wheat improvement will certainly increase in the next couple of years. More markers will be available and more linkages of individual markers to traits of interest will be found. Many breeding companies will use these markers for marker-assisted backcrossing and marker-assisted selection. While up to now molecular markers are in most cases tightly linked to the target gene, there is always the possibility that marker and gene will be separated by a recombination event. Thus, in the long term, the ideal marker is defined by a fixed polymorphism in the target gene itself. Such polymorphism will in most cases be a single nucleotide polymorphism (SNP) caused by a base exchange or even a small deletion. Nowadays, the number of known target genes is still very limited and will remain so in the next couple of years. The situation may improve in the long term, when the molecular structure of more genes will be elucidated by detailed mapping of traits using microsatellite markers, map-based cloning approaches or genomic research. An example for so-called perfect markers are SNP markers derived from the isolated dwarfing genes *Rht-B1b* and *Rht-D1b* on chromosomes 4B and 4D (Ellis et al. 2002). In barley, an example of high practical value was that SNP markers were identified in a thermostable β-amylase which can serve to monitor superior malting quality (Kaneko et al. 2000; Paris et al. 2002).

References

Anderson JA, Stack RW, Liu S, Waldron BL, Fjeld AD, Coyne C, Moreno-Sevilla B, Fetch JM, Song QJ, Cregan PB, Frohberg RC (2001) DNA markers for Fusarium head blight resistance QTLs in two wheat populations. Theor Appl Genet 102:1164–1168

Arraiano LS, Worland AJ, Ellerbrook C, Brown JKM (2001) Chromosomal location of a gene for resistance to septoria tritici blotch (*Mycosphaerella graminicola*) in the hexaploid wheat 'Synthetic 6x'. Theor Appl Genet 103:758–764

Ayala L, van Ginkel M, Khairallah M, Keller B, Henry M (2001) Expression of *Thinopyrum intermedium*-derived barley yellow dwarf virus resistance in elite bread wheat backgrounds. Phytopathology 91:55–62

Bariana HS, Brown GN, Ahmed NU, Khatkar S, Conner RL, Wellings CR, Haley S, Sharp PJ, Laroche A (2002) Characterisation of *Triticum vavilovii*-derived stripe rust resistance using genetic, cytogenetic and molecular analyses and its marker-assisted selection. Theor Appl Genet 104:315–320

Bohn M, Utz HF, Melchinger AE (1999) Genetic similarities among winter wheat cultivars determined on the basis of RFLPs, AFLPs, and SSRs and their use for predicting progeny variance. Crop Sci 39:228–237

Börner A, Röder MS, Unger O, Meinel A (2000) The detection and molecular mapping of a major gene for non-specific adult-plant disease resistance against stripe rust (*Puccinia striiformis*) in wheat. Theor Appl Genet 100:1095–1099

Börner A, Schumann E, Fürste A, Cöster H, Leithold B, Röder M, Weber W (2002) Mapping of quantitative trait loci determining agronomic important characters in hexaploid wheat (*Triticum aestivum* L.). Theor Appl Genet 105:921–936

Brading PA, Verstappen ECP, Kema GHJ, Brown JKM (2002) A gene-for-gene relationship between wheat and *Mycosphaerella graminicola*, the Septoria tritici blotch pathogen. Phytopathology 92:439–445

Bryan GJ, Collins AJ, Stephenson P, Orry A, Smith JB, Gale MD (1997) Isolation and characterisation of microsatellites from hexaploid wheat. Theor Appl Genet 94:557–563

Buerstmayr H, Lemmens M, Hartl L, Doldi L, Steiner B, Stierschneider M, Ruckenbauer P (2002) Molecular mapping of QTLs for Fusarium head blight resistance in spring wheat. I. Resistance to fungal spread (Type II resistance). Theor Appl Genet 104:84–91

Campbell KG, Finney PL, Bergman CJ, Gualberto DG, Anderson JA, Giroux MJ, Siritunga D, Zhu J, Gendre F, Roué C, Vérel A, Sorrells ME (2001) Quantitative trait loci associated with milling and baking quality in a soft × hard wheat cross. Crop Sci 41:1275–1285

Chagué V, Fahima T, Dahan A, Sun GL, Korol AB, Ronin YI, Grama A, Röder MS, Nevo E (1999) Isolation of microsatellite and RAPD markers flanking the *Yr15* gene of wheat using NILs and bulked segregant analysis. Genome 42:1050–1056

Chantret N, Sourdille P, Röder M, Tavaud M, Bernard M, Doussinault G (2000) Location and mapping of the powdery mildew resistance gene *MlRE* and detection of a resistance QTL by bulked segregant analysis (BSA) with microsatellites in wheat. Theor Appl Genet 100:1217–1224

Christiansen MJ, Andersen SB, Ortiz R (2002) Diversity changes in an intensively bred wheat germplasm during the 20th century. Mol Breed 9:1–11

Del Blanco IA, Frohberg RC, Stack RW, Berzonsky WA, Kianian SF (2003) Detection of QTL linked to Fusarium head blight resistance in Sumai 3-derived North Dakota bread wheat lines. Theor Appl Genet 106:1027–1031

Devos KM, Bryan GJ, Collins AJ, Gale MD (1995) Application of two microsatellite sequences in wheat storage proteins as molecular markers. Theor Appl Genet 90:247–252

Donini P, Stephenson P, Bryan GJ, Koebner RMD (1998) The potential of microsatellites for high throughput genetic diversity assessment in wheat and barley. Genet Res Crop Evol 45:415–421

Donini P, Law JR, Koebner RMD, Reeves JC, Cooke RJ (2000) Temporal trends in the diversity of UK wheat. Theor Appl Genet 100:912–917

Ellis MH, Spielmeyer W, Gale KR, Rebetzke GJ, Richards RA (2002) 'Perfect' markers for the *Rht-B1b* and *Rht-D1b* dwarfing genes in wheat. Theor Appl Genet 105:1038–1042

Endo TR, Gill BS (1996) The deletion stocks of common wheat. J Hered 87:295–307

Eujayl I, Sorrells ME, Baum M, Wolters P, Powell W (2002) Isolation of EST-derived microsatellite markers for genotyping the A and B genomes of wheat. Theor Appl Genet 104:399–407

Gupta PK, Varshney RK, Sharma PC, Ramesh B (1999) Molecular markers and their applications in wheat breeding. Plant Breed 118:369–390

Gupta PK, Balyan HS, Edwards KJ, Isaac P, Korzun V, Röder M, Gautier M-F, Joudrier P, Schlatter AR, Dubcovsky J, de la Pena RC, Khairallah M, Penner G, Hayden MJ, Sharp P, Keller B, Wang RCC, Hardouin JP, Jack P, Leroy P (2002) Genetic mapping of 66 new microsatellite (SSR) loci in bread wheat. Theor Appl Genet 105:413–422

Guyomarc'h H, Sourdille P, Charmet G, Edwards KJ, Bernard M (2002) Characterisation of polymorphic microsatellite markers from *Aegilops tauschii* and transferability to the D-genome of bread wheat. Theor Appl Genet 104:1164–1172

Holton TA, Christopher JT, McClure L, Harker N, Henry RJ (2002) Identification and mapping of polymorphic SSR markers from expressed gene sequences of barley and wheat. Mol Breed 9:63–71

Huang XQ, Hsam SLK, Zeler FJ, Wenzel G, Mohler V (2000) Molecular mapping of the wheat powdery mildew resistance gene *Pm24* and marker validation for molecular breeding. Theor Appl Genet 101:407–414

Huang XQ, Börner A, Röder MS, Ganal MW (2002) Assessing genetic diversity of wheat (*Triticum aestivum* L.) germplasm using microsatellite markers. Theor Appl Genet 105:699–707

Huang XQ, Wang LX, Xu MX, Röder MS (2003a) Microsatellite mapping of the powdery mildew resistance gene *Pm5e* in common wheat (*Triticum aestivum* L.). Theor Appl Genet 106:858–865

Huang XQ, Cöster H, Ganal MW, Röder MS (2003b) Advanced backcross QTL analysis for the identification of quantitative trait loci alleles from wild relatives of wheat (*Triticum aestivum* L.). Theor Appl Genet 106:1379–1389

Huguet-Robert V, Dedryver F, Röder MS, Korzun V, Abélard P, Tanguy AM, Jaudeau B, Jahier J (2001) Isolation of a chromosomally engineered durum wheat line carrying the *Aegilops ventricosa Pch1* gene for resistance to eyespot. Genome 44:345–349

Jahier J, Abelard P, Tanguy AM, Dedryver F, Rivoal R, Khatkar S, Bariana HS (2001) The *Aegilops ventricosa* segment on chromosome 2AS of the wheat cultivar 'VPM1' carries the cereal cyst nematode resistance gene *Cre5*. Plant Breed 120:125–128

Järve K, Peusha HO, Tsymbalova J, Tamm S, Devos KM, Enno TM (2000) Chromosomal location of a *Triticum timopheevii*-derived powdery mildew resistance gene transferred to common wheat. Genome 43:377–381

Kaneko T, Kihara M, Ito K (2000) Genetic analysis of β-amylase thermostability to develop a DNA marker for malt fermentability improvement in barley (*Hordeum vulgare*). Plant Breed 119:197–201

Keller M, Keller B, Schachermayr G, Winzeler M, Schmid JE, Stamp P, Messmer MM (1999a) Quantitative trait loci for resistance against powdery mildew in a segregating wheat × spelt population. Theor Appl Genet 98:903–912

Keller M, Karutz Ch, Schmid JE, Stamp P, Winzeler M, Keller B, Messmer MM (1999b) Quantitative trait loci for lodging resistance in a segregating wheat × spelt population. Theor Appl Genet 98:1171–1182

Khan IA, Procunier JD, Humphreys DG, Tranquilli G, Schlatter AR, Marcucci-Poltri S, Frohberg R, Dubcovsky J (2000) Development of PCR-based markers for a high grain protein content gene from *Triticum turgidum* ssp. *dicoccoides* transferred to bread wheat. Crop Sci 40:518–524

Korzun V, Röder M, Worland AJ, Börner A (1997) Mapping of the dwarfing (*Rht12*) and vernalisation response (*Vrn1*) genes in wheat by using RFLP and microsatellite markers. Plant Breed 116:227–232

Korzun V, Röder MS, Ganal MW, Worland AJ, Law CN (1998) Genetic analysis of the dwarfing gene *Rht8* in wheat, part I. Molecular mapping of *Rht8* on the short arm of chromosome 2D of bread wheat (*Triticum aestivum*). Theor Appl Genet 96:1104–1109

Liu SX, Griffey CA, Saghai Maroof MA (2001a) Identification of molecular markers associated with adult plant resistance to powdery mildew in common wheat cultivar Massey. Crop Sci 41:1268–1275

Liu XM, Smith CM, Gill BS, Tolmay V (2001b) Microsatellite markers linked to six Russian wheat aphid resistance genes in wheat. Theor Appl Genet 102:504–510

Liu Z, Sun Q, Ni Z, Nevo E, Yang TM (2002a) Molecular characterization of a novel powdery mildew resistance gene *Pm30* in wheat originating from wild emmer. Euphytica 123:21–29

Liu XM, Smith CM, Gill BS (2002b) Identification of microsatellite markers linked to Russian wheat aphid resistance genes *Dn4* and *Dn6*. Theor Appl Genet 104:1042–1048

Ma J, Zhou R, Dong Y, Wang L, Wang X, Jia J (2001) Molecular mapping and detection of the yellow rust resistance gene *Yr26* in wheat transferred from *Triticum turgidum* L. using microsatellite markers. Euphytica 120:219–226

Manifesto MM, Schlatter AR, Hopp HE, Suarez EY, Dubcovsky J (2001) Quantitative evaluation of genetic diversity in wheat germplasm using molecular markers. Crop Sci 41:682–690

Marino CL, Nelson JC, Lu YH, Sorrells ME, Leroy P, Tuleen NA, Lopes CR, Hart GE (1996) Molecular genetic maps of the group 6 chromosomes of hexaploid wheat (*Triticum aestivum* L. em. Thell.). Genome 39:359–366

Miller CA, Altinkut A, Lapitan NLV (2001) A microsatellite marker for tagging *Dn2*, a wheat gene conferring resistance to the Russian wheat aphid. Crop Sci 41:1584–1589

Mingeot D, Chantret N, Baret PV, Dekeyser A, Boukhatem N, Sourdille P, Doussinault G, Jacquemin JM (2002) Mapping QTL involved in adult plant resistance to powdery mildew in the winter wheat line RE714 in two susceptible genetic backgrounds. Plant Breed 121:133–140

Nelson JC, Sorrells ME, van Deynze AE, Lu YH, Atkinson M, Bernard M, Leroy P, Faris JD, Anderson JA (1995a) Molecular mapping of wheat: Major genes and rearrangements in homoeologous groups 4, 5, and 7. Genetics 141:721–731

Nelson JC, van Deynze AE, Autrique E, Sorrells ME, Lu YH, Merlino M, Atkinson M, Leroy P (1995b) Molecular mapping of wheat: homoeologous group 2. Genome 38:516–524

Nelson JC, van Deynze AE, Autrique E, Sorrells ME, Lu YH, Negre S, Bernard M, Leroy P (1995c) Molecular mapping of wheat: homoeologous group 3. Genome 38:525–533

Paris M, Jones MGK, Eglinton JK (2002) Genotyping single nucleotide polymorphisms for selection of barley β-amylase alleles. Plant Mol Biol Rep 20:149–159

Parker GD, Chalmers KJ, Rathjen AJ, Langridge P (1999) Mapping loci associated with milling yield in wheat (*Triticum aestivum* L.). Mol Breed 5:561–568

Peng JH, Fahima T, Röder MS, Li YC, Dahan A, Grama A, Ronin YI, Korol AB, Nevo E (1999) Microsatellite tagging of the stripe-rust resistance gene *YrH52* derived from wild emmer wheat, *Triticum dicoccoides*, and suggestive negative crossover interference on chromosome 1B. Theor Appl Genet 98:862–872

Perretant MR, Cadalen T, Charmet G, Sourdille P, Nicolas P, Boeuf C, Tixier MH, Branlard G, Bernard S, Bernard M (2000) QTL analysis of bread-making quality in wheat using a doubled haploid population. Theor Appl Genet 100:1167–1175

Pestsova E, Ganal MW, Röder MS (2000) Isolation and mapping of microsatellite markers specific for the D genome of bread wheat. Genome 43:689–697

Pestsova E, Röder MS (2002) Microsatellite analysis of wheat chromosome 2D allows the reconstruction of chromosomal inheritance in pedigrees of breeding programmes. Theor Appl Genet 106:84–91

Plaschke J, Ganal MW, Röder MS (1995) Detection of genetic diversity in closely related bread wheat using microsatellite markers. Theor Appl Genet 91:1001–1007

Plaschke J, Börner A, Wendehake K, Ganal MW, Röder MS (1996) The use of wheat aneuploids for the chromosomal assignment of microsatellite loci. Euphytica 89:33–40

Prasad M, Varshney RK, Kumar A, Balyan HS, Sharma PC, Edwards KJ, Singh H, Dhaliwal HS, Roy JK, Gupta PK (1999) A microsatellite marker associated with a QTL for grain protein content on chromosome arm 2DL of bread wheat. Theor Appl Genet 99:341–345

Prasad M, Varshney RK, Roy JK, Balyan HS, Gupta PK (2000) The use of microsatellites for detecting DNA polymorphism, genotype identification and genetic diversity in wheat. Theor Appl Genet 100:584–592

Prasad M, Kumar N, Kulwal PL, Röder MS, Balyan HS, Dhaliwal HS Gupta PK (2003) QTL analysis for grain protein content using SSR markers and validation studies using NILs in bread wheat. Theor Appl Genet 106:659–667

Raupp WJ, Singh S, Brown-Guedira GL, Gill BS (2001) Cytogenetic and molecular mapping of the leaf rust resistance gene *Lr39* in wheat. Theor Appl Genet 102:347–352

Röder MS, Plaschke J, König SU, Börner A, Sorrells ME, Tanksley SD, Ganal MW (1995) Abundance, variability and chromosomal location of microsatellites in wheat. Mol Gen Genet 246:327–333

Röder MS, Korzun V, Gill B, Ganal MW (1998a) The physical mapping of microsatellite markers in wheat. Genome 41:278–283

Röder MS, Korzun V, Wendehake K, Plaschke J, Tixier MH, Leroy P, Ganal MW (1998b) A microsatellite map of wheat. Genetics 149:2007–2023

Röder MS, Wendehake K, Korzun V, Bredemeijer G, Laborie D, Bertrand L, Isaac P, Rendell S, Jackson J, Cooke RJ, Vosman B, Ganal MW (2002) Construction and analysis of a microsatellite-based database of European wheat varieties. Theor Appl Genet 106:67–73

Roy JK, Prasad M, Varshney RK, Balyan HS, Blake TK, Dhaliwal HS, Singh H, Edwards KJ, Gupta PK (1999) Identification of a microsatellite on chromosomes 6B and a STS on 7D of bread wheat showing an association with preharvest sprouting tolerance. Theor Appl Genet 99:336–340

Salina E, Börner A, Leonova I, Korzun V, Laikova L, Maystrenko O, Röder MS (2000) Microsatellite mapping of the induced sphaerococcoid mutation genes in *Triticum aestivum*. Theor Appl Genet 100:686–689

Sarbarzeh A, Singh H, Dhaliwal HS (2001) A microsatellite marker linked to leaf rust resistance transferred from *Aegilops triuncialis* into hexaploid wheat. Plant Breed 120:259–261

Seyfarth R, Feuillet C, Schachermayr G, Messmer M, Winzeler M, Keller B (2000) Molecular mapping of the adult-plant leaf rust resistance gene *Lr13* in wheat (*Triticum aestivum* L.). J Genet Breed 54:193–198

Singh H, Prasad M, Varshney RK, Roy JK, Balyan HS, Dhaliwal HS, Gupta PK (2001) STMS markers for grain protein content and their validation using near-isogenic lines in bread wheat. Plant Breed 120:273–278

Shariflou MR, Sharp PJ (1999) A polymorphic microsatellite in the 3' end of 'waxy' genes of wheat, *Triticum aestivum*. Plant Breed 118:275–277

Song QJ, Fickus EW, Cregan PB (2002) Characterization of trinucleotide SSR motifs in wheat. Theor Appl Genet 104:286–293

Stachel M, Lelley T, Grausgruber H, Vollmann J (2000) Application of microsatellites in wheat (*Triticum aestivum* L.) for studying genetic differentiation caused by selection for adaptation and use. Theor Appl Genet 100:242–248

Stephenson P, Bryan G, Kirby J, Collins A, Devos K, Busso C, Gale M (1998) Fifty new microsatellite loci for the wheat genetic map. Theor Appl Genet 97:946–949

Tanksley SD, Nelson JC (1996) Advanced backcross QTL analysis: a method for the simultaneous discovery and transfer of valuable QTLs from unadapted germplasm into elite breeding lines. Theor Appl Genet 92:191–203

Van Deynze AE, Dubcovsky J, Gill KS, Nelson JC, Sorrells ME, Dvořák J, Gill BS, Lagudah ES, McCouch SR, Appels R (1995) Molecular-genetic maps for group 1 chromosomes of Triticeae species and their relation to chromosomes in rice and oat. Genome 38:45–59

Varshney RK, Prasad M, Roy JK, Kumar N, Harjit-Singh, Dhaliwal HS, Balyan HS, Gupta PK (2000) Identification of eight chromosomes and a microsatellite marker on 1AS associated with QTL for grain weight in bread wheat. Theor Appl Genet 100:1290–1294

Wang L, Ma J, Zhou R, Wang X, Jia J (2002) Molecular tagging of the yellow rust resistance gene *Yr10* in common wheat, P.I.178383 (*Triticum aestivum* L.). Euphytica 124:71–73

Weng Y, Lazar MD (2002) Amplified fragment length polymorphism- and simple sequence repeat-based molecular tagging and mapping of greenbug resistance gene *Gb3* in wheat. Plant Breed 121:218–223

Worland AJ, Korzun V, Röder MS, Ganal MW, Law CN (1998) Genetic analysis of the dwarfing gene *Rht8* in wheat, part II. The distribution and adaptive significance of allelic variants at the *Rht8* locus of wheat as revealed by microsatellite screening. Theor Appl Genet 96:1110–1120

Xie C, Sun Q, Ni Z, Yang TM, Nevo E, Fahima T (2003) Chromosomal location of a *Triticum dicoccoides*-derived powdery mildew resistance gene in common wheat by using microsatellite markers. Theor Appl Genet 106:341–345

Zanetti S, Winzeler M, Keller M, Keller B, Messmer M (2000) Genetic analysis of pre-harvest sprouting resistance in a wheat × spelt cross. Crop Sci 40:1406–1417

Zhou W, Kolb FL, Bai G, Shaner G, Domier LL (2002) Genetic analysis of scab resistance QTL in wheat with microsatellite and AFLP markers. Genome 45:719–727

II.12 Comparative Genetic Mapping in Trees: The Group of Conifers

D.B. NEALE and K.V. KRUTOVSKY[1]

1 Introduction

The power of comparative genomics is widely accepted and applies to all taxa (Sankoff and Nadeau 2000). The genomes of the model systems *Arabidopsis* (The Arabidopsis Genome Initiative 2000) and rice (Goff et al. 2002; Yu et al. 2002) have been completely sequenced and are used to aid in positional cloning of genes from related species having much larger genomes and lacking a complete genome sequence. Comparative mapping among nonmodel species helps to understand the evolution of plant genomes (Bennetzen and Freeling 1993; Gale and Devos 1998) and can help validate quantitative trait loci (QTL) from one crop species to another (Paterson et al. 1995). Although there is no known small genome model species in conifers to be the equivalent of *Arabidopsis* and rice, conifers would still benefit significantly from an organized comparative mapping effort.

Comparative mapping in plants began with the rather simple demonstration that maps in one species could be constructed using restriction fragment length polymorphism (RFLP) probes from a related species and once such maps were made, they could be compared (Bonierbale et al. 1988; Ann and Tanksley 1993). Loci revealed by RFLP probes are assumed to be orthologous between species, meaning that the gene was present in a common ancestor (Fig. 1). In contrast, paralogous loci result from duplications following speciation. Comparative maps have now been made for species in several important plant families including Brassicaceae (Paterson et al. 2000; Barnes 2002; Hall et al. 2002), Poaceae (Feuillet and Keller 2002; Laurie and Devos 2002; Ware et al. 2002; Ware and Stein 2003) and Solanaceae (Doganlar et al. 2002). The syntenic relationship among species provides insight into the type and number of chromosomal rearrangements that have occurred in the evolution of these plant groups.

The international genome mapping community in forestry is very small and many different tree species are involved. For example, there are probably no more than 20 labs worldwide actively constructing forest tree genetic maps and there are at least an equal number of species being mapped. Forest

---

[1] Institute of Forest Genetics, Pacific Southwest Research Station, USDA Forest Service, Davis, California 95616, USA

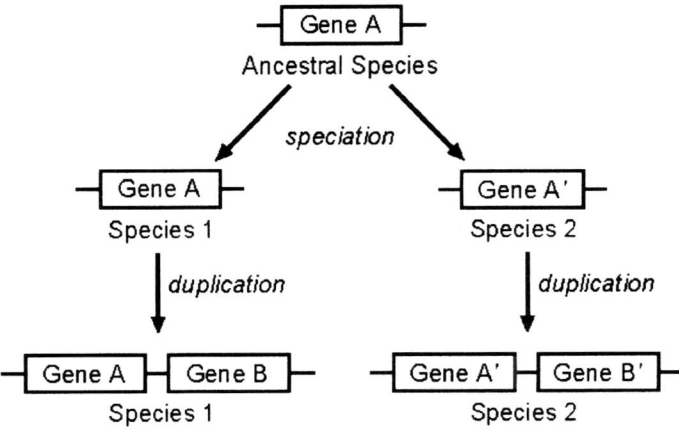

Fig. 1. Gene duplication. *A* and *A'* are orthologs because they are related by descent, whereas *B* and *B'* are paralogs because they are not directly related by descent, but rather are formed from duplication of individual genes in one species following speciation

tree mapping efforts include species from both angiosperms (*Eucalyptus*, *Populus*, *Quercus*) and gymnosperms (*Pinus* and other conifers). Comparative maps have been constructed in *Eucalyptus* (Marques et al. 2002), and between *Quercus* and *Castanea* (Fagaceae) (Barreneche et al. 2004). In this chapter, we review the progress toward constructing comparative genetic maps in conifers.

2 Conifer Genomes

Conifer genomes are among the largest of plant genomes. Genome size estimates range from 5.8–32.2 pg per 1 N nucleus (Leitch et al. 2001; Bennett and Leitch 2003). All conifers are diploids with just a couple of exceptions, such as hexaploid *Sequoia sempervirens* (coast redwood; 2n=6x=66; Ahuja and Neale 2002). It is not known how or why conifer genomes became so large, although it is clear that genes have been amplified by some mechanism, resulting in a large number of pseudogenes (Kinlaw and Neale 1997). Pseudogenes are much more likely to be paralogs than orthologs, thereby making comparative mapping quite challenging. In terms of genetic map distance, however, conifers are not much different from many of the crop species with estimated sizes of around 2000 cM (Echt and Nelson 1997). Except for the Podocarpaceae and *Pseudolarix amabilis*, the base chromosome number in conifers ranges from 11 to 13. In fact, all members of the Pinaceae have 12 pairs of chromosomes, with *Pseudotsuga menziesii* (Douglas fir) being the only exception with 13 pairs (O'Brien et al. 1996). Comparative karyotype analysis suggests that conifer chromosomes have undergone little rearrangement in their evolutionary history (Sax and Sax 1933; Prager et al. 1976).

Genetic maps have been constructed for about 20 conifer species (Cervera et al. 1999; http://dendrome.ucdavis.edu/treegenes). Most are from *Pinus* (*brutia, elliottii, lambertiana, palustris, pinaster, radiata, strobus, sylvestris, taeda*) and from a few other genera (*Cryptomeria, Larix, Picea, Pseudotsuga, Taxus*). Maps have been constructed using an array of genetic marker types [isozymes, RFLPs, random amplified polymorphic DNAs (RAPDs), amplified fragment length polymorphisms (AFLPs), and simple sequence repeats (SSRs)], but in very few cases have the same genetic markers been used on more than one map. Classical (based on visible mutations) or cytogenetic maps have not been constructed for conifers. Thus, the genetic marker maps cannot be assigned to chromosomes nor can the marker maps be compared to one another.

To address this problem in conifer genetics, the Conifer Comparative Genomics Project (CCGP, http://dendrome.ucdavis.edu/ccgp) was established in 1999. The project had two primary goals: (1) to develop and distribute orthologous genetic markers and (2) to construct comparative genetic maps for several of the more important conifer species. These goals are nearly complete and will be summarized in the remainder of this chapter.

3 Loblolly Pine Reference Genetic Map

Pinus taeda (loblolly pine) was chosen for the reference genetic map simply because it possesses the greatest wealth of genetic marker information. RFLP genetic maps were constructed using two 3-generation mapping populations (Devey et al. 1994; Groover et al. 1994). The mapping populations were given the names *base* and *qtl*, respectively. These maps were merged into a single consensus genetic map (Sewell et al. 1999). The *base* and *qtl* reference mapping populations are distributed freely along with framework marker segregation data (http://dendrome.ucdavis.edu/ccgp/refmap.html). This has enabled researchers developing markers in other species to map those markers to the reference genetic map.

4 Genetic Markers for Comparative Mapping in Conifers

Orthologous genetic markers are essential for comparative mapping. RFLPs have been used almost exclusively for comparative mapping in other plant taxa. Ahuja et al. (1994) showed that cDNA RFLP probes derived from *P. taeda* would hybridize to genomic DNA from other species of *Pinus* and even other members of the Pinaceae and Coniferales, suggesting that RFLP probes could be shared among labs for mapping purposes and that comparative maps would result from such exchanges. This has not occurred due to the difficulty in performing RFLP analyses in conifers and the preference for using PCR-based markers. Most genome mapping projects in conifers have used one of the PCR-based marker systems (RAPDs, AFLPs, or SSRs). Unfortunately, these marker types do not have the potential for providing orthologous markers that can be used across many species. Even SSRs can only be used within a narrow range of related species (Echt et al. 1999).

The CCGP has developed and mapped 135 new genetic markers based on expressed sequenced tags (Harry et al. 1998; Temesgen et al. 2000, 2001; Brown et al. 2001). These markers are called expressed sequence tagged polymorphisms (ESTP) because they are derived from expressed genes. Brown et al. (2001) showed that ESTP primers amplify subgenus *Pinus* DNA at nearly a 100% success rate and at about a 50% rate in the subgenus *Strobus*. Furthermore, many of these primers amplified DNA from other genera of the Pinaceae. The orthology of ESTP markers was established through a combination of conserved map location and comparative sequence analysis (Brown et al. 2001). This set of orthologous ESTP markers was used to construct comparative maps between *P. taeda* and several species of *Pinus* and genera of Pinaceae.

5 Comparative Mapping in *Pinus*

The CCGP was formed to construct comparative maps between a few of the more important *Pinus* species found worldwide (Table 1; Fig. 2). Taxonomic representation is limited as only three subsections are included. Nevertheless, the species included are of the highest economic importance and possess significant genomic resources (e.g., maps, gene sequences, phenotypic data).

Table 1. Conifer comparative maps. *Pinus taeda* serves as the reference map for all comparative maps

Genus	Species	Subsection	Homologous LGs	Comparative markers	Reference
Pinus	elliottii	Australes	12	57	Brown et al. (2001)
Pinus	radiata	Attenuatae	12	67	Devey et al. (1999)
Pinus	pinaster	Pinus	10	31	Chagné et al. (2003)
Pinus	sylvestris	Pinus	12	37	Komulainen et al. (2003)
Pseudotsuga	menziesii	–	10	46	Krutovsky et al. (2004)
Picea	abies	–	7	26	Troggio et al. (unpubl.)

Fig. 2. Comparative mapping in the genus *Pinus* (linkage group 6). Orthologous comparative mapping markers are *underlined* and shown in **bold**. The high level of synteny and conservation of gene order allowed homologous linkage groups among pine species to be easily identified

Comparative Genetic Mapping in Trees: The Group of Conifers 271

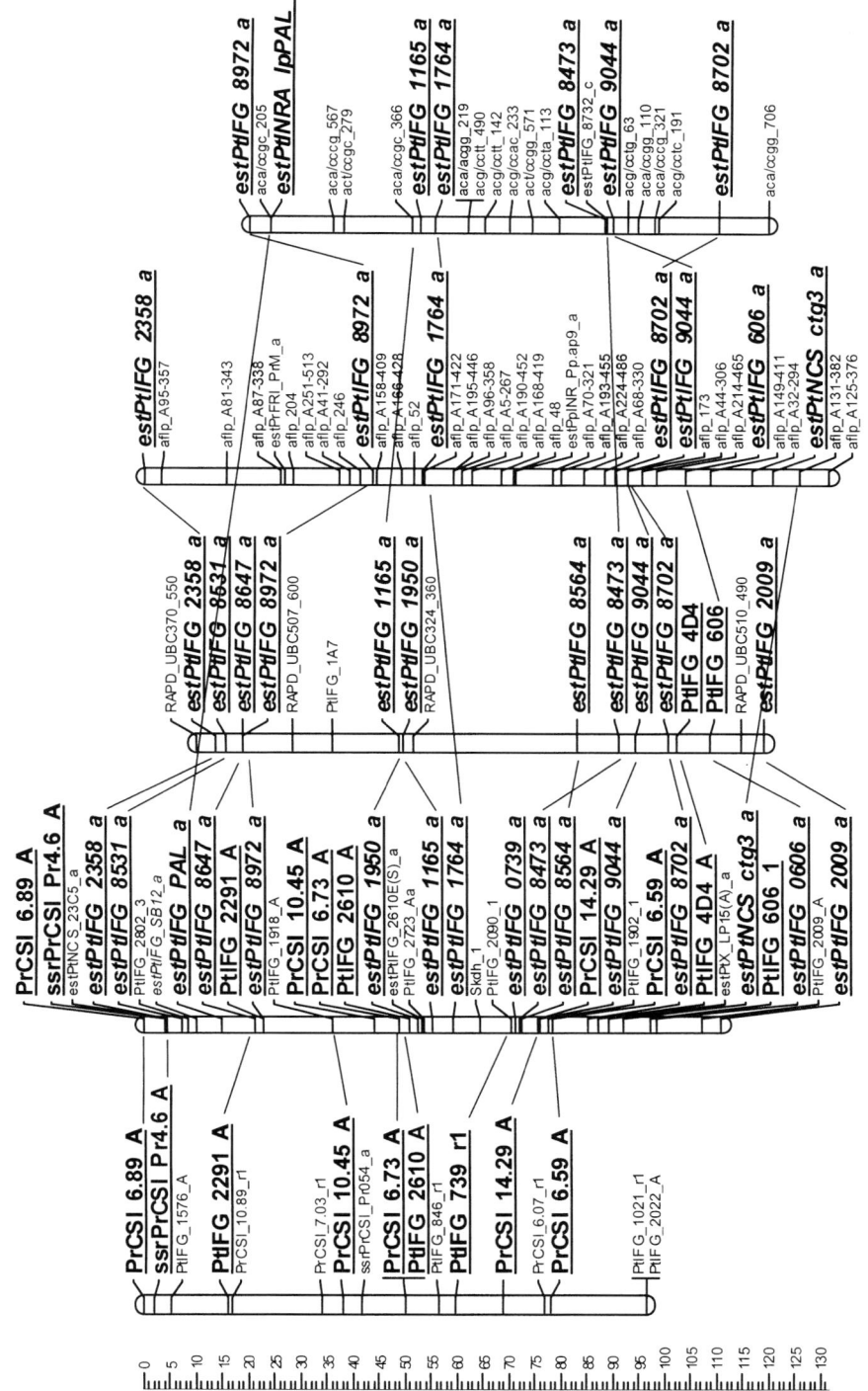

Fig. 2.

The two most closely related species to share a comparative map are *P. taeda* and *P. elliottii* (Brown et al. 2001). The *P. elliottii* reference map was based on RAPDs, RFLP probes from *P. taeda* and isozymes. The map consisted of 15 linkage groups (LG) of three or more markers. Thirteen of the *P. elliottii* linkage groups could be aligned to the homologous linkage groups in *P. taeda*. In three cases (LGs 2, 3 and 8), two unlinked *P. elliottii* LGs were assigned to a single *P. taeda* LG. This illustrates how partial linkage maps can be coalesced by comparative mapping. The *P. taeda* × *P. elliottii* map had ten homologous LGs, the 11th and 12th homologous groups could not be determined. However, by comparing these remaining groups to the *P. taeda* × *P. sylvestris* comparative map (Komulainen et al. 2003), it can be inferred based on common markers that *P. taeda* LGs 11 and 12, and *P. elliottii* LGs 14 and 13, respectively, are homologous groups. Linkage group 11 provides a nice example of how multi-species comparisons can identify and coalesce linkage groups (Fig. 3). A small LG in *P. taeda* that included the marker *estPTIFG-1750-a* was merged with LG 12 based on a comparison to *P. sylvestris*. Next, the homologous LG 11 in *P. elliottii* could be identified based on the three-

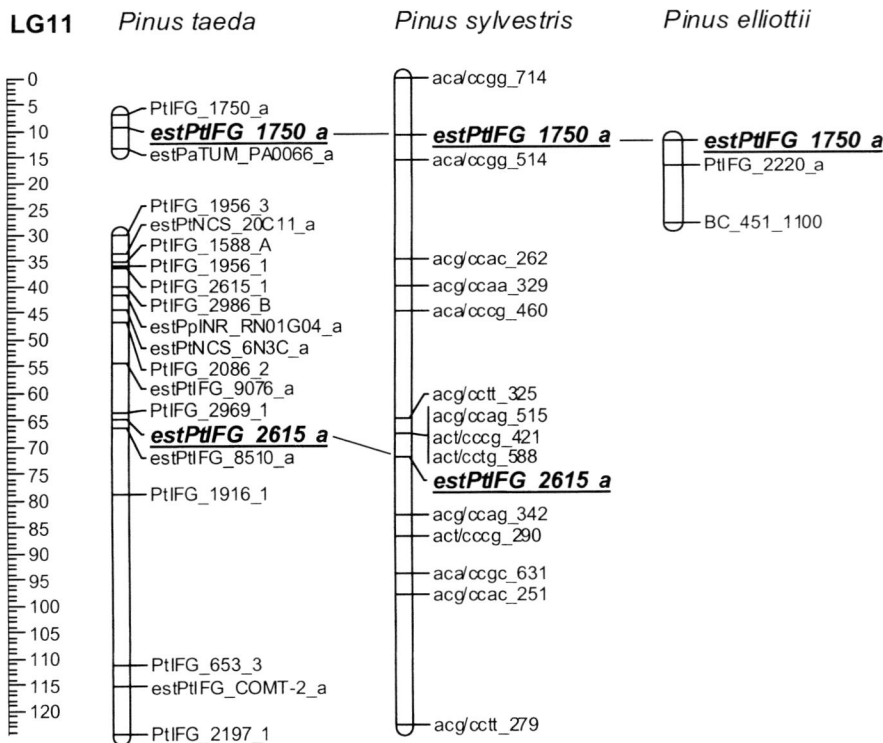

Fig. 3. Comparative mapping among three *Pinus* species (*P. taeda*, *P. sylvestris* and *P. elliottii*; linkage group 11). This example illustrates how multi-species comparisons identify and coalesce homologous linkage groups

way comparison of *P. elliottii, P. sylvestris* and *P. taeda*. This example illustrates the synthesis power of comparative mapping.

Comparative genetic maps have been constructed for two members of the subsection Pinus, *P. pinaster* (Chagné et al. 2003) and *P. sylvestris* (Komulainen et al. 2003). Homologous LGs 1–10 were identified between *P. taeda* and *P. pinaster*, however, orthologous ESTP markers could not be mapped to *P. pinaster* LGs 11 and 12 and could not be assigned to homologous LGs in *P. taeda*. All 12 homologous LGs were identified in the comparative map between *P. taeda* and *P. sylvestris*. A number of very small linkage groups in *P. sylvestris* were not linked to any of the 12 homologous groups in either *P. sylvestris* or *P. taeda*.

The first comparative map constructed in *Pinus* species was between *P. taeda* and *P. radiata*, subsection Attenuatae (Devey et al. 1999). This comparative map was based on RFLP and SSR markers. Establishing the orthology of RFLP and SSR markers in conifers is difficult due to the large number of paralogs revealed by these markers. RFLP probes that revealed just a single locus in both species were assumed to be orthologous, however, there were few examples of this probe type. More frequently, RFLP probes revealed multiple loci in both species, thus making identification of orthologous loci difficult. In cases where a locus was mapped to nearly the same location in both species, it was assumed to be orthologous. This interpretation is reasonable, yet somewhat circular. Comparative maps based on RFLP markers in *Pinus* should be viewed cautiously. Cross-amplification using SSR primers was low and their utility for comparative mapping was limited.

The primary goal of the CCCP was to establish homologous linkage groups among species and assign numbers to each of the 12 chromosomes in *Pinus*. This goal was successful and the first step toward development of a comparative map bioinformatic resource is complete. The comparative maps have been submitted to the genome database for forest trees, TreeGenes (http://dendrome.ucdavis.edu). For the first time it will be possible to compare the map positions of QTLs and expressed genes among species of *Pinus*. The power of comparative genomic analysis can now be realized.

6 Comparative Mapping in Pinaceae

The family Pinaceae includes ten genera (*Abies, Cathaya, Cedrus, Keteleeria, Larix, Picea, Pinus, Pseudolarix, Pseudotsuga, and Tsuga*). Comparative maps between *P. taeda* and *Pseudotsuga menziesii* (Douglas fir) and between *P. taeda* and *Picea abies* (Norway spruce) have been constructed (Table 1; Fig. 4). The *P. taeda* by *Pseudotsuga menziesii* comparative map was based on 46 comparative map markers (Krutovskii et al. 2004) and ten homologous linkage groups were identified. ESTP markers derived from *P. taeda* and RFLP probes from *P. taeda* were used as comparative map markers, just like

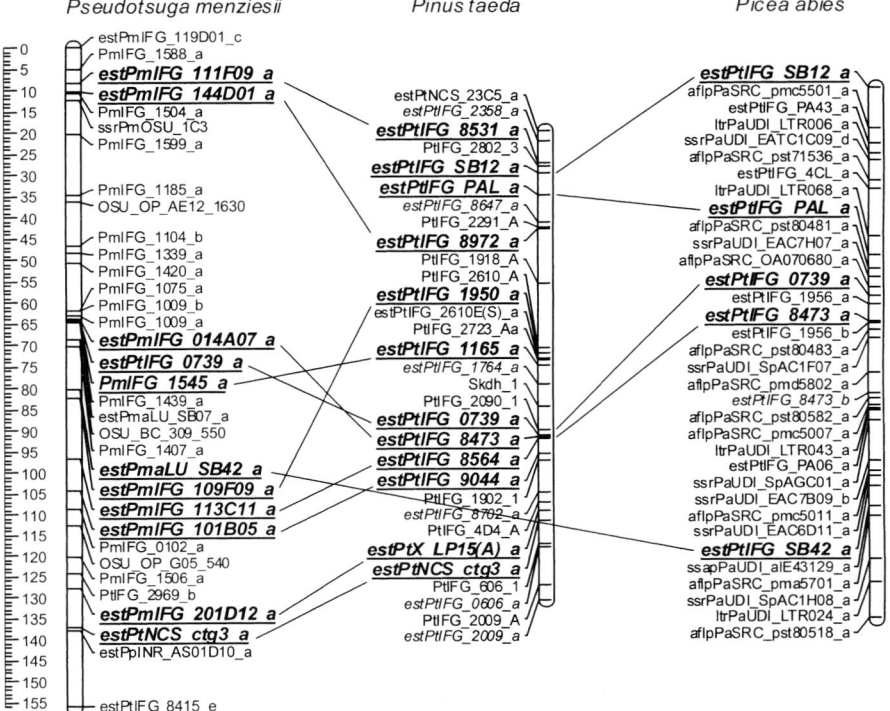

Fig. 4. Comparative mapping in the family Pinaceae (linkage group 6). It has also been possible to map the orthologous comparative mapping markers to other genera of the Pinaceae. The comparative map structure enables the Pinaceae to be viewed as one large genetic system

the cases between *Pinus* species comparative maps. In addition, a third type of comparative map marker was developed for the intergeneric maps. ESTP markers were developed directly from *Pseudotsuga menziesii* based on a database of ~5000 *Pseudotsuga menziesii* ESTs (Krutovskii et al. 2004). Comparative sequence analyses were performed to identify putatively orthologous ESTs from *Pseudotsuga menziesii* to mapped ESTPs in *P. taeda*. Orthology was confirmed based on comparative sequence analyses of ESTP amplicons from both species (Krutovskii et al. 2004). A comparative map between *P. taeda* and *P. abies* has been partially constructed (Troggio et al., unpubl.). It is reasonable to assume that comparative genetic maps could now be constructed for all genera of Pinaceae and that a comparative genomic infrastructure could be built based on the family Pinaceae as a single genetic system.

7 Conclusions: Needs of Linking the Genetic Map to Chromosomes

A future goal for the CCGP is to connect each of the pine genetic linkage groups to each of the 12 chromosomes in pine. Our strategy is to: (1) identify single- or low-copy subfragments of clones from a *P. taeda* bacterial artificial chromosome (BAC) library, (2) genetically map subfragments of these clones to the *P. taeda* consensus map, and (3) physically map the corresponding BAC clones to pine chromosomes using fluorescence in situ hybridization (FISH). A partial (approx. 5%) *P. taeda* BAC library has been constructed and BAC clones have been screened by Southern hybridization to identify low/single-copy subfragments (Kinlaw et al., unpubl.). Single nucleotide polymorphisms (SNPs) have been identified in mapping populations within several subfragments. This approach should eventually lead to the mapping of all 12 *P. taeda* genetic linkage groups to their respective chromosomes. Once completed, this should facilitate the development of chromosome-specific libraries and genomic sequencing of targeted regions of the *P. taeda* genome.

Acknowledgements. We thank all the members of the CCGP for their collaborative efforts to construct comparative maps in conifers. The CCGP was supported by USDA Plant Genome National Research Initiative Grants (USDA NRI 95-37300-1632 and 00-35300-9316).

References

Ahuja MR, Neale DB (2002) Origins of polyploidy in coast redwood (*Sequoia sempervirens* (D. Don) Endl.) and relationship of coast redwood to other genera of Taxodiaceae. Silvae Genet 51:93–100

Ahuja MR, Devey ME, Groover AT, Jermstad KD, Neale DB (1994) Mapped DNA probes from loblolly pine can be used for restriction fragment length polymorphism mapping in other conifers. Theor Appl Genet 88:279–282

Ahn S, Tanksley SD (1993) Comparative linkage maps of the rice and maize genomes. Proc Natl Acad Sci USA 90:7980–7984

Barnes S (2002) Comparing *Arabidopsis* to other flowering plants. Curr Opin Plant Biol 5:128–134

Barreneche T, Casasoli M, Russell K, Akkak A, Meddour H, Plomion C, Villani F, Kremer A (2004) Comparative mapping between *Quercus* and *Castanea* using simple-sequence repeats (SSRs). Theor Appl Genet 108:558–566

Bennett MD, Leitch IJ (2003) Plant DNA C-values database (release 2.0, January 2003), http://www.rbgkew.org.uk/cval/homepage.html

Bennetzen JL, Freeling M (1993) Grasses as a single genetic system: genome composition, collinearity and compatibility. Trends Genet 9:279–282

Bonierbale R, Plaisted RL, Tanksley SD (1988) RFLP maps of potato and tomato based on a common set of clones reveal modes of chromosomal evolution. Genetics 120:1095–1103

Brown GR, Kadel III EE, Bassoni DA, Kiehne KL, Temesgen B, van Buijtenen JP, Sewell MM, Marshall KA, Neale DB (2001) Anchored reference loci in loblolly pine (*Pinus taeda* L.) for integrating pine genomics. Genetics 159:799–809

Cervera MT, Plomion C, Malpica C (1999) Molecular markers and genome mapping in woody plants. In: Jain SM, Minocha SC (eds) Molecular biology of woody plants, vol 1. Kluwer, Dordrecht, pp 375–394

Chagné D, Brown G, Lalanne C, Madur D, Pot D, Neale D, Plomion C (2003) Comparative genome and QTL mapping between maritime and loblolly pines. Mol Breed 12:185–195

Devey ME, Fiddler TA, Liu BH, Knapp SJ, Neale DB (1994) An RFLP linkage map for loblolly pine based on a three-generation outbred pedigree. Theor Appl Genet 88:273–278

Devey ME, Sewell MM, Uren TL, Neale DB (1999) Comparative mapping in loblolly pine and radiata pine using RFLP and microsatellite markers. Theor Appl Genet 93:656–662

Doganlar S, Frary A, Daunay MC, Lester RN, Tanksley SD (2002) Conservation of gene function in the Solanaceae as revealed by comparative mapping of domestication traits in eggplant. Genetics 161:1713–1726

Echt CS, Nelson CD (1997) Linkage mapping and genome length in eastern white pine (*Pinus strobus* L.). Theor Appl Genet 94:1031–1037

Echt CS, Vendramin GG, Nelson CD, Marquart P (1999) Microsatellite DNA as shared genetic markers among conifer species. Can J For Res 29:365–371

Feuillet C, Keller B (2002) Comparative genomics in the grass family: molecular characterization of grass genome structure and evolution. Ann Bot 89:3–10

Gale MD, Devos KM (1998) Plant comparative genetics after 10 years. Science 282:656–659

Goff SA, Ricke D, Lan TH, Presting G, Wang R et al. (2002) A draft sequence of the rice genome (*Oryza sativa* L. ssp. *japonica*). Science 296:92–100

Groover AT, Devey ME, Fiddler TA, Lee JM, Megraw RA, Mitchell-Olds T, Sherman BK, Vujcic SL, Williams CG, Neale DB (1994) Identification of quantitative trait loci influencing wood specific gravity in loblolly pine. Genetics 138:1293–1300

Hall AE, Fiebig A, Preuss D (2002) Beyond the Arabidopsis genome: opportunities for comparative genomics. Plant Physiol 129:1439–1447

Harry DE, Temesgen B, Neale DB (1998) Codominant PCR-based markers for *Pinus taeda* developed from mapped cDNA clones. Theor Appl Genet 97:327–336

Kinlaw CS, Neale DB (1997) Complex gene families in pine genomes. Trends Plant Sci 2:356–359

Komulainen P, Brown GR, Mikkonen M, Karhu A, Garcia-Gil MR, O'Malley D, Lee B, Neale DB, Savolainen O (2003) Comparing EST-based genetic maps between *Pinus sylvestris* and *Pinus taeda*. Theor Appl Genet 107:667–678

Krutovsky KV, Troggio M, Brown GR, Jermstad KD, Neale DB (2004) Comparative mapping in the Pinaceae. Genetics (in press)

Laurie DA, Devos KM (2002) Trends in comparative genetics and their potential impacts on wheat and barley research. Plant Mol Biol 48:729–740

Leitch IJ, Hanson L, Winfield M, Parker J, Bennett MD (2001) Nuclear DNA C-values complete familial representation in gymnosperms. Ann Bot 88:843–849

Marques CM, Brondani RPV, Grattapaglia D, Sederoff R (2002) Conservation and synteny of SSR loci and QTLs for vegetative propagation in four *Eucalyptus* species. Theor Appl Genet 105:474–478

O'Brien IEW, Smith DR, Gardner RC, Murray BG (1996) Flow cytometric determination of genome size in *Pinus*. Plant Sci 115:91–99

Paterson AH, Lin YR, Li Z, Schertz KF, Doebley JF, Pinson SRM, Liu SC, Stansel JW, Irvine JE (1995) Convergent domestication of cereal crops by independent mutations at corresponding genetic loci. Science 269:1714–1718

Paterson AH, Bowers JE, Burow MD, Draye X, Elsik CG, Jiang CX, Katsar CS, Lan TH, Lin YR, Ming R, Wright RJ (2000) Comparative genomics of plant chromosomes. Plant Cell 12:1523–1540

Prager EM, Fowler DP, Wilson AC (1976) Rates of evolution of conifers. Evolution 30:637–649

Sankoff D, Nadeau JH (eds) (2000) Comparative genomics: empirical and analytical approaches to gene order dynamics, map alignment and the evolution of gene families. Computational biology series, vol 1. Kluwer, Dordrecht

Sax K, Sax HJ (1933) Chromosome number and morphology in the conifers. J Arnold Arbor 14:356–375
Sewell MM, Sherman BK, Neale DB (1999) A consensus map for loblolly pine (*Pinus taeda* L.). I. Construction and integration of individual linkage maps from two outbred three-generation pedigrees. Genetics 151:321–330
Temesgen B, Neale DB, Harry DE (2000) Use of haploid mixtures and heteroduplex analysis enhance polymorphisms revealed by denaturing gradient gel electrophoresis. BioTechniques 28:114–116
Temesgen B, Brown GB, Harry DE, Kinlaw CS, Sewell MM, Neale DB (2001) Genetic mapping of expressed sequence tag polymorphism (ESTP) markers in loblolly pine (*Pinus taeda* L.). Theor Appl Genet 102:664–675
The Arabidopsis Genome Initiative (2000) Analysis of the genome sequence of the flowering plant *Arabidopsis thaliana*. Nature 408:796-815
Ware D, Stein L (2003) Comparison of genes among cereals. Curr Opin Plant Biol 6:121–127
Ware D, Jaiswal P, Ni J, Yap IV, Pan X, Clark KY, Teytelman L, Schmidt SC, Zhao W, Chang K, Cartinhour S, Stein LD, McCouch SR (2002) Gramene, a tool for grass genomics. Plant Physiol 130:1606–1613
Yu J, Hu S, Wang J, Wong GK, Li S et al. (2002) A draft sequence of the rice genome (*Oryza sativa* L. ssp. *Indica*). Science 296:79–92

II.13 Markers in Fruit Tree Breeding: Improvement of Peach

E. Dirlewanger[1] and P. Arús[2]

1 Introduction

Peach [*Prunus persica* (L.) Batsch] belongs to the *Prunus* genus, member of the Rosaceae family. The *Prunus* genus, within the subfamily Prunoideae, is characterized by species that produce drupes as fruit (also referred to as stone fruits), and contains a significant number of agriculturally important fruit tree species [i.e., almond *(Prunus dulcis* Mill.), apricot (*Prunus armeniaca* Linn.), sweet cherry (*Prunus avium* L.) and sour cherry (*Prunus cerasus* L.), and plum (*Prunus japonica* and *Prunus domestica*)]. Several other species like myrobalan plum (*Prunus cerasifera* Ehrh.) or Sainte Lucie cherry (*Prunus mahaleb* L.) are mainly used as *Prunus* rootstocks. Although *Prunus* is an economically and biologically important genus, little was known about the genome structure and organization until the breakthrough of DNA marker technologies. Peach has distinct advantages that make it suitable as a model species for comparative and functional genomics. It has a short juvenile phase (2–3 years) compared to many other tree species, and a small genome: 5.9×10^8 bp or 0.61 pg/diploid nucleus (Baird et al. 1994). This is only about twice the genome size of *Arabidopsis thaliana* (Arumuganathan and Earle 1991). All the *Prunus* species have a base chromosome number of x=8. Peach, almond, sweet cherry and myrobalan plum have a diploid genome (2n=2x=16), whereas sour cherry is tetraploid (2n=4x=32) and European plum hexaploid (2n=6x=48). Moreover, peach is genetically the best-characterized *Prunus* species with a fair number of genes controlling important traits and having a Mendelian behavior (Hesse 1975; Monet et al. 1996; Table 1). For all these reasons, peach was chosen as a model for *Rosaceae* and a physical map has been initiated (Fig. 1; Abbott et al. 2002).

The development of molecular markers and linkage maps provides efficient tools to locate genes or quantitative trait loci (QTLs) involved in agronomical characters and could be helpful for monitoring a breeding program through marker-assisted selection (Young 1996). Mapping genes of interest

---

[1] Unité de Recherches sur les Espèces Fruitières et la Vigne, INRA, BP 81, 33883 Villenave d'Ornon, France
[2] Laboratori CSIC-IRTA de Genètica Molecular Vegetal, Departament de Genética Vegetal, Carretera de Cabrils s/n, 08348 Cabrils (Barcelona), Spain

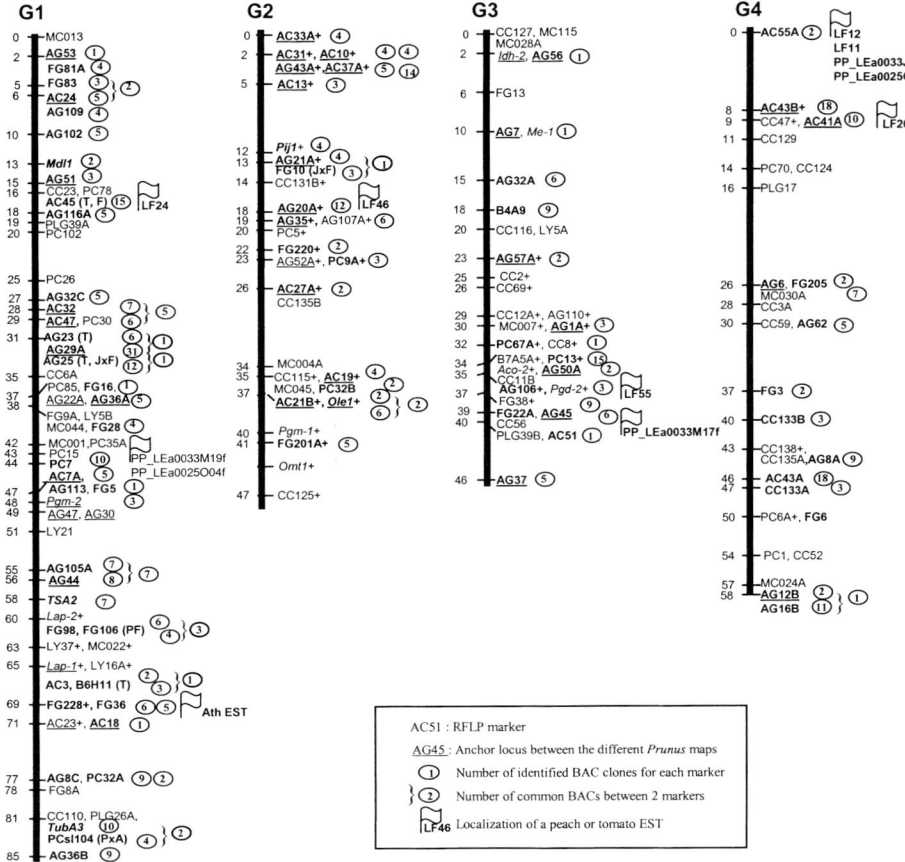

Fig. 1. The *Prunus* reference map TxE (Joobeur et al. 1998) and the genetically anchored physical map development (www.genome.clemson.edu/gdr)

will facilitate the breeding program by quickly combining the best traits isolated in different varieties or in other species from the *Prunus* genus. In *Prunus*, several linkage maps have been obtained and are based on intra- or interspecific *Prunus* crosses (Table 2). One of these maps, entirely constructed with transferable markers [restriction fragment length polymorphisms (RFLPs) plus a few isozyme genes] in an almond 'Texas' × peach 'Earlygold' (TxE) F_2 population (Joobeur et al. 1998), was considered to be saturated and has been taken as a reference by the *Prunus* scientific community. Markers from this map have been studied in many other *Prunus* populations allowing the comparison of their maps and the establishment of a common framework where major genes or QTLs found in different genetic backgrounds could be located.

When compared to other *Prunus* crops, peach has a lower level of variation (Byrne 1990) as a consequence of its self-compatible mating system, in con-

Markers in Fruit Tree Breeding: Improvement of Peach

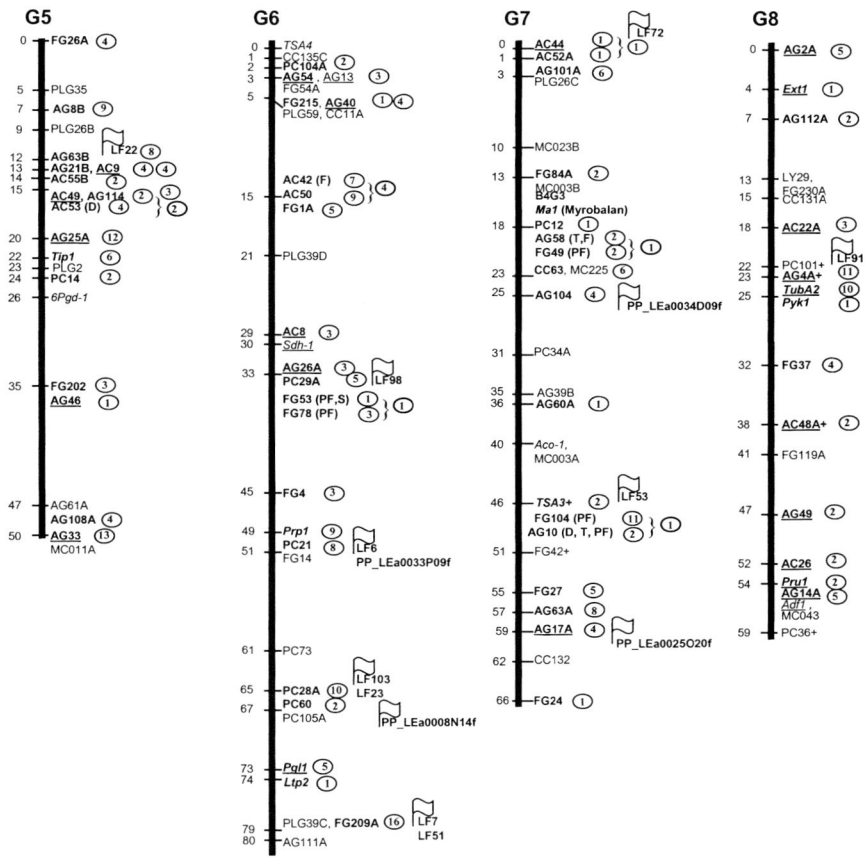

trast to the gametophytic self-incompatibility of most species of this genus, and of the narrow germplasm basis with which most of the modern cultivars have been obtained (Scorza et al. 1985). Low polymorphism is a limitation to marker-assisted selection and to realizing the full potential of the *Prunus* reference map. For example, only 39 (23%) of the 171 RFLP probes mapped in the reference map produced polymorphic loci in an intraspecific peach × peach F_2 progeny (Dirlewanger et al. 1998). This problem has been partly solved with the development of microsatellite (or simple-sequence repeat, SSR) markers (Table 3), most of them obtained from peach DNA sequences. SSRs are highly polymorphic in peach and show a relatively high level of observed heterozygosity (an average 37% of polymorphic SSRs was estimated by Aranzana et al. 2002). Ninety-six of them have been recently used to upgrade the reference map (Aranzana et al. 2003b) and about 30 more have been located in other comparable *Prunus* maps (Joobeur et al. 2000; Dettori et al. 2001; Dirlewanger et al. 2002; Yamamoto et al. 2002). Aranzana et al. (2003b) suggested that the position of 200 or more SSRs should be

Table 1. Documented single gene traits described in peach

Characters	Genes symbols	References
Genes affecting trees		
With anthocyanins/anthocyanins less	An/an	Monet (1967)
Normal/albino	C/c	Bailey and French (1949)
Tall, normal/pillar (broom)	Br/br or Pi/pi	Lammerts (1945)
Tall, normal/ bushy	Bu1/bu1	Lammerts (1945)
	Bu2/bu2	
Normal shape/compact shape	Ct/ct	Mehlenbaker and Scorza (1986)
Tall normal/ brachytic dwarf	Dw/dw	Lammerts (1945)
	Dw2/dw2	Hansche (1988)
	Dw3/dw3	Chaparro et al. (1994)
Normal shape/weeping shape	Pl/pl	Monet et al. (1996)
	We/we	Chaparro et al. (1994)
Genes affecting leaves		
Red leaf/green leaf	Gr/gr	Blake (1937)
Glandular foliage/eglandular foliage	E/r	Connors (1922)
Deciduous/evergreen	Evg/evg	Rodriguez et al. (1994)
Leaf shape (narrow/wide)	Nl/nl	Yamamoto et al. (2001)
Smooth leaf margin/wavy leaf margin	Wa/wa	Scott and Cullinan (1942)
	Wa2/wa2	Chaparro et al. (1994)
Genes affecting flowers		
Single/double flower	D1/d1	Lammerts (1945)
Pollen fertile/pollen sterile	Ps/ps	Scott and Weinberger (1944)
	Ps2/ps2	Chaparro et al. (1994)
Colored/white flower	W/w	Lammerts (1945)
Pink/red flower	R/r	Lammerts (1945)
Dark pink/light pink	P/p	Lammerts (1945)
Pink/pale pink flower color	Fc/fc	Yamamoto et al. (2001)
Large showy flowers/small showy flowers	L/l	Lammerts (1945)
Nonshowy/showy flower	Sh/sh	Bailey and French (1949)
Genes affecting fruits		
Monocarpel/polycarpel	Pcp/pcp	Bliss et al. (2002)
Normal anthocyanin/anthocyanins (blood/flesh)	Bf/bf	Werner et al. (1998)
Sweet fruit/normal fruit	D/d	Monet (1979)
Freestone/clingstone	F/f	Bailey and French (1949)
Pubescent skin/glabrous	G/g	Blake (1932)
Saucer shape/nonsaucer	S/s	Lesley (1939)
Bitter kernel/sweet kernel	Sk/sk	Werner and Creller (1997)
White flesh/yellow flesh	Y/y	Connors (1920)
Red/green skin color	Sc/Sc	Yamamoto et al. (2001)
Red/white flesh color around stone	Cs/cs	Yamamoto et al. (2001)
Melting flesh/nonmelting flesh	M/m	Bailey and French (1949)
Soft melting flesh/stony hard flesh	St/st	Bailey and French (1949)
Disease or pest resistances		
Resistance to *Myzus persicae*/susceptible	Rm1/rm1	Massonié et al. (1982)
Resistance to powdery mildew/susceptible	Sf/sf	Dabov (1983)
Resistance to *M. incognita*/susceptible	Mi/mi	Weinberger et al. (1943)
Resistance to *M. javanica*/susceptible	Mj/mj	Sharp et al. (1970)

Table 2. Intraspecific and interspecific peach crosses used for maps construction

Crosses	Name of pop.	Type of pop.(size)	Nb and type of markers	Size of the map	Location of genes and QTLs	References
P. persica × *P. persica*						
Weeping clone (1161:12×2678:47)1:55 × 'Early Sungrand'		F_2 (270)	52 RAPDs	350 cM Partial map	Pl	Dirlewanger and Bodo (1994)
Peach clones used to generate the F_1: 'DavieII', 'Georgia Belle', 'Honey Glo', 'Marsun', 'Sweet Melody'		11 F_2 (96)	83 RAPDs, 2 isoenzymes	396 cM + 198 cM, Partial maps	Br,Dl, Dw3, G, Gr, Ps2, W, Wa, We, Y	Chaparro et al. (1994)
'Ferjalou Jalousia' × 'Fantasia'	JxF	F_2 (63)	50 RFLPs, 92 RAPDs, 8 ISSRs, 115 AFLPs, 1 isoenzyme, SSRs	712 cM Partial map	D, G, S, Ps, QTLs for fruit quality	Dirlewanger et al. (1998, 1999); Etienne et al. (2002b)
'New Jersey Pillar' × 'KV 77119'	WV	F_2 (71)	46 RFLPs, 12 RAPDs	465 cM Partial map	Dl, Pi, Y, F, tree architecture	Rajapakse et al. (1995); Abbott et al. (1998); Sosinski et al. (1998)
'Suncrest' × 'Bailey'		F2 (48)	51 RFLPs, 12 RAPDs, 82 AFLPs	926 cM Partial map	QTLs for fruit quality	Abbott et al. (1998)
'Lovell' × 'Nemared'		F_2 (55)	153 AFLPs	1297 cM Partial map	Mi, Mij, Sh, F, Gr,	Abbott et al. (1998); Lu et al. (1998)
'Akame' × 'Juseitou'	AxJ	F_2 (126)	35 AFLPs, 31 RAPDs, 11 SSRs, 5 ISSRs, 1 PCR-RFLP	960 cM Partial map	F, Cs, Fc, Sc, Gr, Dw, Nl, Mi, Mj	Yamamoto et al. (2001)

Table 2. (Continue)

Crosses	Name of pop.	Type of pop.(size)	Nb and type of markers	Size of the map	Location of genes and QTLs	References
			P. dulcis × *P. persica*			
'Texas' × 'Earlygold'	TxE	F$_2$ (82)	11 isoenzymes, 235 RFLPs, 96 SSRs	522 cM Saturated map		Joobeur et al. (1998); Aranzana et al. (2002)
'Garfi' × 'Nemared'	GxN	Complex F$_2$ (78)	46 RFLPs, 5 isozymes		*Mi*	Jáuregui et al. (2001)
'Padre' × '54P455'	PMP1	F$_2$ (64)	143 RFLPs, 8 isoenzymes, 2 SSRs, 1 CAP, 1RAPD	1144 cM	*Y, Pcp, D, Sk, St, Dw, Ps*	Foolad et al. (1995); Warburton et al. (1996); Bliss et al. (2002)
			P. cerasifera × (*P. dulcis* × *P. persica*)			
P2175 × ('Garfi' × 'Nemared')22	P2175 × GN22	F$_1$ (83)	69 SSRs, 130 SSRs	438 cM, 742 cM	*Ma Mi, Gr*	Dirlewanger et al. (2003a, b)
			P. persica × *P. ferganensis*			
(IF7310828 × *P. ferganensis*) × IF7310828	PxF	BC$_1$(297)	74 RFLPs, 17 SSRs, 16 RAPDs	521 cM Partial map	*F, E, QTLs for powdery mildew resistance*	Quarta et al. (1998, 2000); Dettori et al. (2001)
			P. davidiana × *P. persica*			
Clone P1908 × 'Summergrand'	SD	F$_1$ (77)	15 RAPDs, 84 RAPDs	83 cM, 536 cM	QTLs for powdery mildew resistance	Dirlewanger et al. (1996); Viruel et al. (1998)
	SD40^2	F$_2$ (99)	66 RFLPs, 24 SSRs, 103 AFLPs	874 cM	QTLs for powdery mildew resistance	Foulongne et al. (2003b)

Table 3. *Prunus* microsatellites

SSR names	*Prunus* species	Repeats	Origins	References
UDP	*P. persica*	CT, GT	Two enriched genomic libraries from 'Redhaven'	Cipriani et al. (1999); Testolin et al. (2000)
CPPCT	*P. persica*	CT	Enriched genomic library from 'O'Henry'	Aranzana et al. (2002)
BPPCT	*P. persica*	CT	Enriched genomic library from 'O'Henry'	Dirlewanger et al. (2002)
pchgms	*P. persica*	CT, CA	Genomic library from 'Bicentennial'	Sosinski et al. (2000)
pchcms	*P. persica*		cDNA library from 'Suncrest'	
MA	*P. persica*	GA	Genomic DNA from 'Akatsuki'	Yamamoto et al. (2002)
M	*P. persica*	CT, GA	cDNA library from 'Akatsuki'	Yamamoto et al. (2002)
pms	*P. avium*	CT, CA, GA	Genomic library from 'Valerij Tschakhalov'	Cantini et al. (2001)
PS	*P. avium*	GA, GT, GTT	Enriched genomic library from 'Napoleon'	Joobeur et al. (2000); Cantini et al. (2001)
PceGA	*P. cerasus*	GA	Genomic library from 'Erdi Botermo'	Downey and Iezzoni (2000); Cantini et al. (2001)
ssrPaCITA	*P. armeniaca*	CT	Genomic library from 'Ungarische Beste'	Lopes et al. (2002)

determined in peach to have a high probability of finding at least one of them heterozygous in each of the 24 bins (approximately 25 cM/bin) in which they divided the *Prunus* genome, in an average genotype. This seems an objective attainable in the short term, given the rapid progress of peach SSR development and mapping.

2 Use of Molecular Markers for Fruit Quality Improvement

Fruit producers must satisfy consumers by producing fruits of good flavor, color and texture and must also provide marketers with fruits resistant to mechanical damage. Among temperate fruit crops, the peach breeding industry is one of the most dynamic (Fideghelli et al. 1998). Peach breeders continuously release new commercial cultivars, most of which are tasty and aromatic if ripened on the tree. However, in the last decade, the consumption of raw peaches and nectarines in the European Union and in the United States has not increased. This trend is largely due to the low quality of fruit resulting from harvesting at an immature stage for storage and shipment reasons. Thus, peach breeding objectives are to find the right compromise between quality and immaturity at harvest.

At the same time, the organoleptic quality has to be associated with a good healthy property of the fruit. The *Rosaceae* family has been increasingly reported to be involved in adverse fruit allergy reactions (Rodriguez et al. 2000), it being one of the most frequent causes of food allergy in Europe (Pastorello et al. 1999). Lowering the allergenicity of peach fruits appears to be one breeding goal for the near future.

2.1 Sugar and Acid Contents

The variation in fruit quality at harvest involves a large number of interrelated factors (Génard and Bruchou 1992). However, organic acid and soluble sugar contents and composition are major determinants of peach quality (Pangborn 1963). The predominant organic acids in ripe peach fruit are malic and citric acids, whilst quinic acid accumulates in lower amounts (Byrne et al. 1991; Moing et al. 1998). The soluble sugars present in peach are sucrose, fructose, glucose and sorbitol. Sucrose is the predominant soluble sugar at maturity while sorbitol accumulates at very low levels.

The 'nonacid' character, of mature fruits with a juice pH >4.0, was first reported to be controlled by a single dominant gene *D* (Monet 1979) located on the genetic linkage map constructed by using the progeny of a cross between two peach varieties, one with the 'nonacid' fruit and the other with normal 'acid' fruit (Dirlewanger et al. 1998). However, the major determinants of fruit flesh quality are usually inherited quantitatively and QTLs involved in those characters were detected.

2.1.1 Quantitative Trait Loci Controlling Peach Fruit Quality

Significant QTLs controlling acidity and sugar composition, fruit size and firmness were detected and mapped on several peach maps (Abbott et al. 1998; Dirlewanger et al. 1999; Quarta et al. 2000; Etienne et al. 2002b). On linkage group 5, QTLs were located near the *D* gene, including QTLs for pH and titratable acidity which correlate to the perception of acidity in the mouth (Dirlewanger et al. 1999; Etienne et al. 2002b). Near the saucer gene *S* of peach fruit shape, i.e., flat or round, QTLs for fresh weight and productivity were detected.

Quantitative trait loci detected on the different maps were compared and putative conservation of QTLs identified (Lecouls et al. 2002). Using amplified fragment length polymorphism (AFLP) markers located in those QTLs and a peach bacterial artificial clone (BAC) library constructed from the rootstock cultivar 'Nemared', BAC contigs around the markers were built by fingerprinting (Georgi et al. 2002). SSR markers derived from the AFLP markers were then located on the different maps in order to do a fine mapping of the region.

A QTL for soluble solid content (SSC) was detected on linkage group 6 (Etienne et al. 2002b), in the same group as a QTL detected in sour cherry, indicating that this QTL might be conserved in both species (Wang et al. 2000).

2.1.2 Candidate Gene Approach

The identification of genes involved in variation of peach fruit quality would assist breeders in creating new cultivars with improved fruit quality. Knowledge of soluble sugar and organic acid accumulation in fleshy fruits has considerably increased over the last decade. Critical steps for this process include: (1) phloem unloading of sucrose, (2) sugar metabolism, (3) organic acid metabolism and, (4) solute accumulation into the vacuole. In addition, other processes that enable cell expansion such as cell wall loosening and water transport may also be crucial. Eighteen peach fruit-related genes encoding enzymes involved in the metabolism or storage of organic acids were cloned and characterized (Rothan et al. 1999; Etienne et al. 2002a). Twelve of them were mapped (Fig. 2) using the Texas × Earlygold (TxE) reference European *Prunus* map (Joobeur et al. 1998). PRUpe;Vp2 encoding a vacuolar pyrophosphatase involved in the establishment of an electrochemical gradient across tonoplast, was co-located with QTLs controlling SSC and sucrose (Fig. 3; Etienne et al. 2002b). This co-location has to be confirmed by the fine mapping of the region.

2.2 Allergens of Peach Fruit

An increasing number of people show diverse reactions to foods often associated with fruit and vegetable consumption. It has been shown that the major allergen of peach is a lipid transfer protein (LTP) with a low molecular weight (9 kDa; (Pastorello et al. 1999; Sánchez-Monge et al. 1999)), stable after heat treatment (Brenna et al. 2000), and under acid and proteasic conditions of the stomach (Asero et al. 2000). Plant LTPs show common features, such as eight conserved cysteines forming disulfide bridges, basic isoelectric point and similar crystallographic structure. A 269-bp cDNA clone corresponding to an LTP gene was isolated from peach fruit and the accumulation of the specific transcript was evaluated in ripe fruit of different peach varieties (Botton et al. 2002). Expression data show that LTPs transcripts are totally absent in the mesocarp of ripe fruit, but strongly accumulate in the epicarp, and the aller-

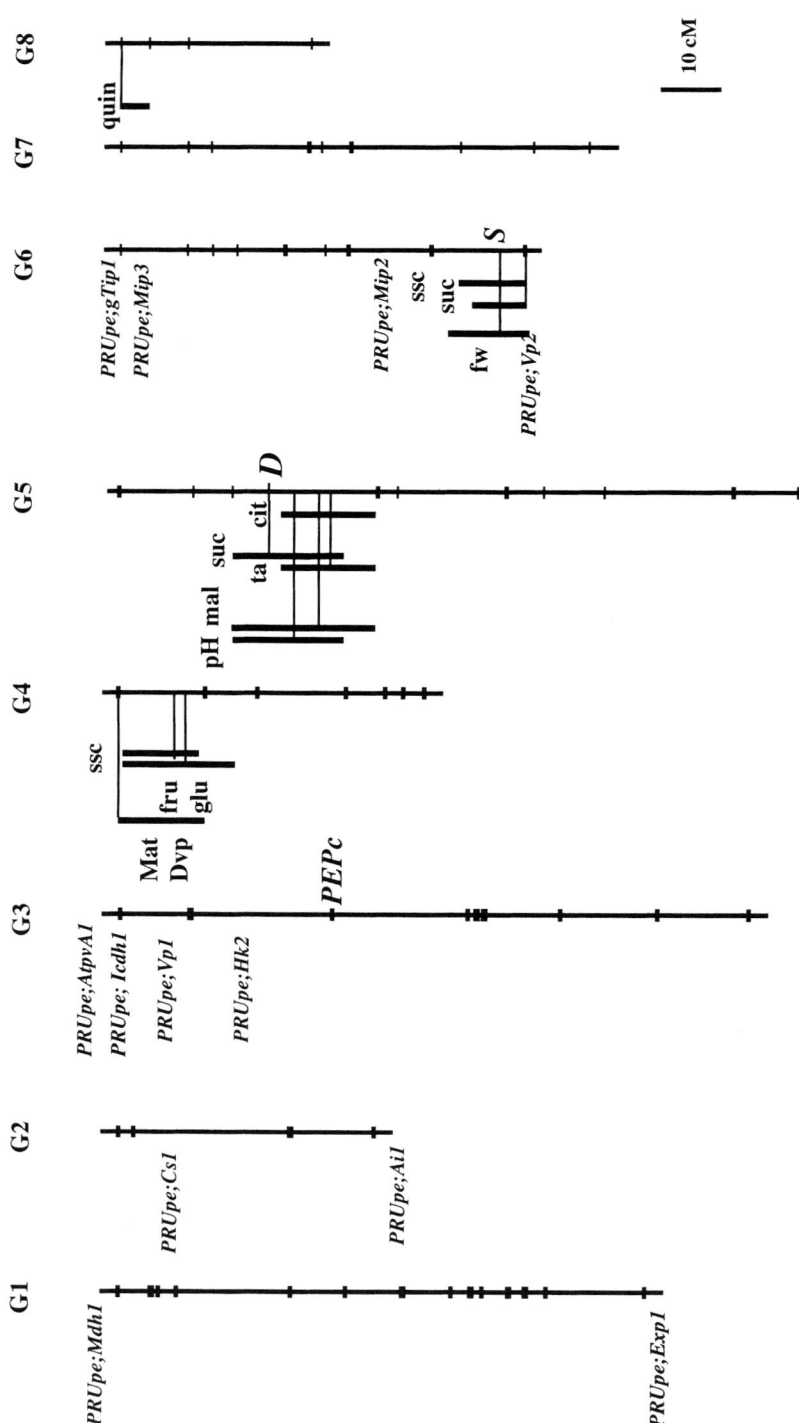

Fig. 2. QTL/candidate gene mapping for fruit sugar and organic acid contents in peach (Etienne et al. 2002b). Loci in *italics* are the Mendelian characters (*D* low-acid, *G* glabrous or pubescent skin, *ps* pollen sterility, *S* saucer shape). Candidate genes are placed on the linkage groups (*right*); they were assigned to by comparative mapping with the TxE map (Joobeur et al. 1998). Most likely positions of QTLs detected by MAPMAKER/QTL for fruit quality components (overall estimate): *cit* citrate, *fru* fructose, *fw* fresh weight, *glu* glucose, *mal* malate, *quin* quinate, *ssc* soluble solid content, *suc* sucrose, *ta* titratable acidity. The most likely position for a QTL is indicated by a *horizontal line* proportional to the effect (R2). The one-LOD support confidence interval is indicated by a *vertical line*

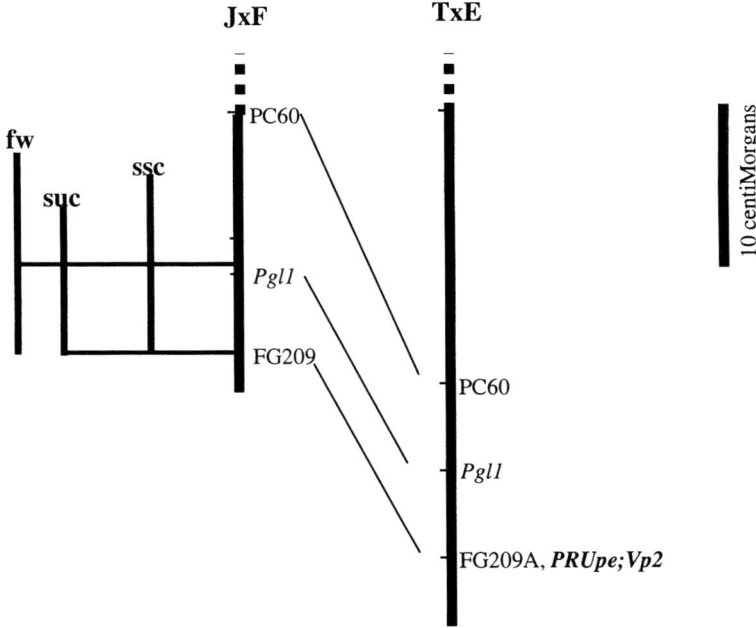

Fig. 3. Co-location of the *PRUpe;Vp2* proton pump and QTLs controlling SSC and sucrose accumulation. Alignment of the homologous fragments at the end of linkage group 6 from the 'Jalousia' × 'Fantasia' (JxF) and the 'Texas' × 'Earlygold' (TxE) maps, shared markers are indicated. Relative position of QTLs are indicated by *vertical bars* (*fw* fresh weight, *suc* sucrose content, *SSC* solid-soluble content). (Etienne et al. 2002b)

genic activity of peaches is largely concentrated in the skin. Different behavior was observed among the peach varieties. The detection of polymorphisms between peach varieties within the LTP clone sequence, and its possible association with different levels of allergenicity may prove to be useful when selecting varieties with low allergen properties.

3 Use of Molecular Markers for Disease Resistance

For environmental reasons, it is essential to decrease the use of chemicals for controlling pests and diseases in orchards. This would respond to the consumer's request for fruit without chemical residues and would help growers to limit toxic residues in soil (Byrne 2002). As a result, breeding programs focusing on fruit quality until recently, may be extended to the improvement of resistance to pests and diseases.

Sharka, caused by the plum pox potyvirus (PPV), powdery mildew, caused by *Spaerotheca pannosa* (Wallr.) var. *persica* and peach leaf curl, caused by the *Taphrina deformans* (Berk.) Tul., are among the most serious diseases in European peach production areas. For peach rootstocks, root-knot nematodes (RKN) are major crop pests in the Mediterranean basin. Most rootstock material is susceptible to RKN. Only the peach 'Shalil' is reported to be resistant to *Meloidogyne arenaria* and *incognita*, but susceptible to *M. javanica* and to an RKN population from Florida (Esmenjaud et al. 1994). The peach 'Nemared' is also resistant to most *M. javanica* populations (Ramming and Tanner 1983), but not to *Meloidogyne* sp. Florida. However, the narrow genetic base of commercial peach varieties results in a low variability for resistance to pests and disease, leaving only a few natural resistance genes for breeding purposes (Scorza et al. 1985). Consequently, additional sources of resistance must be sought in related wild species. *P. davidiana* and *P. ferganensis*, two close relatives of peach resistant to several peach pests and diseases, have already been used as genitors in breeding programs (Verde et al. 2002; Foulongne et al. 2003a). *P. davidiana*, originating from China, was found to be resistant to powdery mildew (Smykov et al. 1982), the green aphid (Massonié et al. 1982; Sauge et al. 1998), plum pox virus (Pascal et al. 1998) and leaf curl (Hesse 1975) and can be used to introgress these resistances into peach (Foulongne et al. 2003a). *P. ferganensis* was reported to carry a source of resistance to powdery mildew (Quarta et al. 2000; Verde et al. 2002). Among myrobalan plum (*P. cerasifera*), often used as *Prunus* rootstock, three clones (P.2175, P.1079 and P.2980) are resistant to *Meloidogyne arenaria, incognita, javanica* and the sp. Florida (Lecouls et al. 1997). Marker-assisted selection (MAS) is already effective for the selection of resistance genes (Lecouls et al. 1999, 2004; Bergougnoux et al. 2002) and markers are currently being used to select in a progeny of interspecific hybrids *P. cerasifera* × (*P. dulcis* × *P. persica*), obtained with the *Meloidogyne*-resistant clones of each species, to develop a new generation of peach rootstocks (Dirlewanger et al. 2003a).

The evaluation of sharka resistance is not easy. PPV-infected trees produce a wide range of reactions and symptoms can take several years to appear. Vilanova et al. (2003), working in apricot, located a major gene involved in sharka resistance in linkage group 1, but results in *P. davidiana* × peach crosses, suggest a more complex pattern of inheritance (Foulongne et al. 2003b). Alternatively, methods for powdery mildew and peach leaf curl assessment are more reliable and QTLs controlling resistance to powdery mildew (Dirlewanger et al. 1996; Quarta et al. 2000; Foulongne et al. 2003b) and peach leaf curl (Viruel et al. 1998) have already been reported allowing MAS of these characters.

3.1 Introgression of Polygenic Resistance to Powdery Mildew

Analysis of segregation data for powdery mildew resistance in the interspecific progenies *P. davidiana* × *P. persica* F_1 (Dirlewanger et al. 1996) and F_2 (Foulongne et al. 2003b), suggested a polygenic inheritance of the resistance carried by *P. davidiana*. The same was reported in a BC1 progeny of a cross with *P. ferganensis* (Verde et al. 2002). A high agreement between QTL positions across generations (F_1 and F_2) were demonstrated and illustrated the feasibility of marker-assisted selection (Foulongne et al. 2003b). The QTL with the highest effect found by Verde et al. (2002) was detected on the same position of a gene that controls the presence/absence and shape of the leaf glands (*E/e*).

3.2 Marker-Assisted Selection Root-Knot Nematode Resistance in Peach Rootstocks

Root-knot nematode (RKN) resistance is one of the characters of interest in breeding new *Prunus* rootstocks. One or a few dominant genes control the resistance found in several species of peach (*P. persica*), myrobalan plum (*P.cerasifera*) and almond (*P. dulcis*). Selection of these genes (R-genes) with the aid of molecular markers tightly linked to them appears as a highly efficient method compared to the complex, slow and space-consuming methods of nematode inoculation.

All genes described so far to determine resistance to RKN coming from the peach were placed on linkage group 2 of the *Prunus* map based on common markers with the TxE map. These include the two genes (*Mi* and *Mij*) found by Lu et al. (1998), the major QTL detected by Jáuregui (1998), the unique gene (M_{iaNem}) found by Arús et al. (2003). Most of these genes are located in the first part (0–35 cM) of the linkage group, but not exactly at the same estimated position.

The *Ma* gene coming from *P. cerasifera* is placed in linkage group 7 of the *Prunus* map (Dirlewanger et al. 2003b). A sequence characterized amplified region (SCAR) marker co-segregating with the gene has already been identified by bulked segregant analysis (BSA; Lecouls et al. 1999) and fine mapping around the gene makes the chromosome walking for gene cloning possible with the use of a BAC library from myrobalon P.2175 (Claverie et al. 2004).

4 Marker-Assisted Selection for Tree Architecture Characters

Several tree characters are controlled by single genes in peach: the evergrowing, also called evergreen (*evg*) gene, the gene for columnar growth, also termed 'pillar' or 'broomy' (*br*), and the gene controlling the wiping shape (*Pl*), dwarf (*Dw*) or compact (*Ct*) habit (Table 1). Molecular markers linked to some of them are already available.

Columnar trees are useful for high-density production systems. Tree fruit breeding requires large numbers of seedlings in the field for selection, thus the use of columnar growth trees would reduce the area occupied by the trees, and management expenses. Selection in the greenhouse would spare much of the expense. A microsatellite marker has already been detected for selecting columnar growth habit in peach (Scorza et al. 2002).

The ever-growing phenotype in peach has been used as a model to study cold hardiness and dormancy in perennial tree species (Wang et al. 2002). The ever-growing peach tree has a substantially different growing pattern in winter to the deciduous peach tree. Most woody plants native to temperate regions have an annual cycle alternating between active growth and winter dormancy. Dormancy is a survival strategy that enables the plant to resist unfavorable environments and is the most adaptable, regulatory function in plant development (Dennis 1996). In temperate regions, the terminal shoots on ever-growing trees keep growing in winter until killed by freezing temperatures, whereas the lateral buds go into dormancy. Cold hardiness is present in both ever-growing and deciduous genotypes, but the maximum hardiness level in deciduous trees is more than twice that of ever-growing trees. This character is controlled by a single recessive allele named *evg* (Rodriguez et al. 1994). AFLP marker fragments linked to the *evg* gene were cloned and used for screening the 'Nemared' BAC library and BAC contigs were identified. Some of these clones were used for developing SSR markers in this region that are used as the starting points of a chromosome walk; this will eventually lead to the map-based cloning of this gene (Wang et al. 2002).

5 Synteny Among *Prunus* Species

Knowledge of the genetic basis of traits and their linkage with molecular markers is important for breeding new varieties at higher speed and lower cost. In *Prunus*, several linkage maps have been developed based on interspecific crosses between peach and almond (Foolad et al. 1995; Joobeur et al. 1998; Jáuregui et al. 2001; Bliss et al. 2002) or intraspecific crosses of peach (Chaparro et al. 1994; Rajapakse et al. 1995; Dirlewanger et al. 1998; Lu et al. 1998; Dettori et al. 2001; Yamamoto et al. 2001), almond (Viruel et al. 1995; Joobeur et al. 2000), apricot (Hurtado et al. 2002; Lambert et al. 2004; Vilanova et al. 2003), and sour cherry (Wang et al. 1998). Most of these maps share a sufficient number of markers with the reference *Prunus* map allowing the marker position on the map and the marker order within each linkage group to be compared.

A high colinearity between markers belonging to homologous linkage groups was observed for all species studied, suggesting that the *Prunus* genome can be treated as a single genetic entity. The linkage group 1, taken as an example, is presented in Fig. 4. Given the colinearity between the maps and the fact that many anchor loci are found between them, it was possible to establish the approximate position of Mendelian agronomic characters from several *Prunus* species (Fig. 5). Only one exception to the *Prunus* highly syn-

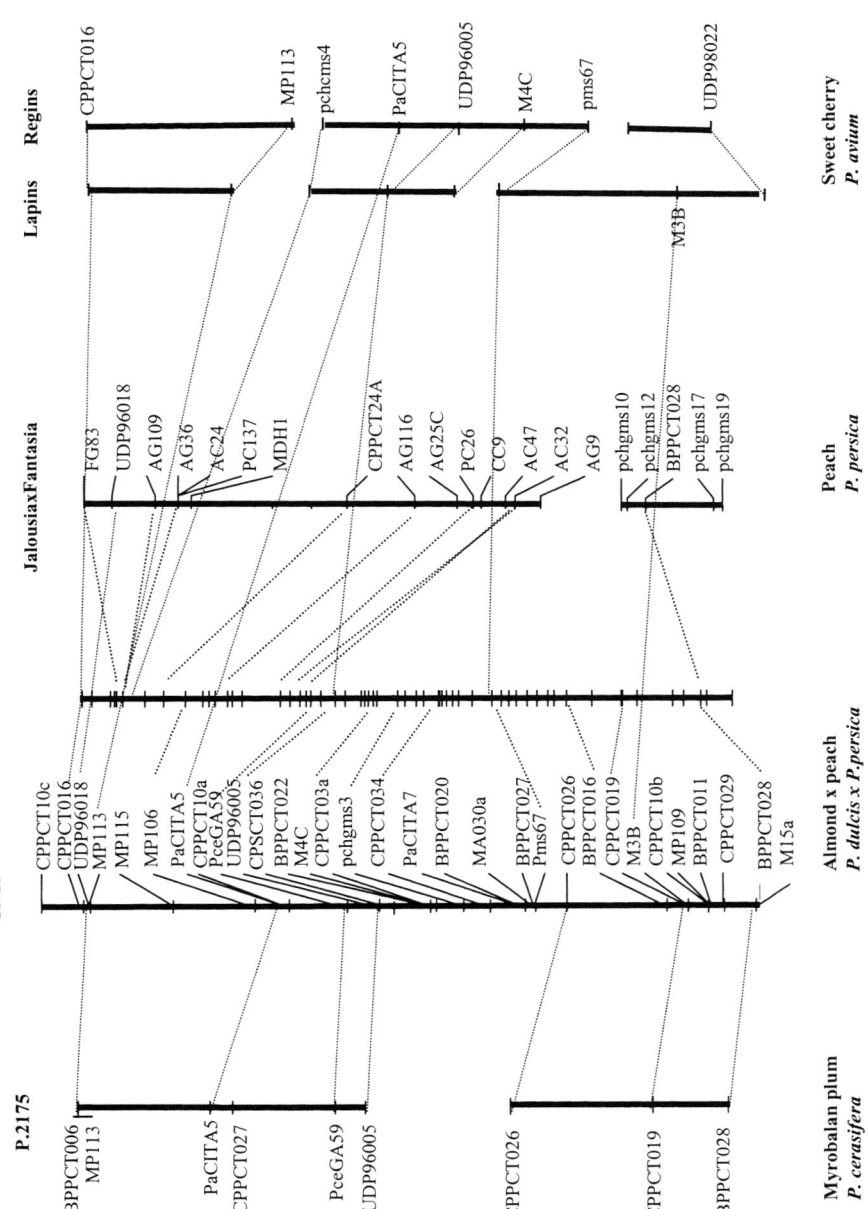

Fig. 4. Comparison of the linkage group 1 of several *Prunus* linkage maps built in different *Prunus* species (*P. persica*, *P. avium*, *P. cerasifera*, *P. dulcis*), see Table 2 for the names of crosses

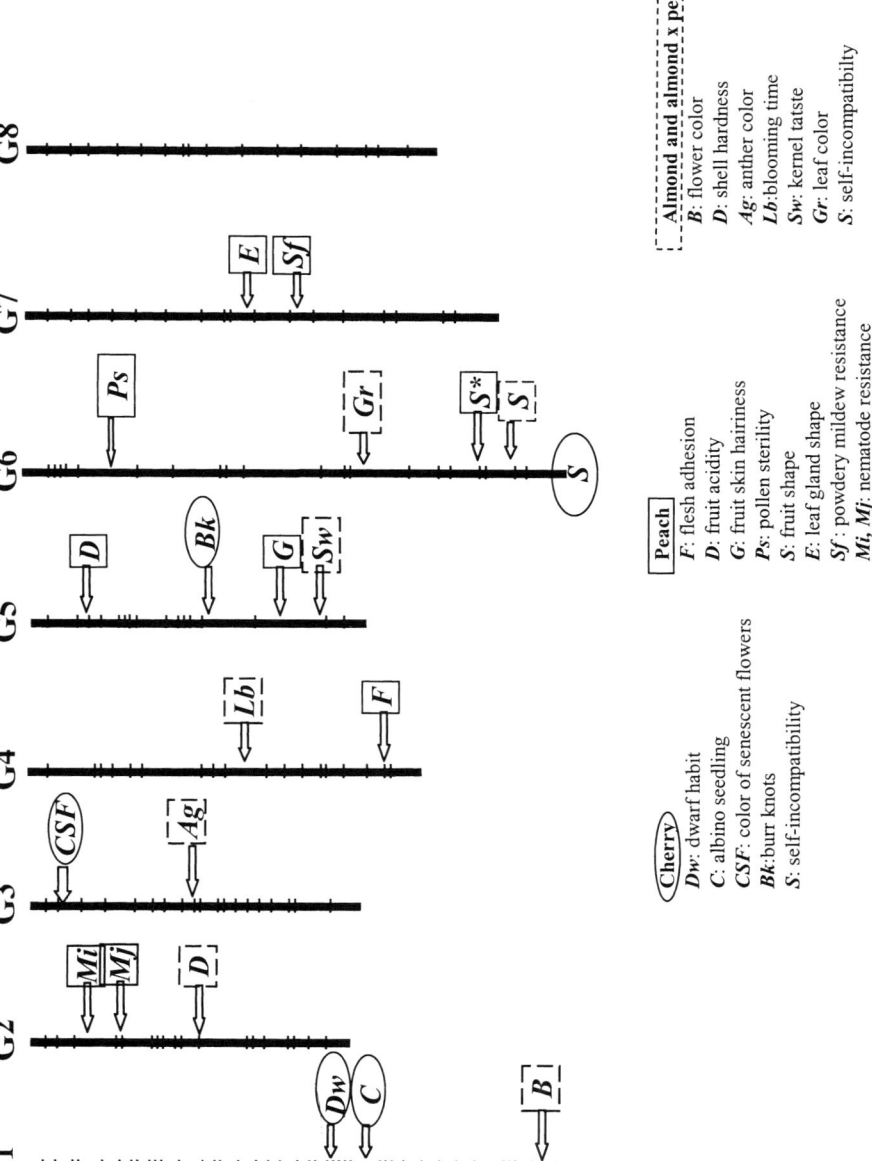

Fig. 5. Peach, cherry and almond monogenic characters located on a single genetic linkage map (TxE map) by using synteny between *Prunus* genus. (Arús et al. 1999)

tenic pattern was found in the cross between 'Garfi' almond and 'Nemared' peach, where a reciprocal translocation occurs between linkage groups 6 and 8 of the the GxN map (Jáuregui et al. 2001). Exactly the same was observed in the myrobalan plum P.2175 × ['Garfi' × 'Nemared'] population (Dirlewanger et al. 2003b). The markers studied in this population mapped to seven linkage groups instead of the expected eight for *Prunus* and markers located in groups 6 and 8 formed a single linkage group (Fig. 6). Pollen viability and meiotic behavior studies confirmed the presence of this chromosome rearrangement. A similar result was observed in the 'Akame' × 'Juseitou' peach intraspecific cross (Yamamoto et al. 2001; Yamamoto, pers. comm.), suggesting that one of the parents of this cross carried the same translocation. It has not been possible so far to establish which parents had the translocation and which had the standard chromosome composition, but the fact that one of the parents of each cross, 'Nemared' and 'Juseitou', is a red-leafed peach cul-

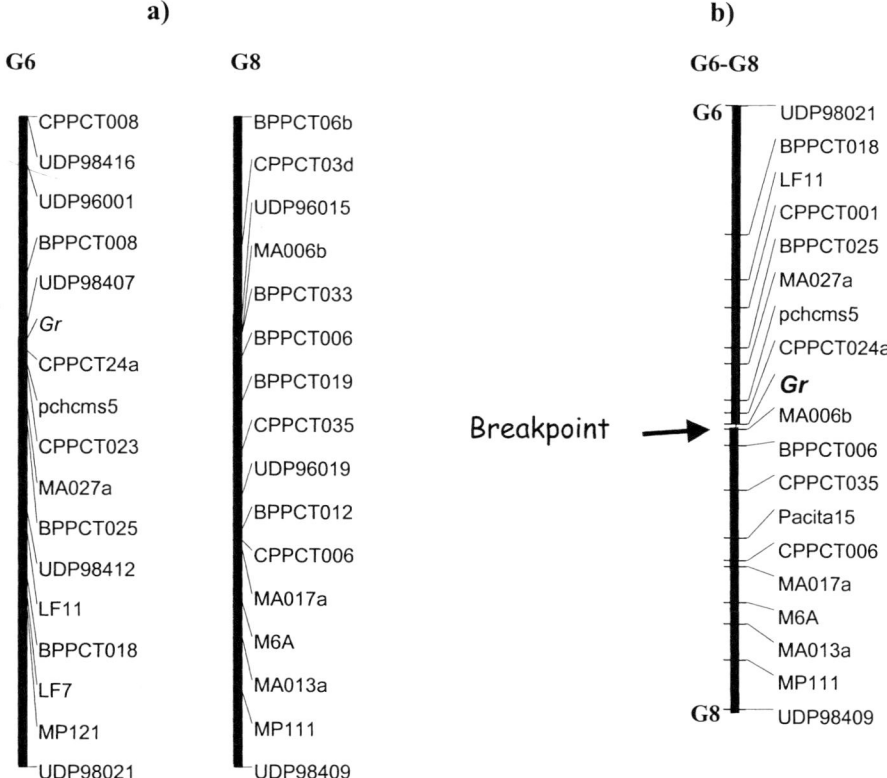

Fig. 6. Reciprocal translocation between the groups 6 and 8 observed within the almond '[Garfi' × peach 'Nemared']22 linkage map built by using the interspecific myrobalan plum 'P.2175' × [GN]22 population (Dirlewanger et al. 2003b). **a** Groups homologous to those obtained in previous *Prunus* maps. **b** Linkage group that includes markers of groups 6 and 8 of other maps. *Gr* is the red/green leaf gene

tivar suggests that the translocation may be associated to genotypes sharing this character.

The use of synteny appears to be a powerful tool for speeding up the construction of genetic maps in different crosses between members of the *Prunus* genus. By choosing a set of well-distributed transferable markers along the linkage groups of the reference map or other pre-existing maps, it is possible to construct an anchored map covering most of the *Prunus* genome. Once major genes or QTLs for the characters of interest are located in the new progeny, the regions of interest may be saturated with markers from other maps. This should allow an optimization of the identification of major genes or QTLs, or markers linked to them, involved in the expression of target characters from the results generated in other breeding populations of *Prunus*.

6 Development of Peach Molecular Markers and Their Use for Fingerprinting and for the Evaluation of Genetic Resources

Peach is the second most important fruit crop in temperate and subtropical zones worldwide after apple. The peach cultivar structure is characterized by a great diversity of cultivars with a fast turnover. Peach is self-compatible and tolerant to inbreeding, which makes the breeding of new cultivars either by outcrossing or by inbreeding possible. Modern peach cultivars are vegetatively propagated, which allows the conservation of their genetic information, but it also makes the breeder's rights more difficult to protect. The high number of existing cultivars and their important economic value has encouraged the development of fast and reliable techniques for peach molecular fingerprinting. Moreover, they may be used, as a complement to morphological evaluations, in the distinct tests needed for the registration of new cultivars into national and international official catalogues and for the recognition of protection status.

Many *Prunus* microsatellites have been recently developed in *Prunus* (Table 3). Many of them were tested for their usefulness to fingerprint peach varieties. Several sets of *Prunus* microsatellite markers were chosen (Dirlewanger et al. 2002; Aranzana et al. 2003a) and proven to be highly efficient for this purpose.

The development of SNPs (single nucleotide polymorphism) in peach is in progress. A new bioinformatic pipeline for automatic SNP detection in EST dataset is available at INRA-Bordeaux (Le Dantec et al. 2004). Those markers, if defined within the sequence of genes, should have the advantage over the SSRs currently in use by being able to detect variability potentially meaningful for agronomic performance, making them particularly valuable in the evaluation of the genetic variation of peach collections and as a means of cultivar characterization.

7 Conclusions

During the last decade, important progress has been achieved in the breeding of peach in order to obtain new cultivars adapted to different biotic or abiotic conditions, as well as responding to consumer demands for quality. The level of knowledge has considerably increased, especially in genetics and biotechnology. Molecular analysis has allowed important progress to be made in the understanding of the genome structure and genetic diversity of *Prunus*. The use of molecular markers is already efficient for the selection of several characters. The adaptation of marker-based breeding strategies developed for annual crops to woody perennials, particularly those that facilitate the introgression of genes from wild or exotic materials, is necessary to take advantage of the enormous genetic variability of the genus *Prunus*. Moreover, a better knowledge of synteny in the Rosaceae will favor the transfer of genetic information among stone fruit species and other important crops like apple, strawberry or rose. This goal may be facilitated by the recent creation by Dr. A. Abbott (www.genome.clemson.edu/projects/peach/gdr) of a Rosaceae Mapping Consortium, aimed at combining the efforts of the major research groups working on the *Rosaceae*. This consortium will focus on (1) structural genomics analysis with the development of a complete physical map of the peach genome, taken as the model species for the family, and the anchoring of the genetic maps of many of the economically important Rosaceae species maps on this physical map and, (2) on functional genomics with the development of an extensive expressed sequence tag (EST) database for fruit, shoot and seed tissues and integration of the unigene set onto the physical and genetic maps of peach. The results should soon be integrated into a '*Prunus* Genomic Database' (PGD) developed at Clemson University (www.genome.clemson.edu/projects/gdr). In June 2003, 10,200 peach ESTs were recorded in the NCBI Genbank database, but this number is expected to quickly increase. This initiative will foster the research on this group of species and lead to results of immediate application in fruit tree breeding.

References

Abbott AG, Rajapakse S, Sosinski B, Lu ZX, Sossey-Alaoui K, Gannavarapu M, Reighard G, Ballard RE, Baird WV, Scorza R, Callahan A (1998) Construction of saturated linkage maps of peach crosses segregating for characters controlling fruit quality, tree architecture and pest resistance. Acta Hortic 465:41–49

Abbott AG, Lecouls AC, Wang Y, Georgi L, Scorza R, Reighard G (2002) Peach: The model genome for *Rosaceae* genomics. Acta Hortic 592:199–203

Aranzana MJ, Garcia-Mas J, Carbo J, Arús P (2002) Development and variability analysis of microsatellite markers in peach. Plant Breed 121:87–92

Aranzana MJ, Carbo J, Arús P (2003a) Microsatellite variability in peach [*Prunus persica* (L.) Batsch]: cultivar identification, marker mutation, pedigree inferences and population structure. Theor Appl Genet 106:1341–1352

Aranzana MJ, Pineda A, Cosson P, Ascasibar J, Dirlewanger E, Ascasibar J, Cipriani G, Ryder CD, Testolin R, Abbott A, King GJ, Iezzoni AF, Arùs P (2003b) A set of simple-sequence repeat (SSR) markers covering the *Prunus* genome. Theor Appl Genet 106:819–825

Arumuganathan K, Earle E (1991) Nuclear DNA content of some important plant species. Plant Mol Biol Rep 9:208–218

Arús P, Dirlewanger E, Quarta R, Tobutt K, J Ballester, Boskovic R, Dettori MT, de Vicente C, Jàuregui B, Joobeur T, Russell K, Verde I, Viruel M (1999) Location of 20 major genes of peach, almond and cherry on the *Prunus* linkage map. Plant and animal genome VII. San Diego, 17–21 Jan 1999)

Arús P, Mnejja M, Dirlewanger E, Esmenjaud D (2003) High marker density around the peach nematode resistance I genes. Proceedings of the 1st international symposium on rootstock for deciduous fruit tree species, ISHS Fruit Section. Acta Hortic (in press)

Asero R, Mistrello G, Roncarolo DC, de Vries S, Gautier MF, Ciurana CLF, Verbeek E, Mohammadi T, Knul-Brettlova V, Akkerdaas JH, Bulder I, Aalberse RC, van Ree R (2000) Lipid transfert protein: a pan allergen in plant-derived foods that is highly resistant to pepsin digestion. Int Arch Allergy Immunol 122:20–32

Bailey JS, French AP (1949) The inheritance of certain fruit and foliage characters in the peach. Mass Agric Expt Sta Bull p 452

Baird WV, Estager AS, Wells J (1994) Estimating nuclear DNA content in peach and related diploid species using laser flow cytometry and DNA hybridization. J Am Soc Hort Sci 119:1312–1316

Bergougnoux V, Claverie M, Bosselut N, Lecouls AC, Esmenjaud D, Dirlewanger E, Salesses G (2002) Marker-assisted selection of the *Ma* gene from Myrobalan Plum for a complete-spectrum root-knot nematode (RKN) resistance in *Prunus* rootstocks. Acta Hortic 592 (ISHS 2002):223–228

Blake MA (1932) The J H Hale peach as a parent in peach crosses. Proc Am Soc Hortic Sci 29:131–136

Blake MA (1937) Progress in peach breeding. Proc Am Soc Hortic Sci 35:49–53

Bliss FA, Arulsekar S, Foolad MR, Becerra V, Gillen AM, Warburton ML, Dandekar AM, Kocsisne GM, Mydin KK (2002) An expanded genetic linkage map of *Prunus* based on an interspecific cross between almond and peach. Genome 45:520–529

Botton A, Begheldo M, Rasori A, Bonghi C, Tonutti P (2002) Factors affecting gene expression of lipid transfert protein (LTP), the major allergen of peach fruit. Acta Hortic 592:237–243

Brenna O, Pompei C, Ortolani C, Pravettoni V, Farioli L, Pastorello EA (2000) Technological processes to decrease the allergenicity of peach juice and nectar. J Agric Food Chem 48:493–497

Byrne DH (1990) Isozyme variability in four diploid stone fruits compared with other woody perennial plants. J Hered 81:68–71

Byrne DH (2002) Peach breeding trends: a world wide perspective. Acta Hortic 592:49–59

Byrne DH, Nikolic AN, Burns EE (1991) Variability in sugars, acids, firmness, and colour characteristics of 12 peach genotypes. J Am Soc Hortic Sci 116:1004–1006

Cantini C, Iezzoni AF, Lamboy WF, Bortizki M, Struss D (2001) DNA fingerprinting of tetraploid cherry germplasm using simple sequence repeats. J Am Soc Hortic Sci 126:205–209

Chaparro JX, Werner DJ, O'Malley D, Sederoff RR (1994) Targeted mapping and linkage analysis of morphological, isozyme, and RAPD markers in peach. Theor Appl Genet 87:805–815

Cipriani G, Lot G, Huang WG, Marrazzo MT, Peterlunger E, Testolin R (1999) AC/GT and AG/CT microsatellite repeats in peach [*Prunus persica* (L) Batsch]: isolation, characterization and cross-species amplification in *Prunus*. Theor Appl Genet 99:65–72

Claverie M, Dirlewanger E, Cosson P, Bosselut N, Lecouls AC, Voisin R, Kleinhentz M, Lafargue B, Caboche M, Chalhoub B, Esmenjaud D (2004) Fine Mapping and Chromosome Landing at the Root-Knot Nematode Resistance Locus *Ma* from Myrobalan Plum Using a Large-Insert BAC DNA Library. Theor Appl Genet (in press)

Connors CH (1920) Peach breeding – a summary of results. Proc Am Soc Hortic Sci 17:108–115

Connors CH (1922) Inheritance of foliar glands of the peach. Proc Am Soc Hortic Sci 18:20–26

Dabov S (1983) Inheritance of peach resistance to powdery mildew. III. Leaf resistance in F1 of J.H. Hale × nectarine Ferganensis 2. Genet Plant Breed 16:146–150

Dennis FG (1996) A physiological comparison of seed and bud dormancy. In: Lang GA (ed) Plant dormancy. CAB International, Wallingford

Dettori MT, Quarta R, Verde I (2001) A peach linkage map integrating RFLPs, SSRs, RAPDs and morphological markers. Genome 44:783–790

Dirlewanger E, Bodo C (1994) Molecular genetic mapping of peach. Euphytica 77:101–103

Dirlewanger E, Pascal T, Zuger C, Kervella J (1996) Analysis of molecular markers associated with powdery mildew resistance genes in peach (*Prunus persica* (L.) Batsch) × *Prunus davidiana* hybrids. Theor Appl Genet 93:909–919

Dirlewanger E, Pronier V, Parvery C, Rothan C, Guy A, Monet R (1998) Genetic linkage map of peach (*Prunus persica* (L.) Batsch) using morphological and molecular markers. Theor Appl Genet 97:888–895

Dirlewanger E, Moing A, Rothan C, Svanella L, Pronier V, Guye A, Plomion C, Monet R (1999) Mapping QTLs controlling fruit quality in peach (*Prunus persica* (L) Batsch). Theor Appl Genet 98:18–31

Dirlewanger E, Cosson P, Tavaud M, Aranzana MJ, Poizat C, Zanetto A, Arús P, Laigret F (2002) Development of microsatellite markers in peach [*Prunus persica* (L.) Batsch] and their use in genetic diversity analysis in peach and sweet cherry (*Prunus avium* L.). Theor Appl Genet 105:127–138

Dirlewanger E, Kleinhentz M, Claverie M, Lecouls AC, Bosselut N, Voisin R, Poessel JL, Faurobert M, Arús P, Gomez-Aparisi J, Xiloyannis C, Di Vito M, Esmenjaud D (2003a) Breeding for a new generation of *Prunus* rootstocks: an example of marker-assisted selection. Proceedings of the 1st international symposium on rootstock for deciduous fruit tree species, ISHS Fruit Section. Acta Hortic (in press)

Dirlewanger E, Poizat C, Cosson P, Lafargue B, Kleinhentz M, Claverie M, Bosselut N, Voisin R, Esmenjaud D, Laigret F (2003b) Genetic linkage maps of myrobalan plum and of an almond-peach hybrid-location of root-knot nematode resistance genes. 7th international congress of plant molecular biology, ISPMB, Barcelona, 23–28 June 2003

Downey SL, Iezzoni AF (2000) Polymorphic DNA markers in black cherry (*Prunus serotina*) are identified using sequences from sweet cherry, peach, and sour cherry. J Am Soc Hortic Sci 125:76–80

Esmenjaud D, Minot JC, Voisin R, Pinochet J, Salesses G (1994) Inter- and intraspecific resistance variability in Myrobalan plum, peach, and peach-almond rootstocks using 22 root-knot nematode populations. J Am Soc Hortic Sci 119:94–100

Etienne C, Moing A, Dirlewanger E, Raymond P, Monet R, Rothan C (2002a) Isolation and characterisation of six peach cDNAs encoding key proteins in organic acid metabolism and solute accumulation: involvement in regulating peach fruit acidity. Physiol Plant 114:259–270

Etienne C, Rothan C, Moing A, Plomion C, Bodénès C, Svanella-Dumas L, Cosson P, Pronier V, Monet R, Dirlewanger E (2002b) Candidate genes and QTLs for sugar and organic acid content in peach [*Prunus persica* (L.) Batsch]. Theor Appl Genet 105:145–159

Fideghelli C, Della Strada G, Grassi F, Morico G (1998) The peach industry in the world: present situation and trend. Acta Hortic 465:29–40

Foolad MR, Arulsekar S, Becerra V, Bliss FA (1995) A genetic map of *Prunus* based on an interspecific cross between peach and almond. Theor Appl Genet 91:262–269

Foulongne M, Pascal T, Arús P, Kervella J (2003a) The potential of *Prunus davidiana* for introgression into peach [*Prunus persica* (L.) Batsch] assessed by comparative mapping. Theor Appl Genet 107:227–238

Foulongne M, Pascal T, Pfeiffer F, Kervella J (2003b) QTLs for powdery mildew resistance in peach × *Prunus davidiana* crosses: consistency across generations and environments. Mol Breed 12:33–50

Génard M, Bruchou C (1992) Multivariate analysis of within-tree factors accounting for the variation of peach fruit quality. Sci Hort 52:37–51

Georgi LL, Wang Y, Yvergniaux D, Ormsbee T, Iñigo M, Reighard G, Abbott AG (2002) Construction of a BAC library and its application to the identification of simple sequence repeats in peach [*Prunus persica* (L.) Batsch]. Theor Appl Genet 105:1151–1158

Hansche P (1988) Two genes that induce brachytic dwarfism in peach. Hortic Sci 23:604–606

Hesse CO (1975) Peach. In: Janick J, Moore JN (eds) Advances in fruit breeding. Purdue Univ Press, West Lafayette, Indiana, pp 325–326

Hurtado MA, Romero C, Vilanova S, Abbott AG, Llácer G, Badenes ML (2002) Genetic linkage maps of two apricot cultivars (*Prunus armaniaca* L.), and mapping of PPV (sharka) resistance. Theor Appl Genet 105:182–191

Jáuregui B (1998) Localizacion de marcadores moleculares ligados a caracteres agronomicos en un cruzamiento interespecifico almendro × melocotonero. PhD Thesis. University of Barcelona, Spain.

Jáuregui B, de Vicente MC, Messeguer R, Felipe A, Bonnet A, Salesses G, Arús P (2001) A reciprocal translocation between 'Garfi' almond and 'Nemared' peach. Theor Appl Genet 102:1169–1176

Joobeur T, Viruel MA, de Vicente MC, Jáuregui B, Ballester J, Dettori MT, Verde I, Truco MJ, Messeguer R, Batlle I, Quarta R, Dirlewanger E, Arús P (1998) Construction of a saturated linkage map for *Prunus* using an almond × peach F_2 progeny. Theor Appl Genet 97:1034–1041

Joobeur T, Periam N, de Vicente MC, King GJ, Arús P (2000) Development of a second generation linkage map for almond using RAPD and SSR markers. Genome 43:649–655

Lambert P, Hagen LS, Arús P, Audergon JM (2004) Genetic linkage maps of two apricot cultivars (*Prunus armeniaca* L.) compared with the almond 'Texas' × peach 'Earlygold' reference map for Prunus. Theor Appl Genet 108:1120–1130

Lammerts WE (1945) The breeding of ornamental edible peaches for mild climates I. Inheritance of tree and flower characters. Am J Bot 30:707–711

Lecouls AC, Salesses G, Minot JC, Voisin R, Bonnet A, Esmenjaud D (1997) Spectrum of the *Ma* genes for resistance to *Meloidogyne* spp. in Myrobalan plum. Theor Appl Genet 95:1325–1334

Lecouls AC, Rubio-Cabetas MJ, Minot JC, Voisin R, Bonnet A, Salesses G, Dirlewanger E, Esmenjaud D (1999) RAPD and SCAR markers linked to the *Ma1* root-knot nematode resistance gene in Myrobalan plum (*Prunus cerasifera* Ehr.). Theor Appl Genet 99:328–335

Lecouls AC, Reighard GL, Abbott AG, Dirlewanger E (2002) Physical mapping and integration of QTL intervals involved in fruit quality on peach fruit variety and rootstock molecular maps. Proc 5th IS Peach Acta Hortic 592 (ISHS 2002):273–278

Lecouls AC, Bergougnoux V, Rubio-Cabetas MJ, Bosselut N, Voisin R, Bonnet A, Salesses G, Dirlewanger E, Esmenjaud D (2004) Marker-assisted selection of *Prunus* rootstocks for the wide-spectrum root-knot nematode resistance conferred by the *Ma* gene from Myrobalan plum (*Prunus cerasifera*). Mol Breed 13:113–124

Le Dantec L, Chagné D, Pot D, Cantin O, Garnier-Géré P, Bedon F, Frigerio JM, Chaumeil P, Léger P, Garcia V, Laigret F, de Daruvar A, Plomion C (2004) Automated SNP Detection in Expressed Sequence Tags: Statistical Considerations and Application to Maritime Pine Sequences. Plant Mol Biol (in press)

Lesley JW (1939) A genetic study of saucer fruit shape and other characters in the peach. Proc Am Soc Hortic Sci 38:218–222

Lopes MS, Sefc KM, Laimer M, da Camara Machado A (2002) Identification of microsatellite loci in apricot. Mol Ecol Notes 2:24–26

Lu ZX, Sosinski B, Reighard GL, Baird WV, Abbott AG (1998) Construction of a genetic linkage map and identification of AFLP markers for resistance to root-knot nematodes in peach rootstocks. Genome 41:199–207

Massonié G, Monet R, Bastard Y, Grasselly C (1982) Résistance au puceron vert du pêcher, *Myzus persicae* Sulzer (*Homoptera aphididae*) chez *Prunus persica* (L.) Batsch et d'autres espèces de *Prunus*. Agronomie 2:63–70

Mehlenbacher SA, Scorza R (1986) Inheritance of growth habit in progenies of compact Redhaven peach. Hortscience 21:124–126

Moing A, Svanella L, Rolin D, Gaudillere M, Gaudillere JP, Monet R (1998) Compositional changes during the fruit development of two peach cultivars differing in juice acidity. J Am Soc Hortic Sci 123:770–775

Monet R (1967) A contribution to the genetics of peaches (in French). Ann Amelior Plant 17:5–11

Monet R (1979) Genetic transmission of the 'non-acid' character. Incidence on selection for quality. Eucarpia symposium tree fruit breeding. INRA, Angers, pp 273–276

Monet R, Guye A, Roy M, Dachary N (1996) Peach mendelian genetics: a short review of results. Agronomie 16:321–329

Pangborn (1963) Relative taste intensities of selected sugars and organic acids. J Food Sci 28:726–733

Pascal T, Kervella J, Pfeiffer F, Sauge MH, Esmenjaud D (1998) Evaluation of the interspecific progeny *Prunus persica* cv Summergrand × *Prunus davidiana* for disease resistance and some agronomic features. Acta Hortic 465:185–191

Pastorello EA, Farioli J, Pravettoni V, Ortolani C, Ispano M, Monza M, Broglio C, Scibola E, Ansaloni R, Incorvaia C, Conti A (1999) The major allergen of peach (*Prunus persica*) is a lipid transfer protein. J Allergy Clin Immunol 103:520–526

Quarta R, Dettori MT, Verde I, Gentile A, Broda Z (1998) Genetic analysis of agronomic traits and genetics linkage mapping in a BC1 peach population using RFLPs and RAPDs. Acta Hortic 465:51–59

Quarta R, Dettori MT, Sartori A, Verde I (2000) Genetic linkage map and QTL analysis in peach. Acta Hortic 521:233–241

Rajapakse S, Belthoff LE, He G, Estager AE, Scorza R, Verde I, Ballard RE, Baird WV, Callahan A, Monet R, Abbott AG (1995) Genetic linkage mapping in peach using morphological, RFLP and RAPD markers. Theor Appl Genet 91:964–971

Ramming DW, Tanner O (1983) Nemared peach rootstock. HortScience 18:376

Rodriguez J, Crespo JF, Lopez-Rubio A, de la Cruz Bertolo J, Ferrando-Vivas P, Vives R, da Roca P (2000) Clinical cross-reactivity among foods of the *Rosaceae* family. J Allergy Clin Immunol 106:183–189

Rodriguez J, Sherman WB, Scorza R, Wisniewski M, Okie WR (1994) 'Evergreen' peach, its inheritance and dormance behavior. J Am Soc Hortic Sci 119:789–792

Rothan C, Etienne C, Moing A, Dirlewanger E, Raymond P, Monet R (1999) Isolation of a cDNA encoding a metallothionein-like protein (Accession N°AJ243532) expressed during peach fruit development. Plant Physiol 121:311 (Electronic plant gene register)

Sánchez-Monge R, Lombardero M, Garcia-Sellé FJ, Barber D, Salcedo G (1999) Lipid-transfer proteins are relevant allergens in fruit allergy. J Allergy Clin Immunol 103:514–519

Sauge MH, Kervella J, Pascal T (1998) Settling behavior and reproductive potential of the green peach aphid *Myzus persicae* on peach varieties and a related wild *Prunus*. Entomol Exp Appl 89:233–242

Scorza R, Mehlenbacher SA, Lightner GW (1985) Inbreeding and coancestry of freestone peach cultivars of the eastern United States and implications for peach germplasm improvement. J Am Soc Hortic Sci 110:547–552

Scorza R, Melnicenco L, Dang P, Abbott AG (2002) Testing a microsatellite marker for selection of columnar growth habit in peach [*Prunus persica* (L.) Batsch]. Acta Hortic 592:285–289

Scott DH, Cullinan FP (1942) The inheritance of wavy-leaf character in the peach. J Hered 33:293–295

Scott DH, Weinberger JH (1944) Inheritance of pollen sterility in some peach varieties. Proc Am Soc Hort Sci 45:229–232

Sharpe RH, Hesse CO, Lownsberry BF, Perry VG, Hansen CJ (1970) Breeding peaches for root knot nematode resistance. J Am Soc Hortic Sci 94:209–212

Smykov VK, Ovcharenko GV, Perfilyeva ZN, Shoferistov EP (1982) Estimation of the peach hybrid resources by its mildew resistance against the infection background. Byull Gos Nikitsh Bot Sada 88:74–80

Sosinski B, Sossey-Alaoui K, Rajapakse S, Glassmoyer K, Ballard R, Abbott A, Lu X, Baird WV, Reighard G, Tabb A, Scorza R (1998) Use of AFLP and RFLP markers to create a combined linkage map in peach (*Prunus persica* (L.) Batsch) for use in marker assisted selection. Acta Hortic 465:61–68

Sosinski B, Gannavarapu M, Hager LD, Beck LE, King GJ, Ryder CD, Rajapakse S, Baird WV, Ballard RE, Abbott AG (2000) Characterization of microsatellite markers in peach [*Prunus persica* (L.) Batsch]. Theor Appl Genet 97:1034–1041

Testolin R, Marrazzo T, Cipriani G, Quarta R, Verde I, Dettori MT, Pancaldi M, Sansavini S (2000) Microsatellite DNA in peach (*Prunus persica* L. Batch) and its use in fingerprinting and testing the genetic origin of cultivars. Genome 43:512–520

Verde I, Quarta R, Cedrola C, Dettori MT (2002) QTL analysis of agronomic traits in a BC1 peach population. Acta Hortic 592:291–297

Vilanova S, Romero C, Abbott AG, Llácer G, Badenes ML (2003) An apricot (*Prunus armeniaca* L.) F2 progeny linkage map based on SSR and AFLP markers, mapping plum pox virus resistance and self-incompatibility traits. Theor Appl Genet 107:239–247

Viruel MA, Messeguer R, de Vicente MC, Garcia-Mas J, Puigdomènech P, Vargas F, Arús P (1995) A linkage map with RFLP and isozyme markers for almond. Theor Appl Genet 91:964–971

Viruel MA, Madur D, Dirlewanger E, Pascal T, Kervella J (1998) Mapping quantitative trait loci controlling peach leaf curl resistance. Acta Hortic 465:79–88

Wang D, Karle R, Brettin TS, Iezzoni AF (1998) Genetic linkage map in sour cherry using RFLP markers. Theor Appl Genet 97:1217–1224

Wang D, Karle R, Iezzoni AF (2000) QTL analysis of flower and fruit traits in sour cherry. Theor Appl Genet 100:535–544

Wang Y, Garay L, Reighard GL, Geargi LL, Abbott AG, Scorza R (2002) Development of bacterial artificial chromosome contigs in the *evergrowing* gene region in peach. Acta Hortic 592:183–189

Warburton ML, Becerra-Velasquez VL, Goffreda JC, Bliss FA (1996) Utility of RAPD markers in identifying genetic linkages to genes of economic interest in peach. Theor Appl Genet 93:920–925

Weinberger JH, Marth PC, Scott DH (1943) Inheritance study of root knot nematode resistance in certain peach varieties. Proc Am Soc Hortic Sci 42:321–325

Werner DJ, Creller MA (1997) Genetic studies in peach: inheritance of sweet kernel and male sterility. J Am Soc Hortic Sci 122:215–217

Werner DJ, Creller MA, Chaparro JX (1998) Inheritance of blood-flesh trait in peach. Hortic Sci 33:1243–1246

Yamamoto T, Shimada T, Imai T, Yaegaki H, Haji T, Matsuta N, Yamaguchi M, Hayashi T (2001) Characterization of morphological traits based on a genetic linkage map in peach. Breed Sci 51:271–278

Yamamoto T, Mochida K, Imai T, Shi Z, Ogiwara I, Hayashi T (2002) Microsatellite markers in peach [*Prunus persica* (L.) Batsch] derived from an enriched genomic and cDNA libraries. Mol Ecol Notes 2:298–301

Young ND (1996) QTL mapping and quantitative disease resistance in plants. Annu Rev Plant Physiol Plant Mol Biol 34:479–501

Section III Breeding Strategies and Silviculture Based on Markers

III.1 General Considerations: Marker-Assisted Selection

V. Mohler and C. Singrün[1]

1 Introduction

Since the first reported linkage of an agronomically important trait (a quantitative trait locus affecting seed weight) to a simply controlled gene (seed color) in common bean by Sax (1923), it has taken more than 60 years for genetic markers to become a qualified tool for widely optimizing genotype building in plant breeding programs. With the advent of molecular marker technology, the identification of genetic markers displaying linkage to any genetically inherited trait became feasible. However, most types of molecular markers, though nowadays PCR-based, are still too impractical to be used in large-scale marker-assisted selection (MAS) schemes due to the complexity of the assay preventing the appropriate automation, insufficient robustness or inadequate level of detected polymorphism (Koebner and Summers 2003). Due to their high polymorphic information content, sequence-tagged microsatellite sites are presently the most appropriate marker class for MAS. The future development of single nucleotide polymorphism (SNP) markers will provide access to affordable and high-throughput genotype determination assays and automated data analyses that are crucial for breeders' acceptance of MAS. MAS will then increasingly be applied to obtain improved efficiency and effectiveness in the selection of genotypes with traits that are difficult and expensive to phenotype, for the pyramiding of disease resistance genes in single genotypes, and for the carefully directed choice of parental lines in crossing programs allowing a controlled combination of alleles targeted for selection.

2 Requirements of Markers for Marker-Assisted Selection

Key issues in successful deployment of molecular markers in MAS are as follows:

1. Markers should co-segregate or map as close as possible to the target gene (within 2 cM), in order to have low recombination frequency between the

---

[1] Department of Plant Sciences, Center of Life and Food Sciences Weihenstephan, Technical University Munich, 85350 Freising-Weihenstephan, Germany

target gene and the marker. A better estimate of map distance between the target gene and the marker will be obtained by analysing further mapping populations which have genotypes in common with those used in the initial mapping population. Accuracy of MAS will be improved if, rather than a single marker, two markers flanking the target gene are used (Peng et al. 2000).
2. For unlimited use in MAS, markers should display polymorphism between genotypes that have and do not have the target gene.
3. Cost-effective, simple PCR markers are required to ensure genotyping power needed for the rapid screening of large populations.

Microsatellite markers, also termed sequence-tagged microsatellite site (STMS) or simple sequence repeat (SSR) markers, which use the high mutation rates of repeated short DNA motifs are presently the most complete tool for MAS. Extensive collections of mapped SSR markers from both the noncoding and expressed portion of the genome are, or will be available in the near future for all major crop species. To allow absolute allele recognition and, consequently, to exploit the full range of marker alleles at a given locus in a panel of breeding lines, SSRs need to be processed on polyacrylamide gel or capillary electrophoresis machines. However, these high demands on fragment detection can be compensated by the simultaneous electrophoresis of different SSR marker samples carrying distinguishable fluorescent dyes in a single lane/capillary.

Development is moving away from anonymous to functional and candidate gene markers as primary MAS tools since linkage relationships which limit the overall applicability of anonymous markers will no longer exist or will be reduced to a minimum. 'Perfect' markers have already been made available for the gibberellin-insensitive semi-dwarfing genes *Rht-B1b* and *Rht-D1b* (Ellis et al. 2002) and the null *Wx-B1* allele of the granule-bound starch synthase I (McLauchlan et al. 2001) of wheat and may be provided by forthcoming map-based cloning experiments and genetic association mapping studies (Rafalski 2002). Resistance gene analogs (RGAs) are a useful resource as candidate gene markers for disease resistance genes (Mohler et al. 2002; Madsen et al. 2003) since RGAs showing close genetic linkage to resistance genes often reflect physical proximity (Leister et al. 1999; Wei et al. 1999). Furthermore, a huge number of candidate gene markers for complex traits will be supplied by investigations directed at the identification of genes differentially expressed among extreme phenotypes, e.g., for potato late blight disease (Ronning et al. 2003).

The marker type by which functional alleles are discriminated from their allelic variants relies on small insertion-deletion (indel) polymorphisms or single nucleotide polymorphisms (SNPs). SNPs are of particular interest for their utilization in crop improvement, since they (1) represent the most frequent variations in the genome of any organism, thus, offering the opportunity to find informative markers for a distinct genomic region in any genetic background and (2) can be simply treated as di-allelic markers making them amenable to automated high-throughput genotyping and data handling.

While indels can be scored by direct sizing on polyacrylamide gels, the determination of SNP genotypes can be preferentially performed using nongel-based technology platforms such as denaturing high-performance liquid chromatography (DHPLC; Oefner and Underhill 1998), the most advanced system for heteroduplex analysis, or by pyrosequencing (Ahmadian et al. 2000) which uses the reaction principle of minisequencing. Sample analyses using DHPLC are carried out sequentially with an autosampler and one analysis takes around 5 min, while minisequencing reactions are performed in a 96-well plate in an automated device and take approximately 15 min.

3 Present Status of Validated Molecular Markers for Molecular Breeding of Important Crops

Markers that have been elaborated and validated for the monitoring of agronomically important traits, most of them determining resistance to disease, and which are (or have been) used in current breeding programs are listed in Table 1. The majority of traits is detectable using simple PCR markers, however, very often they need to be tested for polymorphism between parental lines of breeding programs. SNPs that are assayed via allele-specific PCR (AS-PCR; Ugozzoli and Wallace 1991) and cleaved amplified polymorphic sequences (CAPS; Konieczny and Ausubel 1993) and SSRs provide the prevalent marker classes for MAS. However, the time-consuming post-PCR digestion step limits the application of CAPS markers for small-scale genotyping. Therefore, these markers should be genotyped in the future using more general SNP detection systems in which any sequence polymorphism, irrespective of its location with reference to restriction sites, can be assayed. RFLP markers have only been used as an MAS tool when it was profitable due to the importance of the disease, high costs and unreliability of the bioassays such as in the early-generation selection of the cereal cyst nematode resistance gene *Cre1* in wheat (Ogbonnaya et al. 2001).

4 Marker-Assisted Selection for Quantitative Trait Loci

Most agronomic traits are of a polygenic nature and it is widely accepted that molecular markers are an appropriate tool to identify loci, the so-called quantitative trait loci (QTL), having alleles that differentially affect the expression of a quantitative trait. QTL mapping has been done for yield (Stuber et al. 1987), quality (Igrejas et al. 2002; Tan et al. 2001), tolerance to abi-

Table 1. Validated markers for the monitoring of agronomically important traits in MAS programs

Crop	Trait	Gene(s)	Marker type(s)	References
Wheat	Bread making quality	Glu-$D1$-$x5$	Gene-specific STS	Ahmad (2000)
		Glu-$D1$-$y10$		
		Glu-$B1$-$x7$		
		Glu-$D1$-$x5$	Promotor-specific SNP (DHPLC)	Schwarz et al. (2003)
	Reduced height	Rht-$B1b$	Gene-specific AS-PCRs	Ellis et al. (2002)
		Rht-$D1b$		
		$Rht8$	SSR	Korzun et al. (1998)
	Starch quality	Wx-$B1b$	Gene-specific AS-PCR	McLauchlan et al. (2001)
	Powdery mildew resistance	$Pm1c$	AFLP	Hartl et al. (1999)
		$Pm17$	RFLP-derived STS	Mohler et al. (2001)
		$Pm24$	SSR	Huang et al. (2000)
	Barley yellow dwarf virus resistance	T1 translocation	SSR	Ayala et al. (2001)
			RAPD-derived STS	Stoutjesdijk et al. (2001)
	Cereal cyst nematode resistance	$Cre1$	RGA-RFLP	Ogbonnaya et al. (2001)
		$Cre3$	RGA-based STS	Huang and Gill (2001)
	Leaf rust resistance	$Lr21$	RGA-based CAPS	Helguera et al. (2000)
		$Lr47$	RFLP-derived CAPS[b]	Seah et al. (2001)
	VPM rust resistance	$Sr38/Lr38/Yr17$	RGA-based STS	Chen et al. (2003)
	Stripe rust resistance	$Yr5$	RGA-based CAPS	W.-C. Zhou et al. (2003)
	Fusarium head blight resistance	3BS QTL (Sumai3)	Trait-flanking SSRs	Graner et al. (1999)
Barley	Barley yellow mosaic virus resistance	$rym4/rym5$	SSR	Werner et al. (2000)
		$rym9$	RAPD-derived STS	Paltridge et al. (1998)
	Barley yellow dwarf virus resistance	$Yd2$	AFLP-derived STS	Ford et al. (1998)
	Malting quality	QTL affecting α-amylase activity	CAPS derived from a linked gene	Ayoub et al. (2003)
			Trait-flanking RFLP-derived STS and CAPS	
Rice	Bacterial blight	xa-5	Trait-flanking RFLP-derived STS and CAPS	Huang et al. (1997)
			RFLP-derived CAPS	Sanchez et al. (2000)
		xa-13	RAPD-derived STS	Huang et al. (1997)
		Xa-21	Trait-flanking RFLPs	Chunwongse et al. (1993)
	Blast resistance	$Pi1$	RFLP-derived CAPS	Hittalmani et al. (2000)
		Piz-5		
		$Pita$	Trait-flanking RFLPs	
Potato	Potato virus Y resistance	Ry_{adg}	RGA-based CAPS	Sorri et al. (1999)
			RGA-based STS	Kasai et al. (2000)
Soybean	Soybean cyst nematode	$Rhg4$	TaqMan assay	Meksem et al. (2001)

otic stress (Cattivelli et al. 2002; Price et al. 2002) and durable disease resistance (Lindhout 2002). The sustained utilization of QTL, which is difficult to achieve through conventional breeding, is the principal task of MAS and can be done by selecting for the presence of specific marker alleles that are linked to favorable QTL alleles. Verification of putative QTL is needed prior to application of MAS for QTL because QTL effects found in a single mapping population are generally overestimated (Lande and Thompson 1990). Aspects that should be examined are the magnitude of the bias of estimated QTL effects and the certainty of genetic linkage map positions. The simplest method is to compare phenotypic differences between individuals carrying alternative marker alleles at putative QTL, allowing the detection of significant differences for a true QTL. According to the suggestion from Lande and Thompson (1990), the validation of QTL effects can be done in an independent sample of lines within the same cross (Han et al. 1997; Melchinger et al. 1998; Romagosa et al. 1999; Igartua et al. 2000). However, to fully assess the true breeding value of a QTL, studies validating QTL alleles in different genetic backgrounds and environments have to be carried out. In barley, Toojinda et al. (1998) and Ayoub et al. (2003) succeeded in introgressing QTL alleles that confer resistance to stripe rust and affect α-amylase activity of malt, respectively, into genetic backgrounds other than the original mapping population. Moreover, W.-C. Zhou et al. (2003) validated a major QTL for wheat Fusarium head blight resistance with SSR markers in two different genetic backgrounds, while Yousef and Juvik (2002) reported on the marker-assisted introgression of a beneficial QTL enhancing seedling emergence in three different genotypes of sweet corn. However, the attempt to transfer desired QTL alleles to other genetic backgrounds using MAS can also result in the loss of QTL effects (Sebolt et al. 2001; Reyna and Sneller 2001).

A further aspect of the evaluation and verification of QTL is the complexity of the trait, e.g., grain yield is a more complex trait to handle than disease resistance. Several studies reported that the number of QTL associated with grain yield and yield-related traits depends on the genotypes and the variance created by the cross (e.g., Melchinger et al. 1998; Ajmone Marsan et al. 2001). The effect of an added QTL allele on grain yield in an elite genotype is more difficult to estimate than it is, for example, for a QTL on disease resistance in a susceptible elite genotype. For introgression of a QTL allele influencing grain yield into diverse elite genotypes, it must have superior value to all other alleles at this QTL or to alleles at all grain yield QTL that are present in the gene pool (Reyna and Sneller 2001).

Comparative studies exist about the benefit of MAS versus phenotypic selection (van Berloo and Stam 1999; Yousef and Juvik 2001). The benefit depends on the heritability of the trait and the population size. When the heritability is high, the cost involved in genotyping many plants may not outweigh the expected benefits from phenotypic selection. As calculated for recombinant inbred lines, a benefit can be expected within a range of heritability of 0.1–0.3 (van Berloo and Stam 1998). If the value is less than 0.1, it is not possible to detect the QTL with the accuracy required to rely on flanking markers for selection (van Berloo and Stam 1999).

5 Marker-Assisted Selection in Gene Pyramiding

The great opportunity offered by MAS to select superior lines based on genotype rather than phenotype becomes clearly obvious in the case of combining different simple inherited resistance genes of large effects for a given pathosystem in a single genotype (gene pyramiding), since it is difficult to select plants with multiple resistance genes based on phenotype alone as the action of one gene may mask the action of another. Pyramiding multiple qualitative disease resistance genes with different race specificities has been proposed as a way of achieving more comprehensive resistance (Mundt 1990) due to simultaneous or stepwise mutation of several avirulence genes in the pathogen that is needed to overcome this pyramid. Successful examples for the pyramiding of major genes in single genotypes are given for the pathosystems rice:*Xanthomonas oryzae* pv. *oryzae*, rice:*Magnaporthe grisea*, and wheat:*Blumeria graminis* f. sp. *tritici* (Table 2). Novel approaches deal with the implementation of transgenes in breeding programs such as the pyramiding of *Bt* genes *cry1Ac* and *cry1C* conferring resistance to diamondback moths in broccoli (Cao et al. 2002; Table 2).

Durable disease resistance is not associated with a distinct type or mechanism of resistance, but only refers to the number of genes involved in resistance reaction (Lindhout 2002), for which reason pyramiding multiple quantitative or qualitative and quantitative resistance alleles in single genotypes is

Table 2. MAS in gene pyramiding

Crop	Trait (combination of genes)	References
Rice	Bacterial blight resistance (*xa4+xa5+xa13+Xa21*; *xa5+xa13+Xa21*)	Huang et al. (1997); Sanchez et al. (2000); Singh et al. (2001)
	Blast resistance (*Pi1+Piz-5+Pita*)	Hittalmani et al. (2000)
	(*Pi-tq5, Pi-tq1, Pi-tq6, Pi-lm2*: pyramids of 2 to 4 genes)	Tabien et al. (2000)
	Multiple resistance: bacterial blight (*Xa21*) Sheath blight (*RC7*) Yellow stem borer *Bt* fusion gene (*cry1AB/cry1Ac*)	Datta et al. (2002)
Wheat	Powdery mildew resistance (*Pm2+Pm4a*; *Pm2+Pm21*; *Pm4a+Pm21*)	Liu et al. (2000)
Barley	Stripe rust resistance (3 QTL)	Castro et al. (2003a, b)
Broccoli	Diamondback moths resistance (*cry1Ac+cry1c*)	Cao et al. (2002)
Soybean	Lepidopteran resistance (*cry1Ac*+corn earworm QTL)	Walker et al. (2002)

an approach to increase the level of disease resistance. Castro et al. (2003a, b) reported on the marker-assisted pyramiding of three quantitative resistance loci against barley stripe rust, caused by *Puccinia striiformis* f. sp. *hordei*. Resistance alleles at two QTL were necessary for the seedling resistance phenotype being expressed fitting a complementary gene model, while all three QTL regions were significant determinants of adult plant stripe rust resistance, with an additive effect of existing resistance alleles.

6 Marker-Assisted Selection in Backcross Breeding

The use of molecular markers in improving backcrossing efficiency has been widely accepted and was the subject of studies dealing with the marker-assisted building of disease-resistant, abiotic stress-tolerant and quality-improved genotypes (Table 3).

Conventional backcrossing aims at introgressing a target trait that is controlled by a single gene from a usually exotic donor line into a highly adapted recipient line, the so-called recurrent parent. At each backcross cycle, molecular markers can be used to identify carriers of the target trait (foreground selection) having the closest fit to the recurrent parent genotype (background selection). In order to minimize linkage drag, the selection of lines with the smallest introgressed segment around the target locus is usually done in tandem (Tanksley et al. 1989), i.e., selection for recombination on one side in the first generation and selection for recombination on the other side in the next generation. Although selection for simultaneous recombination events on both sides would save one generation of backcrossing, it is much more cost-effective due to the greater number of individuals that have to be genotyped to obtain one double recombinant. A full informative marker-assisted backcrossing scheme can be performed with markers derived from the DNA sequence of the gene to be introgressed. Chen et al. (2000) reported on the improvement of 'Minghui 63', a restorer line widely used in Chinese hybrid rice production, to bacterial blight resistance, caused by *Xanthomonas oryzae* pv. *oryzae* (*Xoo*), through introgression of *Xa21*, a broad-spectrum bacterial blight resistance gene. The PCR-based foreground selection system consisted of a marker that was part of *Xa21*, a marker located at 0.8 cM from the *Xa21* locus on one side and a marker at 3.0 cM from the gene on the other side, while a total of 128 RFLP markers, evenly distributed throughout the rice genome, was used to recover the genetic background of the recurrent parent in the BC_3F_1. The improved version, 'Minghui 63(*Xa21*)', was exactly the same as the original except for a fragment of less than 3.8 cM in length surrounding the *Xa21* locus. Both 'Minghui 63(*Xa21*)' and its hybrid with 'Zhenshan 97A', 'Shanyou 63(*Xa21*)', showed the same spectrum of bacterial blight resistance as the donor parent. Field examination of a number of agronomic traits showed that the two pairs of versions were identical when there was no disease stress. Under heavily diseased conditions, 'Minghui 63(*Xa21*)' showed significantly higher grain weight and spikelet fertility than 'Minghui 63', and 'Shanyou 63(*Xa21*)' was significantly higher than 'Shanyou 63' in grains per panicle, grain weight, and yield. In a later experiment by Chen et al. (2001), efficiency of background selection was enhanced by using the high-volume amplified fragment length polymorphism (AFLP) marker technique allowing the fast and cost-effective selection of individuals having 99.3% amount of the recurrent parent genome in BC_1F_1.

Nearly isogenic lines for QTL (QTL NILs) were developed by repeated backcrossing of individuals from primary mapping populations carrying the desired QTL genotype to one of the parental lines (Kandemir et al. 2000; Monforte and Tanksley 2000; Yamamoto et al. 2000; Shen et al. 2001; van Berloo et al. 2001; Willcox et al. 2002). To accelerate the creation of QTL NIL, some authors used background selection (Table 3). QTL NILs represent qualified genetic stocks for the validation of QTL effects in different environments. They can be further used to study epistatic interactions among QTL by intercrossing of single QTL NILs and for fine mapping of QTL for map-based cloning.

Advanced backcross QTL (AB-QTL; Tanksley and Nelson 1996) analysis was proposed as a general strategy for the simultaneous detection of QTL qualified for breeding purposes and cultivar development. The delay of QTL analysis until an advanced backcross generation offers advantages for QTL characterization such that the probability is reduced for the detection of QTL

Table 3. MAS in backcross breeding of single genes and QTL alleles

Crop	Trait (gene)	Foreground selection at	Background selection at	References
	Major genes			
Rice	Bacterial blight resistance (*Xa21*)	Each backcross cycle up to BC_3F_1	BC_3F_1 (128 RFLPs)	Chen et al. (2000)
		Each backcross cycle up to BC_3F_1	BC_1F_1 and BC_2F_1 (129 AFLPs)	Chen et al. (2001)
	Cooking and eating quality (*Waxy* gene region)	Each backcross cycle up to BC_3F_1	BC_3F_1 (118 AFLPs)	P.H. Zhou et al. (2003)
Barley	Barley yellow dwarf virus resistance (*Yd2*)	BC_1F_1 and BC_2F_2	Not performed	Jefferies et al. (2003)
	QTL			
Rice	Root depth (1–2 QTL)	Each backcross cycle up to BC_3F_2	BC_3F_2 (60 SSRs)	Shen et al. (2001)
Barley	Leaf rust (*Rphq2*)	Each backcross cycle up to BC_3S_2	Each backcross cycle up to BC_3S_2	van Berloo et al. (2001)
Maize	Southwestern corn borer resistance (3 QTL)	Each backcross cycle up to BC_2F_2	Each backcross cycle up to BC_2F_2	Willcox et al. (2002)

displaying epistatic interactions among donor alleles due to overall lower frequency of donor alleles. In fact, there will be a higher probability of detecting additive QTL which still function in a nearly isogenic background.

7 Conclusions

The future of MAS aims not only at utilizing perfect markers for improving existing breeding schemes, e.g., backcrossing, but also controlling all allelic variation for all genes of agronomic relevance. To build superior genotypes in silico, Peleman and van der Voort (2003) introduced a concept, 'Breeding by Design' that requires the knowledge of the map positions of all loci of agronomic importance, the allelic variation at those loci, and their contribution to the phenotype. Although great efforts have to be made to gather all this information, the starting position looks promising: molecular marker technology is very well developed, precise genetic stocks such as introgression line libraries (Eshed and Zamir 1995) for mapping all relevant traits are available for several crop plants and allelic variation at any locus in the genome can be assessed by establishing haplotypes of multiple tightly linked markers. This all-embracing approach has to be addressed immediately to make molecular markers an accepted and irreplaceable tool for developing better crop plants.

References

Ahmad M (2000) Molecular marker-assisted selection of HMW glutenin alleles related to wheat bread quality by PCR-generated DNA markers. Theor Appl Genet 101:892–896

Ahmadian A, Gharizadeh B, Gustafsson AC, Sterky F, Nyren P, Uhlen M, Lundeberg J (2000) Single nucleotide polymorphism analysis by pyrosequencing. Anal Biochem 280:103–110

Ajmone Marsan P, Gorni C, Chittò A, Redaelli R, van Vijk R, Stam P, Motto M (2001). Identification of QTLs for grain yield and grain-related traits of maize (*Zea mays* L.) using an AFLP map, different testers, and cofactor analysis. Theor Appl Genet 102:230–243

Ayala L, Henry N, González-de-León D, van Ginkel M, Mujeeb-Kazi A, Keller B, Khairallah M (2001) A diagnostic molecular marker allowing the study of *Th. intermedium*-derived resistance to BYDV in bread wheat segregating populations. Theor Appl Genet 102:942–949

Ayoub M, Armstrong E, Bridger G, Fortin MG, Mather DE (2003) Marker-based selection in barley for a QTL region affecting α-amylase activity of malt. Crop Sci 43:556–561

Cao J, Zhao J-Z, Tang JD, Shelton AM, Earle ED (2002) Broccoli plants with pyramided *cry1Ac* and *cry1C* Bt genes control diamondback moths resistant to Cry1A and Cry1C proteins. Theor Appl Genet 105:258–264

Castro AJ, Chen X, Corey A, Filichkina T, Hayes PM, Mundt C, Richardson K, Sandoval-Islas S, Vivar H (2003a) Pyramiding and validation of quantitative trait locus (QTL) alleles determining resistance to barley stripe rust: effects on adult plant resistance. Crop Sci 43:2234–2239

Castro AJ, Chen X, Hayes PM, Johnston M (2003b) Pyramiding quantitative trait locus (QTL) alleles determining resistance to barley stripe rust: effects on resistance at the seedling stage. Crop Sci 43:651–659

Cattivelli L, Baldi P, Crosatti C, di Fonzo N, Faccioli P, Grossi M, Mastrangelo AM, Pecchioni N, Stanca AM (2002) Chromosome regions and stress-related sequences involved in resistance to abiotic stress in Triticeae. Plant Mol Biol 48:649–665

Chen S, Lin XH, Xu CG, Zhang Q (2000) Improvement of bacterial blight resistance of 'Minghui 63', an elite restorer line of hybrid rice, by molecular marker-assisted selection. Crop Sci 40:239–244

Chen S, Xu CG, Lin XH, Zhang Q (2001) Improving bacterial blight resistance of 6078, an elite restorer line of hybrid rice, by molecular marker-assisted selection. Plant Breed 120:133–137

Chen X, Soria MA, Yan G, Sun J, Dubcovsky J (2003) Development of sequence tagged site and cleaved amplified polymorphic sequence markers for wheat stripe rust resistance gene Yr5. Crop Sci 43:2058–2064

Chunwongse J, Martin GB, Tanksley SD (1993) Pregermination genotypic screening using PCR amplification of half seeds. Theor Appl Genet 86:694–698

Datta K, Baisakh N, Maung Thet K, Tu J, Datta SK (2002) Pyramiding transgenes for multiple resistance in rice against bacterial blight, yellow stem borer and sheath blight. Theor Appl Genet 106:1–8

Ellis MH, Spielmeyer W, Gale KR, Rebetzke GJ, Richards RA (2002) "Perfect" markers for the Rht-B1b and Rht-D1b dwarfing genes in wheat. Theor Appl Genet 105:1038–1042

Eshed Y, Zamir D (1995) An introgression line population of Lycopersicon pennellii in the cultivated tomato enables the identification and fine mapping of yield-associated QTL. Genetics 141:1147–1162

Ford CM, Paltridge NG, Rathjen JP, Moritz RL, Simpson RJ, Symons RH (1998) Assays for Yd2, the barley yellow dwarf virus resistance gene, based on the nucleotide sequence of a closely linked gene. Mol Breed 4:23–31

Graner A, Streng S, Kellermann A, Schiemann A, Bauer E, Waugh R, Pellio B, Ordon F (1999) Molecular mapping and genetic fine-structure of the rym5 locus encoding resistance to different strains of the Barley Yellow Mosaic Virus Complex. Theor Appl Genet 98:285–290

Han F, Romagosa I, Ullrich SE, Jones BL, Hayes PM, Wesenberg DM (1997) Molecular marker assisted selection for malting quality traits in barley. Mol Breed 3:427–437

Hartl L, Mohler V, Zeller FJ, Hsam SLK, Schweizer G (1999) Identification of AFLP markers closely linked to the powdery mildew resistance genes Pm1c and Pm4a in common wheat (Triticum aestivum L.). Genome 42:322–329

Helguera M, Khan IA, Dubcovsky J (2000) Development of PCR markers for wheat leaf rust resistance gene Lr47. Theor Appl Genet 101:625–631

Hittalmani S, Parco A, Mew TV, Zeigler RS, Huang N (2000) Fine mapping and DNA marker-assisted pyramiding of the three major genes for blast resistance in rice. Theor Appl Genet 100:1121–1128

Huang L, Gill BS (2001) An RGA-like marker detects all known Lr21 leaf rust resistance gene family members in Aegilops tauschii and wheat. Theor Appl Genet 103:1007–1013

Huang N, Angeles ER, Domingo J, Magpantay G, Singh S, Zhang G, Kumaravadivel N, Bennett J, Khush GS (1997) Pyramiding of bacterial blight resistance genes in rice: marker-assisted selection using RFLP and PCR. Theor Appl Genet 95:313–320

Huang XQ, Hsam SLK, Zeller FJ, Wenzel G, Mohler V (2000) Molecular mapping of the wheat powdery mildew resistance gene Pm24 and marker validation for molecular breeding. Theor Appl Genet 101:407–414

Igartua E, Edney M, Rossnagel BG, Spaner D, Legge WG, Scoles GJ, Eckstein PE, Penner GA, Tinker NA, Briggs KG, Falk DE, Mather DE (2000) Marker-based selection of QTL affecting grain and malt quality in two-row barley. Crop Sci 40:1426–1433

Igrejas G, Leroy P, Charmet G, Gaborit T, Marion D, Branlard G (2002) Mapping QTLs for grain hardness and puroindoline content in wheat (Triticum aestivum L.). Theor Appl Genet 106:19–27

Jefferies SP, King BJ, Barr AR, Warner P, Logue SJ, Langridge P (2003) Marker-assisted backcross introgression of the Yd2 gene conferring resistance to barley yellow dwarf virus in barley. Plant Breed 122:52–56

Kandemir N, Kudrna DA, Ullrich SE, Kleinhofs A (2000) Molecular marker assisted genetic analysis of head shattering in six-rowed barley. Theor Appl Genet 101:203–210

Kasai K, Morikawa Y, Sorri VA, Valkonen JPT, Gebhardt C, Watanabe KN (2000) Development of SCAR markers to the PVY resistance gene Ry_{adg} based on a common feature of plant disease resistance genes. Genome 43:1–8

Koebner RMD, Summers RW (2003) 21st century wheat breeding: plot selection or plate detection? Trends Biotechnol 21:59–63

Konieczny A, Ausubel F (1993) A procedure for mapping *Arabidopsis* mutations using co-dominant ecotype-specific PCR-based markers. Plant J 4:403–410

Korzun V, Röder MS, Ganal MW, Worland AJ, Law CN (1998) Genetic analysis of the dwarfing gene (*Rht8*) in wheat, part I. Molecular mapping of *Rht8* on the short arm of chromosome 2D of bread wheat (*Triticum aestivum* L.). Theor Appl Genet 96:1104–1109

Lande R, Thompson R (1990) Efficiency of marker-assisted selection in the improvement of quantitative traits. Genetics 124:743–756

Leister D, Kurth J, Laurie DA, Yano M, Sasaki T, Graner A, Schulze-Lefert P (1999) RFLP and physical mapping of resistance gene homologues in rice (*O. sativa*) and barley (*H. vulgare*). Theor Appl Genet 98:509–520

Lindhout P (2002) The perspective of polygenic resistance in breeding for durable disease resistance. Euphytica 124:217–226

Liu J, Liu D, Tao W, Li W, Wang S, Chen P, Cheng S, Gao D (2000) Molecular marker-facilitated pyramiding of different genes for powdery mildew resistance in wheat. Plant Breed 119:21–24

Madsen LH, Collins NC, Rakwalska M, Backes G, Sandal N, Krusell L, Jensen J, Waterman EH, Jahoor A, Ayliffe M, Pryor AJ, Langridge P, Schulze-Lefert P, Stougaard J (2003) Barley disease resistance gene analogs of the NBS-LRR class: identification and mapping. Mol Gen Genomics 269:150–161

McLauchlan A, Ogbonnaya FC, Hollingsworth B, Carter M, Gale KR, Henry RJ, Holton TA, Morell MK, Rampling LR, Sharp PJ, Shariflou MR, Jones MGK, Appels R (2001) Development of robust PCR-based DNA markers for each homoeo-allele of granule-bound starch synthase and their application in wheat breeding programs. Aust J Agr Res 52:1409–1416

Meksem K, Ruben E, Hyten DL, Schmidt ME, Lightfoot DA (2001) High-throughput genotyping for a polymorphism linked to soybean cyst nematode resistance gene *Rhg4* by using Taqman probes. Mol Breed 7:63–71

Melchinger AE, Utz HF, Schön CC (1998) Quantitative trait locus (QTL) mapping using different testers and independent population samples in maize reveals low power of QTL detection and large bias in estimates of QTL effects. Genetics 149:383–403

Mohler V, Hsam SLK, Zeller FJ, Wenzel G (2001) An STS marker distinguishing the rye-derived powdery mildew resistance alleles at the *Pm8/Pm17* locus of common wheat (*T. aestivum* L. em Thell.). Plant Breed 120:448–450

Mohler V, Klahr A, Wenzel G, Schwarz G (2002) A resistance gene analog useful for targeting disease resistance genes against different pathogens on group 1S chromosomes of barley, wheat and rye. Theor Appl Genet 105:364–368

Monforte AJ, Tanksley SD (2000) Fine mapping of a quantitative trait locus (QTL) from *Lycopersicon hirsutum* chromosome 1 affecting fruit characteristics and agronomic traits: breaking linkage among QTLs affecting different traits and dissection of heterosis for yield. Theor Appl Genet 100:471–479

Mundt CC (1990) Probability of mutation to multiple virulence and durability of resistance gene pyramids. Phytopathology 80:221–223

Oefner PJ, Underhill PA (1998) DNA mutation detection using denaturing high-performance liquid chromatography (DHPLC). In: Dracopoli NC, Haines JL, Korf BR, Moir DT, Morton CC, Seidman CE (eds) Current protocols in human genetics (supplement 19). Wiley, New York, pp 7.10.1–7.10.12

Ogbonnaya FC, Subrahmanyam NC, Moullet O, de Majnik J, Eagles HA, Brown JS, Eastwood RF, Kollmorgen J, Appels R, Lagudah ES (2001) Diagnostic DNA markers for cereal cyst nematode resistance in bread wheat. Aust J Agric Res 52:1367–1374

Paltridge NG, Collins NC, Bendahmane A, Symons RH (1998) Development of YLM, a codominant PCR marker closely linked to the *Yd2* gene for resistance to barley yellow dwarf disease. Theor Appl Genet 96:1170–1177

Peleman JD, van der Voort JR (2003) Breeding by design. Trends Plant Sci 8:330–334

Peng JH, Fahima T, Röder MS, Li YC, Grama A, Nevo E (2000) Microsatellite high-density mapping of the stripe rust resistance gene *YrH52* region on chromosome 1B and evaluation of its marker-assisted selection in the F_2 generation in wild emmer wheat. New Phytol 146:141–154

Price AH, Cairns JE, Horton P, Jones HG, Griffiths H (2002) Linking drought-resistance mechanisms to drought avoidance in upland rice using a QTL approach: progress and new opportunities to integrate stomatal and mesophyll responses. J Exp Bot 53:989–1004

Rafalski A (2002) Applications of single nucleotide polymorphisms in crop genetics. Curr Opin Plant Biol 5:94–100

Reyna N, Sneller CH (2001) Evaluation of marker-assisted introgression of yield QTL alleles into adapted soybean. Crop Sci 41:1317–1321

Romagosa I, Han F, Ullrich SE, Hayes PM, Wesenberg DM (1999) Verification of yield QTL through realized molecular marker-assisted selection responses in a barley cross. Mol Breed 5:143–152

Ronning CM, Stegalkina SS, Ascenzi RA, Bougri O, Hart AL, Utterbach TR, Vanaken SE, Riedmuller SB, White JA, Cho J, Pertea GM, Lee Y, Karamycheva S, Sultana R, Tsai J, Quackenbush J, Griffiths HM, Restrepo S, Smart CD, Fry WE, van der Hoeven R, Tanksley S, Zhang P, Jin H, Yamamoto ML, Baker BJ, Buell CR (2003) Comparative analyses of potato expressed sequence tag libraries. Plant Physiol 131:419–429

Sanchez AC, Brar DS, Huang N, Li Z, Khush GS (2000) Sequence tagged site marker-assisted selection for three bacterial blight resistance genes in rice. Crop Sci 40:792–797

Sax K (1923) The association of size differences with seed-coat pattern and pigmentation in *Phaseolus vulgaris*. Genetics 8:552–560

Schwarz G, Sift A, Wenzel G, Mohler V (2003) DHPLC scoring of a SNP between promoter sequences of HMW glutenin x-type alleles at the *Glu-D1* locus in wheat. J Agric Food Chem 51:4263–4267

Seah S, Bariana H, Jahier JK, Sivasithamparam K, Lagudah ES (2001) The introgressed segment carrying rust resistance genes *Yr17*, *Lr37* and *Sr38* in wheat can be assayed by a cloned disease resistance gene-like sequence. Theor Appl Genet 102:600–605

Sebolt AM, Shoemaker RC, Diers BW (2001) Analysis of a quantitative trait locus allele from wild soybean that increases seed protein concentration in soybean. Crop Sci 40:1438–1444

Shen L, Courtois B, McNally KL, Robin S, Li Z (2001) Evaluation of near-isogenic lines of rice introgressed with QTLs for root depth through marker-aided selection. Theor Appl Genet 103:75–83

Singh S, Sidhu JS, Huang N, Vikal Y, Li Z, Brar DS, Dhaliwal HS, Khush GS (2001) Pyramiding three bacterial blight resistance genes (*xa5*, *xa13* and *Xa21*) using marker-assisted selection into indica rice cultivar PR106. Theor Appl Genet 102:1011–1015

Sorri VA, Watanabe KN, Valkonen JPT (1999) Predicted kinase-3a motif of a resistance gene-like fragment as a unique marker for PVY resistance. Theor Appl Genet 99:164–170

Stoutjesdijk P, Kammholz SJ, Kleven S, Matsay S, Banks PM, Larkin PJ (2001) PCR-based molecular marker for the Bdv2 *Thinopyrum intermedium* source of barley yellow dwarf virus resistance in wheat. Aust J Agric Res 52:1383–1388

Stuber CW, Edwards MD, Wendel JF (1987) Molecular marker-facilitated investigations of quantitative trait loci in maize. II. Factors influencing yield and its component traits. Crop Sci 27:639–648

Tabien RE, Li Z, Patterson AH, Marchetti MA, Stansel JW, Pinson SRM (2000) Mapping of four major rice bast resistance genes from Lemont and Teqint and evaluation of their combinatorial effect for field resistance. Theor Appl Genet 101:1215–1225

Tan YF, Sun M, Xing YZ, Hua JP, Sun XL, Zhang QF, Corke H (2001) Mapping quantitative trait loci for milling quality, protein content and color characteristics of rice using a recombinant inbred line population derived from an elite rice hybrid. Theor Appl Genet 103:1037–1045

Tanksley SD, Nelson JC (1996) Advanced backcross QTL analysis: a method for the simultaneous discovery and transfer of valuable QTLs from unadapted germplasm into elite breeding lines. Theor Appl Genet 92:191–203

Tanksley SD, Young ND, Paterson AH, Bonierbale MW (1989) RFLP mapping in plant breeding: new tools for an old science. Biotechnology 7:257–264

Toojinda T, Baird E, Booth A, Broers L, Hayes P, Powell W, Thomas W, Vivar H, Young G (1998) Introgression of quantitative trait loci (QTLs) determining stripe rust resistance in barley: an example of marker-assisted development. Theor Appl Genet 96:123–131

Ugozzoli L, Wallace RB (1991) Allele-specific polymerase chain reaction. Methods Enzymol 2:42–48

Van Berloo R, Stam P (1998) Marker assisted selection in autogamous RIL populations: a simulation study. Theor Appl Genet 96:147–154

Van Berloo R, Stam P (1999) Comparison between marker-assisted selection and phenotypical selection in a set of *Arabidopsis thaliana* recombinant inbred lines. Theor Appl Genet 98:113–118

Van Berloo R, Aalbers H, Werkman A, Niks RE (2001) Resistance QTL confirmed through development of QTL-NILs for barley leaf rust resistance. Mol Breed 8:187–195

Walker D, Boerma HR, All J, Parrott W (2002) Combining *cry1Ac* with QTL alleles from PI 229358 to improve soybean resistance to lepidopteran pests. Mol Breed 9:43–51

Wei F, Gobelman-Werner K, Morroll SM, Kurth J, Mao L, Wing R, Leister D, Schulze-Lefert P, Wise RP (1999) The *Mla* (powdery mildew) resistance cluster is associated with three NBS-LRR gene families and suppressed recombination within a 240-kb DNA interval on chromosome 5S (1HS) of barley. Genetics 153:1929–1948

Werner K, Pellio B, Ordon F, Friedt W (2000) Development of an STS marker and SSRs suitable for marker-assisted selection for the BaMMV resistance gene *rym9* in barley. Plant Breed 119:517–519

Willcox MC, Khairallah MM, Bergvinson D, Crossa J, Deutsch JA, Edmeades GO, González-de-León D, Jiang C, Jewell DC, Mihm JA, Williams WP, Hoisington D (2002) Selection for resistance to southwestern corn borer using marker-assisted and conventional backcrossing. Crop Sci 2002 42:1516–1528

Yamamoto T, Lin H, Sasaki T, Yano M (2000) Identification of heading date quantitative trait locus *Hd6* and characterization of its epistatic interactions with *Hd2* in rice using advanced backcross progeny. Genetics 154:885–891

Yousef GG, Juvik JA (2001) Comparison of phenotypic and marker-assisted selection for quantitative traits in sweet corn. Crop Sci 2001 41:645–655

Yousef GG, Juvik JA (2002) Enhancement of seedling emergence in sweet corn by marker-assisted backcrossing of beneficial QTL. Crop Sci 42:96–104

Zhou PH, Tan YF, He YQ, Xu CG, Zhang Q (2003) Simultaneous improvement for four quality traits of Zhenshan 97, an elite parent of hybrid rice, by molecular marker-assisted selection. Theor Appl Genet 106:326–331

Zhou W-C, Kolb FL, Bai G-H, Domier LL, Boze LK, Smith NJ (2003) Validation of a major QTL for scab resistance with SSR markers and use of marker-assisted selection in wheat. Plant Breed 122:40–46

III.2 Breeding Strategies: Optimum Design of Marker-Assisted Backcross Programs

M. Frisch[1]

1 Introduction

Recurrent backcrossing is used to transfer the genes underlying agronomically important traits from a donor into the genetic background of a recipient genotype, usually an inbred line (Allard 1960). In a backcross program, molecular markers can be used for indirect selection for the presence of a favorable allele (Tanksey 1983) and for selection against the undesired genetic background of the donor genotype (Tanksley et al. 1989). Selection against the genetic background of the donor ('background selection') allows us to reduce the number of backcross generations required for gene introgression from six to three (Frisch et al. 1999a). Due to this time saving and the possibility to monitor the donor genome content of the converted line, background selection has become a standard tool in plant breeding, as demonstrated by the example of the introgression of a gene coding for the *Bacillus thuringiensis* toxin into a maize inbred line (Ragot et al. 1995). However, the cost of a breeding program applying background selection is determined by an optimum allocation of resources, because the price of marker analyses is still high. In this chapter, principles for the optimum design of backcross programs for introgression of qualitatively inherited traits with marker-assisted background selection are described. Considered topics are (1) introgression of a dominant gene, (2) introgression of a recessive gene, and (3) simultaneous introgression of two genes.

2 Introgression of One Dominant Gene

2.1 Minimum Population Size Required for Finding Recombinant Plants

Theory (Stam and Zeven 1981) and experimental results (Young and Tanksley 1989) show that the intact donor chromosome segment around a target gene in backcrossing remains large, even in advanced backcross generations. It can

---

[1] Institute of Plant Breeding, Seed Science, and Population Genetics, University of Hohenheim, 70593 Stuttgart, Germany

form the major part of the carrier chromosome of the target gene in a backcross product, which is responsible for the transfer of undesired traits from the donor into the recipient parent (Zeven et al. 1983). By monitoring markers flanking the target locus and selecting individuals carrying the donor allele at the target locus and the recipient alleles at the flanking markers, the length of the intact donor chromosome segment around the target gene can be reduced efficiently (Tanksley et al. 1989). This rationale can be used to determine the population size in a backcross program such that recombinants between the target gene and flanking markers can be found with a high probability.

The population size required to generate in one backcross generation with a high probability at least one plant recombinant between the target gene and both flanking markers is greater than the multiplication rate of most crop species. For example, for a flanking marker distance of 5 cM on each side of the target gene, about 4000 individuals are required to find a double recombinant with a probability of 0.99; even for the rather large flanking marker distance of 25 cM at least 300 individuals are required (Frisch et al. 1999b). Therefore, we recommend a sequential strategy to find an individual with recombination between the target gene and one flanking marker in generation BC_1, and a recombinant between the target gene and the second flanking marker in generation BC_2 (Frisch et al. 1999b).

To calculate the minimum population size required in a backcross program applying this approach, we consider a chromosome on which positions are denoted in map distance from the telomere. The target locus is located at position x and two flanking markers at positions y_l and y_r such that $y_l < x < y_r$. Let $d_1 = x - y_l$ and $d_2 = y_r - x$ denote the lengths of the chromosome intervals between the target locus and its flanking markers. Without loss of generality, we assume $d_1 \leq d_2$. We denote z^- as the genotype of an individual homozygous for the recipient allele and z^+ as the genotype of a heterozygous individual at the locus at position $z \in \{y_l, x, y_r\}$.

If the probability that a plant has a desired genotype is p, then the minimum population size n required to find with probability q at least one plant, which has a desired genotype, can be obtained from the probability function of the binomial distribution as

$$n \geq \ln(1-q)/\ln(1-p). \tag{1}$$

Probabilities p of obtaining single or double recombinant plants between the target gene and the flanking markers are listed in Table 1. Probabilities p of obtaining plants, which are not only defined by conditions concerning the target gene and its flanking markers, but also by the condition that the complete chromosome region between a flanking marker and the nearest telomere consists entirely of the recurrent parent genome were given by Frisch et al. (1999b).

Applying these results, a simple method to carry out a two-generation backcross program designed to find with probability q_2 at least one BC_2 plant of genotype $y_l^- x^+ y_r^-$ can be conducted as follows:

Table 1. Transition probabilities p between genotypes in backcrossing

Genotype in generation BC_s	Genotype in generation BC_{s+1}	Transition probability
$y_l^+ x^+ y_r^+$ [a]	$y_l^- x^+ y_r^-$	$p = (1 - e^{-2d_1})(1 - e^{-2d_2})/8$ [b]
$y_l^+ x^+ y_r^+$	$y_l^- x^+ y_r^+$ or $y_l^+ x^+ y_r^-$	$p = (1 - e^{-2(d_1 + d_2)})/4$
$y_l^- x^+ y_r^+$	$y_l^- x^+ y_r^-$	$p = (1 - e^{-2d_2})/4$
$y_l^+ x^+ y_r^-$	$y_l^- x^+ y_r^-$	$p = (1 - e^{-2d_1})/4$

[a] The symbols y_l and y_r denote the background selection markers and x the target locus. A superscript + or − indicates that the locus is heterozygous or homozygous for the recurrent parent allele, respectively
[b] d_1 and d_2 denote the map distances between the target gene and two flanking markers

1. Choose the desired probability of success q_2. Set the probability of finding at least one BC_1 individual of type $y_l^- x^+ y_r^+$ or $y_l^+ x^+ y_r^-$ to $q_1 = q_2$
2. Carry out BC_1 with population size n_1 such that at least one individual of genotype $y_l^- x^+ y_r^+$ or $y_l^+ x^+ y_r^-$ is generated with probability q_1.
3. Select a BC_1 individual according to $(d_1 \leq d_2)$

$$y_l^- x^+ y_r^- > y_l^- x^+ y_r^+ > y_l^+ x^+ y_r^- > y_l^+ x^+ y_r^+.$$

(The symbol $>$ denotes that the genotype on the left-hand side is preferred over the genotype on the right-hand side.)

4. Carry out generation BC_2 with n_2 such that at least one individual with genotype $y_l^- x^+ y_r^-$ is generated with probability q_2.

An optimization of this scheme is possible by choosing $q_1 \neq q_2$ in step 1. A certain probability of success q_2 can be reached irrespective of the chosen probability q_1 because the population size n_2 can be chosen in step 4 such that a desired level of q_2 is reached irrespective of the genotype of the plant selected in step 3. In consequence, the choice $q_1 = q_2$ is arbitrary, and an optimum criterion for q_1 can be defined such that q_1 is optimal if the expected total number of individuals required for the backcross program is minimized: $E(n) = n_1 + E(n_2) \rightarrow$ min. A mathematical description of this optimum criterion is given by Eq. (35) of Frisch et al. (1999b), a graphic illustration of the effect of q_1 on the expected total number of individuals required is shown in Fig. 1. For example, for two flanking markers 5 cM distant from the target gene ($d_1 = d_2 = 0.05$) and a desired probability $q_2 = 0.99$ of finding in generation BC_2 at least one double recombinant individual, choosing $q_1 = 0.90$ results in an expected total number of individuals required of $E(n) > 400$ (Fig. 1). In contrast, choosing $q_1 \approx 0.995$ results in a minimum of the expected total number of individuals of $E(n) = 222$.

Calculation of the optimum values q_1 for different flanking marker distances and probabilities of success is numerically demanding, therefore, we tabulated the population size n_1 corresponding to the optimum value of q_1 for flanking marker distances of 4, 6, 8, 12, and 16 cM in Table 2.

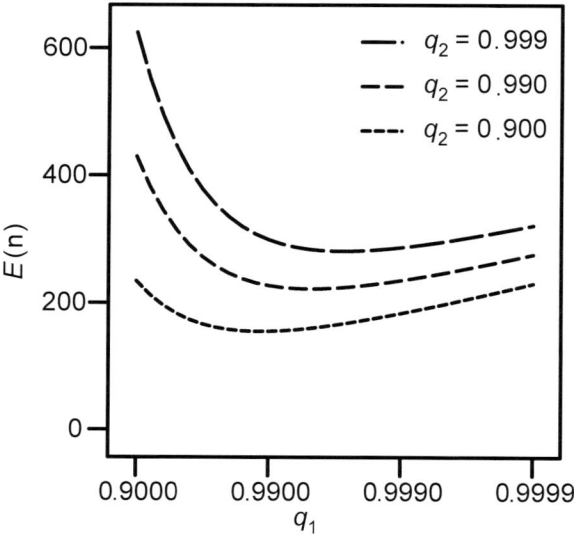

Fig. 1. Expected total number $E(n)$ of individuals required to reach probabilities of success $q_2=0.900$, 0.990 or 0.999 depending on the value chosen for q_1. The flanking marker distances are $d_1=d_2=0.05$ M

Table 2. Optimum population size n_1 in generation BC_1 and corresponding expected population size $E(n_2)$ in generation BC_2 such that the expected total number of individuals $E(n)=n_1+E(n_2)$ required to introgress one gene with a minimum number of individuals in a two-generation backcross program is minimized. The values depend on the map distances d_1 and d_2 between the target gene and two flanking markers

	d_2 (M)				
d_1 (M)	0.04	0.06	0.08	0.12	0.16
			$n_1/E(n_2)$		
0.04	143/252	136/186	130/155	123/128	117/117
0.06		91/167	88/135	83/105	79/93
0.08			66/125	63/94	60/80
0.12				43/83	41/68
0.16					32/62

In the above approach, the population size of generation BC_2 is determined after learning the result of generation BC_1 (a posteriori) such that the overall success is reached independent of the outcome of generation BC_1. Hospital and Charcosset (1997) presented an approach to determine the population sizes for all generations of a backcross program before starting the breeding program (a priori). Their approach applies constant population sizes over generations and, furthermore, does not favor individuals of type $y_l^- x^+ y_r^+$ over those of type $y_l^+ x^+ y_r^-$ even if $d_1 < d_2$. In comparison with the a priori approach, determining the population size n_2 a posteriori has the advantages that (1) only the number of individuals actually required to reach q_2 are generated, and (2) the probability q_2 can be reached for all outcomes of generation BC_1.

2.2 Reducing the Number of Backcross Generations

The above approach for calculating the population size focuses on reduction of the length of the intact donor chromosome segment around the target gene. Alternatively, the population size of a backcross program can be targeted to save a defined number of backcross generations in the breeding program. The population size required for this approach depends on the number and positions of the markers and the selection intensity. There are no analytical solutions to determine population sizes, but simulations can be used to solve the problem.

In such a simulation, first a reference breeding plan is simulated which describes the breeding program as it would be carried out without using markers. Usually the reference plan consists of six (Allard 1960) to eight (Fehr 1987) generations of backcrossing. With this simulation, the reference value reached for the recurrent parent genome is determined. In subsequent simulations of alternative scenarios with marker-assisted backcrossing, the parameters of the backcross program are varied until the reference value for the recurrent parent genome content is reached in the desired number of generations.

For a linkage map of maize, Frisch et al. (1999a) used the simulation software Plabsim (Frisch et al. 2000). We found that the recurrent parent genome content of 96.8%, which was reached after six backcross generations without marker-assisted background selection, was reached after three backcross generations when using 80 markers and a population size of 100 individuals in each of generations BC_1 to BC_3.

2.3 Marker Positions

If only two background selection markers on the target chromosome are used (assuming direct selection for the target gene), the distances d_1 and d_2 between target gene and markers can be chosen such that the expected donor chromosome content on the target chromosome is minimized if both markers are fixed for the recipient allele (Hospital et al. 1992) by applying

$$d_1 = d_2 = \frac{1}{2} \ln\left(1 + 2\sqrt{s}\right), \qquad (2)$$

where s is the proportion of selected BC_1 individuals. This approach is based upon the assumption of an infinite population size and the optimum properties only hold true if exactly two markers on the carrier chromosome of the target gene are used.

An alternative method of calculating of d_1 and d_2 is based on the rationale that in a population with given size n, at least one single or double recombinant individual is found with a given probability q (Frisch et al. 1999b). To determine n, the probabilities p given in Table 1 are inserted in

$$p = 1 - (1 - q)^{1/n} \tag{3}$$

and the resulting equation is solved for d_1 and d_2. Tabulated results are given by Frisch et al. (1999b).

The positions of markers used for background selection on noncarrier chromosomes can be determined such that the correlation between the molecular marker estimate of the donor genome content and the true donor genome content is maximized (Visscher 1996). In this approach, the distance between the telomere and the first marker is determined by numerical comparisons of alternative map positions in order to optimize the correlation. For the remaining markers, Visscher (1996) showed that the maximum correlation is reached if these are evenly spaced. The optimum distance between telomere and the first marker is different for each backcross generation, which makes it difficult to choose marker distances for a backcross program with subsequent generations of background selection.

An approach which takes selection over several generations into account was presented by Servin and Hospital (2002). They suggest choosing the positions of markers such that after all markers have been fixed for the recurrent parent allele, the expected donor genome content on the chromosome is maximized. As with Visscher's approach, the distance between the telomere and the first marker is determined with numerical evaluations and the remaining markers are evenly spaced.

2.4 Selection Strategies

We consider here a marker-assisted backcross program consisting of $s=1...t$ generations, where in total $n=n_1+...+n_t$ plants are employed and the population size n_s per generation is considerably greater than the minimum population size required to find at least one recombinant between target gene and flanking markers. The goal of marker-assisted background selection is to reduce the recurrent parent genome across all chromosomes. A straightforward design to accomplish this goal is to generate in each generation $n_s=n/t$ plants and to apply a two-stage selection strategy, consisting of selection for the target gene and one marker-assisted background selection step. For the background selection step, in generation BC_1 m markers with a good coverage of the entire genome are analyzed and an individual which carries the target gene and the recurrent parent alleles at most of the m markers is selected as parent for producing the next backcross generation. In subsequent backcross generations, selection is carried out according to the same scheme, but only those markers are analyzed which have not been fixed for the recurrent parent allele in the preceding generation.

When at least three generations of marker-assisted backcrossing have been carried out, the efficiency can be enhanced considerably by: (1) employing a small population size in generation BC_1 and increasing the population size in subsequent backcross generations, or (2) employing three- or four-stage selection strategies, emphasizing selection for recombinants on the carrier chromosome of the target gene during the first generations (Frisch et al. 1999a).

Employing increasing, constant, or decreasing population sizes from generations BC_1 to BC_3 in a simulation study had little effect on the recurrent parent genome values of the selected BC_3 plants (Frisch et al. 1999a). For example, allocating a total of $n=300$ plants such that 100 plants are generated in each of generations BC_1 to BC_3 (ratio $n_1:n_2:n_3=1:1:1$) resulted in a lower 10% percentile of the recurrent parent genome (Q10) of 97.4%, while various ratios from 3:2:1 on the one extreme to 1:3:9 on the other resulted in Q10 values of 97.3 or 97.4%. In contrast, employing a large population size in generation BC_1 multiplied the number of marker data points required for the marker-assisted backcrossing program. For example, only 2650 marker data points were required for $n_1:n_2:n_3=1:3:9$, while 5000 or even 7250 marker data points were required for ratios of 1:1:1 and 3:2:1, respectively.

The constant recurrent parent genome values for different ratios $n_1:n_2:n_3$ are in contrast to what is expected in multi-stage selection for a quantitative character. There, large populations in early generations are advantageous, because when high selection intensity is applied, a large selection gain is expected due to the large segregation variance. However, in marker-assisted backcrossing the increase in recurrent parent genome is not only driven by selection, but also by the backcross process itself. It is to be expected that backcrossing reduces the donor genome content by one half in each generation, irrespective of the amount present in the nonrecurrent parent. This implies that the selection gain attained in a certain backcross generation is halved by each additional backcross. Only the selection gain attained in the last backcross generation is fully recovered in the final product of the backcross program. Consequently, if high selection pressure (i.e., selection of one individual from a large population) is applied at the beginning of a marker-assisted backcrossing program, then a high absolute value for the selection gain is reached, but it is halved with each additional backcross generation. In contrast, if high selection pressure is applied in advanced backcross generations the selection gain is smaller, but the rate of recovering it in the final product of the breeding program is greater. A compensation between both effects explains why the ratio of dividing a constant number of individuals amongst the backcross generations hardly influences the recurrent parent genome in the plants selected at the end of a marker-assisted backcrossing program.

In generation BC_1, all markers are analyzed at the plants carrying the target gene; an approximation of the required number of marker data points is $mn_1/2$. With no marker-assisted selection, in each subsequent backcross generation the number of heterozygous markers is expected to be halved, therefore, a rough approximation of the portion of markers which are still heterozygous in generation BC_s, is $mn_s/2^s$. With marker-assisted background selection, the actual number of heterozygous markers which need to be analyzed in generation BC_s is below this approximation, because in addition to the effect of the backcrossing per se, homozygosity is increased by marker-assisted selection. These approximations illustrate that the expected portion of markers which need to be analyzed is greater in early than in advanced backcross generations. Therefore, large populations in early generations of a marker-assisted backcross program require more marker data points than large populations in late backcross generations. In conclusion, increasing the population size reduces the number of marker data points in a two-stage selection program compared to applying constant population sizes, but reaches comparable percentages of recurrent parent genome in the final breeding product.

Saving marker data points by using small BC_1 populations can be successfully applied without a linkage map of the markers. If linkage information is available, a sequential three- or four-stage selection strategy is another option to increase efficiency of a marker-assisted background selection program.

A three-stage selection strategy, consisting of one foreground selection step and two background selection steps, can be conducted as follows (Frisch et al. 1999a): after preselecting all individuals carrying the target gene, these are analyzed for the two markers flanking the target gene. On the basis of the result of this analysis, a selection index is constructed for each individual, taking the value 2, if both flanking markers are fixed for the recurrent parent allele and the value 1, if one out of the two flanking markers is fixed for the recurrent parents' allele. If both flanking markers are still heterozygous, the index takes the value 0. Subsequently, all individuals for which this selection index takes the largest observed value are analyzed for the remaining $m-2$ markers. Out of these individuals, the one carrying the recurrent parent allele at the largest number of markers is selected as parent for the next backcross generation.

An additional selection step extends three-stage selection to four-stage selection. After preselecting the individuals having the best selection index with respect to the flanking markers, these individuals are analyzed for all markers on the carrier chromosome of the target gene. Then a selection index is constructed, reflecting the number of markers on the carrier chromosome which were fixed for the recurrent parent allele. The plants for which this selection index takes the largest observed value are analyzed for the markers on the remaining chromosomes and the one carrying the recurrent parent allele at most markers is selected as parent for the next backcross generation.

In a simulated backcross experiment with maize, using 100 plants per backcross generation, two-stage selection reached a Q10 value of 97.4% in generation BC_3 while three- and four-stage selection reached Q10 values 97.2 and 96.8%, respectively (Frisch et al.1999a). However, while for two-stage selection 5430 marker data points were required, three-stage and four-stage selection required only 1810 and 1390 marker data points, respectively. These results show that three- and four-stage selection provide an option which significantly reduces the number of marker data points required compared to two-stage selection, with only a minor reduction of the recurrent parent genome percentage reached.

3 Introgression of a Recessive Gene

Introgression of a recessive gene by recurrent backcrossing without the aid of molecular markers requires progeny tests in each backcross generation in order to determine whether a plant is a heterozygous carrier of the recessive gene or not. In addition to background selection, molecular markers can be used to indirectly select for the target gene such that progeny tests are not required in each backcross generation, but only at the end of the backcross program. In this section, we focus on the carrier chromosome of the target gene and outline a strategy to answer the following questions which describe a two-generation backcross program for introgression of a recessive gene by applying combined foreground and background selection (Frisch and Melchinger 2001a): (1) What is the necessary population size n_1 in BC_1? (2) Suppose the marker genotypes of the n_1 BC_1 individuals are known. Which marker genotypes g and how many individuals i_g of each should be selected as parents for further backcrossing? (3) What should be the size f_g of a BC_2 family produced from a selected BC_1 individual of genotype g?

We consider a chromosome with a sequence of loci at positions y_l, m_l, x, m_r, y_r. The target locus is located at position x and two flanking markers, used for foreground selection, are located at positions m_l and m_r. Two markers located at positions y_l and y_r are used for background selection. We denote z^- as the genotype of an individual homozygous for the recipient allele and z^+ as the genotype of a heterozygous individual at the locus at position $z \in \{y_l, m_l, x, m_r, y_r\}$.

To calculate the population size for generation BC_1, we define a set G of multi-locus genotypes with respect to map positions y_l, m_l, m_r, y_r, which comprises the marker genotypes of all plants considered as possible parents for generation BC_2. It contains all multi-locus marker genotypes with at least one background selection marker homozygous for the recurrent parent allele and at least one heterozygous foreground selection marker. Marker genotypes with no foreground selection marker carrying the donor allele are not included in G because with high probability they do not carry the target gene. Likewise, genotypes with only heterozygous background selection markers are excluded from G because, with respect to the goal of reducing the donor genome around the target gene, they show no improvement compared with F_1 individuals.

To calculate the selection parameters for generation BC_2, we define a set T of multi-locus genotypes with respect to map positions y_l, m_l, m_r, y_r comprising the marker genotypes of all plants, which are considered as a successful outcome of generation BC_2. It contains all multi-locus marker genotypes with both background selection markers homozygous for the recurrent parent allele and at least one heterozygous foreground selection marker. Marker genotypes with no foreground selection marker carrying the donor allele are not included in T because with high probability they do not carry the target gene. Likewise, genotypes with only one homozygous background selection marker are excluded because the goal of reducing the donor genome around the target gene has not been reached.

Using these definitions, the population size for generation BC_1 can be determined by inserting into Eq. (1) the probability

$$p = \sum_{g \in G} p_{0,g+} \qquad (4)$$

that a BC_1 individual belongs to the set of genotypes G and carries the target gene. For each marker genotype $g \in G$, the probability $p_{0,g+}$ that a BC_1 individual has marker genotype g and carries the target gene is given in Table 3.

After producing the BC_1 generation and analyzing the plants with markers at positions y_l, m_l, m_r, y_r, the selection parameters for producing generation BC_2 need to be determined. We denote o_g as the number of individuals with genotype g observed in BC_1, i_g as the number of BC_1 individuals with genotype g used for further backcrossing, and f_g as the size of a BC_2 family produced from a BC_1 individual with genotype g. A certain parameter setting for generating the BC_2 generation, consisting of the number of individuals i_g to be backcrossed and the respective family size f_g for each marker genotype g, is denoted by S. The set of all admissible parameter settings, denoted by A, is determined by the following three conditions: (1) $0 \leq i_g \leq o_g$ for all $g \in G$, i.e., the number of selected individuals of genotype g cannot exceed the number of observed individuals, (2) $0 \leq f_g \leq m$ for all $g \in G$, i.e., the number of progenies generated from one plant cannot exceed the maximum possible family size m (which can be determined either by the multiplication rate of the species or the resources of the breeder), and (3) $q(S) \leq q_2$, i.e., the desired probability of success q_2 must be reached by the parameter combination S.

The probability $q(S)$ of recovering at least one BC_2 plant of marker genotype $t \in T$ carrying the target gene when using the parameter setting S is calculated as

$$q(S) = 1 - \prod_{g \in G} [1 - q_g(i_g, f_g)], \qquad (5)$$

where $q_g(i_g, f_g)$ is the probability of finding among the i_g backcross families of size f_g at least one carrier of the target gene with genotype $t \in T$

Table 3. Formulas to calculate the probabilities $p_{0,g+}$ (probability that a BC$_1$ individual has marker genotype g and carries the target gene), $p_{g+|0,g}$ (probability that a BC$_1$ individual with marker genotype g carries the target gene), and $p_{g+,T+}$ (probability that a BC$_1$ individual with marker genotype g, which carries the target gene, generates a BC$_2$ individual with marker genotype $t \in T$, which carries the target gene.)

Marker genotype $g \in G$	$p_{0,g+}$	$p_{g+|0,g}$	$p_{g+,T+}$
$y_l^- m_l^+ m_r^+ y_r^-$ [a]	$p_b(1-p_c)(1-p_a)p_e/2$ [b]	$(1-p_c)(1-p_a)(1-p_h)$	$(1-p_c p_a)/2$
$y_l^- m_l^- m_r^+ y_r^-$	$(1-p_b)p_c(1-p_a)p_e/2$	$p_c(1-p_a)/p_h$	$(1-p_d)/2$
$y_l^- m_l^+ m_r^- y_r^-$	$p_b(1-p_c)p_d(1-p_e)/2$	$(1-p_c)p_d/p_h$	$(1-p_c)/2$
$y_l^+ m_l^+ m_r^+ y_r^-$	$(1-p_c)(1-p_c)(1-p_a)p_e/2$	$(1-p_c)(1-p_a)/(1-p_h)$	$[p_b(1-p_c)+(1-p_b)p_c(1-p_d)]/2$
$p_b(1-p_c)(1-p_a)(1-p_e)/2$	$(1-p_c)(1-p_a)(1-p_h)$	$[(1-p_a)p_e+(1-p_c)p_d(1-p_e)]/2$
$y_l^- m_l^+ m_r^+ y_r^+$	$p_b p_c(1-p_a)p_e/2$	$p_c(1-p_a)/p_h$	$p_a(1-p_a)/2$
$y_l^+ m_l^- m_r^+ y_r^-$	$(1-p_b)(1-p_c)p_d(1-p_e)/2$	$(1-p_c)p_d/p_h$	$p_b(1-p_c)/2$
$y_l^- m_l^+ m_r^- y_r^+$	$(1-p_b)p_c(1-p_a)(1-p_e)/2$	$p_c(1-p_a)/p_h$	$(1-p_d)p_d/2$
$y_l^+ m_l^- m_r^- y_r^-$	$p_b(1-p_c)p_d p_e/2$	$(1-p_c)p_d/p_h$	$(1-p_c)p_f/2$

[a] The symbols y_l and y_r denote the background selection markers, m_l and m_r the foreground selection markers, and x the target locus. A superscript + or − indicates that the locus is heterozygous or homozygous for the recurrent parent allele, respectively.

[b] The probabilities p_a to p_h are the recombination frequencies between the loci delimiting the intervals $[y_l, x]$, $[y_l, m_l]$, $[m_l, x]$, $[x, m_r]$, $[m_l, y_r]$, $[x, y_r]$, $[y_l, m_r]$, and $[m_l, m_r]$, respectively. They can be obtained from $p = (1-e^{-2d})/2$ by inserting the corresponding map distance d between the loci delimiting the interval

$$q_g(i_g, f_g) = \sum_{s=1}^{i_g} [B(i_g, s, p_{g+|0,g})\{1 - B(sf_g, 0, p_{g+,T,+})\}]. \tag{6}$$

For each marker genotype $g \in G$, the probabilities $p_{g+|0,g}$ and $p_{g+,T+}$ are given in Table 3. $p_{g+|0,g}$ denotes the probability that a BC_1 individual with marker genotype g carries the target gene. $p_{g+,T+}$ denotes the probability that a BC_1 individual with marker genotype g, which carries the target gene, generates a BC_2 individual with marker genotype $t \in T$, which carries the target gene.

$B(n, m, p) = \binom{n}{m} p^m (1-p)^{n-m}$ is the probability function of the binomial distribution. If a particular genotype occurs with probability p, the number m of individuals of this type in a sample of size n is binomially distributed with probability $B(n,m,p)$.

The number of individuals required for the parameter setting S is

$$n_2(S) = \sum_{g \in G} i_g f_g \tag{7}$$

and the optimum parameter setting S^* is the one requiring the smallest number of individuals among all elements in A.

$$n_2(S^*) = \min_{S \in A} n_2(S). \tag{8}$$

There is no closed analytical solution for the minimization problem in Eq. (8). To find a suitable parameter setting, we propose to calculate the probability of success $q(S)$ for various parameter settings S and choose the one which is an element of A and requires the smallest number of individuals.

4 Introgression of Two Dominant Genes

Alternative breeding schemes exist for the simultaneous introgression of two genes into the genetic background of an inbred line (Frisch and Melchinger 2001a). They differ in the generation in which a plant carrying both target genes is generated for the first time. The two genes can be merged into one individual before starting the backcross program by crossing the donors of the target genes and using the resulting F_1 as the nonrecurrent parent for backcrossing. Alternatively, the two genes can be introgressed in two separate branches of the breeding program into the recipient and only when the introgression is finished after t generations of backcrossing, the two converted BC_t individuals are crossed in order to merge the target genes. Between these two extremes, breeding plans for a t-generation backcross program can be applied, in which the target genes are merged into one individual in generation BC_s ($s<t$). These alternative breeding plans differ with respect to: (1) the minimum population size required for finding, with a given probability of

success, carriers of both target genes in different types of populations and (2) the selection intensity which has an effect on the percentage of the recurrent parent genome reached and the number of marker data points required in the backcross program.

The minimum population size required to recover carriers of both target genes depends on the degree of linkage between them and whether they are in coupling or repulsion phase linkage in the crossing or selfing parent. The required population size can be calculated by inserting the respective probabilities given by Frisch and Melchinger (2001b) into Eq. (1). Special attention is required for breeding programs in which linked target genes are merged into one individual by crossing two BC_t plants, followed by a selfing generation to generate homozygous carriers of the target genes. In the selfing parents, the target genes occur in repulsion phase, i.e., they are located on different homologous chromosomes, one originating from the male, the second from the female BC_t plant. To generate a plant which carries both target genes homozygous, it is therefore required that recombination between the target genes occurs during the formation of both parental gametes. The probability $p=(1-r)^2/4$ that such a plant occurs is lower, the tighter linkage is. This results in large populations being required for tightly linked target genes.

When two unlinked target genes are merged into one plant before the first backcross generation and a total of n individuals are generated per backcross generation, then about $n/4$ plants are expected to be subjected to marker-assisted background selection. In contrast, when each target genes is introgressed in a separate branch of the breeding program with a population size of $n/2$, then out of the n plants employed in total for a certain generation, about $n/2$ are expected to be subjected to marker-assisted background selection. In consequence, the intensity of selection for the recurrent parent genome in a breeding plan in which the target genes are merged in a later generation is greater than in a breeding plan with early merging of the target genes. The greater selection intensity is accompanied by greater values of the recurrent parent genome reached, but also by larger numbers of marker data points required. This was demonstrated numerically in a simulation study based on a model of the maize genome (Frisch and Melchinger 2001b).

5 Length of the Intact Donor Chromosome Segment Around the Target Gene

For recurrent backcrossing with selection for the presence of a target gene, the expected length of the intact donor chromosome segment attached on one side of the target gene was derived by Hanson (1959) as $(1-e^{-tl})/t$, where t is the number of backcrosses carried out and l the map distance between the target gene and the end of the chromosome. Stam and Zeven (1981) extended

Hanson's approach and derived the expected donor genome content on the carrier chromosome. Their approach includes chromosome segments not directly attached to the target gene and averages over all possible map positions of the target gene on the chromosome. For background selection with exactly two markers on the carrier chromosome of the target gene, Hospital et al. (1992) extended the approach of Stam and Zeven (1981) and determined numerically the expected donor genome content on the target chromosome.

The probability distribution of the intact donor chromosome segment around the target gene in backcrossing with selection for the presence of the target gene and selection for the recipient alleles at flanking markers was investigated by Hospital (2001) and Frisch and Melchinger (2001c). For various situations relevant for practical backcross programs in plant breeding, they derived density functions, expectations, and the variances of the lengths of the attached donor chromosome segment.

A numerical illustration of their results presented in Table 4 shows the expected length $E(X)$ of the intact chromosome segment attached to one side of the target gene in generations BC_t ($t=1...5, 6, 8, 10, 15$) for backcross programs with and without background selection at flanking markers in generation BC_1. The target locus is positioned at distance $l=1.0$ M from the chromosome end and the flanking marker is located at distance 0.1, 0.2, 0.3, 0.4, 0.5 M from the target locus. In generation BC_1, the expected length of the intact chromosome segment is 0.25 M, when selecting for a flanking marker at 0.5 M distance. Without marker-assisted selection, a value of 0.24 M is reached only in generation BC_4. With an increasing number of backcrosses, the differences between applying background selection and not applying

Table 4. Expected length $E(X)$ of the intact chromosome segment attached on one side of the target gene in generations BC_s ($s=1...5, 6, 8, 10, 15$) in the absence of marker-assisted selection (none) and with selection for recombinants at a flanking marker at varying distances ($d=0.1,...0.5$ M) in generation BC_1. The map distance between the target gene and the telomere is 1 M

s		Flanking marker distance (M)				
	None	0.5	0.4	0.3	0.2	0.1
			$E(X)[M]$			
1	0.63	0.24	0.20	0.15	0.01	0.05
2	0.43	0.21	0.17	0.14	0.09	0.05
3	0.32	0.18	0.15	0.12	0.09	0.05
4	0.25	0.16	0.14	0.11	0.08	0.05
5	0.20	0.14	0.12	0.10	0.08	0.04
6	0.17	0.12	0.11	0.10	0.07	0.04
8	0.12	0.10	0.09	0.08	0.07	0.04
10	0.10	0.09	0.08	0.07	0.06	0.04
15	0.07	0.06	0.06	0.05	0.05	0.03

background selection becomes smaller. However, an expected length of the attached chromosome segment of 0.05 M, as reached in generation BC_1 with a flanking marker distance of 0.1 M, is not reached even after 15 backcross generations without background selection.

These results show that selection for recombinants between the target gene and a flanking marker is highly effective even when the marker is fairly distant from the target gene. For example, a saving of three backcross generations concerning the expected length of the linked chromosome segment is realized with a marker distance of 0.5 M. Because recombinants between the target gene and fairly distant flanking markers occur with a high probability even in small backcross generations (Frisch et al. 1999b), marker-assisted background selection can be used to avoid large intact donor chromosome segments around the target gene, even with limited resources for the population size and marker analyses.

Acknowledgment The author thanks A.E. Melchinger for helpful discussions and comments on the manuscript.

References

Allard RW (1960) Principles of plant breeding. Wiley, New York
Fehr WR (1987) Principles of cultivar development, vol 1. Theory and technique. Macmillan, New York
Frisch M, Melchinger AE (2001a) Marker-assisted backcrossing for introgression of a recessive gene. Crop Sci 41:1485–1494
Frisch M, Melchinger AE (2001b) Marker-assisted backcrossing for simultaneous introgression of two genes. Crop Sci 41:1716–1725
Frisch M, Melchinger AE (2001c) The length of the intact chromosome segment around a target gene in marker-assisted backcrossing. Genetics 157:1343–1356
Frisch M, Bohn M, Melchinger AE (1999a) Comparison of selection strategies for marker-assisted backcrossing of a gene. Crop Sci 39:1295–1301
Frisch M, Bohn M, Melchinger AE (1999b) Minimum sample size and optimal positioning of flanking markers in marker-assisted backcrossing for transfer of a target gene. Crop Sci 39:967–975, Erratum: Crop Sci 39:1903
Frisch M, Bohn M, Melchinger AE (2000) Plabsim: software for simulation of marker-assisted backcrossing. J Hered 91:86–87
Hanson WD (1959) Early generation analysis of lengths of heterozygous chromosome segments around a locus held heterozygous with backcrossing or selfing. Genetics 44:833–837
Hospital F (2001) Size of donor chromosome segments around introgressed loci and reduction of linkage drag in marker-assisted backcross programs. Genetics 158:1363–1379
Hospital F, Charcosset A (1997) Marker-assisted introgression of quantitative trait loci. Genetics 147:1469–1485
Hospital F, Chevalet C, Mulsant P (1992) Using markers in gene introgression breeding programs. Genetics 132:1199–1210
Ragot M, Biasiolli M, Delbut MF, Dell'Orco A, Malgarini L, Thevenin P, Vernoy J, Vivant J, Zimmermann R, Gay G (1995) Marker-assisted backcrossing: a practical example. In: INRA (ed) Techniques et utilisations des marqueurs moléculaires. Montepellier, France, 29–31 March 1994

Servin B, Hospital F (2002) Optimal positioning of markers to control genetic background in marker-assisted backcrossing. J Hered 93:214–217

Stam P, Zeven AC (1981) The theoretical proportion of the donor genome in near-isogenic lines of self-fertilizers bred by backcrossing. Euphytica 30:227–238

Tanksley SD (1983) Molecular markers in plant breeding. Plant Mol Biol Rep 1:1–3

Tanksley SD, Young ND, Patterson AH, Bonierbale MW (1989) RFLP mapping in plant breeding: New tools for an old science. Bio/Technology 7:257–263

Visscher PM (1996) Proportion of the variation in genomic composition in backcrossing programs explained by molecular markers. J Hered 87:136–138

Young ND, Tanksley SD (1989) RFLP analysis of the size of chromosomal segments retained around the *tm*-2 locus of tomato during backcross breeding. Theor Appl Genet 77:353–359

Zeven AC, Knott DR, Johnson R (1983) Investigation of linkage drag in near isogenic lines of wheat by testing for seedling reaction to races of stem rust, leaf rust and yellow rust. Euphytica 32:319–327

III.3 From Theory to Practice: Marker-Assisted Selection in Maize

D.A. Hoisington[1] and A.E. Melchinger[2]

1 Introduction

For thousands of years, farmers and, much later, breeders have improved crops through plant selection. These efforts have been based on accumulating favorable alleles, found in individual plants from diverse origins, through recurrent cycles of recombination and selection. As a result, every improved variety corresponds to a unique collection of thousands of alleles, accumulated over the years, the combination of which explains the plant phenotype.

A long-standing goal of plant scientists has been to create more efficient methods and approaches to plant selection first by better understanding the basis of phenotypic variation and then by developing more efficient selection methodologies. Defining the relationship between the presence of a given allele and a phenotypic response relies on classical genetic analyses. During the last two decades, technological innovations and developments in the area of molecular genetics have greatly spurred the utilization and "power" of genetics.

One of the major innovations has been the ability to detect differences in the DNA between individuals. Such differences, detected by various types of molecular markers, can be used to dissect polygenic traits into their Mendelian components or quantitative trait loci (QTL), thus increasing our understanding of the inheritance and gene action for such traits. When tightly linked to genes of interest, the markers can be used to indirectly select for the desirable allele, and represents the simplest form of marker-assisted selection (MAS), whether used to accelerate the backcrossing of such an allele or in pyramiding several desirable alleles. Molecular markers can also be used to probe the level of genetic diversity among different cultivars, within populations, and among related species. The applications of such evaluations are many, including varietal fingerprinting for identification and protection and understanding relationships among germplasm collections. In addition, markers and comparative mapping of various species have been very valuable

---

[1] International Maize and Wheat Improvement Center, Applied Biotechnology Center and Bioinformatics, Apdo Postal 6-641, COL. Juárez, 06600 Mexico, D.F. Mexico
[2] Institute of Plant Breeding, Seed Science and Population Genetics, University of Hohenheim, 70593 Stuttgart, Germany

for improving our understanding of genome structure and function and have allowed the isolation of genes of interest via map-based cloning.

1.1 The Importance of Maize

While other plant species (e.g., rice and *Arabidopsis*) have been considered model species for modern biotechnology, maize (*Zea mays* L.) has also been a major focus of biotechnology research for several reasons. First, maize is important globally for feeding the world. Maize is the third most important food crop (behind rice and wheat), and is expected to increase in importance as the world's population continues to demand maize as a major food and feed source. Maize has also been the focus of many commercial plant breeding and biotechnology companies. The presence of hybrid technology enables the private sector to capitalize on the sale of hybrid seed and derive benefits from their investments in research and development. This, in turn, presents a major dilemma for maize researchers. Cutting-edge maize biotechnology research is concentrated in a handful of industrialized countries and commercial companies, particularly the United States and Europe. Although some of the technology is being made available (with limitations) for public use, and some research is still being conducted by public organizations, the private sector has much greater resources to devote to research and development of the tools and techniques needed to advance maize improvement.

Maize also offers a number of significant scientific advantages. Classical genetic studies have evolved to the point that the collection of known loci and genetic/cytogenetic stocks are enormous. This, coupled with the ease with which many molecular studies can be accomplished – both genetic and biological – has lead to a wealth of investigations and ultimately, the understanding of the maize genome (Hoisington 1992). Efforts to develop the tools and techniques for expanded identification of genes and gene functions via EST databases, reverse genetics and functional genomics promise to maintain the position of maize as a leading genetic organism, while providing powerful approaches for enhancing maize productivity (Stuber et al. 1999; Coe et al. 2002; Lee et al. 2002).

1.2 Molecular Marker Development

Molecular marker technology has evolved from hybridization-based detection to new sequence-based systems; each having their advantages and disadvantages. Restriction fragment length polymorphisms (RFLPs) were the first to be developed (some 15 years ago) and have been widely and successfully used to construct linkage maps of various species, including many of the major cereals. With the development of the polymerase chain reaction (PCR) technology, several marker types emerged. The first of those were RAPD markers (random amplified polymorphic DNA) that gained popularity due to the simplicity and lower costs of the assay. However, most researchers now realise the weaknesses of RAPDs and use them less frequently. Microsatellite markers or simple sequence repeats (SSRs), combine the power of RFLPs (co-dominant markers, reliable, specific genome location) with the ease of RAPDs and have the advantage of detecting higher levels of polymorphism. The AFLP (amplified fragment length polymorphism) approach takes advantage of the PCR technique to selectively amplify DNA fragments previously digested with one or two restriction enzymes. Altering the number of selective bases of the primers and considering the number of amplification products per primer pair, this approach is certainly powerful in terms of polymorphisms identified per reaction. Most recently, systems that detect single base pair changes (termed single nucleotide polymorphisms, SNPs) are becoming available. While being fairly expensive to develop, requiring sequencing of several alleles, they do detect high levels of polymorphism and can be detected with simple and automated technology.

Maize was one of the first major crop species for which a complete molecular marker map was developed (Helentjaris et al. 1986). Since the first publications, many other maps have been produced and are now consolidated into a consensus map using a 'bin' allocation to chromosome segment (Gardiner et al. 1993). Given the high level of polymorphism found even between highly related lines, this consensus map allows one to rapidly identify possible markers for use in further saturating a region of interest, or for developing alternative (e.g., PCR-based) marker systems. Efforts are underway to develop saturated microsatellite marker maps. Most recently, efforts are focused on sequencing alleles at numerous loci to develop the information necessary for SNP analyses (see http://www.agron.missouri.edu and http://www.cerealsdb.uk.net/cgi-bin/maize-snip.pl).

Since these markers also reveal a high level of polymorphism in maize, all prerequisites are fulfilled to supplement or even substitute conventional phenotypic selection (PS) by marker-assisted selection (MAS). In view of the potential advantages of MAS in plant breeding, it is not surprising that the number of publications on this topic has increased exponentially since 1990. A closer look, however, reveals that most reports were quantitative trait loci (QTL) mapping studies, which often arrive at an optimistic assessment about the prospects of MAS. The objectives of this review are to present an overview about the theoretical foundation of MAS as well as simulations studies and first experimental results on the use of MAS in maize breeding.

1.3 Quantitative Trait Loci Mapping

One of the areas in molecular breeding receiving attention is the mapping of chromosomal regions influencing qualitative or quantitative traits of interest. Polygenic characters that are often difficult to manipulate using conventional phenotypic selection can now be tagged using molecular markers. A QTL can be defined as the location of a gene or a cluster of tightly linked genes affecting a trait, and the position of such QTL are statistically inferred. The basic idea of QTL mapping has been known for nearly four decades (Thoday 1961). If genetic markers are scattered throughout the genome of an organism of interest, the segregation of these markers can be used to directly estimate the effects of linked QTL, making possible the mapping and characterization of underlying QTL. Modern QTL mapping involves searching for associations between the segregating molecular markers and the character of interest in a segregating population. Experimental populations such as F_2, backcross (BC), recombinant inbred lines (RILs), and double haploid (DH) lines are commonly used as mapping populations in plants. In the case of F_2 mapping populations, F_2 plants are usually used to genotype, and F_2 families to phenotype. Near isogenic lines (NILs) are used for fine mapping and determining specific QTL effects. RIL and DH populations are permanent populations that permit replicated evaluation of the phenotype, and are useful for mapping traits that are difficult to measure.

Besides localization of polygenes, QTL mapping also allows estimation of the effects of individual QTL as well as their joint effects (epistasis). QTL mapping has been increasingly employed in recent years, primarily because of the availability of PCR-based markers and powerful statistical packages. A number of methods for mapping QTL and estimating their effects have been suggested and investigated (Edwards et al. 1987; Lander and Botstein 1989; Haley and Knott 1992; Jansen and Stam 1994; Zeng 1994; Jiang and Zeng 1995). Genetic dissection of diverse agronomically important traits in maize using QTL mapping has been carried out (Stuber 1995; Stuber et al. 1999, for reviews); comprehensive information about such experiments in maize can be obtained from the MaizGDB (http://www.maizegdb.org).

1.4 Theoretical Aspects of Marker-Assisted Selection

Dekkers and Hospital (2002) distinguished two approaches for implementation of MAS in breeding. In the first approach, markers are treated as secondary traits and used to calculate a marker index score (Lande and Thompson 1990) based on classical results from selection theory. This score can be used as a selection criterion in both recurrent selection as well as recycling breeding.

The second approach, denoted as "gene stacking" or "genotype construction", aims at combining the favorable alleles from all parents in a single new genotype. If markers tagging the desirable genes or QTL regions are known, they are used as discrete building blocks for a systematic construction of the optimal genotype. Potential areas of application are recycling breeding and introgression programs. Marker-assisted backcrossing for transfer of one or several target genes represents a special case of "genotype construction".

1.4.1 Theory and Simulations of Marker-Assisted Selection Efficiency for One Selection Cycle

QTL analyses provide the necessary prerequisites for calculating marker index scores by detecting QTL-marker associations and estimating QTL effects in segregating populations with high linkage disequilibrium. In its simplest form, the marker index score is calculated as follows:

$$M = \sum_{QTL^*} \hat{a}_i x_i.$$

Here, QTL^* is the set of all detected QTL, $\hat{a}_i$ the estimated additive effect of the ith QTL and x_i ($x_i = 0,1,2$) the number of favorable alleles at the ith locus. In "pure" MAS, M is the only selection criterion. In "combined" MAS (cMAS), the following index is calculated (Lande and Thompson 1990):

$$I = b_m M + b_p P.$$

Here, P denotes the phenotypic value and the optimum weights b_m und b_p have the following ratio:

$$b_m/b_p = (1/h^2 - 1)/(1-p).$$

Comparisons of the efficiency of different selection schemes are usually based on the genetic gain ΔG per cycle. For PS, the well-known formula applies (Hallauer and Miranda 1981):

$$\Delta G_{PS} = i_{PS}\, h\, \sigma_G\, c_{PS}.$$

For MAS and cMAS, the following formulas apply (Lande and Thompson 1990):

$$\Delta G_{MAS} = i_{MAS}\, \sigma_G\, p \text{ and } \Delta G_{MAS} = i_{cMAS}\, \sigma_G\, [p + (1-p)^2/(h^2-p)]^{1/2}.$$

Here, i denotes the selection intensity, σ_G the genetic standard deviation in the primary trait, h^2 the heritability, c_{PS} the parental control, and p the proportion of the genetic variance σ_G^2 explained by the detected QTL-markers associations. Assuming identical selection intensities ($i_{MAS}=i_{PS}$), we obtain for the relative efficiency (RE):

$$RE_{MAS} = \Delta G_{MAS}/\Delta G_{PS} = (p/h^2)^{1/2}/c_{PS} \text{ and}$$
$$RE_{cMAS} = \Delta G_{cMAS}/\Delta G_{PS} = [(p/h^2 + (1-p)^2/(1-h^2 p)]^{1/2}/c_{PS}.$$

Consequently, MAS is superior to PS under the following scenarios (Dudley 1993).

1. Direct evaluation of the trait is either difficult or expensive, and thus, MAS allows a higher selection intensity than PS.
2. The trait can be evaluated only after flowering. Therefore, $c_{PS}=0.5$, but $c_{MAS}=1$, because MAS selection can be performed during the seedling stage.

3. The trait can be evaluated only in the target environment, but not in off-season nurseries. Hence, MAS allows selection during all generations and shortens the cycle length.

The dilemma is that for a given sample size of the mapping population (N), the power of QTL detection decreases with h^2 (Lande and Thompson 1990). Thus, RE is optimal for intermediate values of h^2 depending on N. Moreover, p and RE strongly depend on the genetic architecture of the trait. If only a few QTL explain a large proportion of σ_G^2, then RE is high (Moreau et al. 1998). By comparison, the density of markers has only a minor influence beyond a certain coverage (~20 cM).

The formulas illustrate that p is fundamental for evaluating RE of MAS and cMAS compared with PS. In the literature, p is commonly estimated by $p=R^2/h^2$ (Schön et al. 1994) or $p=R^2_{adj.}/h^2$ (Moreau et al. 1998). It is common practice to estimate R^2 or $R^2_{adj.}$ from the same data set that was used for QTL detection (i.e., in model selection). For small N (<500), this causes a considerable inflation of the estimates (Melchinger et al. 1998). To solve this problem, Utz et al. (2000) proposed cross-validation as a tool for obtaining unbiased estimates of p and, thus, obtaining realistic predictions about the prospects of RE.

Comparisons of cMAS and PS that take into account the additional costs of marker assays were reported by Utz et al. (1994) and Moreau et al. (2000). Given a fixed total budget and the same effective population size (N_e) of the selected candidates, the optimum allocation of resources was calculated for each method. As expected, RE is strongly influenced by the cost ratio R_c=marker costs per genotype/cost per plot. RE of MAS increases with an increasing budget, because only under these circumstances, the additional costs of QTL mapping are counterbalanced by a significantly higher selection response. Likewise, RE increases with decreasing heritability (h^2_{plot}) up to a certain point beyond which QTL detection becomes rather inefficient.

Cycle length is another important criterion. In maize, the cycle length is approximately 6 months for MAS, but 2 years for half-sib selection. With this in mind, Hospital et al. (1997) proposed an alternative breeding scheme, in which one cycle of cMAS (combined with QTL mapping) is alternated by two cycles of "pure" MAS. In simulation studies, they found a considerably higher selection response compared with PS and cMAS.

1.4.2 Simulation Results of Marker-Assisted Selection After Several Selection Cycles

The RE of MAS and cMAS with PS was investigated by several simulation studies (for review see Whittaker 2001). The following conclusions can be drawn from the results. (1) For traits influenced by 10 or fewer QTL, cMAS is superior to PS for about 10–20 generations. However, the long-term selection response is smaller due to a rapid fixation of unfavorable alleles. (2) QTL-marker associations must regularly be re-estimated, because MAS becomes rather ineffective after four to six generations due to a rapid decline of linkage disequilibrium. (3) The biometric method for detection and estimation of QTL-marker associations has only a minor effect on an improvement of cMAS. (4) Likewise, increasing the number of markers beyond a 10-cM density results only in a marginal improvement of cMAS. (5) During the initial cycles, the RE of cMAS compared with PS increases with N and h^2 up to an optimum, which depends on N. (6) The RE of cMAS versus PS is higher for QTL in coupling than in repulsion phase.

1.4.3 Prospects of Marker-Assisted Selection in Recycling Breeding

Van Berloo and Stam (1998) proposed a scheme for implementation of MAS in the framework of "genotype construction" (Fig. 1). Here, the goal is to construct transgressive genotypes, which possess the favorable allele at (nearly) all QTL. In a first step, a segregating population of recombinant inbred lines (RIL) is generated from a bi-parental cross, which after phenotyping and genotyping is used for QTL mapping. With commonly employed sample sizes N, the chances of obtaining the optimum genotype are extremely low, hence, the best candidates are recombined in a further round. With PS, only the phenotypic value P of the candidates can be chosen as selection criterion, whereas with MAS or cMAS, in addition to the marker index score I, one can choose the candidates such that they complement each other with regard to the positive alleles at different loci. As criterion for comparing both schemes, van Berloo und Stam (1998) used the genotypic value of the best F_2 progeny in the segregating generation of the second cycle relative to the mean of the unselected RIL population. In simulation studies, they found a superiority of MAS compared to PS, which decreased with increasing h^2 and uncertainties in the estimated QTL-marker associations.

1.4.4 Experimental Results on the Efficiency of Marker-Assisted Selection in Recycling Breeding

Following the procedure in Fig. 1, van Berloo and Stam (1999) compared the efficiency of MAS versus PS for flowering date, a highly heritable trait in *Arabidopsis*. They detected a total of eight QTL in the cross Landsberg × Columbia and conducted divergent selection for early and late flowering. Deviating from the expected superiority of MAS based on simulation results, both procedures were about equally effective disregarding the costs.

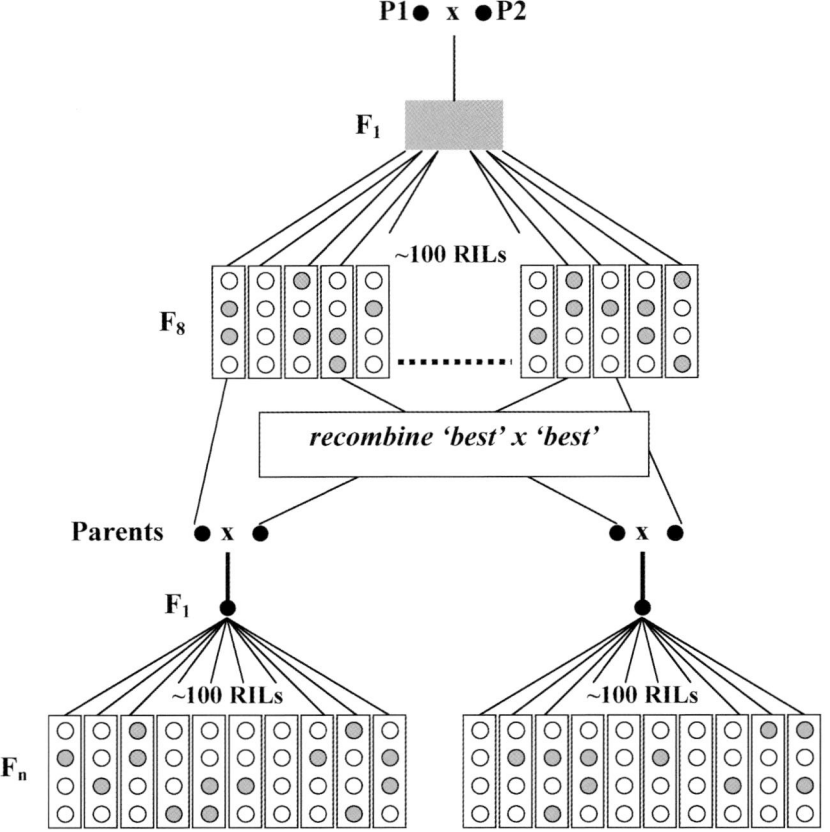

Fig. 1. Schematic representation of the procedure proposed by van Berloo and Stam (1998) for "genotype construction" in recycling breeding

2 Practical Examples of Marker-Assisted Selection

While a number of theoretical and simulation studies have been conducted, there are also a few examples of MAS being applied in a practical sense. These have evaluated the use of MAS to transfer single genes or QTLs for a single trait from a donor line into one or more recipients lines. While many more examples, involving selection of multiple traits, will be needed before broad statements about the general effectiveness of MAS can be made, these do provide concrete examples of how MAS can be applied and the results of such. Those known to the authors will be briefly described. The reader is referred to the actual publication for further details.

2.1 Transgene Backcross-Marker-Assisted Selection

One of the first practical demonstrations of the use of MAS involved the backcrossing of a Bt transgene into a commercial maize variety (Ragot et al. 1994). RFLPs were used to select for the transgene and maximal recurrent parent genome at each generation. After four backcrosses (BC_3) and in less than 2 years, the lines were determined to be equivalent to those normally produced by conventional backcrossing at the BC_6 generation. From a commercial perspective, such a rapid conversion of elite varieties is very attractive.

2.2 Marker-Assisted Selection for Improvement of Insect Resistance in Maize

Willcox et al. (2002) conducted an experimental comparison of MAS and PS for improving the resistance against various stem borers in tropical maize (Fig. 2). MAS was based on mapping of a backcross population between a susceptible and resistant inbred line using a total of 277 BC_1S_1 families, in which three QTL were detected. Following MAS, three BC_2S_2 families were developed, which carried all three target regions of the donor in homozygous state, but had otherwise only a small fraction of the donor genome. In paral-

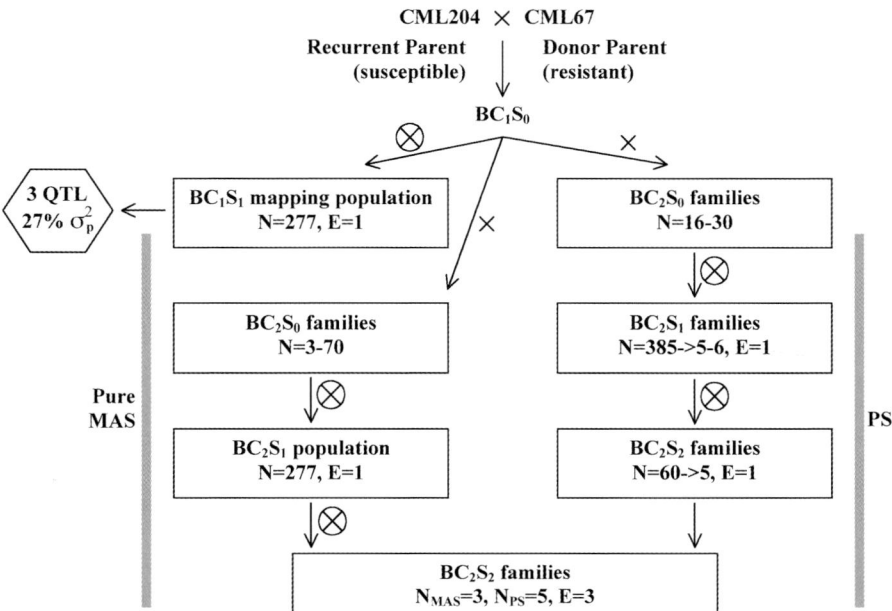

Fig. 2. Breeding scheme of Willcox et al. (2002) for PS and MAS for improvement of insect resistance in tropical maize. N Sample size, E number of environments

lel, PS with artificial infestation was carried out at one location per generation. The comparison of three MAS- and five PS-selected families yielded a significant improvement for leaf damage ratings, with some advantages of PS over MAS. As expected, lines selected by MAS showed a significantly higher proportion of the genome from the recurrent parent (78.3%) than those selected by PS (70.4%).

2.3 Marker-Assisted Selection for Improvement of Important Traits in Sweet Corn

Yousef und Juvik (2001) reported a comparison of MAS and PS for improvement of seedling emergence and important quality traits (kernel sucrose content, kernel tenderness) in three crosses of sweet corn. After QTL mapping in three populations of $F_{2:3}$ lines, five markers were chosen for the traits or combination of traits. Besides divergent selection with MAS and PS, a randomly chosen sample was taken as a control. Each of these fractions was intermated for generating the next breeding cycle (Fig. 3). In all instances, MAS was superior over PS for traits selected either individually or simultaneously (Fig. 4).

2.4 Marker-Assisted Selection for Grain Yield and Earliness in Elite Maize Germplasm

Recently, Bouchez et al. (2002) reported the marker-assisted introgression of favorable alleles at three QTL for earliness and grain yield in the cross of two elite maize lines ($F_2 \times$ Io; Fig. 5). Based on QTL mapping for testcross perfor-

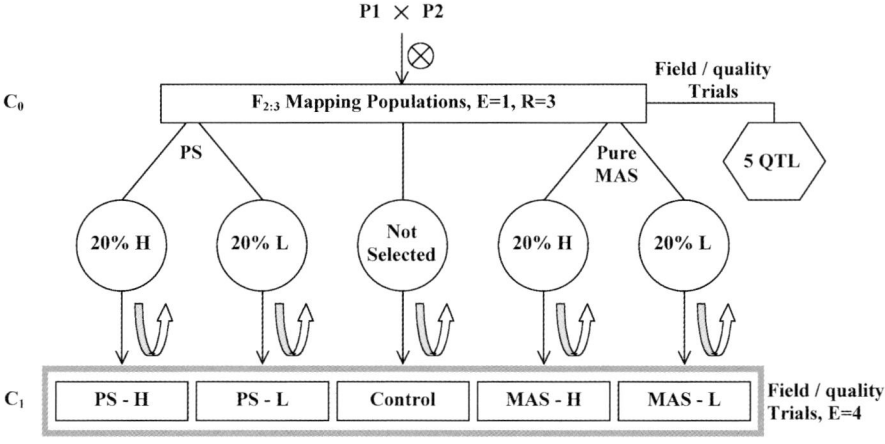

Fig. 3. Breeding scheme of Yousef and Juvik (2001) for comparing the efficiency of PS and MAS with selection for high (*H*) or low (*L*) seedling emergence or quality traits of sweet corn

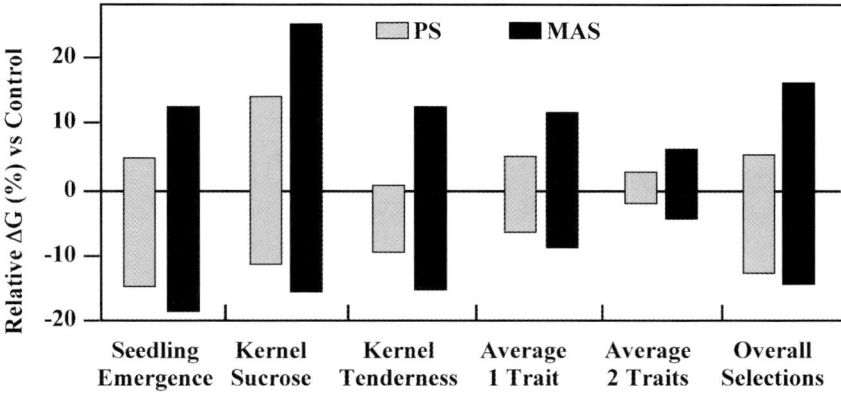

Fig. 4. Observed efficiency of PS and MAS with high (*H*) and low (*L*) selection (relative to the unselected control) for seedling emergence, kernel sucrose content and kernel tenderness. (Yousef and Juvik 2001)

mance of 96 RILs, they detected several QTL regions and chose three of them with fairly large effects on earliness and/or grain yield for MAS. One RIL containing all favorable QTL alleles was backcrossed twice to the one of the parents (Io). MAS was performed to (1) transfer the target QTL regions from the other parent (F_2), (2) recover the Io genome on the noncarrier chromosomes, and (3) reduce the linkage drag attached to the target QTL regions. At the end of the selection program, the effects of introgression were re-evaluated phenotypically in field trials and QTL estimation was repeated with 217 BC_3S_1

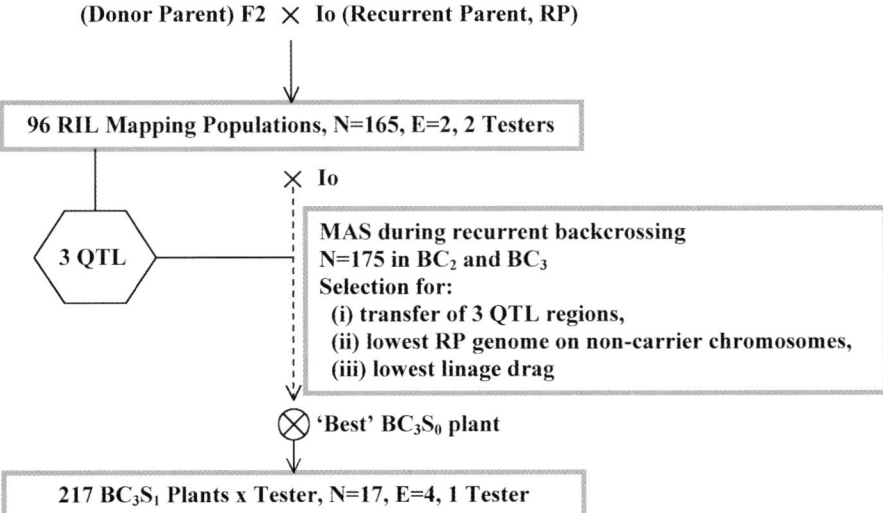

Fig. 5. Scheme of Bouchez et al. (2002) for QTL mapping and MAS for grain yield and earliness in the cross of two elite maize inbreds ($F_2 \times$ Io)

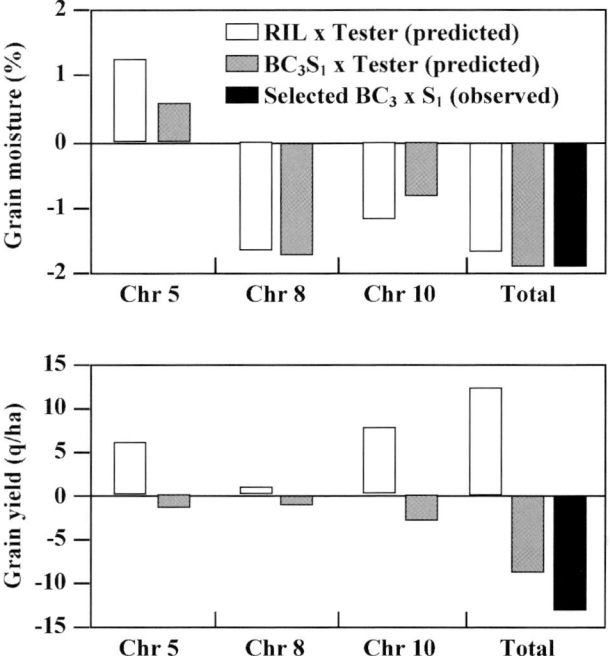

Fig. 6. Effects of three QTL regions on testcross performance for grain yield and earliness predicted from the original mapping population of 96 RILs (*white columns*), estimated from a mapping population of 215 BC$_3$S$_1$ families (*shaded columns*) and the observed performance of a selected BC$_3$S$_1$ line (*black columns*). (Bouchez et al. 2002)

(testcross) families. Clearly, MAS accelerated recovery of the recurrent parent genome. QTL positions and effects originally determined in the RIL population could be confirmed only for earliness (Fig. 6). However, for grain yield, important discrepancies were found in the magnitude and sign of QTL after introgression in generation BC$_3$S$_1$ as compared with the QTL effects estimated from the initial QTL analysis of RILs. The authors explained these discrepancies by genotype × environment interactions and epistatic interactions with the genetic background. In conclusion, these results support the findings of Melchinger et al. (1998) in that estimated QTL positions and effects should be regarded with caution for complex traits, especially for small values of N. Thus, if the mapping population is not sufficiently large enough for obtaining reliable QTL estimates, the final result of MAS can be rather disappointing.

2.5 Drought

Research at the International Maize and Wheat Improvement Center (CIMMYT) for the past few years has focused on the genetic dissection of drought tolerance in maize. Following the QTL mapping of an initial $F_{2:3}$ population between a drought-tolerant and -susceptible inbred (Ribaut et al. 1996, 1997), a BC-MAS project was initiated. The drought-tolerant parent in the mapping study was used as the donor for backcrossing drought-tolerant QTL alleles into a susceptible, but good combiner and high yielding, elite inbred (CML247). Five genomic regions were transferred using flanking PCR-based markers. Following two BCs and two self-pollinations (Fig. 7), the 70 best BC_2F_3 lines were identified and crossed with two testers, CML254 and CML274. The hybrids and the selected lines were evaluated under various water regimes for 3 years in CIMMYT's field station in Tlaltizapan, Mexico. The results indicated that the mean yield under drought stress of the 70 MAS lines was significantly higher than the controls. In addition, the best MAS lines were two to four times higher yielding under drought. No yield reductions were observed under nonstress conditions, and a few MAS lines performed better than the controls even under nonstress conditions.

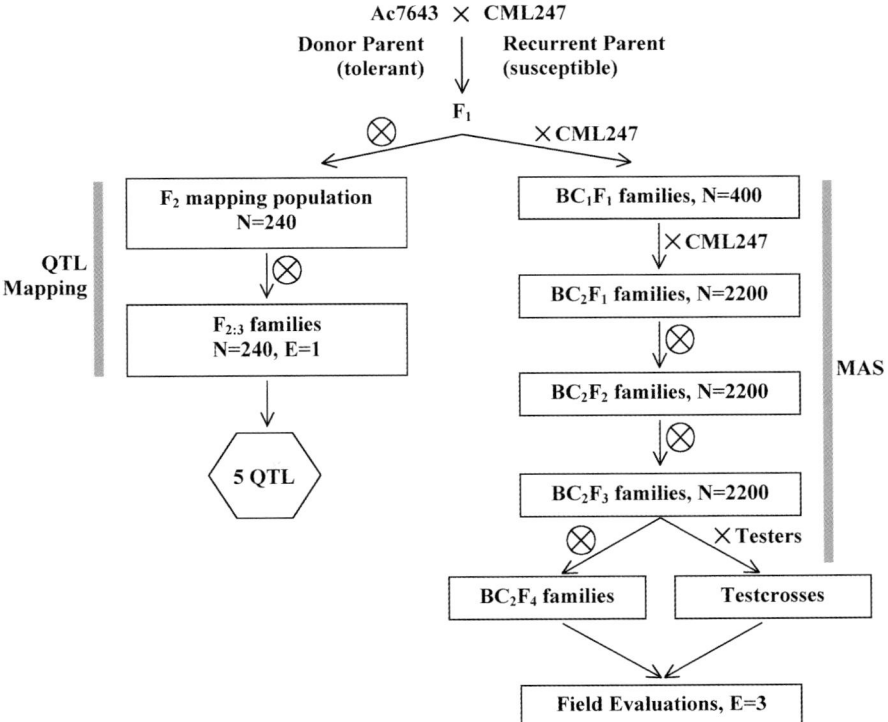

Fig. 7. Scheme of Ribaut et al. (2002) for QTL mapping and BC-MAS for transferring drought tolerance to the elite inbred CML247

3 Economics of Marker-Assisted Selection

Very few studies, either theoretical or actual, have been reported on the economics of MAS as compared to PS. Ragot and Hoisington (1993) presented one of the first economic analyses of MAS, comparing the relative costs of RFLP and RAPD markers. More recently, reports that have calculated the technical efficiency of MAS relative to PS have concluded that it is difficult to make meaningful comparisons without empirical data (Moreau et al. 2000; Yu et al. 2000). One factor that limited the usefulness of these previous studies has been the lack of actual cost data of the PS program.

Dreher et al. (2003) presented the results of a detailed cost analysis study of the field and laboratory procedures associated with conventional (PS) and MAS breeding. The studies were based on strategies and actual costs employed by the maize breeding program at CIMMYT in Mexico. The conversion of lines to quality protein maize (QPM) was used as a specific case study. The results indicated that MAS (using SSR markers) was cost-effective compared to PS for QPM and that adoption of new screening procedures would reduce the cost differential between MAS and PS. The authors concluded that such detailed budget analyses are critical in determining the cost effectiveness of MAS and that such analyses are extremely useful in making informed decisions about the choice of methods.

In a companion paper, Morris et al. (2003) compared the cost effectiveness of various PS strategies and MAS for introgressing a single dominant gene into an elite maize line (genotype conversion). Using costs at CIMMYT in Mexico, neither method showed clear superiority in terms of both cost and speed. PS schemes were less expensive, but MAS required less time. Thus, which method is most cost-effective will depend greatly on the availability of operating capital in the breeding program. If operating capital is abundant, MAS would be most cost-effective; if capital is limited, PS would be the method of choice.

The above studies indicate that there is no single, nor simple, answer as to when to apply MAS versus PS in a breeding program. The answer will require a detailed analysis of the specific field and laboratory costs. The overall cost-effectiveness of MAS will depend on four major parameters: (1) the relative cost of phenotypic versus genotypic screening, (2) the time savings achieved by MAS, (3) the benefits associated with an accelerated release of improved germplasm, and (4) the amount of operating capital available to the breeding program.

4 New Marker-Assisted Selection Strategies

It is not surprising that using an indirect selection tool such as a molecular marker linked to a QTL results in the successful transfer of the QTL (and associated phenotype) to a new variety. However, MAS is still limited by the current requirement to map the QTLs whenever a new source of germplasm is involved in the MAS program. This greatly increases the costs and time, thus decreasing the efficiency and usefulness of MAS in a breeding program.

Two strategies have been proposed to overcome these limitations. One strategy involves selecting plants at an early generation that possess a fixed, desirable genotype at specific loci by conducting a single, large-scale, marker-assisted selection (SLS-MAS; Ribaut and Hoisington 1998; Ribaut and Betrán 1999). MAS is conducted only once and only for the specific target loci, thus greatly reducing the costs and resources required. In addition, since the strategy employs crosses of elite by elite germplasm and no selection outside the target loci, the resulting selected lines contain the best alleles from both parents while maintaining optimal variation. Breeders can then continue to select the best lines while maintaining the target trait without further MAS. CIMMYT molecular geneticists are in the final stages of conducting SLS-MAS for drought tolerance in tropical maize with the national programs in Kenya and Zimbabwe. The only limitation of this approach is the requirement to map the QTL to be selected.

A second approach to overcome the requirement for re-mapping QTLs is to develop a drought consensus map for the trait(s) of interest (Ribaut et al. 2002). The strategy is based on the observation that certain genes form clusters within the maize genome (Khavkin and Coe 1997). By combining into a single consensus map all information regarding the genetic location of multiple components involved in a trait, for example, drought tolerance, it may be possible to identify genomic regions that are common across a range of germplasm and traits. If such regions can be identified, MAS for these regions could be used directly without requiring the construction of linkage maps. It is quite possible that not all common regions will be required in each cross, but the savings from not having to develop a new linkage map will greatly outweigh the small costs of performing MAS for these regions.

5 Conclusions

Maize is extremely rich in the number and type of molecular markers available. These are being used to map many genes and traits of interest. Thus, some of the early predictions of the usefulness of molecular markers have already been demonstrated. Theoretical studies, simulations and experimental results show that the efficiency of MAS and its relative efficiency compared with phenotypic selection depend primarily on the following factors: (1) the heritability of the trait, (2) the population size of the mapping population employed in QTL mapping, (3) the genetic architecture of the trait, (4) the relative costs of marker assays compared to phenotypic trait evaluation, and (5) the total budget of the breeding program. By comparison, marker coverage of the genome beyond a certain marker density and the biometric procedure employed for QTL mapping are only of secondary importance. While MAS for complex traits like yield have been mostly disappointing, encouraging results have been reported for quantitative traits with a simpler genetic basis and for qualitative tolerance to abiotic and biotic stresses. Hence, the choice of suitable traits and the appropriate strategy (Ribaut et al. 2002) is crucial for the prospects to successfully integrate MAS in practical breeding programs. Furthermore, integration of new MAS strategies into ongoing breeding programs is essential for bridging the gap between QTL mapping in special populations and transfer of these results to applied maize breeding.

References

Bouchez A, Hospital F, Caussee M, Gaillais A, Charcosset A (2002) Marker-assisted introgression favorable alleles at quantitative trait loci between maize elite lines. Genetics 162:1945–1959

Coe E, Cone K, McMullen M, Chen S, Davis G, Gardiner J, Liscum E, Polacco M, Paterson A, Sanchez-Villeda H, Soderlund C, Wing R (2002) Access to the maize genome: an integrated physical and genetic map. Plant Physiol 128:9–12

Dekkers JCM, Hospital F (2002) The use of molecular genetics in the improvement of agricultural populations. Nat Genet Rev 3:22–32

Dreher K, Khairallah M, Ribaut J-M, Morris M (2003) Money matters (I): costs of field and laboratory procedures associated with conventional and marker-assisted maize breeding at CIMMYT. Mol Breed 11:221–234

Dudley JW (1993) Molecular markers in plant improvement: Manipulation of genes affecting quantitative traits. Crop Sci 33:660–668

Edwards MD, Stuber CW, Wendel JF (1987) Molecular-marker-facilitated investigations of quantitative trait loci in maize. I. Numbers, genomic distribution, and types of gene action. Genetics 116:113–125

Gardiner J, Melia-Hancock S, Hoisington DA, Chao S, Coe EH (1993) Development of a core RFLP map in maize using an immortalized-F2 population. Genetics 134:917–930

Haley CS, Knott SA (1992) A simple regression method for mapping quantitative trait loci in line crosses using flanking markers. Heredity 69:315–324

Hallauer AR, Miranda JB (1981) Quantitative genetics in maize breeding. Iowa State University Press, Ames

Helentjaris T, Slocum M, Wright S, Schaefer A, Nienhuis J (1986) Construction of genetic linkage maps in maize and tomato using restriction fragment length polymorphisms. Theor Appl Genet 72:761–769

Hoisington D (1992) Maize as a model system. In: Chapman GP (ed) Grass evolution and domestication. Cambridge Univ Press, London

Hospital F, Moreau L, Lacoudre F, Charcosset A, Gallais A (1997) More on the efficiency of marker-assisted selection. Theor Appl Genet 95:1181–1189

Jansen RC, Stam P (1994) High resolution of quantitative traits into multiple loci via interval mapping. Genetics 136:1447–1455

Jiang C, Zeng Z-B (1995) Multiple trait analysis of genetic mapping for quantitative trait loci. Genetics 140:1111–1127

Khavkin E, Coe E (1997) Mapped demonic locations for developmental functions and QTLs reflect concerted groups in maize (*Zea mays* L.). Theor Appl Genet 95:343–352

Lande R, Thompson R (1990) Efficiency of marker-assisted selection in the improvement of quantitative traits. Genetics 124:743–756

Lander ES, Botstein D (1989) Mapping Mendelian factors underlying quantitative traits using RFLP linkage maps. Genetics 121:185–199

Lee J-M, Williams ME, Tingey SV, Rafalski JA (2002) DNA array profiling of gene expression changes during maize embryo development. Funct Integr Genom 2:13–27

Melchinger AE, Utz HF, Schön CC (1998) QTL mapping using different testers and independent population samples in maize reveals low power of QTL detection and large bias in estimates of QTL effects. Genetics 149:383–402

Moreau L, Charcosset A, Hospital F, Gallais A (1998) Marker-assisted selection efficiency in populations of finite size. Genetics 148:1353–1365

Moreau L, Lemarié S, Charcosset A, Gallais A (2000) Economic efficiency of one cycle of marker-assisted selection. Crop Sci 40:329–337

Morris M, Dreher K, Ribaut J-M, Khairallah M (2003) Money matters (II): costs of maize inbred line conversion schemes at CIMMYT using conventional and marker-assisted selection. Mol Breed 11:235–247

Ragot M, Hoisington DA (1993) Molecular markers for plant breeding: comparisons of RFLP and RAPD genotyping costs. Theor Appl Genet 86:975–984

Ragot M, Biasiolli M, Delbut MF, Dell'orco A, Malgarina L, Thevenin P, Vernoy J, Vivant J, Zimmermann R, Gay G (1994) Marker-assisted backcrossing: a practical example. In: Bervillé A, Tersac M (eds) Techniques et utilisations des marqueurs moléculaires. Les colloques 72. INRA, Versailles, France, pp 45–56

Ribaut J-M, Hoisington DA (1998) Marker-assisted selection: new tools and strategies. Trends Plant Sci 3:236–239

Ribaut J-M, Betrán J (1999) Single large-scale marker-assisted selection (SLS-MAS). Mol Breed 5:531–541

Ribaut J-M, Hoisington DA, Deutsch JA, Jiang C, González-de-León D (1996) Identification of quantitative trait loci under drought conditions in tropical maize I. Flowering parameters and the anthesis-silking interval. Theor Appl Genet 92:905–914

Ribaut J-M, Jiang C, González-de-León D, Edmeades GO, Hoisington DA (1997) Identification of quantitative trait loci under drought conditions in tropical maize II. Yield components and marker-assisted selection strategies. Theor Appl Genet 94:887–896

Ribaut J-M, Jiang C, Hoisington D (2002) Simulation experiments on efficiencies of gene introgression by backcrossing. Crop Sci 42:557–565

Schön CC, Melchinger AE, Boppenmaier J, Brunklaus-Jung E, Herrmann RG, Seitzer JF (1994) RFLP mapping in maize: Quantitative trait loci affecting testcross performance of elite European flint lines. Crop Sci 34:378–389

Stuber CW (1995) Mapping and manipulating quantitative traits in maize. Trends Genet 11:477–481

Stuber CW, Polacco M, Senior ML (1999) Synergy of empirical breeding, marker-assisted selection, and genomics to increase yield potential. Crop Sci 39:1571–1583

Thoday JM (1961) Location of polygenes. Nature 191:368–370

Utz HF, Schön CC, Melchinger AE (1994) Markergestützte Selektion auf Qualitätsmerkmale mittels RFLP in einem Körnermaisexperiment. Arbeitstagung der Arbeitsgemeinschaft der Saatzuchtleiter in Gumpenstein 1993, pp 69–74

Utz HF, Melchinger AE, Schön CC (2000) Bias and sampling error of the estimated proportion of genotypic variance explained by QTL determined from experimental data in maize using cross validation and validation with independent samples. Genetics 154:1839–1849

Van Berloo R, Stam P (1998) Marker-assisted selection in autogamous RIL populations: a simulation study. Theor Appl Genet 96:147–154

Van Berloo R, Stam P (1999) Comparison between marker-assisted selection and phenotypical selection in a set of *Arabidopsis thaliana* recombinant inbred lines. Theor Appl Genet 98:113–118

Whittaker JC (2001) Marker-assisted selection and introgression. In: Balding DJ, Bishop M, Cannings C (eds) Handbook of statistical genetics, Wiley, New York, pp 673–693

Willcox MC, Khairallah MM, Bergvinson D, Crossa J, Deutsch JA, Edmeades GO, González-de-León D, Jiang C, Jewell DC, Mihm JA, Williams WP, Hoisington D (2002) Selection for resistance to Southwestern Corn Borer using marker-assisted and conventional backcrossing. Crop Sci 42:1516–1528

Yousef GG, Juvik JA (2001) Comparison of phenotypic and marker-assisted selection for quantitative traits in sweet corn. Crop Sci 41:645–655

Yu K, Park SJ, Poysa V (2000) Marker-assisted selection of common bean for resistance to common bacterial blight: efficacy and economics. Plant Breed 119:411–415

Zeng ZB (1994) Precision mapping of quantitative trait loci. Genetics 136:1457–1468

III.4 Molecular Markers for Disease Resistance: The Example Wheat

C. Feuillet and B. Keller[1]

1 Introduction

Bread wheat (*Triticum aestivum* L.) is attacked by a large number of different pathogen species. Many diseases caused by these pathogens result in severe reductions of yield and a decreased bread-making quality. Depending on the specific environment of wheat production, the relevant pathogens and the extent of the damage can vary. The wheat pathogens include obligate biotrophic fungi such as leaf rust and powdery mildew, necrophytic species such as *Stagonospora nodorum* as well as nematodes and viruses. The wheat gene pool contains a large variety of resistant germplasm against specific pathogens and improving disease resistance is a highly relevant goal in many breeding programs. Naturally occurring resistance, particularly complete resistance against biotrophic pathogens, is frequently based on single, race-specific resistance (*R*) genes which can be durable, but are frequently broken by new, virulent pathogen races. A second type of resistance is called partial resistance: it is quantitative, depends on two or more genes and it is usually durable. In quantitative resistance, the genes which show relatively major effects are particularly interesting. Recently, several quantitative trait loci (QTL) with large effects have been described, e.g., for Fusarium resistance (a quantitative trait locus on 3BS; Zhou et al. 2002) or for resistance against leaf rust (*Lr34*; Nelson et al. 1997; Schnurbusch et al. 2004).

Given the difficulty and the considerable resources needed to score for disease resistance in wheat breeding programs, molecular markers for specific traits can significantly contribute to an improved efficiency of resistance breeding. In the last decade, a large number of such markers have been developed. International collaborations in the Wheat Microsatellite Consortium (WMC) and the International Triticeae EST Cooperative (ITEC) have greatly contributed to the development of markers. Closely linked markers are particularly useful for marker-assisted selection (MAS) of resistance traits that are difficult to score and which are genetically based on a single gene. In wheat, selection for cereal cyst nematode resistance based on the *cre1* and *cre3* resistance loci has become an integral part of some Australian wheat breeding programs (Ogbonnaya et al. 2001a). Markers are equally well suited

---

[1] Institute of Plant Biology, University of Zürich, Zollikerstrasse 107, 8008 Zürich, Switzerland

for the pyramiding of resistance genes, e.g., against leaf rust or powdery mildew diseases for which many different resistance genes are available. Increasingly, markers have also been identified for genomic regions contributing to quantitative resistance. Most markers for such regions have not been extensively tested in breeding material and their applicability outside the population where they were identified remains to be determined. In general, the validation of the markers in a large gene pool for wrong positive or wrong negative results has been a neglected area of research. However, there is an increased awareness of this problem (Sharp et al. 2001). In addition to application in MAS, markers closely linked to a particular gene are a prerequisite for map-based cloning. The recent isolation of the fungal disease resistance genes *Lr10* (Feuillet et al. 2003), *Lr21* (Li et al. 2003) and *Pm3b* (Yahiaoui et al. 2004) has shown that map-based cloning is now feasible from hexaploid wheat.

Excellent reviews have recently been published on the use of molecular markers for wheat breeding (Gupta et al. 1999; Langridge et al. 2001) and a list of molecular markers for specific genes can be found in the gene catalogue of (McIntosh et al. 1998) and its updates (http://wheat.pw.usda.gov/). Here, we focus on the specific markers available for wheat disease resistance breeding, the status of their practical application as well as the problems of MAS in resistance breeding.

2 Development of Molecular Markers

2.1 First Generation of Genetic Markers

Different technologies have been used to develop markers for *R* genes in wheat. A very limited number corresponds to morphological markers such as leaf chlorosis or pseudo black chaff for *Sr2* (Brown 1997) or leaf tip necrosis for *Lr34* and *Yr18* (Singh 1992a, b). This type of marker is usually of limited use as it is often affected by environmental conditions or developmental growth stages. However, in some cases phenotypic markers such as the leaf tip necrosis associated with *Lr34* and *Yr18* have been intensively used in selection for durable resistance in wheat (Rajaram et al. 1988). A number of biochemical markers, particularly isoenzymes, have also been developed as markers for leaf rust and eyespot disease resistance genes in wheat (McMillin et al. 1986; Winzeler et al. 1995). However, this type of marker requires protein extraction, is labor-intensive and not well adapted to automation and high-throughput analysis for breeding.

Molecular markers are based on the detection of polymorphisms in the DNA sequence. Their number is theoretically almost unlimited and, they are not affected by environmental conditions and plant growth stages. The first molecular markers for disease resistance genes in wheat were developed at the beginning of the 1990s and they were mainly restriction fragment length polymorphism (RFLP) markers. A large number of RFLP probes has been generated from *T. aestivum* and *Ae. tauschii* libraries (Liu et al. 1990; Anderson et al. 1992; Devos and Gale 1993); for more details see the review of Gupta et al. (1999). They have been extensively used to establish genetic maps (for review, see Langridge et al. 2001) and the development of markers for agronomically important traits in wheat. For many years, marker development in wheat has relied on RFLP probes directly originating from wheat. This has resulted in low-density maps which did not always allow the efficient development of markers for target genes. The situation dramatically improved with the discovery of the conservation of the marker order (colinearity) at the genetic map level on homoeologous chromosomes of grass genomes (Moore et al. 1995; Keller and Feuillet 2000). As colinearity mainly concerns genes and most of the RFLP probes correspond to cDNAs, RFLP markers from one species have been successfully used for mapping in other grass species. This has allowed the increase of genetic map densities at resistance gene loci in wheat and the development of tightly linked markers. To date, more than 36 RFLP markers have been developed for monogenic pest and disease resistance genes in wheat (Tables 1, 2) and they still represent a large part of the markers used to identify QTLs for quantitative disease resistance (Table 3).

Table 1. Markers for fungal disease resistance genes in wheat

1. Leaf rust (*Puccinia triticina* Eriks.)

Gene	Location	Source	Linkage[a] comments	Marker	Reference
Lr1	5DL	T. aestivum	<1 cM	RFLP/STS, RGA[b]	Feuillet et al. (1995); Ling et al. (2002)
Lr3	6BL	T. aestivum	<1 cM	RFLP	Sacco et al. (1998)
Lr9	6BL	Ae. umbellulata	introgression	RFLP/STS	Schachermayr et al. (1994); Autrique et al. (1995)
Lr10	1AS	T. aestivum	<1 cM	RFLP/STS	Schachermayr et al. (1997); Feuillet et al. (2003)
Lr13	2B	T. aestivum	10 cM	RFLP/SSR	Seyfarth et al. (2000)
Lr18	1BL	T. timopheevi	Introgression	N-Band	Yamamori (1994)
Lr19	7DL	Ae. elongatum	Introgression	Endopeptidase, RFLP/STS	Autrique et al. (1995); Winzeler et al. (1995); Prins et al. (2001)
Lr20	7AL	T. aestivum	<1 cM, introgression?	RAPD/STS	Neu et al. (2002)
Lr21/Lr40	1DS	Ae. tauschii	<1 cM	RFLP/STS	Spielmeyer et al. (2000b); Huang and Gill (2001); Li et al. (2003)
Lr23	2BS	T. turgidum	20 cM	RFLP	Nelson et al. (1997)
Lr24	3DL	Ae. elongatum	Introgression	SCAR/STS	Autrique et al. (1995); Schachermayr et al. (1995); Dedryver et al. (1996)

Gene	Location	Source	Linkage[a] comments	Marker	Reference
Lr25	1BS	S. cereale	Introgression	RAPD	Procunier et al. (1995)
Lr27/Lr31	3BS/4BL	T. aestivum	Complementary genes	RFLP	Nelson et al. (1997)
Lr28	4AL	T. speltoides	Introgression	RAPD/STS	Naik et al. (1998)
Lr29	7DS	Ae. elongatum	Introgression	RAPD	Procunier et al. (1995)
Lr32	3DS	Ae. tauschii	3.6 cM	RFLP	Autrique et al. (1995)
Lr35	2B	T. speltoides	<1 cM	RFLP/STS	Seyfarth et al. (1999)
Lr37	2AS	Ae. ventricosa	Introgression	RFLP/STS	Bonhomme et al. (1995); Seah et al. (2001)
Lr39 (=Lr41)	2DS	Ae. tauschii	10.7 cM	SSR	Raupp et al. (2001)
Lr47	7A	T. speltoides	Introgression	RFLP/CAPS	Dubcovsky et al. (1998); Helguera et al. (2000)

2. Yellow rust (*Puccinia striiformis* Westend.)

Gene	Location	Source	Linkage[a] comments	Marker	Reference
Yr5	2B	T. spelta album	<1 cM	RGA[b]	Yan et al. (2003)
Yr7	2BL	T. durum	<3 cM	AFLP	Bariana et al. (2001)
Yr9	1B/1R	Rye	1 cM	RGA[b]	Shi et al. (2001)
Yr10	1B	T. aestivum	<1 cM	RGA[b]/SSR	Spielmeyer et al. (2000a); Bariana et al. (2002); Smith et al. (2002); Wang et al. (2002)
Yr10vav		T. vavilovii	1.5%	SSR	
YrMoro		T. aestivum	<1 cM	STS	
Yr15	1B	T. dicoccoides	2.6 cM	RAPD/RFLP, SSR	Sun et al. (1997); Peng et al. (2000a)
Yr17	2A	Ae. ventricosa	Introgression	RAPD/RFLP, RFLP/STS	Robert et al. (1999); Seah et al. (2001)
Yr26	1B	T. turgidum	1.9 cM	SSR	Ma et al. (2001)
YrH52	1B	T. dicoccoides	1.3 cM	SSR	Peng et al. (2000b)
Yrns-B1	3BS	T. aestivum	20.5 cM	SSR	Borner et al. (2002)

3. Stem rust (*Puccinia graminis* Pers.: Pers.)

Gene	Location	Source	Linkage[a] comments	Marker	Reference
Sr2	3BS	T. dicoccum	6.9 cM	Leaf chlorosis, pseudo-black chaff, RFLP/STS	Rajaram et al. (1988); Brown (1997); Sharp et al. (2001); Spielmeyer et al. (2003)
Sr22	7A	T. monococcum	Introgression	RFLP	Paull et al. (1994)
Sr30	5DL	T. aestivum	>10 cM	AFLP/RFLP	Bariana et al. (2001)
Sr36	2B	T. timopheevii	<1 cM	SSR	Bariana et al. (2001)
Sr38	2AS	Ae. ventricosa	Introgression	RFLP/STS	Seah et al. (2001)

4. Powdery mildew (*Erysiphe graminis* DC. f. sp. *tritici* Em. Marchal)

Gene	Location	Source	Linkage[a] comments	Marker	Reference
Pm1	7AL		<1 cM	RAPD/RFLP, AFLP	Ma et al. (1994); Hartl et al. (1995); Hu et al. (1997); Neu et al. (2002)
Pm2	5DS		3.5 cM	RFLP	Ma et al. (1994); Hartl et al. (1995); Mohler and Jahoor (1996)
Pm3	1AS		1.3 cM	RFLP, SSR	Hartl et al. (1993); Ma et al. (1994); Bougot et al. (2002); Yahiaoui et al. (2004)
Pm4	2AL	*T. dicoccum/ T. cartlicum*	<1 cM	RFLP, RAPD, AFLP	Ma et al. (1994); Hartl et al. (1999)
Pm5e	7BL	*T. aestivum*	6.6 cM	SSR	Huang et al. (2003)
Pm6	2BL	*T. timopheevii*	1.6 cM	RFLP	Tao et al. (2000)
Pm8/Pm17	1BL	Rye 1R	For 1R translocation	STS, AFLP	Mohler et al. (2001, 2002)
Pm12	6B	*Ae. speltoides*	Introgression	RFLP	Jia et al. (1996)
Pm13	3DS	*Ae. longissima*	Introgression	RFLP/STS	Cenci et al. (1999)
Pm18			4.4 cM	RFLP	Hartl et al. (1995)
Pm21	6AL	*H. villosa*	Introgression	RAPD, SCAR	Qi et al. (1996); Liu et al. (1999)
Pm24	1D	*T. aestivum*	2.4 cM	SSR, AFLP	Huang et al. (2000)
Pm25	1A	*T. monococcum*	12.8 cM	RAPD	Shi et al. (1998)
Pm26	2BS	*T. dicoccoides*	<1 cM	RFLP	Rong et al. (2000)
Pm27	6B	*T. timopheevii*	<1 cM	SSR	Jarve et al. (2000)
Pm29	7D	*Ae. ovata*	<1 cM	RFLP, AFLP	Zeller et al. (2002)
Pm30	5BS	*T. dicoccoides*	5–6 cM	SSR	Z.Y. Liu et al. (2002)

5. Common bunt (*Tilletia tritici*), loose smut (*Ustilago tritici* (Pers) Rostr.), eyespot (*Pseudocercosporella herpotricoides* (Fron)

Gene	Location	Source	Linkage[a] comments	Marker	Reference
Bt-10		*T. aestivum*	1 cM	RAPD, STS	Demeke et al. (1996); Laroche et al. (2000)
T10	2BS		10 cM	SCAR	Procunier et al. (1997)
Pch-1	7DL	*Ae. ventricosa*	Introgression	Ep-D1b, RFLP	McMillin et al. (1986)
Pch-2	7AL	*T. aestivum*	11 cM	RFLP	delaPena et al. (1997)

[a] Linkage <1 cM corresponds to markers either completely linked or less than 1 cM. When several studies were performed on the same gene, the closest marker distance is indicated
[b] Markers derived from resistance gene analogs (RGA)

Table 2. Markers for resistance against nematodes, pests and viruses

1. Nematodes: Cereal cyst nematode (*Heterodera avenae*), root lesion nematode (*Pratylenchus neglectus*), root-knot nematode (*Meloidogyne naasi*)

Gene	Location	Source	Linkage[a] comments	Marker	Reference
Cre1	2B	T. aestivum	<1 cM	RFLP/STS	Williams et al. (1994, 1996)
Cre3	2DL		<1 cM	RAPD, RGA[b]	Eastwood et al. (1994); Lagudah et al. (1997)
Cre5	2AS	Ae. ventricosa	Linked to Yr17, Lr37, Sr38		Jahier et al. (2001)
Cre6	5 N^v	Ae. ventricosa		RGA[b]	Ogbonnaya et al. (2001b)
Rlnn1	7AL		9.1 cM	RFLP	Williams et al. (2002)
Rkn-mn1	3BL	Ae. variabilis	<1 cM	RAPD	Dweikat et al. (1994, 1997); Barloy et al. (2000)

2. Insects: Hessian fly (*mayetiola destructor*), Russian wheat aphid [*Diuraphis noxia* (Mordvilko)]

Gene	Location	Source	Linkage[a] comments	Marker	Reference
H3, H6, H9, H10, H12, H16, H17	5A	T. aestivum	<1 cM	RAPD	Dweikat et al. (1994)
H5, H11, H13, H14	1A				Dweikat et al. (1997)
H21		T. aestivum	<1 cM	RAPD	Seo et al. (1997)
H23	6D	Ae. tauschii	6.9 cM	RFLP	Ma et al. (1993)
H24	3D		5.9 cM		
Dn1	7D	T. aestivum	3.8 cM	SSR	X.M. Liu et al. (2001)
Dn2	7D	T. aestivum	2.8 cM	RFLP, SCAR, SSR	Ma et al. (1998); Myburg et al. (1998); X.M. Liu et al. (2001); Miller et al. (2001)
Dn5	7D	T. aestivum	<3.2 cM	SCAR, SSR	Venter and Botha (2000); X.M. Liu et al. (2001)
Dn4	1DS	T. aestivum	7.4 cM	RFLP	Ma et al. (1998);
Dn6	7DS		3 cM	SSR	X.M. Liu et al. (2002)
Dn8	7DS	T. aestivum	<3.2 cM	SSR	X.M. Liu et al. (2001)
Dn9	1DL		<3.2 cM		

3. Viruses: barley yellow dwarf virus, wheat spindle streak mosaic bymovirus, wheat streak mosaic virus

Gene	Location	Source	Linkage[a] comments	Marker	Reference
BYDV (Bvd2)	7D	*Ae. intermedium*	Introgression	RAPD/SCAR, SSR	Ayala et al. (2001); Stoutjesdijk et al. (2001)
WSSMV	2DL	*T. aestivum*	79%	RFLP	Khan et al. (2000)
Wms1	4D	*Ae. intermedium*	Translocation	RAPD/STS	Talbert et al. (1996)

[a] Linkage <1 cM corresponds to markers either completely linked or less than 1 cM. When several studies were performed on the same gene, the closest marker distance is indicated
[b] Markers derived from resistance gene analogs (RGA)

Table 3. Markers for quantitative disease resistance in wheat

QTL	Location	Contribution (R^2)	Marker	Reference
Leaf rust	1B, 7B	18–34	RAPD/RFLP	William et al. (1997)
		14–34		
	1B, 3A, 4B, 4D, 7B			Messmer et al. (2000)
Lr34	7DS	16–42	RFLP/SSR	Nelson et al. (1997); Suenaga (2002); Schnurbusch et al. (2004)
		52		
Lr46	1BL	45	SSR/AFLP	Suenaga (2002); William et al. (2003)
		49		
Fusarium head blight[a]	3BS	15–56	RFLP/AFLP/ SSR/STS	For review: Anderson et al. (2001); Kolb et al. (2001) Borner et al. (2002); Buerstmayr et al. (2002); Zhou et al. (2002); Gervais et al. (2003); Guo et al. (2003)
	2A			
	6B			
	5A			
	7A, B			
Powdery mildew	5A, 7B, 3D	15–32	RFLP, SSR	Keller et al. (1999)
	5D	20–39		Chantret et al. (2000)
	1B, 2A, 2B	11–29		S.X. Liu et al. (2001)
Yellow rust	7D	15	SSR/RFLP/ AFLP	Bariana et al. (2001)
	1B			
	2D			
Yr29	1BL	24–31	AFLP	William et al. (2003)
Yr18	7DS, linked to Lr34	32	RFLP	Singh et al. (2000)
Yr 28	4DS		RFLP	Singh et al. (2000)
Stem rust	3B		SSR/RFLP/ AFLP	Bariana et al. (2001)
	6A			

[a] Only major QTLs that have been found in more than one study are given for FHB resistance

However, RFLPs are not well suited for marker-assisted selection as they are labor-intensive, time-consuming, require large amounts of DNA and often have to be radioactively labeled. In addition, because they only rely on sequence differences in restrictions sites, they show a limited amount of polymorphism in wheat. For these reasons, they are not used routinely in marker-assisted selection programs.

2.2 Polymerase Chain Reaction-Based Markers: High-Throughput for Marker-Assisted Selection

The discovery of the polymerase chain reaction (PCR) has revolutionized the development of molecular markers because it only requires very low amounts of DNA which can be rapidly extracted from different plant material with high-throughput methods (Kang et al. 1998; Paris and Carter 2000; Stein et al. 2001). The first type of PCR-based markers were random amplified polymorphic DNA (RAPD) markers. Their main advantage is that RAPDs do not require any knowledge of the target sequence as single random primers of 9–10 mers are used for PCR. To date, 17 RAPDs have been developed for wheat disease *R* genes (Tables 1–3). However, the low temperature of amplification used in this technique makes RAPD markers not very robust and difficult to reproduce in different laboratories using different thermocyclers. For these reasons, in many cases sequence tagged site (STS) markers, which correspond to the specific amplification of a target DNA sequence at stringent temperature, have been derived either from low-copy RFLP or from RAPD markers.

The second generation of PCR-based markers consisted of microsatellites or simple sequence repeats (SSR) and amplified fragment length polymorphisms (AFLP). SSRs which comprise short repeat units of 1–6 nucleotides are very abundant and dispersed throughout the genome. AFLPs which combine the use of restriction site polymorphisms and specific PCR amplifications have a high multiplex ratio compared to the other marker systems. Both techniques benefit from the advantages of PCR and have a higher marker index (calculated on the information content and multiplex ratio of the marker) than RFLP and RAPD (Powell et al. 1996). Microsatellites, which are mainly codominant, are more robust than RAPD and easier to transfer between populations than AFLPs. The main disadvantage of SSRs resides in their high developmental costs which cannot be supported by every laboratory. However, initiatives such as the Wheat Microsatellite Consortium have helped to share the costs and to provide SSRs developed in different labs to the entire wheat community. Nineteen SSR and seven AFLP markers have been published for single pest and disease *R* genes in wheat mainly in the last 3 years (Tables 1, 2). Combined with RFLP, they represent major sources of markers to establish the genetic maps necessary for QTL analysis for disease resistance in wheat (Table 3).

2.3 Markers for Single Traits vs. Markers to Dissect Complex Traits

So far, most of the markers have been developed for monogenic *R* genes (Tables 1, 2), which when used as single genes are not very durable in the field. Even if molecular markers allow the combination of several single *R* genes (pyramidization) to increase durability, breeders are most interested in targeting quantitative forms of resistance. Quantitative resistance can either be due to the combined action of several minor genes (QTLs) such as the resistance to Fusarium head blight or to single loci which are strongly influenced by the environment such as the slow rusting genes *Lr34* and *Lr46*. The increase in the type and number of molecular markers in wheat in the last decade has allowed a more efficient dissection of quantitative disease resistances into single QTLs. This is demonstrated by the release of more than 20 publications since 2000 for QTLs for disease resistance in wheat (Table 3). Quantitative resistance is very difficult to select for in conventional breeding programs. With the comparison of QTLs obtained in different environments and populations, it is now possible to identify major QTLs which can be targeted by markers and integrated in MAS schemes.

2.4 Markers Derived by Homology with Known Plant Disease Resistance Genes

Since 1998, a number of plant *R* genes have been cloned (Richter and Ronald 2000; Hulbert et al. 2001). A majority of them belong to the NBS-LRR class and contain short conserved domains. So-called resistance gene analogs (RGAs) have been isolated from wheat and barley using degenerated primers corresponding to very conserved regions (Ploop, kinase2 and GLPLAL) within or close to the NBS domain (Leister et al. 1998; Seah et al. 1998; Spielmeyer et al. 1998; Collins et al. 2001), or using the resistance gene analog polymorphism (RGAP) technique (Chen et al. 1998; Shi et al. 2001; Yan et al. 2003). A number of these RGAs were mapped at known disease resistance loci and represent good markers for wheat disease resistance genes. In a few cases, RFLP or RAPD markers associated with wheat *R* genes were also found to correspond to RGAs (Lagudah et al. 1997; Frick et al. 1998; Huang and Gill 2001; Ling et al. 2002). So far, six of the RGAs are used as markers for *R* genes in wheat (Tables 1–3).

2.5 Third Marker Generation Derived from Large-Scale Analysis

Single nucleotide polymorphisms in allelic sequences have recently been investigated as a new source of markers in plant breeding, especially in maize (Rafalski 2002). SNPs can be discovered either by sequencing a number of PCR products amplified from specific target sequences in different genotypes or by in silico analysis of genomic or cDNA sequences. In wheat, a tremendous effort of cDNA sequencing has been undertaken and coordinated by the International Trititiceae EST Cooperative (ITEC; http://wheat.pw.usda.gov/genome/) in the last 3 years. This has resulted in the best collection of expressed sequence tags (ESTs) among plants, with more than 549,000 ESTs in the public database to date (http://www.ncbi.nlm.nih.gov/dbEST/dbEST--summary.html). These sequences represent a valuable resource for SNP detection as well as for SSR detection which are also found in noncoding regions of cDNAs. SNPs can be used as single genetic markers which may be identified in the vicinity of target genes, but there is also a great potential in using the association of SNP haplotypes with particular traits such as disease resistance. Association studies can be of particular help in analyzing quantitative traits with a much higher resolution than QTL analysis performed in F_2 populations or recombinant inbred lines (Buckler and Thornsberry 2002; Rafalski 2002). Different technologies are now available to assay SNPs (Langridge et al. 2001) and there is no doubt that in the future such markers will also be integrated in molecular breeding programs.

3 Use of Molecular Markers in Marker-Assisted Selection for Disease Resistance

To date, molecular markers have been developed for more than 85 different monogenic disease resistance genes and for five different quantitative disease resistances. A key question lies in the validity of these markers for MAS. Langridge et al. (2001) have distinguished different steps in the validation of a marker for MAS. First of all, markers which are often developed in one population have to be tested in other populations originating from crosses with one parent of the original mapping population. In many cases, the markers developed in one or two germplasms will not be found in others originating from different breeding programs. Therefore, it is very important to saturate the resistance locus with about ten markers within a genetic distance of less than 10 cM (ideally with markers located at less than 1 cM from the target genes). This increases the chance of finding at least one marker useful for any breeding population. For these reasons, the high level of SSR polymorphism makes them ideal for MAS. So far, very few markers have been thoroughly tested in practical breeding programs. Recently, Sharp et al. (2001) have evaluated the usefulness of four sets of markers for important rust resistance genes. Validation was done using other mapping populations than the original one and a wide range of International Maize and Wheat Improvement Center (CIMMYT) and Australian breeding lines. Except in one case where linkage was not sufficient, the analysis showed that the published markers are robust and valid for MAS. Another example of the effort done for the application of markers in breeding programs is given by the National Wheat Marker

Assisted Selection Consortium, a USDA founded project which includes 12 wheat-breeding and research programs across the US. This project aims at bringing "genomics to the wheat field" and provides protocols on line for markers for mosaic viruses, Hessian fly and rust resistance genes (see link at http://maswheat.ucdavis.edu/protocols/protocols.htm).

4 Conclusions

Despite these efforts, there is generally a lack of reports concerning MAS and the usefulness of molecular markers in breeding. There are probably several reasons for this. Molecular breeding is still a new field and there is a need for a better transfer of the molecular tools to the breeding programs. There is still a lot of debate around the advantages of MAS over conventional breeding and the wheat MAS consortium described above addresses this question. Another point of debate is the economic return of MAS and very few studies have been conducted to estimate it. They only analyze individual cases and it is very difficult to draw general conclusions from them (Young 1999; Dreher et al. 2000). In a very complete study, Dreher et al. (2000) compared the cost of MAS and conventional breeding methods in the maize program at CIMMYT. They conclude that neither conventional breeding nor MAS offers an unequivocal cost advantage under all circumstances and that the success of MAS lies in the identification of applications where markers offer a real advantage in terms of cost and time savings over classical breeding. Markers will be most useful when the phenotyping is expensive, when multiple genes are involved in a trait, when traits are expressed only at certain times of the year, or under some environmental conditions and for recessive genes. A very good example of this is illustrated in breeding for resistance against the barley yellow dwarf virus (BYDV). Testing for BYDV resistance, which has been introgressed into wheat from *Thinopyrum intermedium*, is very laborious and technically difficult. As a consequence, the efficient transfer of resistance to elite germplasms was very slow until molecular markers which can efficiently assess the presence of the introgressed *Thinopyrum intermedium* fragment have been developed. The SSR marker *gwm37* (Ayala et al. 2001) is now routinely used in the CIMMYT wheat breeding program (M. van Ginkel, pers. comm.) and together with the STS marker *csTiB1* (Stoutjesdijk et al. 2001), it has been validated in the Australian wheat breeding program (Zhang et al. 2001). It is also often neglected that additional technological costs brought by the application of markers are largely compensated by the time saved in the breeding programs. For these reasons, the future of MAS looks promising and as the technologies become cheaper there is no doubt that MAS will be more and more integrated into classical breeding programs.

References

Anderson JA, Ogihara Y, Sorrells ME, Tanksley SD (1992) Development of a chromosomal-arm map for wheat based on RFLP markers. Theor Appl Genet 83:1035–1043

Anderson JA, Stack RW, Liu S, Waldron BL, Fjeld AD, Coyne C, Moreno-Sevilla B, Fetch JM, Song QJ, Cregan PB, Frohberg RC (2001) DNA markers for Fusarium head blight resistance QTLs in two wheat populations. Theor Appl Genet 102:1164–1168

Autrique E, Singh RP, Tanksley SD, Sorrells ME (1995) Molecular markers for 4 leaf rust resistance genes introgressed into wheat from wild relatives. Genome 38:75–83

Ayala L, Henry M, Gonzalez-de-Leon D, van Ginkel M, Mujeeb-Kazi A, Keller B, Khairallah M (2001) A diagnostic molecular marker allowing the study of Th. intermedium-derived resistance to BYDV in bread wheat segregating populations. Theor Appl Genet 102:942–949

Bariana HS, Hayden MJ, Ahmed NU, Bell JA, Sharp PJ, McIntosh RA (2001) Mapping of durable adult plant and seedling resistances to stripe rust and stem rust diseases in wheat. Aust J Agric Res 52:1247–1255

Bariana HS, Brown GN, Ahmed NU, Khatkar S, Conner RL, Wellings CR, Haley S, Sharp PJ, Laroche A (2002) Characterisation of *Triticum vavilovii*-derived stripe rust resistance using genetic, cytogenetic and molecular analyses and its marker-assisted selection. Theor Appl Genet 104:315–320

Barloy D, Lemoine J, Dredryver F, Jahier J (2000) Molecular markers linked to the *Aegilops variabilis*-derived root-knot nematode resistance gene *Rkn-mn1* in wheat. Plant Breed 119:169–172

Bonhomme A, Gale MD, Koebner RMD, Nicolas P, Jahier J, Bernard M (1995) RFLP Analysis of an *Aegilops ventricosa* chromosome that carries a gene conferring resistance to leaf rust (*Puccinia recondita*) when transferred to hexaploid wheat. Theor Appl Genet 90:1042–1048

Borner A, Schumann E, Furste A, Coster H, Leithold B, Roder MS, Weber WE (2002) Mapping of quantitative trait loci determining agronomic important characters in hexaploid wheat (*Triticum aestivum* L.). Theor Appl Genet 105:921–936

Bougot Y, Lemoine J, Pavoine MT, Barloy D, Doussinault G (2002) Identification of a microsatellite marker associated with *Pm3* resistance alleles to powdery mildew in wheat. Plant Breed 121:325–329

Brown GN (1997) The inheritance and expression of leaf chlorosis associated with gene *Sr2* for adult plant resistance to wheat stem rust. Euphytica 95:67–71

Buckler ES, Thornsberry JM (2002) Plant molecular diversity and applications to genomics. Curr Opin Plant Biol 5:107–111

Buerstmayr H, Lemmens M, Hartl L, Doldi L, Steiner B, Stierschneider M, Ruckenbauer P (2002) Molecular mapping of QTLs for Fusarium head blight resistance in spring wheat. I. Resistance to fungal spread (type II resistance). Theor Appl Genet 104:84–91

Cenci A, D'Ovidio R, Tanzarella OA, Ceoloni C, Porceddu E (1999) Identification of molecular markers linked to *PM13*, an *Aegilops longissima* gene conferring resistance to powdery mildew in wheat. Theor Appl Genet 98:448–454

Chantret N, Sourdille P, Roder M, Tavaud M, Bernard M, Doussinault G (2000) Location and mapping of the powdery mildew resistance gene MIRE and detection of a resistance QTL by bulked segregant analysis (BSA) with microsatellites in wheat. Theor Appl Genet 100:1217–1224

Chen XM, Line RF, Leung H (1998) Genome scanning for resistance-gene analogs in rice, barley, and wheat by high-resolution electrophoresis. Theor Appl Genet 97:345–355

Collins N, Park R, Spielmeyer W, Ellis J, Pryor AJ (2001) Resistance gene analogs in barley and their relationship to rust resistance genes. Genome 44:375–381

Dedryver F, Jubier MF, Thouvenin J, Goyeau H (1996) Molecular markers linked to the leaf rust resistance gene *Lr24* in different wheat cultivars. Genome 39:830–835

DelaPena RC, Murray TD, Jones SS (1997) Identification of an RFLP interval containing *Pch2* on chromosome 7AL in wheat. Genome 40:249–252

Demeke T, Laroche A, Gaudet DA (1996) A DNA marker for the *Bt-10* common bunt resistance gene in wheat. Genome 39:51–55

Devos K, Gale M (1993) The genetic maps of wheat and their potential in plant-breeding. Outlook Agric 22:93–99

Dreher K, Morris M, Khairallah M, Ribaut JM, Pandey S, Srinivasan G (2000) Is marker-assisted selection cost-effective compared to conventional plant breeding methods? The case of quality protein maize. International Consortium on Agricultural Biotechnology Research, Ravello, Italy

Dubcovsky J, Lukaszewski AJ, Echaide M, Antonelli EF, Porter DR (1998) Molecular characterization of two *Triticum speltoides* interstitial translocations carrying leaf rust and greenbug resistance genes. Crop Sci 38:1655–1660

Dweikat I, Ohm H, Mackenzie S, Patterson F, Cambron S, Ratcliffe R (1994) Association of a DNA marker with Hessian fly resistance gene *H9* in wheat. Theor Appl Genet 89:964–968

Dweikat I, Ohm H, Patterson F, Cambron S (1997) Identification of RAPD markers for 11 Hessian fly resistance genes in wheat. Theor Appl Genet 94:419–423

Eastwood RF, Lagudah ES, Appels R (1994) A directed search for DNA sequences tightly linked to cereal cyst-nematode resistance genes in *Triticum tauschii*. Genome 37:311–319

Feuillet C, Messmer M, Schachermayr G, Keller B (1995) Genetic and physical characterization of the *Lr1* leaf rust resistance locus in wheat (*Triticum aestivum* L). Mol Gen Genet 248:553–562

Feuillet C, Travella S, Stein N, Albar L, Nublat A, Keller B (2003) Map-based isolation of the leaf rust disease resistance gene *Lr10* from the hexaploid wheat (*Triticum aestivum* L.) genome. Proc Natl Acad Sci USA 100:15253–15258

Frick MM, Huel R, Nykiforuk CL, Conner RL, Kuzyk A, Laroche A (1998) Molecular characterization of a wheat stripe rust resistance gene in Moro wheat. 9th international wheat genetics symposium, Saskatoon, Canada, University Extension Press, University of Saskatchewan, pp 181–182

Gervais L, Dedryver F, Morlais J-Y, Bodusseau V, Negre N, Bilous M, Groos C, Trottet M (2003) Mapping of quantitative trait loci for field resistance to Fusarium head blight in an European winter wheat. Theor Appl Genet 106:961–970

Guo P-G, Bai GH, Shaner GE (2003) AFLP and STS tagging of a major QTL for Fusarium head blight resistance in wheat. Theor Appl Genet 106:1011–1017

Gupta PK, Varshney RK, Sharma PC, Ramesh B (1999) Molecular markers and their applications in wheat breeding. Plant Breed 118:369–390

Hartl L, Weiss H, Zeller FJ, Jahoor A (1993) Use of RFLP markers for the identification of alleles of the *Pm3* locus conferring powdery mildew resistance in wheat (*Triticum aestivum* L). Theor Appl Genet 86:959–963

Hartl L, Weiss H, Stephan U, Zeller FJ, Jahoor A (1995) Molecular identification of powdery mildew resistance genes in common wheat (*Triticum aestivum* L). Theor Appl Genet 90:601–606

Hartl L, Mohler V, Zeller FJ, Hsam SLK, Schweizer G (1999) Identification of AFLP markers closely linked to the powdery mildew resistance genes *Pm1c* and *Pm4a* in common wheat (*Triticum aestivum* L.). Genome 42:322–329

Helguera M, Khan IA, Dubcovsky J (2000) Development of PCR markers for the wheat leaf rust resistance gene *Lr47*. Theor Appl Genet 100:1137–1143

Hu XY, Ohm HW, Dweikat I (1997) Identification of RAPD markers linked to the gene *PM1* for resistance to powdery mildew in wheat. Theor Appl Genet 94:832–840

Huang L, Gill BS (2001) An RGA-like marker detects all known *Lr21* leaf rust resistance gene family members in *Aegilops tauschii* and wheat. Theor Appl Genet 103:1007–1013

Huang XQ, Hsam SLK, Zeller FJ, Wenzel G, Mohler V (2000) Molecular mapping of the wheat powdery mildew resistance gene *Pm24* and marker validation for molecular breeding. Theor Appl Genet 101:407–414

Huang XQ, Wang LX, Xu MX, Röder MS (2003) Microsatellite mapping of the powdery mildew resistance gene *Pm5e* in common wheat (*Triticum aestivum* L.). Theor Appl Genet 106:858–865

Hulbert SH, Webb CA, Smith SM, Sun Q (2001) Resistance gene complexes: evolution and utilization. Annu Rev Phytopathol 39:285–312

Jahier J, Abelard P, Tanguy AM, Dedryver F, Rivoal R, Khatkar S, Bariana HS (2001) The *Aegilops ventricosa* segment on chromosome 2AS of the wheat cultivar 'VPM1' carries the cereal cyst nematode resistance gene *Cre5*. Plant Breed 120:125–128

Jarve K, Peusha HO, Tsymbalova J, Tamm S, Devos KM, Enno TM (2000) Chromosomal location of a *Triticum timopheevii*-derived powdery mildew resistance gene transferred to common wheat. Genome 43:377–381

Jia J, Devos KM, Chao S, Miller TE, Reader SM, Gale MD (1996) RFLP-based maps of the homoeologous group-6 chromosomes of wheat and their application in the tagging of *Pm12*, a powdery mildew resistance gene transferred from *Aegilops speltoides* to wheat. Theor Appl Genet 92:559–565

Kang HW, Cho YG, Yoon UH, Eun MY (1998) A rapid DNA extraction method for RFLP and PCR analysis from a single dry seed. Plant Mol Biol Rep 16:1–9

Keller B, Feuillet C (2000) Colinearity and gene density in grass genomes. Trends Plant Sci 5:246–251

Keller M, Keller B, Schachermayr G, Winzeler M, Schmid JE, Stamp P, Messmer MM (1999) Quantitative trait loci for resistance against powdery mildew in a segregating wheat × spelt population. Theor Appl Genet 98:903–912

Khan AA, Bergstrom GC, Nelson JC, Sorrells ME (2000) Identification of RFLP markers for resistance to wheat spindle streak mosaic bymovirus (WSSMV) disease. Genome 43:477–482

Kolb FL, Bai GH, Muehlbauer GJ, Anderson JA, Smith KP, Fedak G (2001) Host plant resistance genes for fusarium head blight: mapping and manipulation with molecular markers. Crop Sci 41:611–619

Lagudah ES, Moullet O, Appels R (1997) Map-based cloning of a gene sequence encoding a nucleotide binding domain and a leucine-rich region at the *Cre3* nematode resistance locus of wheat. Genome 40:659–665

Langridge P, Lagudah ES, Holton TA, Appels R, Sharp PJ, Chalmers KJ (2001) Trends in genetic and genome analyses in wheat: a review. Aust J Agric Res 52:1043–1077

Laroche A, Demeke T, Gaudet DA, Puchalski B, Frick M, McKenzie R (2000) Development of a PCR marker for rapid identification of the *Bt-10* gene for common bunt resistance in wheat. Genome 43:217–223

Leister D, Kurth J, Laurie DA, Yano M, Sasaki T, Devos K, Graner A, Schulze-Lefert P (1998) Rapid reorganization of resistance gene homologues in cereal genomes. Proc Natl Acad Sci USA 95:370–375

Li H, Brooks SA, Li W, Fellers JP, Trick HN, Gill BS (2003) Map-based cloning of leaf rust resistance gene *Lr21* from the large and polyploid genome of bread wheat. Genetics 164:655–664

Ling H-Q, Zhu Y, Keller B (2002) High-resolution mapping of the leaf rust disease resistance gene *Lr1* in wheat and characterization of BAC clones from the *Lr1* locus. Theor Appl Genet 106:875–882

Liu SX, Griffey CA, Maroof MAS (2001) Identification of molecular markers associated with adult plant resistance to powdery mildew in common wheat cultivar Massey. Crop Sci 41:1268–1275

Liu XM, Smith CM, Gill BS, Tolmay V (2001) Microsatellite markers linked to six Russian wheat aphid resistance genes in wheat. Theor Appl Genet 102:504–510

Liu XM, Smith CM, Gill BS (2002) Identification of microsatellite markers linked to Russian wheat aphid resistance genes *Dn4* and *Dn6*. Theor Appl Genet 104:1042–1048

Liu YG, Mori N, Tsunewaki K (1990) Restriction-fragment-length-polymorphism (RFLP) analysis in wheat 1. Genomic DNA library construction and RFLP analysis in common wheat. Jpn J Genet 65:367–380

Liu Z, Sun Q, Ni Z, Yang T (1999) Development of SCAR markers linked to the *Pm21* gene conferring resistance to powdery mildew in common wheat. Plant Breed 118:215–219

Liu ZY, Sun QX, Ni ZF, Nevo E, Yang TM (2002) Molecular characterization of a novel powdery mildew resistance gene *Pm30* in wheat originating from wild emmer. Euphytica 123:21–29

Ma JX, Zhou RH, Dong YS, Wang LF, Wang XM, Jia JZ (2001) Molecular mapping and detection of the yellow rust resistance gene *Yr26* in wheat transferred from *Triticum turgidum* L. using microsatellite markers. Euphytica 120:219–226

Ma ZQ, Gill BS, Sorrells ME, Tanksley SD (1993) RFLP markers linked to 2 Hessian fly resistance genes in wheat (*Triticum aestivum* L) from *Triticum tauschii* (Coss) Schmal. Theor Appl Genet 85:750–754

Ma ZQ, Sorrells ME, Tanksley SD (1994) RFLP markers linked to powdery mildew resistance genes *Pm1*, *Pm2*, *Pm3*, and *Pm4* in wheat. Genome 37:871–875

Ma ZQ, Saidi A, Quick JS, Lapitan NLV (1998) Genetic mapping of Russian wheat aphid resistance genes *Dn2* and *Dn4* in wheat. Genome 41:303–306

McIntosh RA, Hartl GE, Devos KM, Gale MD, Rogers WJ (1998) Catalogue of gene symbols for wheat. 9th international wheat genetics symposium, vol 5. University Extension Press, Saskatoon, Canada

McMillin DE, Allan RE, Roberts DE (1986) Association of an isozyme locus and strawbreaker foot rot resistance derived from *Aegilops ventricosa* in wheat. Theor Appl Genet 72:743–747

Messmer MM, Seyfarth R, Keller M, Schachermayr G, Winzeler M, Zanetti S, Feuillet C, Keller B (2000) Genetic analysis of durable leaf rust resistance in winter wheat. Theor Appl Genet 100:419–431

Miller CA, Altinkut A, Lapitan NLV (2001) A microsatellite marker for tagging *Dn2*, a wheat gene conferring resistance to the Russian wheat aphid. Crop Sci 41:1584–1589

Mohler V, Jahoor A (1996) Allele-specific amplification of polymorphic sites for the detection of powdery mildew resistance loci in cereals. Theor Appl Genet 93:1078–1082

Mohler V, Hsam SLK, Zeller FJ, Wenzel G (2001) An STS marker distinguishing the rye-derived powdery mildew resistance alleles at the *Pm8/Pm17* locus of common wheat. Plant Breed 120:448–450

Mohler V, Klahr A, Wenzel G, Schwarz G (2002) A resistance gene analog useful for targeting disease resistance genes against different pathogens on group 1S chromosomes of barley, wheat and rye. Theor Appl Genet 105:364–368

Moore G, Devos KM, Wang Z, Gale MD (1995) Cereal genome evolution – grasses, line up and form a circle. Curr Biol 5:737–739

Myburg AA, Cawood M, Wingfield BD, Botha AM (1998) Development of RAPD and SCAR markers linked to the Russian wheat aphid resistance gene *Dn2* in wheat. Theor Appl Genet 96:1162–1169

Naik S, Gill VS, Rao VSP, Gupta VS, Tamhankar SA, Pujar S, Gill BS, Ranjekar PK (1998) Identification of a STS marker linked to the *Aegilops speltoides*-derived leaf rust resistance gene *Lr28* in wheat. Theor Appl Genet 97:535–540

Nelson JC, Singh RP, Autrique JE, Sorrells ME (1997) Mapping genes conferring and suppressing leaf rust resistance in wheat. Crop Sci 37:1928–1935

Neu C, Stein N, Keller B (2002) Genetic mapping of the *Lr20-Pm1* resistance locus reveals suppressed recombination on chromosome arm 7AL in hexaploid wheat. Genome 45:737–744

Ogbonnaya FC, Seah S, Delibes A, Jahier J, Lopez-Brana I, Eastwood RF, Lagudah ES (2001b) Molecular-genetic characterisation of a new nematode resistance gene in wheat. Theor Appl Genet 102:623–629

Ogbonnaya FC, Subrahmanyam NC, Moullet O, de Majnik J, Eagles HA, Brown JS, Eastwood RF, Kollmorgen J, Appels R, Lagudah ES (2001a) Diagnostic DNA markers for cereal cyst nematode resistance in bread wheat. Aust J Agric Res 52:1367–1374

Paris M, Carter M (2000) Cereal DNA: a rapid high-throughput extraction method for marker assisted selection. Plant Mol Biol Rep 18:357–360

Paull JG, Pallotta MA, Langridge P, The TT (1994) RFLP markers associated with *Sr22* and recombination between chromosome 7A of bread wheat and the diploid species *Triticum boeoticum*. Theor Appl Genet 89:1039–1045

Peng JH, Fahima T, Roder M, Huang QY, Dahan A, Li YC, Grama A, Nevo E (2000a) High-density molecular map of chromosome region harboring stripe-rust resistance genes *YrH52* and *Yr15* derived from wild emmer wheat, *Triticum dicoccoides*. Genetica 109:199–210

Peng JH, Fahima T, Roder MS, Li YC, Grama A, Nevo E (2000b) Microsatellite high-density mapping of the stripe rust resistance gene *YrH52* region on chromosome 1B and evaluation of its marker-assisted selection in the F-2 generation in wild emmer wheat. New Phytol 146:141–154

Powell W, Morgante M, Andre C, Hanafey M, Vogel J, Tingey S, Rafalski A (1996) The comparison of RFLP, RAPD, AFLP and SSR (microsatellite) markers for germplasm analysis. Mol Breed 2:225–238

Prins R, Groenewald JZ, Marais GF, Snape JW, Koebner RMD (2001) AFLP and STS tagging of *Lr19*, a gene conferring resistance to leaf rust in wheat. Theor Appl Genet 103:618–624

Procunier J, Townley-Smith TF, Fox S, Prashar S, Gray M, Kim W, Czarnecki E, Dyck P (1995) PCR-based RAPD/DGGE markers linked to leaf rust resistance genes *Lr29* and *Lr25* in wheat (*Triticum aestivum* L.). J Genet Breed 49:87–92

Procunier JD, Knox RE, Bernier AM, Gray MA, Howes NK (1997) DNA markers linked to a *T10* loose smut resistance gene in wheat (*Triticum aestivum* L). Genome 40:176–179

Qi LL, Cao MS, Chen PD, Li WL, Liu DJ (1996) Identification, mapping, and application of polymorphic DNA associated with resistance gene *Pm21* of wheat. Genome 39:191–197

Rafalski A (2002) Applications of single nucleotide polymorphisms in crop genetics. Curr Opin Plant Biol 5:94–100

Rajaram S, Singh RP, Torres E (1988) Current CIMMYT approaches in breeding wheat for rust resistance: breeding strategies for resistance to the rusts of wheat. CIMMYT, Mexico, pp 101–118

Raupp WJ, Sukhwinder-Singh, Brown-Guedira GL, Gill BS (2001) Cytogenetic and molecular mapping of the leaf rust resistance gene *Lr39* in wheat. Theor Appl Genet 102:347–352

Richter TE, Ronald PC (2000) The evolution of disease resistance genes. Plant Mol Biol 42:195–204

Robert O, Abelard C, Dedryver F (1999) Identification of molecular markers for the detection of the yellow rust resistance gene *Yr17* in wheat. Mol Breed 5:167–175

Rong JK, Millet E, Manisterski J, Feldman M (2000) A new powdery mildew resistance gene: Introgression from wild emmer into common wheat and RFLP-based mapping. Euphytica 115:121–126

Sacco F, Suarez EY, Naranjo T (1998) Mapping of the leaf rust resistance gene *Lr3* on chromosome 6B of Sinvalocho MA wheat. Genome 41:686–690

Schachermayr G, Siedler H, Gale MD, Winzeler H, Winzeler M, Keller B (1994) Identification and localization of molecular markers linked to the *Lr9* leaf rust resistance gene of wheat. Theor Appl Genet 88:110–115

Schachermayr G, Feuillet C, Keller B (1997) Molecular markers for the detection of the wheat leaf rust resistance gene *Lr10* in diverse genetic backgrounds. Mol Breed 3:65–74

Schachermayr GM, Messmer MM, Feuillet C, Winzeler H, Winzeler M, Keller B (1995) Identification of molecular markers linked to the *Agropyron elongatum*-derived leaf rust resistance gene *Lr24* in wheat. Theor Appl Genet 90:982–990

Schnurbusch T, Paillard S, Schori A, Messmer M, Schachermayr G, Winzeler M, Keller B (2004) Dissection of quantitative and durable leaf rust resistance in Swiss winter wheat reveals a major resistance QTL in the *Lr34* chromosomal region. Theor Appl Genet 108:477–484

Seah S, Sivasithamparam K, Karakousis A, Lagudah ES (1998) Cloning and characterisation of a family of disease resistance gene analogs from wheat and barley. Theor Appl Genet 97:937–945

Seah S, Bariana H, Jahier J, Sivasithamparam K, Lagudah ES (2001) The introgressed segment carrying rust resistance genes *Yr17*, *Lr37* and *Sr38* in wheat can be assayed by a cloned disease resistance gene-like sequence. Theor Appl Genet 102:600–605

Seo YW, Johnson JW, Jarret RL (1997) A molecular marker associated with the H21 Hessian fly resistance gene in wheat. Mol Breed 3:177–181

Seyfarth R, Feuillet C, Schachermayr G, Winzeler M, Keller B (1999) Development of a molecular marker for the adult plant leaf rust resistance gene *Lr35* in wheat. Theor Appl Genet 99:554–560

Seyfarth R, Feuillet C, Schachermayr G, Messmer M, Winzeler M, Keller B (2000) Molecular mapping of the adult plant leaf rust resistance gene *Lr13* in wheat (*Triticum aestivum* L.). J Genet Breed 54:193–198

Sharp PJ, Johnston S, Brown G, McIntosh RA, Pallotta M, Carter M, Bariana HS, Khartkar S, Lagudah ES, Singh RP, Khairallah M, Potter R, Jones MGK (2001) Validation of molecular markers for wheat breeding. Aust J Agric Res 52:1357–1366

Shi AN, Leath S, Murphy JP (1998) A major gene for powdery mildew resistance transferred to common wheat from wild einkorn wheat. Phytopathology 88:144–147

Shi ZX, Chen XM, Line RF, Leung H, Wellings CR (2001) Development of resistance gene analog polymorphism markers for the *Yr9* gene resistance to wheat stripe rust. Genome 44:509–516

Singh RP (1992a) Association between gene *Lr34* for leaf rust resistance and leaf tip necrosis in wheat. Crop Sci 32:874–878

Singh RP (1992b) Genetic association of leaf rust resistance gene *Lr34* with adult plant resistance to stripe rust in bread wheat. Phytopathology 82:835–838

Singh RP, Nelson JC, Sorrells ME (2000) Mapping *Yr28* and other genes for resistance to stripe rust in wheat. Crop Sci 40:1148–1155

Smith PH, Koebner RMD, Boyd LA (2002) The development of a STS marker linked to a yellow rust resistance derived from the wheat cultivar Moro. Theor Appl Genet 104:1278–1282

Spielmeyer W, Sharp PJ, Lagudah ES (2003) Identification and validation of markers linked to broad-spectrum stem rust resistance gene *Sr2* in wheat (*Triticum aestivum* L.). Crop Sci 43:333–336

Spielmeyer W, Huang L, Bariana H, Laroche A, Gill BS, Lagudah ES (2000a) NBS-LRR sequence family is associated with leaf and stripe rust resistance on the end of homoeologous chromosome group 1S of wheat. Theor Appl Genet 101:1139–1144

Spielmeyer W, Moullet O, Laroche A, Lagudah ES (2000b) Highly recombinogenic regions at seed storage protein loci on chromosome 1DS of *Aegilops tauschii*, the D-genome donor of wheat. Genetics 155:361–367

Spielmeyer W, Robertson M, Collins N, Leister D, Schulze-Lefert P, Seah S, Moullet O, Lagudah ES (1998) A superfamily of disease resistance gene analogs is located on all homoeologous chromosome groups of wheat (*Triticum aestivum*). Genome 41:782–788

Stein N, Herren G, Keller B (2001) A new DNA extraction method for high-throughput marker analysis in a large-genome species such as *Triticum aestivum*. Plant Breed 120:354–356

Stoutjesdijk P, Kammholz SJ, Kleven S, Matsay S, Banks PM, Larkin PJ (2001) PCR-based molecular marker for the *Bdv2 Thinopyrum intermedium* source of barley yellow dwarf virus resistance in wheat. Aust J Agric Res 52:1383–1388

Suenaga K (2002) Tagging of leaf rust resistance genes, *Lr34* and *Lr46*, using microsatellite markers in wheat. Japan International Research Centre for Agricultural Sciences Newsletter 31 http://ss.jircas.affrc.go.jp/kanko/newsletter/nl2002/No.31/HTML/05.htm

Sun GL, Fahima T, Korol AB, Turpeinen T, Grama A, Ronin YI, Nevo E (1997) Identification of molecular markers linked to the *Yr15* stripe rust resistance gene of wheat originated in wild emmer wheat, *Triticum dicoccoides*. Theor Appl Genet 95:622–628

Talbert LE, Bruckner PL, Smith LY, Sears R, Martin TJ (1996) Development of PCR markers linked to resistance to wheat streak mosaic virus in wheat. Theor Appl Genet 93:463–467

Tao W, Liu D, Liu J, Feng Y, Chen P (2000) Genetic mapping of the powdery mildew resistance gene *Pm6* in wheat by RFLP analysis. Theor Appl Genet 100:564–568

Venter E, Botha AM (2000) Development of markers linked to *Diuraphis noxia* resistance in wheat using a novel PCR-RFLP approach. Theor Appl Genet 100:965–970

Wang LF, Ma JX, Zhou RH, Wang XM, Jia JZ (2002) Molecular tagging of the yellow rust resistance gene *Yr10* in common wheat, PI178383 (*Triticum aestivum* L.). Euphytica 124:71–73

William HM, Hoisington D, Singh RP, GonzalezdeLeon D (1997) Detection of quantitative trait loci associated with leaf rust resistance in bread wheat. Genome 40:253–260

William M, Singh RP, Huerta-Espino J, Islas SO, Hoisington D (2003) Molecular marker mapping of leaf rust resistance gene *Lr46* and its association with stripe rust resistance gene *Yr29* in wheat. Phytopathology 93:153–159

Williams KJ, Fisher JM, Langridge P (1994) Identification of RFLP markers linked to the cereal cyst-nematode resistance gene (Cre) in wheat. Theor Appl Genet 89:927–930

Williams KJ, Fisher JM, Langridge P (1996) Development of a PCR-based allele-specific assay from an RFLP probe linked to resistance to cereal cyst nematode in wheat. Genome 39:798–801

Williams KJ, Taylor SP, Bogacki P, Pallotta M, Bariana HS, Wallwork H (2002) Mapping of the root lesion nematode (*Pratylenchus neglectus*) resistance gene *Rlnn1* in wheat. Theor Appl Genet 104:874–879

Winzeler M, Winzeler H, Keller B (1995) Endopeptidase polymorphism and linkage of the *Ep-D1c* null allele with the *Lr19* leaf rust resistance gene in hexaploid wheat. Plant Breed 114:24–28

Yahiaoui N, Srichumpa P, Dudler R, Keller B (2004) Genome analysis at different ploidy levels allows cloning of the powdery mildew resistance gene *Pm3b* from hexaploid wheat. Plant J 37:528–538

Yamamori M (1994) An N-band marker for gene *Lr18* for resistance to leaf rust in wheat. Theor Appl Genet 89:643–646

Yan GP, Chen XM, Line RF, Wellings CR (2003) Resistance gene-analog polymorphism markers co-segregating with the *Yr5* gene for resistance to wheat stripe rust. Theor Appl Genet 106:636–643

Young ND (1999) A cautiously optimistic vision for marker-assisted breeding. Mol Breed 5:505–510

Zeller FJ, Kong L, Hartl L, Mohler V, Hsam SLK (2002) Chromosomal location of genes for resistance to powdery mildew in common wheat (*Triticum aestivum* L. em Thell.) 7. Gene *Pm29* in line Pova. Euphytica 123:187–194

Zhang W, Carter M, Matsay S, Stoutjesdijk P, Potter R, Jones MGK, Kleven S, Wilson RE, Larkin PJ, Turner M, Gale KR (2001) Implementation of probes for tracing chromosome segments conferring barley yellow dwarf virus resistance. Aust J Agric Res 52:1389–1392

Zhou WC, Kolb FL, Bai GH, Shaner G, Domier LL (2002) Genetic analysis of scab resistance QTL in wheat with microsatellite and AFLP markers. Genome 45:719–727

III.5 Application of DNA Markers: Soybean Improvement

M.J. IQBAL and D.A. LIGHTFOOT[1]

1 Introduction

Soybean geneticists and breeders aim to improve harvestable yield, reduce crop losses, and reduce grower inputs by re-assortment among favorable haplotypes and the incorporation of new genes for new traits. Many uses exist for soybean-derived products in the food and feed industries, but soybean remains largely a commodity crop (for a review, see Liu 1997). However, with the advent of biotechnology and molecular markers, the creation of specialty soybeans with modified seed composition is a tractable target for future soybean improvement. In this model, some soybeans will be grown as a value-added crop and some soybeans grown as a commodity. Health benefits from specialty soybeans with modified seed compositions present an attractive new target for crop improvement (Bringe 2001; Meksem et al. 2000; Kassem et al. 2003).

There is a widely accepted belief that all the species within *Glycine* are polyploid or paleopolyploid (Lackey 1980; Doyle 1991). However, in soybean [*G. max* (L.) Merr.] the genome is duplicated, but inheritance is diploid with $2n=40$ (Singh and Hymowitz 1988). About 250 morphological, pigmentation and isozyme markers have been identified, but only 63 were assigned to a classical genetic map (Palmer and Hedges 1993). The classical genetic map of soybean consists of 19 linkage groups and is composed of two or three point linkages or clusters thereof that have one marker in common. In contrast to the classical linkage map, Shoemaker and Olson (1993) positioned nearly 490 restriction fragment length polymorphism (RFLP) and random amplified polymorphic DNA (RAPD) markers on 20 linkage groups encompassing a recombination genetic distance of nearly 3000 cM. With the development of simple sequence repeat (SSR) markers, Cregan et al. (1999a) incorporated 606 SSR loci in three different soybean populations. With the efforts of various research groups, we have a core-consensus soybean map of 20 linkage groups that contains a total of 1488 markers, including 1006 SSRs, 723 RFLPs, 73 RAPDs, 23 classic traits and 19 others (Perry Cregan 2003 Soybean Breeder's

---

[1] Center of Excellence in Soybean Research, Teaching and Outreach, Department of Plant, Soil and Agriculture Systems, 176 Ag. Building, MC 4415, Southern Illinois University at Carbondale, Carbondale, Illinois 62901–4415, USA

Workshop). The consensus map length has been reduced to 2478 cM by rigorous elimination of scoring errors. There is a continued distributed community effort to add new markers to the existing linkage map (Cregan et al. 1999b; Meksem et al. 2001b) particularly in regions that do not have many markers currently. In future, this will rely on the physical map (Lightfoot et al. 2003) and the haplotype map that will be developed from it (Zhu et al. 2003).

Markers have been used to anchor fingerprint contigs to a deep draft physical map (Shultz et al. 2001; Lightfoot et al. 2003; Wu et al., unpubl.). Precisely 469 microsatellite markers and 105 RFLP markers have been anchored to contigs in cooperation between two NSF projects, No. 9872565 "A functional genomics program for soybean" and No. 9872635 "An integrated genetic and physical map of soybean". The genetic map location for all markers and plate addresses for all clones can be viewed at www.siu.edu/~pbgc/. The BAC end sequences are deposited at NCBI along with BAC subclone sample sequences. There are useful fingerprints for 78,001 BACs, about 8.7-fold the soybean genome. Wu et al. (unpubl.) have assembled and edited the map contigs from the fingerprint database resulting in two data releases (Table 1; http://hbz.tamu.edu). Distributed community-based contig editing and emergence will refine the map further using downloadable FPC database available at www.siu.edu/~pbgc/ and the genome browser at http://soybeangenome.siu.edu/.

To support functional genomics, there is a best tiling path of 9600 clones (25 plates) that encompasses 99% of the cloneable genome. There is a renewable collection of 100 recombinant inbred lines derived from the cross of Essex and Forrest (Lightfoot et al. 2003; V.N. Njiti, pers. comm.) that are available from SIUC on request. An immortal collection of 100 near isogeneic

Table 1. Progress in the soybean physical map builds (Wu et al. 2004).

	Automated build (August 2002)	Manually edited build (September 2001)
BACs used in contig assembly	75,568	78,001
Number of singletons	5884	4954
Clones in contigs (fold genome)	69,684 (8.7×)	73,047 (9.1×)
Anchored markers	175	459
Number of contigs	5488	2905
Contigs contain:		
>25 clones	220	921
10–25 clones	3038	920
3–9 clones	1845	850
2 clones	385	216
Unique bands of the contigs	396,843	345,457
Length of the contigs (Mb)	1667a	1408[a]

[a] Based on 4.0 kbp per unique band

line pairs, each pair capturing the heterogeneous regions derived from a single RIL (Njiti et al. 1998, 2004), are available.

Sequence tag sites (STS) are available for every marker-anchored clone (Iqbal et al. 2001; Marek et al. 2001) and every contig that is currently not anchored by a genetic marker (Dr. Chris Town, TIGR, pers. comm.). These sequences provide a resource for the generation of new markers and the linking together of contigs.

2 Choice of Markers

The choice of marker depends upon the objective of the selection program, tools available to the individual/group researcher, markers available for a particular region or trait if any, information content associated with a particular marker system etc.

RFLP. The first marker system to be used was RFLP. However, the tetraploid origin of soybean (Hymowitz and Singh 1987) contributed to the detection of multiple DNA fragments with all RFLP probes. Single RFLP probes have been used to detect up to 19 independent loci (Mansur et al. 1996). The multiplicity of RFLP loci can make the locus identity ambiguous. Other factors that prevent the use of RFLP in mapping and marker-assisted breeding are the low levels of polymorphisms observed (Shoemaker and Specht 1995) and limitations of the automation procedure for high throughput screening.

Microsatellite. PCR-based microsatellite DNA markers that contain 10–20 simple tri- or di-nucleotide repeats have been shown to be highly polymorphic in soybean (Akkaya et al. 1992; Morgante and Olivieri 1993). These were called simple sequence repeats (SSRs) by Cregan et al. (1999a). Some highly polymorphic microsatellite loci can have as many as 26 alleles (Maughan et al. 1995; Rongwen et al. 1995; Powell et al. 1996; Diwan and Cregan 1997). Such a high level of allelic diversity increases the possibility of detecting polymorphism between parents of populations derived from the hybridization of adopted soybean genotypes. SSRs seem to be distributed fairly randomly throughout the soybean genome, with a minimum evidence of clustering (Akkaya et al. 1995). There are about two SSRs (as defined by Akkaya et al. 1995) per 100 kbp of soybean sequence. Analysis of soybean genomic DNA has shown shorter microsatellites that contain less than ten simple tri- or di-nucleotide repeats (small simple sequence repeats or SSSRs) are more common than SSRs and frequently polymorphic in soybean (Meksem et al. 1998, 2001b; Triwitayakorn 2002; Triwitayakorn et al. 2004). There are about ten such DNA markers (as defined) per 100 kbp of soybean sequence. In addition, complex microsatellites that contain two different tri- or di-nucleotide repeats are very common and frequently polymorphic in soybean (Witsenboer et al. 1998). Selection for SSRs, SSSRs and SAMPLs (selectively amplified microsatellite polymorphic loci) in soybean can be made with standard gel or capillary electrophoresis (Cregan et al. 1999a; Prabhu et al. 1999), or electrophoresis-free assays like MALDI-TOF MS (Chen et al. 2003).

Therefore, microsatellite markers are currently the method of choice in most of the public laboratories based on their distribution in the soybean genome, the level of polymorphism information content attached to each identified locus and the ease of its automation for high throughput screening programs.

SNP. Single nucleotide polymorphisms (SNP) are single base pair changes (and 1–3 bp indels) between individuals, in this case soybean cultivars. They occur about twice as often in noncoding compared to coding DNA. In coding DNA, about one quarter to one half of SNPs alter amino acid sequence depending on the genes examined (Meksem et al. 2001b; Zhu et al. 2003). The observed frequency in soybean (0.053 and 0.1%) is about twofold lower than *Arabidopsis thaliana* considering known genes and cDNAs (Zhu et al. 2003). However, the SNP frequency is fourfold higher than expected considering AFLP bands as the sequence dataset (Meksem et al. 2001a, b). Linkage blocks in soybean (350 Kbp) are about twofold larger than in *A. thaliana* on average (Zhu et al. 2003). The linkage blocks appear to derive from 3–4 ancestral cultivars. However, genes under intense selection like the major resistance genes *Rhg1*, *Rhg4* (resistance to *Heterodera glycines*), *Rfs1* and *Rfs2* (resistance to *Fusarium solani* f. sp. *glycines*) are encoded by 6–8 alleles (Meksem et al. 1999; Hauge et al. 2001; Triwitayakorn et al. 2004) likely reflecting the extra diversity imported to cultivated soybean by gene introgression. Selection for SNPs in soybean can be made with standard gel or capillary electrophoresis, or electrophoresis-free assays like TaqMan (Meksem et al. 2001a, b).

Deletions and Insertions. Deletions and insertions (indels) are multi-base pair changes between soybean cultivars. They occur in noncoding DNA more often than coding DNA. In coding DNA they alter amino acid sequence (Meksem et al. 2001b). The observed frequency in soybean is about fivefold lower than for SNPs even considering AFLP bands as the sequence dataset (Meksem et al. 2001a, b). Some genes under intense selection like the major resistance gene *Rhg1*, bear indels in one allele (Meksem et al. 1999; Hauge et al. 2001; Triwitayakorn et al. 2003) likely reflecting the extra diversity imported to cultivated soybean by gene introgression. Selection for indels in soybean can be made by non-PCR methods (Baner et al. 2001) and isothermic assays like Invader (Table 2; Mein et al. 2000; Lightfoot, unpubl.). The standard PCR-based gel or capillary electrophoresis, or electrophoresis-free assays like TaqMan (Meksem et al. 2001a, b) or MALDI-TOF MS (Chen et al. 2003) can also be used (Fig. 1). The ease of selection makes indels the most versatile DNA marker available.

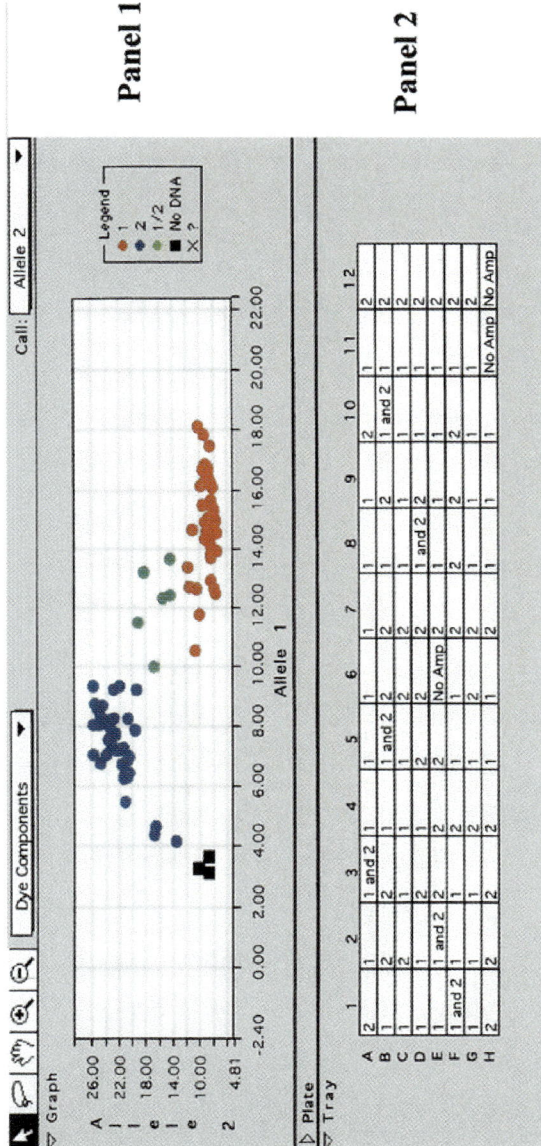

Fig. 1. With permission from Meksem et al. (2001b). Detection of the TMA5 ($E_{ATG}M_{CGA}87$) marker polymorphism by the TaqMan method allows allelic discrimination of soybean genotypes with manual selection of genotypes. Data for 90 individuals from an F_5-derived population of RILs from the cross Essex × Forrest that segregate for resistance to SCN are shown. *Panel 1* The fluorescent signals were viewed under the "Dye Component" field of the sequence detection software and the TMA5 genotypes were manually selected based on the ratio of FAM to TET signals. Allele 1, homozygous Forrest type: TET/FAM greater than 2. Allele 2, homozygous Essex type: FAM/TET greater than 2. If the TET level was less than twofold greater or lesser than the FAM level, the individual was scored as heterozygous Essex/Forrest (1/2). *Panel 2* The Excel spreadsheet shows the scores for the samples as they were arranged in the 96-well plate. There was no DNA in wells H11 and H12. Essex DNA was placed in well A1, Forrest DNA in well B1, The RIL DNA was arrayed in the rest of the 96-well plate (C1 to G12). *1* Resistant allele; *2* susceptible allele; *1* and *2* heterozygous lines

Table 2. An Excel output from an invader assay for marker-assisted selection. Marker representing parent 1 (P_1) has an insertion and parent 2 (P_2) has a deletion in the DNA fragment. In a Biplex Invader Assay the results are fluorescent signal that is detected for each dye – signal is generated if the allele is present. The results are analyzed by looking at the fold over zero (sample signal/no target signal) and then subtracting 1 (to normalize for the inherent background signal from each probe set). This is listed as FOZ-1. These levels are then compared and the genoytping calls are made. If we see (roughly) an equal signal from both dyes, the sample is heterozygous (het), otherwise it is homozygous for one or the other allele

Sample	Average signal Ins probe	Average signal Del probe	Average Ins net counts	Average Del net counts	Fold over 0 Ins probe	Fold over 0 Del probe	FOZ-1 Ins probe	FOZ-1 Del probe	FOZ-1 ratio	Genotype
Blank	118.00	121.75	1.00	1.00	1	1	0.01	0.01	No signal	–
P_1	450.50	119.00	332.50	1.00	3.82	0.98	2.82	0.01	281.78	Ins
P_2	111.50	362.50	1.00	240.75	0.94	2.98	0.01	1.98	0.01	Del
5	120.50	591.50	2.50	469.75	1.02	4.86	0.02	3.86	0.01	Del
6	383.00	123.50	265.00	1.75	3.25	1.01	2.25	0.01	156.24	Ins
7	140.00	429.00	22.00	307.25	1.19	3.52	0.19	2.52	0.07	Del
8	406.50	120.50	288.50	1.00	3.44	0.99	2.44	0.01	244.49	Ins
9	643.50	127.50	525.50	5.75	5.45	1.05	4.45	0.05	94.30	Ins
10	108.50	775.00	1.00	653.25	0.92	6.37	0.01	5.37	0.00	Del
11	311.50	294.50	193.50	172.75	2.64	2.42	1.64	1.42	1.16	Ins/del
12	114.00	290.50	1.00	168.75	0.97	2.39	0.01	1.39	0.01	Del
13	117.50	720.50	1.00	598.75	1.00	5.92	0.01	4.92	0.00	Del
14	384.00	123.50	266.00	1.75	3.25	1.01	2.25	0.01	156.83	Ins
15	311.50	421.00	193.50	299.25	2.64	3.46	1.64	2.46	0.67	Ins/del
16	307.00	126.00	189.00	4.25	2.60	1.03	1.60	0.03	45.88	Ins
17	117.00	1078.50	1.00	956.75	0.99	8.86	0.01	7.86	0.00	Del
19	734.50	176.00	616.50	54.25	6.22	1.45	5.22	0.45	11.73	Ins
20	105.00	1411.50	1.00	1289.75	0.89	11.59	0.01	10.59	0.00	Del
21	115.50	353.00	1.00	231.25	0.98	2.90	0.01	1.90	0.01	Del
22	468.00	115.00	350.00	1.00	3.97	0.94	2.97	0.01	296.61	Ins
23	110.50	437.00	1.00	315.25	0.94	3.59	0.01	2.59	0.00	Del
24	116.50	295.00	1.00	173.25	0.99	2.42	0.01	1.42	0.01	Del
25	110.50	791.50	1.00	669.75	0.94	6.50	0.01	5.50	0.00	Del
26	114.00	688.50	1.00	566.75	0.97	5.66	0.01	4.66	0.00	Del

Application of DNA Markers: Soybean Improvement

Table 2. (Continue)

Sample	Average signal Ins probe	Average signal Del probe	Average Ins net counts	Average Del net counts	Fold over 0 Ins probe	Fold over 0 Del probe	FOZ-1 Ins probe	FOZ-1 Del probe	FOZ-1 ratio	Genotype
27	129.50	132.00	11.50	10.25	1.10	1.08	0.10	0.08	No signal	-
28	326.00	133.50	208.00	11.75	2.76	1.10	1.76	0.10	18.26	Ins
29	505.00	118.00	387.00	1.00	4.28	0.97	3.28	0.01	327.97	Ins
30	522.50	119.50	404.50	1.00	4.43	0.98	3.43	0.01	342.80	Ins
31	316.50	453.50	198.50	331.75	2.68	3.72	1.68	2.72	0.62	Ins/del
32	200.50	272.50	82.50	150.75	1.70	2.24	0.70	1.24	0.56	Ins/del
33	124.50	366.00	6.50	244.25	1.06	3.01	0.06	2.01	0.03	Del
34	382.00	231.50	264.00	109.75	3.24	1.90	2.24	0.90	2.48	Ins/del
35	709.50	131.50	591.50	9.75	6.01	1.08	5.01	0.08	62.59	Ins
36	174.50	365.50	56.50	243.75	1.48	3.00	0.48	2.00	0.24	Del
37	123.00	160.50	5.00	38.75	1.04	1.32	0.04	0.32	0.13	Del
38	117.50	831.50	1.00	709.75	1.00	6.83	0.01	5.83	0.00	Del
39	120.50	397.50	2.50	275.75	1.02	3.26	0.02	2.26	0.01	Del

3 Identification of Polymorphism

The first step in marker-assisted selection (MAS) in breeding programs is the identification of polymorphic loci between the parents that can be traced in the segregating population. The frequency of polymorphism is related to genetic distance between the parents and the type of marker used (Zhu et al. 2003). When low diversity or a paucity of markers is a problem, physical map-based marker identification is an efficient approach (Meksem et al. 1998; Cregan et al. 1999b).

4 Marker-Assisted Recovery of Recurrent Parent Genome

In plant breeding, the backcross procedure is often used to transfer favorable alleles from a donor genotype, which has mostly poor agronomic properties, into a recipient elite genotype. In backcross gene introgression programs, marker-assisted breeding methodologies have a great impact on the genome recovery of recurrent parents. The recovery of recurrent parent genotype (RPG) can be accelerated by marker-assisted strategy as devised by Tanksley et al. (1989). Using this approach, individuals that are homozygous for the alleles of the recurrent parent at a large number of marker loci covering the entire genome are selected. Population size and marker density required in a background selection program can be determined (Openshaw et al. 1994). They recommended the use of four markers per chromosome (of 200-cM length) and a selection strategy for proximal recombinants of the target allele. In the case of soybean, having 20 linkage groups and an estimated map length of 2478 cM, 80 uniformly distributed markers will be very effective in recovering a sufficiently high proportion of RPG in three generations. In the literature, the number of markers used for RPG recovery varies, about a 20-cM marker density represented an optimal trade-off between percentage recurrent parent recovery and management of data-point throughput requirements (Frisch et al. 1999). Based on a marker every 20 cM and the total map distance of 2478 cM, a total of 124 markers are needed for marker-assisted recovery of RPG in soybean. Uniformly distributed markers are more effective than equal numbers per chromosome (Hospital et al. 1992), because smaller population sizes and fewer data points are required with equally spaced markers (Frisch et al. 1999). In soybean, 60 markers, three evenly distributed per linkage group (Vince Pantalone; Soybean breeders Workshop 2003) are reasonably adequate.

5 Marker-Assisted Selection in Recurrent Cross-Populations

The majority of plant breeding programs make advances by re-assorting the entire genomes of two elite lines with favorable and complementary characteristics. The goal of this "shuffling of the decks" is to recover transgressive segregants that yield more than either parent by 1–2% per year for each year necessary to develop the line. Traditional recurrent selection cycles last from 5–7 years from cross to release. The average cultivar is sold for 2–3 years before becoming obsolete. Therefore, marker selection that reduces development time by stacking selections into a shorter time frame are very valuable.

MAS for yield and drought tolerance is not currently very effective. The traits are polygenic and environmentally dependent so that the penetrance of markers (association with traits across years and populations) is limited. MAS for other traits (disease resistance, composition) compete with greenhouse assays and laboratory assays for price and effectiveness. This competition is not always market driven, as investments in infrastructure made in the past determine economics in the present. Over time though, we expect MAS to be widely adopted by all effective soybean breeders. MAS as employed today usually focuses on two to four genes per population with 25 to 6% of lines "passing the test". Samples are selected after preliminary visual selection for yield potential (at the F_3–F_5), rarely are F_2 selections made. Hence, about 10% of a program's recombinants will be tested in the MAS lab. Since a single lab will serve dozens of breeders each with a hundred thousand lines, the task can quickly reach millions of selections per year. Automation is very important in this situation.

Selections are made by a variety of methods often driven more by due diligence of patent holders' rights than technical efficiency. However, with the exception of a few markers linked to five genes for resistance to SCN (Webb et al. 1995) fair-market licensing deals can be negotiated for trait marker associations.

6 Marker-Assisted Selection for Targeted Genes/Traits

Soybean importance in US agriculture has played a significant role in the generation of large number of markers for qualitative and quantitative traits, both by the public and private sector. According to SoyBase (http://soybase.agron.iastate.edu/), a USDA-ARS plant genome funded program, 961 QTL have been identified for 55 agronomic traits in soybean. Moreover, the database also contains 466 genes representing gene class, locus, alleles, phenotypes, and two-point data. The availability of all this information and commercial drive to release new and improved varieties has made soybean one of the best examples where MAS is playing a significant role in new and improved variety development.

A. SCN: One such example of MAS in soybean is selection for resistance to soybean cyst nematode (SCN). Soybean cyst nematode, *Heterodera glycines*, is a small plant-parasitic roundworm that attacks the roots of soybeans and causes significant crop losses in the infected fields. Two QTL significantly contributing to soybean resistance to *H. glycines*, *Rhg*1 and *Rhg*4 have been mapped on linkage groups G and A2. However, varieties selected for *Rhg*1 provide good resistance to *H. glycines* hgtype 7 (race 3). SSR marker Satt309 has been mapped at a 0.5–2 cM distance from *Rhg*1 (42 kbp in physical distance) and is being used for MAS for resistance to SCN race3. With the sequencing of the whole region of linkage group G of soybean genome by Monsanto (St. Louis, Mo), and the SIUC patent for perfect markers for SCN (Meksem et al. 1999; Hauge et al. 2001), it is now possible to select six of eight *Rhg*1 alleles for resistance with 100% certainty. Twenty-six markers are directly located on the gene associated with resistance to SCN race 3, six alter protein code. There is one indel in the single intron.
B. SDS: We have identified six resistance genes associated with resistance to sudden death syndrome of soybean (SDS), three are clustered with SCN resistance *Rhg*1 (Meksem et al. 1999; Iqbal et al. 2001). One gene, *Rfs*1, is responsible for root resistance to *Fusarium solani* f. sp. *glycines* and can be selected for by BAC-derived SSSR (Meksem et al. 1998), a RAPD-derived SCAR (Iqbal, unpubl.) or a gene sequence (Lightfoot, unpubl.).
C. Nutraceuticals: Soybeans and soy products contain non-nutritive, bioactive compounds that may influence, negatively and positively, physiological responses in animals and humans that ingest these products (Meksem et al. 2000). The amount of phytoestrogens in soybean seed can vary up to fivefold. Phytoestrogen content and profile can vary with year, environment, and genotype. Hence, genetic markers closely linked to genes controlling these soy phytoestrogens (DNA fingerprinting) may be used to indirectly select for favorable alleles and complement direct phenotype selection. Marker-assisted breeding of soybeans for phytoestrogen content would also be attractive because the plant could be improved with this value-added trait without genetic engineering techniques, producing so-called GMOs (genetically modified organisms).

7 Methods for Marker-Assisted Selection

Screening for markers linked to a trait or recovery of RPG can be carried out with leaf or seed samples. A high throughput method of MAS is described (Fig. 1). DNA can be isolated from small leaf discs collected from fields as well as from the seeds. For sample collection from plants, a 1-cm^2 leaf disk from a young leaf is removed and stored in small note pads and kept on ice or dry ice. Alternately, seeds are placed on small filter paper discs soaked

Application of DNA Markers: Soybean Improvement 381

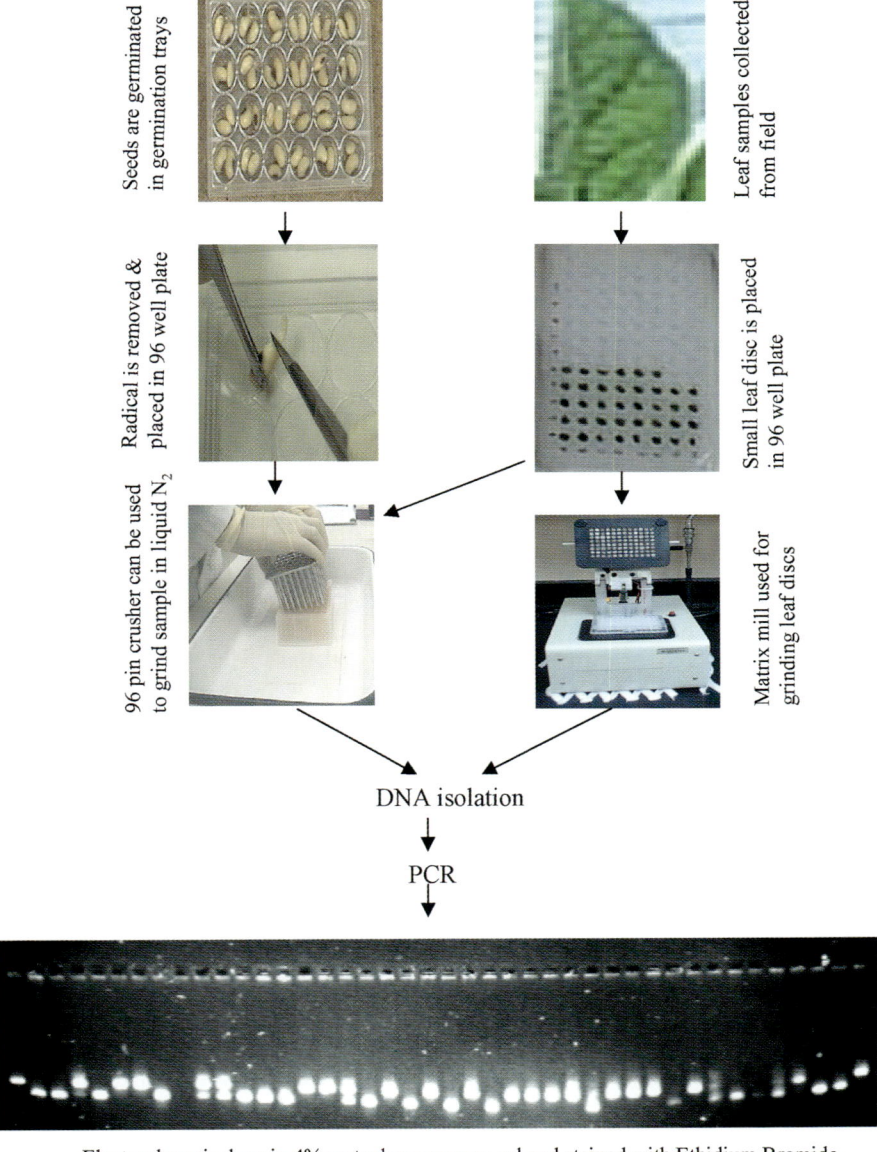

Fig. 2. A schematic presentation of steps involved in MAS. Gel electrophoresis can be replaced by other capillary-based electrophoresis techniques or simply a fluorescence- based assay like TaqMan or invader can be used for scoring polymorphism at the target allele

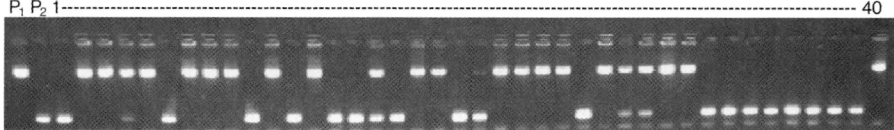

Fig. 3. Segregation of SSR locus, Satt424 in a population. Individuals can be clearly defined as carrying the alleles of either parent 1 (P_1), parent 2 (P_2) or heterozygous. The 10-µl PCR reaction was electrophoresed on a 4% metaphore agarose gel and stained with ethidium bromide. DNA was isolated from radical tissue in a 96-well plate. PCR was carried out in a 96 well plate and samples were loaded on gel with an 8-channel pipette

with sterile dH_2O for 48 h for germination. Radicals are carefully removed with sharp scalpel blades. Leaf disks or cotyledons are placed in 96-well plates. Samples can be ground either by a matrix mill in the presence of 100 µl of 1–0.5 N sodium hydroxide or by 96-pin crusher in the presence of liquid N_2 (Fig. 2). Once the DNA is isolated (Xin et al. 2003 for an alternate method for DNA isolation), PCR with the selected primers is carried out in 96- or 384-well plates and electrophoresed in Metaphore agarose gel/regular agarose gel (Fig. 3) or acrylamide gel based on the size difference of the target alleles. For a higher throughput, SSR or SCAR primers can be labeled with fluorescence tags and multiplexed in the PCR reaction or at the electrophoresis stage. The samples can then be electrophoresed using capillary-based or gel-based DNA sequencers and analyzed by fragment analysis software as provided by the fragment analysis equipment used. The shortcomings of gel electrophoresis can be avoided by developing TaqMan-TM allelic discrimination, PCR-OLA, molecular beacons, padlock probes and well fluorescence assay (Landegren et al. 1998).

8 Conclusions

The main aim of MAS is the selection of highly desirable lines among thousands of genotypes and this can only be achieved by high-throughput, rapid, automated procedures for the detection of DNA polymorphisms attached with the desirable traits. With the availability of capillary electrophoresis-based genetic analyzers, thousands of samples can be screened per day per machine. However, the availability of new nonelectrophoresis-based tools such as TaqMan (Landegren et al. 1998), invader assay (Mein et al. 2000) for the detection of polymorphisms further simplifies the screening procedures for large-scale MAS procedures. Moreover, the availability of gene markers or perfect markers such as SNPs directly on the genes of interest makes the MAS procedures more desirable. With the availability of over 300,000 soybean EST and genomic sequences and 9000 uni-gene sequences in the GenBank, soybean has probably the best genomic resources ready to be used in the development of new markers and their application in marker-assisted breeding of new and improved varieties.

References

Akkaya MS, Bhagwat AA, Cregan PB (1992) Length polymorphism of simple sequence repeat DNA in soybean. Genetics 132:1131–1139

Akkaya MS, Shoemaker RC, Specht JE, Bhagwat AA, Cregan PB (1995) Integration of simple sequence repeat DNA markers into a soybean linkage map. Crop Sci 35:1439–1445

Baner J, Nilsson M, Isaksson A, Mendel-Hartvig M, Antson DO, Landegren U (2001) More keys to padlock probes: mechanisms for high-throughput nucleic acid analysis. Curr Opin Biotechnol 12:11–15

Bringe NA (2001) High beta-conglycinin products and their use. US patent #6,171,640

Chen CH, Potter NT, Taranenko NT (2003) Detection of trinucleotide repeat containing genes by matrix-assisted laser desorption/ionization (MALDI) mass spectrometry. Methods Mol Biol 217:91–100

Cregan PB, Jarvik T, Bush AL, Shoemaker RC, Lark KG, Kahler AL, Kaya N, VanToai TT Lohnes DG, Chung J, Specht JE (1999a) An integrated genetic linkage map of the soybean genome. Crop Sci 39:1464–1490

Cregan PB, Mudge J, Fickus EW, Danesh D, Denny R, Young ND, (1999b) Two simple sequence repeat markers to select for soybean cyst nematode resistance conditioned by the rhg1 locus. Theor Appl Genet 99:811–818

Diwan N, Cregan PB (1997) Automated sizing of fluorescent-labeled simple sequence repeat (SSR) markers to assay genetic variation in soybean. Theor Appl Genet 95:723–733

Doyle JJ (1991)The pros and cons of DNA systematic data: studies of wild perennial relatives of soybean. Evol Trends Plants 5:99–104

Frisch M, Bohn M, Melchinger AE (1999) Comparison of selection strategies for marker-assisted backcrossing of a gene. Crop Sci 39:1295–1301

Hauge BM, Wang ML, Parsons JD, Parnell LD (2001) Nucleic acid molecules and other molecules associated with soybean cyst nematode resistance. US patent: WO 0151627-A 19-JUL-2001

Hospital F, Chevalet C, Mulsant P (1992) Using markers in gene introgression breeding programs. Genetics 132:1199–1210

Hymowitz T, Singh RJ (1987) Taxonomy and speciation. In: Wicox RJ (ed) Soybeans: improvement, production and uses, 2nd edn. Agron Monogr 16. ASA CSSA, Madison, WI, pp 23–48

Iqbal MJ, Meksem K, Njiti VN, Kassem MyA, Lightfoot DA (2001) Microsatellite markers identify three additional quantitative trait loci for resistance to soybean sudden death syndrome (SDS) in Essex × Forrest RILs. Theor Appl Genet 102:187–192

Jamai A, Meksem K, Ishihara H, Bensmail I, Arelli P, Lightfoot DA (2002) Glycine max receptor-like kinase RHG1 and RHG4, mRNA, gene and genomic DNA, complete cds. GenBank AF506516–26; AY163905; AF526257–61

Kassem A, Nitji VN, Meksem K, Banz WJ, Winters TA, Lightfoot DA (2004) Two additional genomic regions that underlie soybean seed isoflavone content. J Biomed Biotech 2004 (1):52–60

Lackey JA (1980) Chromosome number in the Phaseoleae (Fabaceae: Faboideae) and their relation to taxonomy. Am J Bot 67:595–602

Landegren U, Nilsson M, Kwok PW (1998) Reading bits of genetic information: methods for single nucleotide polymorphism analysis. Genome Res 8:769–776

Lightfoot DA, Meksem K, Zhang H-B (2003) An integrated physical and genetic map for the soybean genome. In: Van den Bosch K, Stacey G (eds) Summaries of legume genomics projects from around the globe. Community resources for crops and models. Plant Physiol 131:840–865

Liu KS (ed) (1997) Soybeans: chemistry, technology and utilization. Chapman and Hall, New York

Mansur LM, Orf JH, Chase K, Jarvick T, Cregan PB, Lark KG (1996) Genetic mapping of agronomic traits using recombinant inbred lines of soybean. Crop Sci 36:1327–1336

Marek LF, Mudge J, Darnielle L, Grant D, Hanson N, Paz M, Huihuang Y, Denny R, Larson K, Foster-Hartnett D, Cooper A, Danesh D, Larsen D, Schmidt T, Staggs R, Crow JA, Retzel E, Young ND, Shoemaker RC (2001) Soybean genomic survey: BAC-end sequences near RFLP and SSR markers. Genome 44:572–581

Maughan PJ, Saghai Maroof MA, Buss GR (1995) Microsatellite and amplified sequence length polymorphisms in cultivated and wild soybean. Genome 38:715–723

Mein CA, Barratt BJ, Dunn MG, Siegmund T, Smith AN, Esposito L, Nutland S, Stevens HE, Wilson AJ, Philips MS, Jarvis N, Law S, de Arruda M, Todd JA (2000) Evaluation of single nucleotide polymorphism typing with invader on PCR amplification and its automation. Genome Res 3:330–343

Meksem K, Zhang H-B, Lightfoot DA (1998) A plant transformation ready bacterial artificial chromosome library for soybean: Applications in chromosome walking and genome wide physical mapping. Soybean Genet Newslett 25:108–111

Meksem K, Doubler TW, Chang SJC, Chancharoenchai K, Suttner R, Cregan P, Rao Arelli P, Gibson PT, Lightfoot DA (1999) Clustering among genes underlying QTL for field resistance to Sudden Death Syndrome and cyst nematode race 3. Theor Appl Genet 99:1131–1142

Meksem K, Njiti VN, Banz WJ, Iqbal MJ, Hyten D, Kassem M, Yuan J, Winters TA, Lightfoot DA (2000) Genomic regions that underlie soybean seed phytoestrogen content. J Biomed Biotechnol 2:1–8

Meksem K, Hyten D, Ruben E, Lightfoot DA (2001a) High-throughput genotyping for a polymorphism linked to soybean cyst nematode resistance gene *Rhg*4 by using Taqman probes. Mol Breed 77:63–71

Meksem K, Ruben E, Hyten D, Triwitayakorn K, Lightfoot DA (2001b) Conversion of AFLP bands into high-throughput DNA markers. Mol Genet Genom 265:207–214

Morgante M, Olivieri AM (1993) PCR-amplified microsatellites as markers in plant genetics. Plant J 3:175–182

Njiti VN, Doubler TW, Suttner RJ, Gray LE, Gibson PT, Lightfoot DA (1998) Resistance to soybean sudden death syndrome and root colonization by *Fusarium solani* f. sp. *glycine* in near-isogenic lines. Crop Sci 38:472–477

Njiti VN, Kassem MA Meksem K, Gibson PT, Iqbal MJ, Lightfoot DA (2003) Registration of near isogeneic lines derived from a cross of Essex and Forrest. Crop Sci (in press)

Openshaw SJ, Jarboe SG, Beavis WD (1994) Marker assisted selection in backcross breeding. Proceedings of the symposium "Analysis of molecular marker data", Corvallis, OR 5–6 Aug 1994, Am Soc Hortic Sci, Crop Sci Soc Am

Palmer RG, Hedges BR (1993) Linkage map of soybean (*Glycine max* L. Merr). In: O'Brien SJ (ed) Genetic maps: locus maps of complex genomes. Cold Spring Harbor Laboratory Press, New York, pp 6.139–6.148

Powell W, Morgante M, Andre C, Hanafey M, Vogel J, Tingey S, Rafalski A (1996) The comparison of RFLP, RAPD, AFLP, and SSR (microsatellite) markers for germplasm analysis. Mol Breed 2:225–238

Prabhu RR, Njiti VN, Bell-Johnson B, Johnson JE, Schmidt ME, Klein JH, Lightfoot DA (1999) Selecting soybean cultivars for dual resistance to soybean cyst nematode and sudden death syndrome using two DNA markers. Crop Sci 39:982–987

Rongwen J, Akkaya MS, Bhagwat AA, Lavi U, Cregan PB (1995) The use of microsatellite DNA markers for soybean genotype identification. Theor Appl Genet 90:43–48

Shoemaker RC, Olson TC (1993) Molecular linkage map of soybean (*Glycine max* L. Merr). In: O'Brien SJ (ed) Genetic maps: locus maps of complex genomes. Cold Spring Harbor Laboratory Press, New York, pp 6.131–6.138

Shoemaker RC, Specht JE (1995) Integration of the soybean molecular and classical genetic linkage groups. Crop Sci 35:436–446

Shultz J, Wu C, Santos FA, Nimmakayala P, Springman R, LaMontague C, Zobrist K, Meksem K, Zhang H-B, Lightfoot DA (2001) A physical map for the soybean genome. Soybean Genet Newslett 28:5–10

Singh RJ, Hymowitz T (1988) The genomic relationship between *Glycine max* (L.) Merr. and *G. soja* Sieb. and Zucc. As revealed by pachytene chromosome analysis. Theor Appl Genet 76:705–711

Tanksley SD, Young ND, Paterson AH, Bonierbale MW (1989) RFLP mapping in plant breeding: new tools for an old science. Bio/technology 7:257–264

Triwitayakorn K (2002) Positional cloning of the *Rfs* loci. PhD Thesis, SIUC Carbondale, 232 pp

Triwitayakorn K, Njiti VN, Meksem K, Iqbal MJ, Yaegashi S, Jamai A, Town C, Lightfoot DA (2004) Genomic analysis of a region encompassing QRfs1 and QRfs2: genes that underlie soybean resistance to sudden death syndrome. Genome (in press)

Webb DM, Baltazar BM, Rao-Arelli AP, Schupp J, Keim P, Clayton K, Ferreira AR, Owens T, Beavis WD (1995) QTLs affecting soybean cyst-nematode resistance. Theor Appl Genet 91:574–581

Witsenboer H, Vogel J, Michelmore RW (1998) Identification, genetic localization and allelic diversity of selectively amplified microsatellite polymorphic loci (SAMPL) in lettuce and wild relatives (*Lactuca* spp.). Genome 40: 923–936

Wu C, Sun S, Nimmakayala P, Santos FA, Springman R, Ding K, Meksem K, Lightfoot DA, Zhang HB (2004) A BAC and BIBAC-based physical map of the soybean genome. Genome Res 14:319–326

Xin Z, Velten JP, Oliver MJ, Burke JJ (2003) High-throughput DNA extraction method suitable for PCR. BioTechniques 34:820–826

Zhu YL, Song QJ, Hyten DL, van Tassell CP, Matukumalli LK, Grimm DR, Hyatt SM, Fickus EW, Young ND, Cregan PB (2003) Single-nucleotide polymorphisms in soybean. Genetics 163:1123–1134

III.6 Forest Management and Conservation Using Microsatellite Markers: The Example of *Fagus*

Y. Tsumura[1], M. Takahashi[2], T. Takahashi[3], N. Tani[1], Y. Asuka[4], and N. Tomaru[4]

1 Introduction

The genetic diversity and mating systems of tree species have been studied extensively during the last two decades by allozyme analysis (Hamrick 1989). The resulting information is important for forest management and conservation. In particular, information on genetic differentiation among populations of forestry species is necessary for the conservation of genetic resources. Highly polymorphic genetic markers such as microsatellite markers (Litt and Luty 1989; Weber and May 1989), are highly sensitive at detecting the dynamics of gene flow within and among populations (Dow and Ashley 1996). These markers give us information on selfing rate and biparental inbreeding (Kelly and Willis 2002; Obayashi et al. 2002), the differential paternal contribution of each individual tree to future populations, and the fine-scale genetic structure within forests (Ueno et al. 2000). As we can understand pollen flow within forests when we use these markers, we can determine the best management system to maintain genetic diversity among fragmented natural populations and man-made populations for the purposes of forestry and conservation.

Fagus crenata and *F. japonica* are monoecious, long-lived, woody angiosperm species with an outcrossing breeding system based on wind pollination, with gravity- and animal-dispersed seeds (Kitamura and Murata 1979). Both grow in Japan. *F. crenata* is considered to be important as an ecosystem component and for the conservation of bio-diversity; the World Heritage listed Mt. Shirakami is dominated by *F. crenata* forests (UNESCO 2002). However, since the 1950s, many areas of beech forest have been logged and converted to coniferous forests for timber production. Coniferous forests planted in high-altitude or heavy-snowfall regions sometimes are not adapted to the severe environmental conditions and do not grow well. Forest rehabilitation

---

[1] Department of Forest Genetics, Forestry and Forest Products Research Institute, Tsukuba, Ibaraki 305-8687, Japan
[2] Forest Tree Breeding Center, Juo, Ibaraki 319-1301, Japan
[3] Graduate School of Science and Technology, Niigata University, Niigata 950-2181, Japan
[4] Laboratory of Forest Ecology and Physiology, Graduate School of Bioagricultural Sciences, Nagoya University, Nagoya 464-8601, Japan

has been used to recover such forests by the planting of native species such as *F. crenata*. To maintain the genetic diversity of planted populations and to adapt them to the prevailing conditions, superior tree selection of *F. crenata*, clonal propagation by grafting, and construction of an experimental seed orchard have been started in the Forest Tree Breeding Center of Japan. Microsatellite markers are a very powerful and sensitive tool for evaluating the genetic diversity of collected superior trees in a seed orchard, and, unlike allozyme markers, can detect subtle changes in genetic diversity.

In this chapter, we describe an effective method for developing microsatellite markers, and discuss forest conservation and management based on information such as the genetic structure and gene flow of *Fagus* populations. We demonstrate that the pattern of gene flow is influenced by several factors – reproductive system, mating system, environment, and others – by using data obtained from fine-scale genetic structure studies between closely related species in different environmental conditions, and from a gene flow study in a seed orchard.

2 Development and Evaluation of Microsatellite Markers in *Fagus*

Several improvements in methodology have been reported recently for efficient development of simple-sequence repeat (SSR) markers, including the vectorette polymerase chain reaction (PCR) strategy (Lench et al. 1996), the random amplified hybridization microsatellites (RAHM) method (Cifarelli et al. 1995), and the library enrichment method for SSR regions (Ostrander et al. 1992; Karagyozov et al. 1993; Lyall et al. 1993; Kirkpatrick et al. 1995; Takahashi et al. 1996). The library enrichment method for the development of microsatellite markers is much more efficient and less labor-intensive than nonenrichment methods. Several procedures have been developed for enrichment of DNA libraries, among which the magnetic bead method is well established and has a very high efficiency of enrichment. We adopted two modified methods to develop microsatellite markers in *Fagus*, the RAHM method (Cifarelli et al. 1995) and the enrichment method using magnetic beads (Fischer and Bachmann 1998; Hamilton et al. 1999), and compared their efficiency.

The RAHM procedure, based on PCR, is very convenient because it is not necessary to prepare high-quality genomic DNA to make a genomic DNA library (Cifarelli et al. 1995), so we merely selected the RAPD fragments containing SSR regions. The screening efficiency is very high compared with that of the colony hybridization method, because one random primer amplifies many DNA fragments, which can be screened at the same time. We detected 60 positive fragments by using 38 out of 360 random primers; thus, 10.6% of primers yielded positive fragments (Tanaka et al. 1999). Some plant species have high contents of secondary metabolites in their cells, so if we use the colony hybridization method, we have to exclude the components during the DNA extraction before we can prepare the genomic library. We developed nine polymorphic microsatellite markers in *F. crenata* by using the RAHM method, eight of which are available also in *F. japonica*. The polymorphic level was extremely high in both species; the average heterozygosity was 0.615 and 0.660, respectively.

Our method for enriching a genomic DNA library containing SSR regions is based on the methods of Fischer and Bachmann (1998) and Hamilton et al. (1999). We used the following procedure to develop SSR markers in *F. crenata*. Genomic DNA of *F. crenata* was extracted by using a modified CTAB method (Murray and Thompson 1980) and purified by ultracentrifugation. Ten micrograms of DNA was digested with *Nde*II and electrophoretically separated on a 1.2% agarose gel. DNA fragments ranging from 300 to 1000 bp were recovered. Approximately 600 ng of the fragments was ligated to 5 pmol of *Sau*3AI cassette (TaKaRa). The nick between the genomic DNA and the cassette sequence was filled by using DNA polymerase I. One hundred ng of DNA fragments with *Sau*3AI cassettes were denatured at 95°C for 15 min and hybridized at 70°C overnight to 2 pmol of biotinylated oligonucleotides, $(CT)_{15}$, in 100 ml of buffer containing 6×SSC and 0.05% SDS at 55°C. These hybrids were captured with 20 mg of pre-washed streptavidin-coated magnetic beads (Dynal), and microsatellite-containing fragments were enriched and recovered in eluate as described by Hamilton et al. (1999). Double-stranded conformation was performed by PCR with the Primer C1 (Takara). The PCR products were digested with *Sau*3AI to remove the cassette, ligated into pUC118 (Takara) plasmid vectors, and cloned into competent cells of *E. coli*. Plasmid DNA was extracted from positive clones and sequenced on a 3100 Genetic Analyzer by using the Big Dye Terminator Cycle Sequencing Kit (Applied Biosystems). PCR primer pairs for microsatellites were designed by using OLIGO software (National Biosciences). After PCR optimization, successful forward primers were fluorescently labeled, and amplifications were carried out in 10-μl reactions containing 1×PCR buffer (10 mM Tris·HCl (pH 8.3), 50 mM KCl, 100 mM of each dNTP), 1.5 mM $MgCl_2$, 0.25 U *Taq* polymerase, 0.2 mM of each primer, and 5–10 ng of template DNA. The PCR conditions were 3 min at 94°C; 30 cycles of 30 s at 94°C, 30 s at primer-specific annealing temperature, 30 s at 72°C; and final extension at 72°C for 7 min. The PCR products were run on a 3100 Genetic Analyzer with GeneScan software (Applied Biosystems), and genotypes were determined. We obtained a highly enriched DNA library containing SSR regions. The proportion of SSR-containing clones was about 61%: sequence data were obtained from 806 clones, of which 496 contained SSR regions. PCR primers to amplify SSR regions were designed by using sequence data that had pure repeats after redundancy was excluded. Finally, we designed PCR primers for at least 96 loci in *F. crenata*; 16 of the markers showed clear patterns and high polymorphisms (Asuaka et al. 2004a).

Both methods are effective for developing SSR markers. The RAHM method is much simpler than the enrichment method using magnetic beads, but the latter is much more efficient; in particular, many markers are needed for construction of the linkage map. The RAHM procedure is a step-by-step method that can be followed by an operator who is not a specialist in molecular biology techniques. The genomic library is not necessary in this method because of the PCR based-method, thus, high quality DNA is also not needed. Conversely, the enrichment method using magnetic beads requires a relatively high level of skill in molecular biology techniques. For the purpose of genetic monitoring of a forest population, fewer than ten loci of SSR markers are probably adequate if the markers show high polymorphism because we can determine the pollen and seed dispersal using these markers. Therefore, both methods can be used for the purpose of genetic monitoring, but, for the construction of a genetic linkage map, the latter method is much more effective.

3 Spatial Analysis of Genetic Structure Within Forests by Microsatellite Markers

3.1 Differences in Fine-Scale Genetic Structure Between *F. crenata* and *F. japonica*

We compared the spatial genetic structures of *F. crenata* and *F. japonica* by using four microsatellite markers (Takahashi et al. 2004, in press). The study site was a 2-ha plot within a mixed population on Mt. Takahara, central Honshu, Japan. Two statistics, genetic relatedness and number of alleles in common, were used to detect spatial genetic structure. A significant negative correlation between genetic relatedness and spatial distance was detected among all individuals in each species. However, this correlation was weak; the genetic structures likely resulted from extensive pollen flow caused by wind pollination. Similarly, Merzeau et al. (1994) and Streiff et al. (1998) also detected weak genetic structure in one of three *Fagus sylvatica* stands and in *Quercus petraea* and *Q. robur*, which are wind-pollinated species whose seeds are gravity-dispersed, like those of *Fagus*. Spatial genetic clustering in *F. japonica* was stronger than in *F. crenata* over short distance classes. The presence of self-incompatibility may also influence genetic structure (Loveless and Hamrick 1984; Doligez et al. 1998). Self-incompatibility can induce decreases in spatial genetic structure within populations (Doligez et al. 1998). Furthermore, if self-incompatibility combines with family structure, pollen dispersal becomes large and genetic subdivision is less likely (Loveless and Hamrick 1984). Although only two studies investigated self-incompatibility in *F. crenata* (Kouno and Mukouda 1985) and *F. japonica* (Igarashi 1996), they suggested that the ratios of mature and immature seeds in *F. japonica* were higher than in *F. crenata* in controlled pollination experiments. The reproductive system might also influence the difference in spatial genetic structure between the two species. The regeneration of *F. crenata* depends mainly on the growth of seedlings or saplings under canopy gaps (e.g., Nakashizuka 1987). On the other hand, *F. japonica* forms stools by vigorous sprouting, and the stools help to maintain trees (Ohkubo 1992). Thus, an *F. japonica* stool can reach a substantial age – about 1000 years, against about 200–300 years for *F. crenata*. Differences in genetic structure between those species would be caused by different periods of generation overlap.

3.2 Influence of Environmental Differences and Forest History on Spatial Genetic Structure

The spatial genetic structure of *F. crenata* in a 4-ha plot (200×200 m^2) of an old-growth beech forest was analyzed by using microsatellite markers (Asuka et al. 2004b). Two types of coefficient were used to assess the genetic structure: Moran's *I* spatial autocorrelation coefficients and genetic relatedness. The correlation between spatial distance separating individuals and genetic relatedness was tested by a Mantel test. Correlograms of both Moran's *I* and Mantel's *r* values showed significant positive values for short distance classes, indicating weak genetic structure, the same as in the previous studies (Kitamura et al. 1997; Takahashi et al. 2000). The genetic structuring within the population is probably created by limited seed dispersal, but likely weakened by extensive pollen flow and overlapping seed shadows. Genetic structure was detected in an eastern subplot of 1 ha (50×200 m^2) with immature soils and almost no dwarf bamboo (*Sasa* spp.), but none was found in a western subplot of the same size with mature soils and *Sasa* cover. The apparent genetic structure detected in the 4-ha plot was, therefore, due to the structure in the western portion of the plot. The heterogeneity of genetic structure presumably reflects variation in regeneration, which is strongly influenced by heterogeneity of environmental conditions.

Takahashi et al. (2000) examined the effect of logging on within-population genetic structure by comparing two forests (selectively logged and unlogged) by allozyme analysis. They found that logging slightly, but significantly decreased the genetic variability and reinforced the spatial genetic structure by reducing the mixing of half-sib progeny derived from a limited number of reproductive trees. They also found that linkage disequilibrium was higher in the logged forest than in the unlogged forest, and suggested that this value might be a good indicator of forest decline. However, linkage disequilibrium is influenced by population history and natural selection as well. Therefore, by considering the history and natural selection of each forest, we will be able to use the value of linkage disequilibrium for suitable management.

3.3 Gene Flow Within Seed Orchard Revealed by Microsatellite Analysis

To understand gene flow within a seed orchard of *F. crenata*, we investigated seedlings derived from open-pollinated seeds of six clones by using four microsatellite loci (Tanaka et al. 1999). The seed orchard consisted of 38 clones, which were established by grafting in 1979. We searched for pollen donor candidates for each seedling. When we found a single match between seedling haplotype and pollen donor haplotype, we could determine the pollen donor.

Finally, we could determine the pollen donor for 172 out of the 217 seedlings. The assigned paternity rate for each parent clone ranged from 71.4 to 100.0%, with an average of 79.3% (Table 1; "1 male parent" column). Thirty-one out of the remaining 45 seedlings had more than one candidate paternal clone in the seed orchard ("2 possible male parents" and "3 or more possible male parents" columns). We could not find a candidate within the six parent clones in the seed orchard for 13 seedlings. The pollen responsible would have traveled a long distance, because no reproductive mature trees grow within 500 m of the seed orchard. The degree of long-distance pollen transport is generally high in wind-pollinated species (contamination rate; 69–71% in *Picea abies*, Pakkanen et al. 2000; 48% in *Pinus taeda*, Friedman and Adams 1985; about 70% in *Quercus robur*, Buiteveld et al. 2001). Thus, this seed orchard seems to be isolated from beech forests, because the pollen contamination was only 6%. We also found only one selfed seedling, and thus the outcrossing rate is very high (99.5%), which reconfirms that *Fagus* is an allogamous species.

The contribution of the parent clones to seedling paternity differed greatly: Sanbongi-103 fathered 43.4% of all seedlings (Table 2), followed by Fukaura-101 (17.3%), Ajigasawa-102 (7.5%), and the rest (<5%). Sanbongi-

Table 1. Paternity analysis of seed orchard of *Fagus crenata*

Parent clone	No. seedlings investigated	No. seedlings with paternity assigned			No. selfed seedlings	Male parent not in nursery
		One male parent	Two possible male parents	Three or more possible male parents		
Ajigasawa 102	40	29	3	0	1	7
Fukaura 101	14	14	0	0	0	0
Fukaura 102	42	30	8	3	0	1
Hirosaki 103	40	35	3	0	0	2
Iwaizumi 103	49	37	8	2	0	2
Tayama 104	32	27	4	0	0	1
Total	217	172	26	5	1	13
Proportion (%)		79.3	12.0	2.3	0.5	6.0

Table 2. Paternity contribution of parent clones in the seed orchard

Female clone	Male clone							
	Sanbongi 103	Fukaura 101	Ajigasawa 102	Tayama 102	Tayama 104	Hirosaki 102	Tohno 101	Other 15 clones
Ajigasawa 102	15	5	1	0	0	1	0	8
Fukaura 101	7	0	2	0	0	0	0	5
Fukaura 102	6	16	5	0	1	2	0	0
Hirosaki 103	15	7	1	0	3	1	0	8
Iwaizumi 103	19	0	1	0	1	0	5	11
Tayama 104	13	2	3	5	0	1	0	3
Total	75	30	13	5	5	5	5	35
Proportion (%)	43.35	17.34	7.51	2.89	2.89	2.89	2.89	20.23

103 has the highest pollen fecundity in the orchard. Pollen fecundity and flowering phenology might be highly related to success of mating.

Microsatellite analysis showed that *F. crenata* is an almost completely outcrossing species that *F. crenata* trees with high pollen fecundity may contribute disproportionately to future generations, and that *F. crenata* pollen travels long distances. In our seed orchard, three clones fathered 68.2% of all seedlings. However, to maximize genetic diversity, it is important to select several clones with synchronous flowering and comparable pollen fecundity. With regard to long-distance gene flow, allozyme study showed that genetic differentiation between populations is very low (G_{ST}=0.038; Tomaru et al. 1997), because gene migrants from neighboring populations frequently come into a population in pollen. Pollen fecundity, flowering phenology, and pollen contamination are critical issues for genetic diversity and production of high-quality seedlings in seed orchards.

4 Genetic Management of *Fagus* Forests for Conservation and Sustainable Use

Forests frequently experience fragmentation or isolation by human disturbance and thus suffer genetic bottlenecks. Severe bottlenecks, such as drastic reduction in population size, result in genetic erosion and loss of adaptation to environmental change. In a small and isolated population, inbreeding depression as a result of mating between relatives reduces the variability and viability of forest. These forces cause forests to decline and can change the forest environment. To maintain relatively high genetic diversity within forests, long-term genetic monitoring is needed. Microsatellite markers are very suitable for this purpose due to their high polymorphism. This marker can be traced to the pollen and seed movements. This kind of information is necessary to maintain adequate genetic diversity within forests especially for the conservation and management of forest tree species.

Fagus forests are widely distributed in Europe, eastern Asia, and eastern North America (Peters 1997), but their area has been decreased and fragmented by exploitation for land and timber. In particular, suburban spread has promoted fragmentation. For the conservation of such forests, maintenance of genetic diversity is very important. Data on the heterozygosity, allelic richness, genetic differentiation between local populations, outcrossing rate, and genetic structure of those forests are important for conservation. Gene flow through pollen is not strongly restricted in *Fagus*, because this species is wind-pollinated. An allozyme study of 23 populations of *Fagus* forest showed that the genetic differentiation between populations was low (Tomaru et al. 1997). However, gene flow through seeds is strongly restricted: genetic differentiation revealed by both mtDNA and cpDNA polymorphisms between populations was extremely high (Tomaru et al. 1998; Fujii et al. 2002; Okaura and Harada 2002). Guidelines for gene conservation within populations can be based on information on maternal inherited DNA markers such as mtDNA and cpDNA. However, an understanding of gene flow within a forest through pollen and seed is necessary for conservation and sustainable use. Fine-scale genetic structure is influenced by the regeneration system, forest history, and microenvironment heterogeneity, such as forest floor conditions. Generally, long-lived, wind-pollinated, dominant species such as *Fagus*, *Quercus*, and conifers in temperate regions show weak spatial genetic structure (Merzeau et al. 1994; Streiff et al. 1998; Takahashi et al. 2000; Epperson and Chung 2001). If the structure becomes strong, this indicates that a forest is declining.

Tree density is an important factor in maintaining a high outcrossing rate and genetic diversity (Rajora et al. 2000; Obayashi et al. 2002). Inbreeding increases in an isolated or low-density forest, and the forest declines owing to inbreeding depression. Most forest tree species are predominantly allogamous, and some are self-incompatible.

The outcrossing rate and a fixation index can be used to assess the integrity of a forest. Takahashi et al. (2000) suggested that reduced genetic variability and linkage disequilibrium would have a significant influence over several generations. Reductions in genetic variability imply a higher potential for inbreeding depression, and the existence of linkage disequilibrium means distortions in the composition of the gene set in the population. If the natural composition of the gene set is assumed to be the most highly adapted to a given environment, linkage disequilibrium also implies reductions in the adaptability of populations in succeeding generations, which could be detrimental to the conservation of important genetic resources. For rehabilitation programs, indicators such as spatial genetic structure, outcrossing rate, fixation index, and linkage disequilibrium can be used to assess the integrity of planted populations.

5 Conclusions

We could develop a sufficient number of microsatellite markers in *F. crenata* using the enrichment method of the microsatellite region. These markers will provide important information for the conservation and management of *F. crenata* forests.

Fine-scale genetic structure was influenced by life history such as the regeneration system, microenvironment, and forest history. *Fagus crenata* had a weaker fine-scale genetic structure than that of *F. japonica*, probably due to the different level of self-incompatibility and longevity. The genetic structure of *F. crenata* has also been changed by their microenvironment such as soil type and forest floor, which are closely related to forest history. The parameter of linkage disequilibrium could be one of better indicators to understand the forest history, which might show the maturity and stability of forest population. Combining ecological and environment data together with genetic data, we can understand the integrity of forest population and may take a suitable strategy for conservation and management. For the rehabilitation program of *Fagus* forest, seed sources with high genetic diversity and genetic similarity to introduced population is critically important to maintain original genetic component of the population. For that purpose, the suitable seed orchard is necessary to supply such seedlings for the plantation.

We believe that microsatellite markers are the best markers to monitor the gene flow within a forest, fine-scale genetic structure, and to investigate genetic diversity and similarity between the seed source and the rehabilitation forests for conservation and management of *Fagus* populations.

References

Asuka Y, Tani N, Tsumura Y, Tomaru N (2004a) Development and characterization of microsatellite markers for *Fagus crenata* Blume. Mol Ecol Note 4:101–103

Asuka Y, Tomaru N, Nishimura N, Tsumura Y, Yamamoto S (2004b) Spatial genetic structure of *Fagus crenata* (Fagaceae) in an old-growth beech forest revealed by microsatellite markers. Mol Ecol 13:1241–1250

Buiteveld J, Bakker EG, Bovenschen J, de Vries SMG (2001) Paternity analysis in a seed orchard of *Quercus robur* L. and estimation of the amount of background pollination using microsatellite markers. For Genet 8:331–337

Cifarelli RA, Gallitelli M, Cellini F (1995) Random amplified hybridization microsatellites (RAHM): isolation of a new class of microsatellite-containing DNA clones. Nucleic Acid Res 23:3802–3803

Doligez A, Baril C, Joly HI (1998) Fine-scale spatial genetic structure with nonuniform distribution of individual. Genetics 148:905–919

Dow BD, Ashley MV (1996) Microsatellite analysis of seed dispersal and parentage of sapling in bur oak, *Quercus macrocarpa*. Mol Ecol 5:615–627

Epperson BK, Chung MG (2001) Spatial genetic structure of allozyme polymorphisms within populations of *Pinus strobus* (Pinaceae). Am J Bot 88:1006–1010

Fischer D, Bachmann K (1998) Microsatellite enrichment in organisms with large genomes (*Allium cepa* L.). Biotechniques 24:796–802

Friedman ST, Adams WT (1985) Estimation of gene flow into two seed orchards of loblolly pine (*Pinus taeda* L.). Theor Appl Genet 69:609–615

Fujii N, Tomaru N, Okuyama K, Koike T, Mikami T, Ueda K (2002) Chloroplast DNA phylogeography of *Fagus crenata* (Fagaceae) in Japan. Plant Systemat Evol 232:21–33

Hamilton MB, Pincus EL, di Fiore A, Fleischer RC (1999) Universal linker and ligation procedures for construction of genomic DNA libraries enriched for microsatellites. Biotechniques 27:500–507

Hamrick JL (1989) Isozymes and the analysis of genetic structure in plant populations. In: Soltis DE, Soltis PS (eds) Isozymes in plant biology. Dioscorides Press, Portland, Oregon, USA, pp 87–105

Igarashi T (1996) The relationship between variation of fructification and efficiency of pollination in *Fagus crenata* and *F. japonica*. Msc Diss, University of Tokyo

Karagyozov L, Kalcheva ID, Chapman VM (1993) Construction of random small-insert genomic libraries highly enriched for simple sequence repeats. Nucleic Acids Res 21:3911–3912

Kelly JK, Willis JH (2002) A manipulative experiment to estimate biparental inbreeding in monkeyflowers. Int J Plant Sci 163:575–579

Kirkpatrick BW, Bradshaw M, Barendse W, Dentine MR (1995) Development of bovine microsatellite markers from a microsatellite-enriched library. Mamm Genome 6:526–528

Kitamura S, Murata G (1979) Colored illustrations of woody plants of Japan, vol II. Hoikusha Publ, Osaka, Japan

Kitamura K, Shimada K, Nakashima K, Kawano S (1997) Demographic genetics of the Japanese beech, *Fagus crenata*, at Ogawa forest preserve,Ibaraki, central Honshu, Japan. I. Spatial genetic substructuring in local population. Plant Species Biol 12:107–136

Kouno K, Mukouda M (1985) Flowering and seed-setting traits of three broadleaf trees, *Fagus crenata*, *Cornus controvera* and *Aesculus turbinata*. Bull Tohoku For Tree Breed Ctr 25:74–76

Lench NJ, Norris A, Bailey A, Booth A, Markham AF (1996) Vectorette PCR isolation of microsatellite repeat sequence using anchored dinucleotide repeat primers. Nucleic Acids Res 24:2190–2191

Litt M, Luty JA (1989) A hypervariable microsatellite revealed by in vitro amplification of a dinucleotide repeat within the cardiac muscle actin gene. Am J Hum Genet 44:397–401

Loveless MD, Hamrick JL (1984) Ecological determinants of generic structure in plant populations. Annu Rev Ecol Sys 15:65–95

Lyall JEW, Brown GM, Furlong RA, Ferguson-Smith MA, Affara NA (1993) A method for creating chromosome-specific plasmid libraries enriched in clones containing [CA]n microsatellite repeat sequences directly from flow-sorted chromosome. Nucleic Acids Res 21:4641–4642

Merzeau D, Comps B, Thiebaut B, Cuguen J, Letouzey J (1994) Genetic structure of natural stands of *Fagus sylvatica* L. (beech). Heredity 72:269–277

Murray MG, Thompson WF (1980) Rapid isolation of high molecular weight plant DANN. Nucleic Acids Res 8:4321–4325

Nakashizuka T (1987) Regeneration dynamic of beech forests in Japan. Vegetatio 69:169–175

Obayashi K, Tsumura Y, Ihara-Ujino T, Niiyama K, Tanouchi H, Suyama Y, Washitani I, Lee C-T, Lee S-L, Muhammad N (2002) Genetic diversity and outcrossing rate between undisturbed and selectively logged forests of *Shorea curtisii* (Dipterocarpaceae) using microsatellite DNA analysis. Int J Plant Sci 163:151–158

Ohkubo T (1992) Structure and dynamics of Japanese beech (*Fagus japonica* Maxim.) stools and sprouts in the regeneration of the natural forests. Vegetatio 101:65–80

Okaura T, Harada K (2002) Phylogeographical structure revealed by chloroplast DNA variation in Japanese beech (*Fagus crenata* Blume). Heredity 88:322–329

Ostrander EA, Jong PM, Rine J, Duyk G (1992) Construction of small-insert genomic DNA libraries highly enriched for microsatellite repeat sequences. Proc Natl Acad Sci USA 89:3419–3423

Pakkanen A, Nikkanen T, Pulkkinen P (2000) Annual variation in pollen contamination and outcrossing in a *Picea abies* seed orchard. Scand J For Res 15:399–404

Peters R (1997) Beech forests. pp. 169, Kluwer, Dordrecht

Rajora OP, Rahman MH, Buchert GP, Dancik BP (2000) Microsatellite DNA analysis of genetic effects of harvesting in old-growth eastern white pine (*Pinus strobus*) in Ontario, Canada. Mol Ecol 9:339–348

Streiff R, Labbe T, Bacilieri R, Steinkellner H, Glossl J, Kremer A (1998) Within-population genetic structure in *Quercus robur* L. and *Quercus petraea* (Matt.) Liebl. assessed with isozymes and microsatellites. Mol Ecol 7:317–328

Takahashi H, Nirawasa N, Furukawa T (1996) An efficient method to clone chicken microsatellite repeat sequences. Jpn Poultry Sci 33:292–299

Takahashi M, Mukouda M, Koono K (2000) Differences in genetic structure between two Japanese beech (*Fagus crenata* Blume) stands. Heredity 84:103–115

Takahashi T, Konuma A, Ohkubo T, Taira H, Tsumura Y (2004) Comparison of spatial genetic structures in *Fagus crenata* and *F. japonica* by the use of microsatellite markers. Silvae Genet (in press)

Tanaka K, Tsumura Y, Nakamura T (1999) Development and polymorphism of microsatellite markers for *Fagus crenata* and closely related species, *F. japonica*. Theor Appl Genet 99:11–15

Tomaru N, Mitsutsuji T, Takahashi M, Tsumura Y, Uchida K, Ohba K (1997) Genetic diversity in *Fagus crenata* (Japanese beech): influence of the distributional shift during the late-Quaternary. Heredity 78:241–251

Tomaru N, Takahashi M, Tsumura Y, Takahashi M, Ohba K (1998) Intraspecific variation and phylogeographic patterns of *Fagus crenata* (Fagaceae) mitochondrial DNA. Am J Bot 85:629–636

Ueno S, Tomaru N, Yoshimaru H, Manabe T, Yamamoto S (2000) Genetic structure of *Camellia japonica* L. in an old-growth evergreen forest, Tsushima, Japan. Mol Ecol 9:647–656

UNESCO (2002) Properties inscribed on the word heritage list world heritage centre. UNESCO, Paris

Weber JL, May PE (1989) Abundant class of human DNA polymorphisms which can be typed using the polymerase chain reaction. Am J Hum Genet 44:388–396

III.7 Molecular Markers in Tree Improvement: Characterisation and Use in *Eucalyptus*

M. SHEPHERD and M.E. JONES[1]

1 Introduction

Eucalypts are the most widely planted hardwood trees in the world, occupying a global estate of around 12 million ha (Turnbull 1999). The genus comprises over 700 species, most of which are endemic to Australia, and its diverse membership offers species with adaptability to a range of exotic tropical and temperate conditions with high growth rates on productive sites (Eldridge et al. 1994). They are a major source of wood for paper pulp and construction timber, as well as fuelwood for industrial and domestic purposes in many developing countries (Eldridge et al. 1994).

Domestication of eucalypts is still at an early stage, with most breeding populations only several generations removed from wild populations (Eldridge et al. 1994). The challenge for eucalypt breeders, as with most tree crops, is to make genetic gains in the face of long generations and delays in selecting mature traits, which can be as long as 20–30 years for wood properties. Eucalypts, having mixed mating systems, are predominantly outcrossing (rates between 0.7 and 0.92), and are thought to possess high levels of genetic load and exhibit deleterious effects when inbred (Eldridge 1970; Potts and Reid 1990; Myburg et al. 2000). Consequently, breeders tend to avoid inbreeding, instead they manage broadly based breeding as well as specialty populations to select for genetic gains (Eldridge et al. 1994). Eucalypts, as a group, are recognised as being promiscuous, with weak reproductive barriers amongst taxa (Pryor 1976), and a frequency of natural and artificial hybridisation that declines as taxonomic distance between parents increases (Griffin et al. 1988; Potts et al. 2001). Interspecific F_1 hybrids feature in a number of breeding programs, often because they combine desirable characteristics from the parental taxa, but also for hybrid superiority imparted through heterosis, epistasis or trait complementarity (Nikles and Griffin 1992). Many tropical eucalypts are amenable to vegetative propagation, hence, one strategy for improvement is intensive within family selection followed by mass clonal reproduction of elite hybrid trees (Eldridge et al. 1994).

---

[1] Centre for Plant Conservation Genetics, Southern Cross University, P.O. Box 157, Lismore, NSW 2480 Australia

Molecular markers have been embraced enthusiastically in the face of these complex and varied challenges for eucalypt breeders, as they offer hope of circumventing some restrictions to or accelerating improvement. In this review, we provide representative or unique case studies where applications of molecular markers are benefiting eucalypt breeders. Initially, we consider how gene resource (base) populations can be managed and best exploited with the aid of molecular markers. Then, a study to determine whether anthropological factors influence the genetic diversity of base populations of a eucalypt is reviewed. Next, we consider how markers are helping to define the gene pool available for breeding eucalypts and how they are revealing new understanding about the genetic mechanisms of hybrid inviability. We review two recent accounts of gene flow and paternity analysis that are providing data to optimise seed orchard design and finally, perhaps the area where markers will ultimately have the greatest impact on breeding, molecular breeding. In recent years, research into the molecular breeding of eucalypts has entered a new era with an emphasis on multi-allelic markers and candidate gene mapping. This second generation of experiments addresses short falls and builds on the discoveries of a first generation of genetic mapping and quantitative trait loci (QTL) experiments in the earlier years of the last decade. On the subject of molecular breeding, our review updates an excellent recent review in this area (Grattapaglia 2000). Grattapaglia (2000) also provides a review of applications of DNA fingerprinting and germplasm management not considered in this review. For a comprehensive review of eucalypt genetics and genecology, see Potts and Wiltshire (1997), and for tree improvement of eucalypts in general, Eldridge et al. (1994).

2 Base Population Characterisation – *Eucalyptus globulus* is Geographically Structured with Chloroplast Haplotypes Coincident with Quantitative Genetic Variation

Our understanding of the natural population structure of *E. globulus* is perhaps the most detailed of any eucalypt. It is a major plantation species for temperate regions in Australia, Chile, Portugal, Spain and China (Eldridge et al. 1994) and has been subject to a concerted research effort over the past decade that has revealed a detailed picture of the relationships and evolutionary forces shaping the quantitative and molecular variation in this species.

Eucalyptus globulus has a broad natural distribution in south-eastern Australia with populations in Victoria, Tasmania and on the islands of Bass Strait. It is currently recognised as having four geographical subspecies, spp. *globulus*, spp. *biocostata*, spp. *pseudoglobulus* and spp. *maidenii* (Chippendale 1988). A major latitudinal cline was evident in random amplified polymorphic DNA (RAPD) markers from the northern mainland Australian localities to the southern Tasmanian localities (Nesbitt et al. 1995). Like

RAPD, restriction fragment length polymorphism (RFLP) variation of *E. globulus* chloroplast revealed lineages that showed strong geographic structuring rather than alignment with taxonomic subspecies boundaries (Jackson et al. 1999). This study also found that chloroplast haplotypes transcended species boundaries, as haplotypes were shared with co-occurring endemic eucalypts. Hybridisation and introgression of chloroplasts is believed to be the most likely explanation for this, emphasising the importance of reticulate evolution in the eucalypt group.

Chloroplast lineages within Tasmania were also found to be coincident with patterns of quantitative genetic variation that had been used to establish a racial classification for *E. globulus* (Dutkowski and Potts 1999; Jackson et al. 1999). Traits including bark thickness, wood basic density and flowering precocity exhibited strong spatial patterns. Racial groups are of interest to breeders as they summarise complex patterns of variation and may improve prediction of breeding values (Dutkowski and Potts 1999).

The development of microsatellite markers in *E. globulus* (Steane et al. 2001) has enabled further molecular characterisation of this species. These markers were found to be highly polymorphic across the geographic range of *E. globulus*. Microsatellite analysis found significant differentiation across the geographic range of ssp. *globulus* and was consistent with the previous RAPD and quantitative genetic studies, providing a further tool for race identification and exploring relationships within and between races.

3 Effect of Utilisation on the Base Population Resource – Influence of Silvicultural and Harvesting on *Eucalyptus sieberi* Genetic Diversity

The impact of native forest management, including logging, upon genetic diversity of *E. sieberi*, a eucalypt from lowland mixed-species forests of south-eastern Australia, was recently examined (Glaubitz et al. 1999; Moran et al. 2000). Three silvicultural treatments that were commonly practiced in this region were investigated; clearfelling with aerial resowing, regeneration from seed trees following burning, and regeneration from seed trees following mechanical disturbance. Several measures of genetic diversity at RFLP and microsatellite loci indicated there were no significant differences amongst treatments or within unharvested controls. As *E. sieberi* is abundant in the region, it was thought that the diversity in regenerated areas was maintained in this system by a high number of pollen parents contributing to offspring in harvested coupes. Although the silvicultural practices apparently do not significantly affect diversity for common widespread species over one generation, it remains to be seen whether these practices are suitable for more localised species or for longer time frames (Moran et al. 2000).

4 Defining the Gene Pool for Breeding – Hybridisation

Eucalypt hybrids have been an important part of plantation forestry, particularly in the tropics (Potts and Dungey 2001). DNA markers are helping breeders to characterise the gene pool available to create hybrid eucalypts and the genetic causes of hybrid inviability.

4.1 Resolving Anomalies in the Success of Wide Hybrids

Hybridisation between subgenera for a long time was thought to delimit the extent of hybridisation in eucalypts (Pryor and Johnson 1981). Reports of an occurrence of a putative natural hybrid between *E. cloeziana* and *E. acmenodies*, members of the *Idiogenes* and *Monocalyptus* subgenera, respectively, were thought to be the single exception (Brooker and Kleinig 1994). Analysis of morphological characters as well as the chloroplast J_{LA} region verified hybridisation and established the direction of crossing between these two species (Stokoe et al. 2001). Genealogical analysis of hybrid offspring based on microsatellite markers indicated that the extent of hybridisation was likely to be restricted to an F_1 generation (Stokoe 2002). Phylogenetic analysis of the J_{LA} region *E. cloeziana* and 20 subgenus *Monocalyptus* species, however, found no supporting evidence for a division between the single monotypic subgenus *Idiogenes* and *Monocalyptus* (Stokoe et al. 2002). Molecular evidence, therefore, suggests there is no basis for a transgression of the subgenera rule.

4.2 Genetic Causes of Hybrid Inviability

A glimpse of the importance and complexity of genic mechanisms maintaining species and defining patterns of hybridisation was obtained from comparisons of transmission ratio distortion (TRD) in genetic maps from interspecific crosses of *E. globulus* and *E. grandis* (Myburg et al. 2003, 2004). In this study, AFLP markers were used to generate maps of an interspecific F_1 individual and two backcross parents. The F_1 *E. globulus* × *E. grandis* cross exhibits high levels of hybrid inviability and breakdown (Griffin et al. 2000). As with many wide crosses, a high proportion (27%) of markers were distorted, but these loci mapped to a few regions and marker alleles were biased to either of the two parents, supporting biological rather than methodological causes of distortion. Multiple putative TRD loci (TRDL) were found in each map, but there was surprisingly little evidence of epistasis between these loci. Those interactions that were detected tended to be between donor and recurrent parent alleles and they tended to be positive, increasing hybrid fitness. Remarkably, the donor genes were not predominantly selected against in the recurrent background, as might be expected with dysfunctional homologous recombination. Preferential survival of donor alleles suggest either the presence of "selfish" genes that enhance the success of gametes or that there was alleviation of genetic load, which is a major factor influencing fitness in forest trees (Williams and Savolainen 1996). Comparison of TRDL between maps revealed some loci were fixed, whereas other loci segregated between the two species.

5 Direct Measures of Gene Flow – Implications for Orchard Design

Gene flow analysis in eucalypts is increasingly relying on direct measures derived from molecular markers because they provide absolute measures, high resolution paternity analysis and enable the separate contributions from pollen and seed to be distinguished (reviewed in Potts and Wiltshire 1997). Isozyme analysis of gene flow in a 10-year-old seed orchard of the insect-pollinated *E. regnans* indicated the likely importance of floral phenology, tree spacing, provenance, and within-orchard position in determining complex mating patterns (Burczyk et al. 2002). Asynchronicity in floral phenology between provenances, for example, was thought to largely explain why intra-provenance crosses were three times more likely to occur than inter-provenance crosses. Despite a preference for within-provenance mating, pollen dispersal was found to be extensive throughout the orchard with 50% of all effective pollen travelling a distance of at least 40 m and bypassing large numbers of nearer neighbours.

Microsatellite markers and paternity assignment were used to study pollen flow in a seed orchard of *E. grandis* (Jones et al., in prep.). Nearly half of the progeny analysed (46%) was found to be the result of pollen originating from trees located outside the seed orchard, probably from nearby surrounding *E. grandis* plantations (Fig. 1). The high level of pollen contamination from outside the *E. grandis* orchard suggested that longer-range pollen dispersal vectors such as flying fox bats may be important for pollen dispersal as they are in some south-eastern, coastal eucalypts (House 1997). Pollination distance within the seed orchard was also extensive with 48% of identified parent pairs located more than 50 m apart, and pollen travelling distances up to the maximum detectable distance of ∼192 m. Of the pollen parents identified within the orchard, the majority were from a provenance that was different to the mother tree (inter-provenance) and low intra-provenance and within-family crossing was observed.

The low level of intra-provenance provenance pollination in the *E. grandis* study contrasted with the preferential intra-provenance mating exhibited in the *E. regnans* orchard (Burczyk et al. 2002). This difference between the two studies may be accounted for by the greater breadth of material studied in the

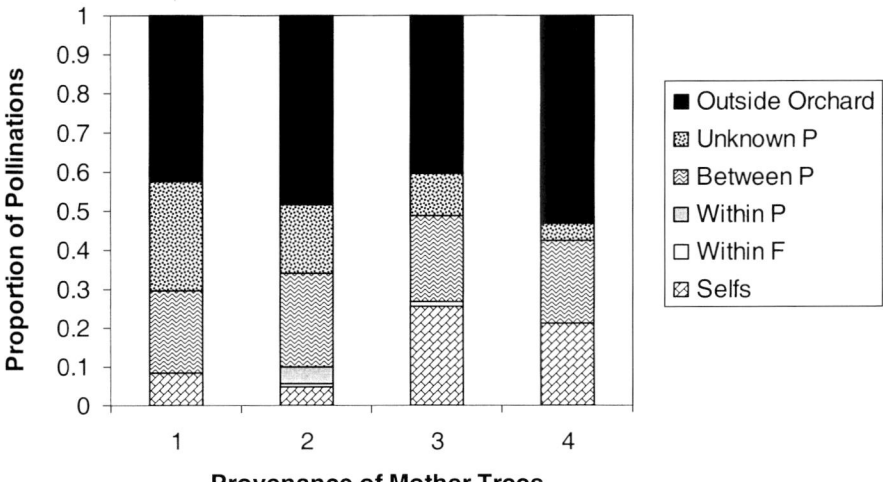

Fig. 1. Pollination events categorised by the degree of relationship between the pollen donor and the mother tree for six *E. grandis* trees. The mother trees are grouped into four provenances: *1* Boambee SF, *2* Pine Creek plantation 3, *3* Newry plantation, *4* Pine Creek plantation 17. Pollination events were classified as; self fertilisation (*Selfs*), within-family (*Within F*), within-provenance (*Within P*), between provenance (*Between P*), unknown provenance (*Unknown P*; orchard trees for which no provenance data is available) and contamination from outside the orchard (*Outside Orchard*)

E. regnans orchard compared to the *E. grandis* orchard. The *E. regnans* orchard was derived from two distinct provenances that differed in flowering time, whereas the *E. grandis* orchard was based on material from the Coffs Harbour region of New South Wales. Although the Coffs Harbour region has been recognised as encompassing multi-provenance sources (Burgess et al. 1996), it is unlikely this material would exhibit the marked differences in flowering phenology observed in the *E. regnans* study, as *E. grandis* is known for high flowering synchronicity across provenances compared to other species (Law et al. 2000).

Among the six mother trees studied in the Coffs Harbour seed orchard, the proportion of progeny derived from self-fertilisation ranged between zero and 36% with an overall average of 13%, consistent with the high degree of between-tree and within-canopy variability in self fertilisation rates of eucalypts (Potts and Wiltshire 1997). This observed outcrossing rate compared closely with an 'effective' multi-locus outcrossing rate (t) of 0.89 based on eight microsatellite markers generated using MLTR v2.4 (Ritland 2002) and was similar to the rate typical for natural populations of this species (0.84; J.C. Bell in Eldridge et al. 1994 p. 194). The level of selfing was apparently higher, however, than that detected in an *E. grandis* orchard in Uruguay, which had low levels of selfing for this species (5%; Russell et al. 2001). At present, we may only speculate on possible causes for apparent differences in breeding system parameters between these studies because of the many sources of variation influencing fecundity that prevent critical comparison. Nonetheless, it seems that the planting of eucalypts as exotics does not necessarily lead to large increases in selfing rates as might be expected with some "offsite" plantings due to poor or sporadic flowering.

6 Genetic Architecture of Commercial Traits – Quantitative Trait Loci and Candidate Gene Mapping

6.1 First Generation Genetic Mapping and Quantitative Trait Loci Studies – Detection of Major Effect Genes and Quantitative Trait Loci Stability Across Physiological Age Classes

Genetic mapping and QTL detection studies in eucalypts over the early part of the last decade were characterised by the use of unplanned crosses, typically F_1 families from interspecific matings and the use of the pseudo-testcross mapping strategy (e.g., Grattapaglia and Sederoff 1994; Grattapaglia et al. 1996; Verhaegen and Plomion 1996; Shepherd et al. 1999). Hybrid families were targeted for these studies because of the favourable prospects for gains from early within-family selection using marker-aided selection (MAS) directly in populations used for deployment (Bradshaw and Grattapaglia 1994). Gains can be realised even earlier in many tropical eucalypts because of their amenability to vegetative propagation (Bradshaw and Foster 1992; Eldridge et al. 1994). There were important exceptions to this approach, however that were based on multi-generation intraspecific crosses or that used family arrays or factorial designs in attempts to identify QTL of average effect (O'Malley and McKeand 1994; Byrne et al. 1995; Grattapaglia et al. 1996; Verhaegen and Plomion 1996; Squilassi and Grattapaglia 1998). A thorough review of earlier work has been recently published (Grattapaglia 2000). Here, we report only the major outcomes, as a prelude to a review of more recent reports.

A major theme that emerged from early genetic mapping and QTL detection experiments was the high proportion of variation in many traits that were apparently controlled by a few major genes (Grattapaglia 2000). The detection of major gene effects for traits believed to have simple underlying genetic control was expected, and recently it has been possible to validate some of these putative QTL for vegetative propagation characteristics (Marques et al. 2002; also see below).

The detection of major gene effects for traits considered as quantitative, however, such as diameter or volume stem growth, was less expected. This outcome may have been a consequence of the predominance of studies based on wide hybrids where atypical large effects segregated as a consequence of hybrid incompatibility. Alternatively, there are confounding issues of limited experimental power and potential problems with sampling effects due to small population sizes, which may have led to detection of false QTL or inflated QTL parameters (Beavis 1998). This issue needs to be resolved by large-scale QTL detection and validation experiments before too much emphasis can be placed on these early estimates of QTL effects (Grattapaglia 2000).

Another outcome from early QTL detection studies was the discovery that the stability of QTL across different physiological ages may be higher than initially thought (Campinhos et al. 1997; Verhaegen et al. 1997; Grattapaglia 2000; P. Bundock, pers. comm.). This was important, as one difficulty in tree improvement is the need to establish reliable correlations between a tree's performance at a young age and that at harvest, as different genes may be influencing a trait at different ages in a tree's life. These studies suggested, however, that some genes (or at least their effect), are important for growth for long periods in a tree's life time.

It was also clear from early QTL studies that there was a need to investigate QTL variability across populations to identify the most useful QTL for breeding (Grattapaglia 2000). Several studies had indicated that genetic background significantly affects QTL detection (Grattapaglia et al. 1996; Verhaegen et al. 1998). Furthermore, little was known about QTL × site interaction (Bradshaw and Grattapaglia 1994; Grattapaglia 2000). These early studies also highlighted the need for more informative multi-allelic marker types, such as microsatellites to facilitate exchange of genetic information within and across species in the genus (Grattapaglia 2000).

6.2 Second Generation Experiments – Multi-Allelic Markers, Quantitative Trait Loci × Site Effects, Candidate Gene Mapping

6.2.1 Microsatellites and Genus-Wide Maps

A genus-wide map for *Eucalyptus* is now feasible with the availability of an abundance of highly variable, multi-allelic microsatellite markers and their ready transfer amongst related species (Brondani et al. 1998, 2002; Byrne et al. 1996; Glaubitz et al. 2001; Jones et al. 2001; Steane et al. 2001). A genetic map with 240 microsatellite loci was developed from a pool of over 500 microsatellite markers (Brondani et al. 2001). Many of these loci exhibited high levels of polymorphism, with average expected heterozygosities in the range of 0.82–0.87 (Brondani et al. 2002). Transferability of microsatellites was high amongst eucalypts with around 80–90% of loci transferring amongst the key commercial species belonging to the *Symphyomyrtus* group, *E. globulus*, *E. grandis*, *E. urophylla* and *E. tereticornis* (Brondani et al. 2001). Similarly high levels of transfer were found for microsatellite markers developed from *Corymbia variegata* (spotted gum; formerly *Eucalyptus*; Jones et al. 2001). All 14 loci tested, transferred to another species within the same

Corymbia section (*Politaria*). Transfer to species belonging to the *Eucalyptus* subgenus *Symphyomyrtus* was also high (40–50%), with lower transfer to more distal species in the subgenus *Monocalyptus* (21%) and *E. cloeziana* (29%; subgenus *Idiogenes*).

6.2.2 Transfer of Genetic Information Across Populations and Species – Quantitative Trait Loci Stability Across Genetic Backgrounds and Environments

The development of highly informative, transferable microsatellite markers has been a major step towards studies of QTL diversity at a breeding population level in eucalypts, and the exchange of genetic information amongst pedigrees and species. The advantage in exchanging genetic information across species was demonstrated in the validation of QTL for vegetative propagation characteristics (Marques et al. 2002). Comparative mapping of microsatellite loci and QTL influencing sprouting and adventitious rooting ability was possible in four species, *E. grandis*, *E. urophylla*, *E. tereticornis* and *E. globulus*. Using a set of 40 microsatellites, many homeologous linkage groups were identified amongst these species and in most cases, locus order was conserved. Putative QTL for adventitious root formation were located on homeologous linkage groups of two species, providing independent validation that genes influencing rooting were located in this region of the eucalypt genome. Putative QTL for sprouting were located on a homeologous linkage group of a third species, indicating that a cluster of genes influencing different aspects of vegetative propagation could be located in this region, or that there was pleiotropy of the same major gene.

Highly variable co-dominant markers in mapping experiments will also assist to transfer genetic information amongst pedigrees in efforts to introgress early flowering genes into elite clones in eucalypts (Missiaggia et al. 2002). Genes for early flowering could be important in a eucalypt breeding program to allow accelerated breeding cycles and the potential to develop inbred lines (Missiaggia et al. 2002). Linkage between a mutant early flowering phenotype and a microsatellite marker was established by selective genotyping in one family and it is anticipated that once other markers are found to bracket the QTL, it will be transferred into elite clones.

A recent QTL detection study in *E. globulus* was encouraging in that at least some QTL for wood density were stable across environments (P. Bundock, pers. comm.). In this experiment, a single family was grown across seven sites in Australia. There was no evidence of QTL × site interaction for pilodyn penetration, an indirect measure of wood density, yet volume growth exhibited large QTL × site interaction.

6.2.3 Gene Discovery and Candidate Gene Mapping

There has been an intensive effort over the past few decades to understand the cellular processes and more recently the molecular aspects of wood formation (Jain and Minocha 2000; Savidge et al. 2000). As a result, many of the key features of the development and function of the vascular cambium, the differentiation and control of xylem (wood) cell formation, and the biosynthesis of the major components of their cell walls, cellulose and lignin, have been elucidated. Some of the genes involved in these processes are known and cloned as a result of functional or mutation analysis (Bossinger and Leitch 2000; Hertzberg et al. 2001; Whetten et al. 2001). Other genes with unknown function, but believed to be involved in the processes of cell wall formation, developmental regulation, signal transduction and hormone biosynthesis have also been cloned (Sterky et al. 1998). Currently, there are over 273 eucalypt sequences in the genetic database GenBank (March 2002), with about 50% of these associated or likely to be associated with wood formation. Recently, a major genomics initiative in eucalypts, the GENOLYPTUS project has commenced in Brazil (Grattapaglia 2002; D. Grattapaglia, pers. comm.). The objective of this project is to discover, map, validate and characterise genes of economic importance in eucalypts. A key aspect of this work will be the establishment of segregating reference populations for genetic mapping experiments similar to those available in humans.

In the first step toward understanding the relationship between variation in gene sequence and its phenotypic effect, candidate genes for wood quality and other traits were mapped in eucalypts. In one example, six genes of known function involved in lignin biosynthesis or the common phenylpropanoid pathway, and which may be related to QTL controlling wood quality, as well as two genes implicated in morphological formation in roots, were mapped using polymorphism detected by single strand conformation polymorphism (SSCP) in a *E. grandis* × *E. urophylla* family (Gion et al. 1999). In a second study, genes of known function involved in monolignol biosynthesis and floral expression in addition to 31 cambium-specific expressed sequence tags (EST), were mapped by RFLP on to a genetic map for *E. globulus* (Thamarus et al. 2002). The next crucial step in this area will be the linking of candidate genes with phenotypic values of characteristics of economic importance by QTL or association studies (Brown et al. 2001; Thamarus et al. 2002).

Acknowledgements. The authors thank B. Potts, R. Vaillancourt, P. Bundock, R. Griffin and Z. Myburg for helpful comments on the manuscript.

References

Beavis WD (1998) QTL analysis: power, precision, and accuracy. In: Paterson AH (ed) Molecular dissection of complex traits. CRC Press, Boca Raton, pp 145–162

Bossinger G, Leitch M (2000) Isolation of cambium-specific genes from *Eucalyptus globulus* Labill. In: Savidge R, Barnett J Napier (eds) Cell and molecular biology of wood formation. BIOS Scientific, Oxford, pp 203–207

Bradshaw HD, Foster GS (1992) Marker aided selection and propagation systems in trees. Advantages of cloning for studying quantitative inheritance. Can J For Res 22:1044–1049

Bradshaw HD, Grattapaglia D (1994) QTL mapping in interspecific hybrids of forest trees. For Genet 1:191–196

Brondani RPV, Brondani C, Tarchini R, Grattapaglia D (1998) Development, characterisation and mapping of microsatellite markers in *Eucalyptus grandis* and *E. urophylla*. Theor Appl Genet 97:816–827

Brondani R, Kirst M, Ribeiro V, Gaiotto F, Marques C, Nichols D, Williams E, Grattapaglia D (2001) Fingerprinting and mapping *Eucalyptus* with large batteries of microsatellite loci. Paper presented at the plant and animal genome IX conference, 13–17 Jan 2001, San Diego

Brondani RPV, Brondani C, Grattapaglia D (2002) Towards a genus-wide reference linkage map for *Eucalyptus* based exclusively on highly informative microsatellite markers. Mol Genet Genom 267:338–347

Brooker MIH, Kleinig DA (1994) Field guide to eucalypts, vol. 3. Northern Australia. Inkata Press, Sydney

Brown GR, Gill GP, Sewell MM, Wheeler NC, Megraw RA, Neale DB (2001) Towards association studies in forest trees: wood property QTL verification, candidate genes, and SNPs in Loblolly pine (*Pinus taeda* L.). Paper presented at the international conference on wood, breeding, biotechnology and industrial expectations, 11–14 June 2001, Bordeaux, France

Burczyk J, Adam W, Moran G, Griffin A (2002) Complex patterns of mating revealed in a *Eucalyptus regnans* seed orchard using allozyme markers and the neighbourhood model. Mol Ecol 11:2379–2391

Burgess IP, Williams ER, Bell JC, Harwood CE, Owen JV (1996) The effect of outcrossing rate on the growth of selected families of *Eucalyptus grandis*. Silvae Genet 45(2–3):97–101

Byrne M, Murrell JC, Allen B, Moran G (1995) An integrated genetic linkage map for eucalypts using RFLP, RAPD and isozyme markers. Theor Appl Genet 91:869–875

Byrne M, Marquez-Garcia MI, Uren T, Smith DS, Moran GF (1996) Conservation and genetic diversity of microsatellite loci in the genus *Eucalyptus*. Aust J Bot 44:331–341

Campinhos EN, Grattapaglia D, Alfenas AC, Bertolucci FL (1997) Stability of expression of QTL alleles controlling growth across variable genetic backgrounds in *Eucalyptus*. Paper presented at the proceedings of the IUFRO conference on silviculture and improvement of eucalypts, 24–29 Aug 1997, Salvador, Brazil

Chippendale GM (1988) *Eucalyptus*, *Angophora* (Myrtaceae). Flora of Australia, vol 19. Australian Govt. Publishing Service, Canberra

Dutkowski GW, Potts BM (1999) Geographic patterns of genetic variation in *Eucalyptus globulus* spp. *globulus* and a revised racial classification. Aust J Bot 47:237–263

Eldridge K (1970) Breeding system of *Eucalyptus regnans*. Paper presented at the proceedings of IUFRO section 22 working group on sexual reproduction of forest trees, Varparanta, Finland

Eldridge K, Davidson J, Harwood C, van Wyk G (1994) Eucalypt domestication and breeding, 1st edn. Oxford University Press, Oxford

Gion J-M, Rech P, Grima-Pettenati J, Verhaegen D, Plomion C (1999) Mapping candidate genes in *Eucalyptus* with emphasis on lignification genes. Mol Breed 6:441–449

Glaubitz J, Strk J, Moran G (1999) Genetic impact of different silvicultural practices in native eucalypt forests. Paper presented at the forest genetics and sustainability; Proceeding of IUFRO Conference, Beijing, China

Glaubitz J, Emebiri, Moran G (2001) Dinucleotide microsatellites in *Eucalyptus sieberi*: inheritance, diversity and improved scoring of single base differences. Genome 44:1041–1045

Grattapaglia D (2000) In: Jain SM, Minocha SC (Eds.) Molecular biology of wood plants, vol. 1. Kluwer Academic, Dordrecht, pp 451–474

Grattapaglia D (2002) Integrating genomic biotechnologies into genetic improvement of *Eucalyptus*: the GENOYPTUS project in Brazil. Paper presented at the Simposio Internacional sobre socioeconomia, tecnologia, patologia y sostenibilidad del eucalipto, Pontevedra, Spain, 29–31 May 2002, 16 pp

Grattapaglia D, Sederoff R (1994) Genetic linkage maps of *Eucalyptus grandis* and *Eucalyptus urophylla* using a pseudo-testcross: mapping strategy and RAPD markers. Genetics 137:1121–1137

Grattapaglia D, Bertolucci FLG, Penchel R, Sederoff R (1996) Genetic mapping of quantitative trait loci controlling growth and wood quality traits in *Eucalyptus grandis* using a maternal half-sib family and RAPD markers. Genetics 144:1205–1214

Griffin AR, Burgess IP, Wolf L (1988) Patterns of natural and manipulated hybridisation in the genus *Eucalyptus* L'Herit – a review. Aust J Bot 36:41–66

Griffin A, Harbard J, Centurion C, Santini P (2000) Breeding *Eucalyptus grandis* × *globulus* and other inter-specific hybrids with high inviability – problem analysis and experience with Shell forestry projects in Uruguay and Chile. Hybrid Breeding and Genetics Symposium, 9–14 April 2000, Noosa, Queensland, Australia, pp 1–13

Hertzberg M, Aspeborg H, Schrader J, Andersson A, Erlandsson R, Blomqvist K, Bhalerao R, Uhlen M, Teeri TT, Lundeberg J, Sundberg B, Nilsson P, Sandberg G (2001) A transcriptional roadmap to wood formation. Proc Natl Acad Sci USA 98:14732–14737

House SM (1997) Reproductive biology of eucalypts. In: Williams J, Woinarski J (eds) Eucalypt ecology. Cambridge Univ Press, Cambridge, pp 30–55

Jackson HD, Steane DA, Potts BM, Vaillancourt RE (1999) Chloroplast DNA evidence for reticulate evolution in *Eucalyptus* (Myrtaceae). Mol Ecol 8:739–751

Jain SM, Minocha SC (2000) Molecular biology of woody plants, vol 1. Kluwer Academic, Dordrecht

Jones M, Stokoe R, Cross M, Scott L, Maguire T, Shepherd M (2001) Isolation of microsatellite loci from spotted gum (*Corymbia variegata*), and cross-species amplification in *Corymbia* and *Eucalyptus*. Mol Ecol Notes 1:276–278

Law B, Mackowski C, Schoer L, Tweedie T (2000) Flowering phenology of myrtaceous trees and their relation to climatic, environmental and disturbance variables in northern New South Wales. Aust Ecol 25:160–178

Marques C, Brondani R, Grattapaglia D, Sederoff R (2002) Conservation and synteny of SSR loci and QTLs for vegetative propagation in four *Eucalyptus* species. Theor Appl Genet 105:474–478

Missiaggia M, Piaceza A, Grattapaglia D (2002) A major effect QTL for early flowering in *Eucalyptus* mapped by selective genotyping of microsatellite markers detected in fluorescent multiplexes. Paper presented at the Proceedings of the 48th Brazilian Congress of Genetics, 17–20 Sept 2002, Aguas de Lindoir

Moran GF, Butcher PA, Glaubitz JC (2000) Application of genetic markers in the domestication, conservation and utilisation of genetic resources of Australasian tree species. Aust J Bot 48:313–320

Myburg AA, Griffin R, Sederoff RR, Whetten R (2000) Genetic analysis of interspecific backcross families of a hybrid of *Eucalyptus grandis* and *Eucalyptus globulus*. Paper presented at the Hybrid Breeding and Genetics of Forest Trees Proceedings of QFRI/CRC-SPF Symposium, 9–14 April 2000, Noosa, Queensland, Australia, pp 462–467

Myburg AA, Griffin RA, Sederoff RR, Whetten RW (2003) Comparative genetic linkage maps of *Eucalyptus grandis*, *Eucalyptus globulus* and their F-1 hybrid based on a double pseudo-backcross mapping approach Theor Appl Genet 107:1028–1042

Myburg AA, Vogl C, Griffin RA, Sederoff RR, Whetten RW (2004) Genetics of postzygotic isolation in *Eucalyptus* II. Whole-genome analysis of barriers to introgression in a wide interspecific cross of *E. grandis* and *E. globulus*. Genetics 166:1405–1418

Nesbitt KA, Potts BM, Vaillancourt RE, West AK, Reid JB (1995) Partitioning and distribution of RAPD variation in a forest tree species, *Eucalyptus globulus* (Myrtaceae). Heredity 74:628–637

Nikles DG, Griffin AR (1992) Breeding hybrids of forest trees: definitions, theory, some practical examples and guidelines on strategy with tropical acacias. ACIAR Proc 37:101–109

O'Malley DM, McKeand SE (1994) Marker-assisted selection for breeding value in forest trees. For Genet 1:207–218

Potts BM, Reid JB (1990) The evolutionary significance of hybridisation in *Eucalyptus*. Evolution 44:2151–2152

Potts BM, Wiltshire RJE (1997) Eucalypt genetics and genecology. In: Williams J, Woinarski J (eds) Eucalypt ecology: individuals to ecosystems. Cambridge Univ Press, Cambridge, pp 56–91

Potts BM, Dungey HS (2001) Hybridisation of *Eucalyptus*: key issues for breeders and geneticists. Paper presented at the IUFRO Conference: Developing the eucalypt of the future, Valdivia, Chile.

Potts BM, Barbour RC, Hingston AB (2001) Genetic pollution from farm forestry using eucalypt species and hybrids. A report for RIRDC/L&WA/FWPRDC joint venture agroforestry program, Sept 2001, RIRDC publ no 01/114

Pryor LD (1976) The biology of eucalypts. Arnold, London

Pryor LD, Johnson LAS (1981) *Eucalyptus*, the universal Australian. In: Keast A (ed) Ecological biogeography of Australia. Junk, The Hague, pp 499–536

Ritland K (2002) Extensions of models for the estimation of mating systems using n independent loci. Heredity 88:221–228

Russell J, Marshall D, Griffin R, Harbard J, Powell W (2001) Gene flow in South American *Eucalyptus grandis* and *Eucalyptus globulus* seed orchards. In: Barros S (ed) "Developing the eucalypt of the future." Proceedings of IUFRO international symposium, Valdivia, Chile

Savidge R (2000) Biochemistry of seasonal cambial growth and wood formation – an overview of the challenges. In: Savidge R, Barnett J, Napier R (eds) Cell and molecular biology of wood formation. BIOS Scientific Ltd, Oxford, pp 1–28

Shepherd M, Chaparro JX, Teasdale R (1999) Genetic mapping of monoterpene composition in an interspecific eucalypt hybrid. Theor Appl Genet 99: 1207–1215

Squilassi M, Grattapaglia D (1998) Mapping QTL using linkage disequilibrium and efficiency of early marker-assisted selection in *Eucalyptus*. Paper presented at the plant and animal genome VI, Jan 1998, San Diego

Steane DA, Vaillancourt RE, Russell J, Powell W, Marshall D, Potts BM (2001) Development and characterisation of microsatellite loci in *Eucalyptus globulus* (Myrtaceae). Silvae Genet 50:89–91

Sterky F, Regan S, Karlsson J, Hertzberg M, Rohde A, Holmberg A, Amini B, Bhalerao R, Larsson M, Villarroel R, Vanmontagu M, Sandberg G, Olsson O, Teeri TT, Boerjan W, Gustafsson P, Uhlen M, Sundberg B, Lundeberg J (1998) Gene discovery in the wood-forming tissues of poplar – analysis of 5,692 expressed sequence tags. Proc Natl Acad Sci USA 95:13330–13335

Stokoe RL (2002) Pattern of genetic diversity and hybridisation of *Eucalyptus cloeziana* F. Muell (Myrtaceae), PhD Thesis, Southern Cross University, Lismore

Stokoe RL, Shepherd M, Lee D, Nikles DG, Henry RJ (2001) Natural interspecific hybridisation between *Eucalyptus acmenoides* Schauer and *E. cloeziana* F. Muell (Myrtaceae). Ann Bot 88:563–570

Thamarus K, Groom K, Murrell J, Byrne M, Moran G (2002) A genetic linkage map for *Eucalyptus globulus* with candidate loci for wood, fibre and floral traits. Theor Appl Genet 104:379–387

Turnbull J (1999) Eucalypt plantations. New For 17:37–52

Verhaegen D, Plomion C (1996) Genetic mapping in *Eucalyptus urophylla* and *E. grandis* using RAPD markers. Genome 39:1051–1061

Verhaegen D, Plomion C, Gion JM, Poitel M, Costa P, Kremer A (1997) Quantitative trait dissection analysis in *Eucalyptus* using RAPD markers 1. Detection of QTL in interspecific hybrid progeny, stability of QTL expression across different ages. Theor Appl Genet 95:597–608

Verhaegen D, Plomion C, Poitel M, Costa P, Kremer A (1998) Quantitative trait dissection analysis in *Eucalyptus* using RAPD markers 2. Linkage disequilibrium in a factorial design between *E. urophylla* and *E. grandis*. For Genet 5:61–69

Whetten R, Sun Y, Zhang Y, Sederoff R (2001) Functional genomics and cell wall biosynthesis in loblolly pine. Plant Mol Bio 47:275–291

Williams CG, Savolainen O (1996) Inbreeding depression in conifers. For Sci 41:1–20

III.8 DNA Markers for Identification and Evaluation of Genetic Resources in Forest Trees: Case Studies in *Abies, Picea* and *Populus*

B. Ziegenhagen[1] and M. Fladung[2]

1 Introduction

1.1 Concept and History of Genetic Markers in Forest Trees

The underlying principle of genetic markers is the biological variation in the organism systems being analysed. Biological variation as a result of, and at the same time, pre-requisite of evolution is the focus of interest in modern plant sciences:

1. It is analysed for conservation purposes
2. It is the basis for selecting or recombining traits in breeding programmes and gene technology, and with special regard to the present chapter
3. It is the source of markers for tracing different states and processes that organisms undergo at different levels of their organisation.

Once the latter processes are understood, a sustainable management of biodiversity and crop or tree improvement may be strongly enhanced. The development and usage of markers is, therefore, a desirable goal in research and practice.

By definition, genetic markers involve evolutionary principles: a variable trait or phenotype is defined as a genetic marker when the relationship between phenotype and the underlying genotype is clearly determined by inheritance analysis (Gillet 1999). Consequently, a marker variation relying on a single locus of the nuclear genome needs to follow Mendelian segregation. A marker variation relying on a single locus of the organelle genome needs to be validated for uniparental inheritance. A genetic marker in sensu strictu is a phenotype or trait that is defined by just one unambiguous genotype or just one unambiguous haplotype.

One of the first genetic markers used in population genetics of forest trees was an *aurea* mutant phenotype in Norway spruce (*Picea abies* Karst L.). *Aurea* and the wild-type phenotype were found to segregate according to a

---

[1] Philipps-University of Marburg, Faculty of Biology, Conservation Biology, Karl-von-Frisch-Strasse, 35032 Marburg, Germany
[2] Federal Research Centre for Forestry and Forest Products, Institute for Forest Genetics and Forest Tree Breeding, Sieker Landstrasse 2, 22927 Grosshansdorf, Germany

codominant mode of gene action at a single gene locus (Langner 1953). The *aurea* phenotype, therefore, was a genetic marker in sensu strictu and allowed the tracing of gene flow by pollen and seeds within the study area in the Arboretum Tannenhöft (Institute for Forest Genetics and Forest Tree Breeding, Grosshansdorf, Germany, Langner 1953). The usage of mutant phenotypes as genetic markers, however, is necessarily restricted by low numbers of known mutants.

A new source of markers was exploited in the early 1970s. Isoforms of constitutively expressed proteins were introduced as genetic markers in forest trees (Bartels 1971a, b; Bergmann 1973, 1974a, b). Since then, about 20 isozyme gene systems have been validated by means of inheritance analysis and used in population and conservation genetics of forest trees (for a general review, see Hamrick and Godt 1990; for European forest trees, see review in Müller-Starck and Ziehe 1992).

A novel generation of genetic markers was provided with the use of DNA. In forest trees, DNA markers were developed and applied comparably late at the beginning of the 1980s (Neale et al. 1992). In contrast to the previous markers, the putative number of DNA markers is nearly infinite and only restricted by the respective genome sizes. Furthermore, a different origin of the markers from either the nuclear or organelle genome allows differentiation among gene flow via pollen and seeds. Depending on the genomic position of the marker loci, the underlying variation may vary by mutation processes and mutation rates (e.g., Hewitt 2000). Using the respective markers, it is, therefore, possible to reconstruct different processes ranging from higher taxonomic levels down to the individual level (e.g., Ziegenhagen and Fladung 1997a, b; Gillet 1999). Using DNA markers, in principle, it is feasible to separately access non-coding or coding regulating regions of the genome. This is potentially useful for estimating the strength and direction of different evolutionary drivers like genetic drift or selection and adaptation. Techniques of identifying and routinely monitoring DNA polymorphism have dramatically improved. The former laborious RFLP analyses (restriction fragment length polymorphism), including radioactively labelled probes, have been replaced by PCR routines. The latter have become a key technology in the detection and diagnostics of DNA polymorphism and thus, in marker application. Marker application in population genetics or conservation genetics of forest trees mainly includes large sample sizes. Thus, automated technologies are increasingly covering the whole procedure from DNA extraction to multiplex gel electrophoresis that either serves for automated detection and scoring of fragment length polymorphism, or for high-throughput sequence analysis (overview in Glaubitz and Moran 2000).

1.2 Genetic Resources and Their Management in Forest Trees

In contrast to most crops, forest tree species have not experienced a long breeding history. Most of them are still regarded as undomesticated wild populations of comparably high levels of genetic diversity. The future effects of global change and/or the present over-exploitation of the tropical and boreal forest ecosystems, however, are a severe threat to these natural forest genetic resources. Young et al. (2000) consider forest trees as paradigms of conservation genetics due to extreme life history traits (longevity, accumulation of mutations, large ranges of mating and dispersal) and the variety of stresses they are forced to adapt to. In contrast to crop genetic resources, the main efforts in forest trees have been put into in situ conservation strategies. In addition, regarding the extreme features, it is a challenge to understand the spatio-temporal distribution of their genetic resources and even more, the underlying processes. Genetic markers greatly facilitate the analysis of patterns and dynamics of genetic diversity in forest tree species (Young et al. 2000). Ex situ conservation practices are a different topic which is not dealt within this chapter.

Trees grow and/or are cultivated in more or less natural forest stands. Some species, however, like poplar, radiata pine, eucalyptus or teak, are also cultivated in high-yield plantations. The management policy behind this is to exploit these plantations to the greatest possible extent and thereby protect the natural resources (Gladstone and Ledig 1990).

In Germany, tree plantations are being considered for the reforestation of former agriculturally used areas. In this context, the questions regarding genetic resources are: from where to select the pheno-/genotypes, and if clones are used: how many of them are needed to guarantee an adaptive potential against pests and abiotic stress (for 'clone mixtures', see FSaatG, National German Regulation on the trade with forest reproductive material, 1994). Genetic markers are useful to identify clones or to verify genomic stability of material originating from micro-propagation or manipulation of the ploidy level. Plantation forestry is furthermore practiced in field release trials of transgenic trees. Here, molecular markers may assist in risk assessment scenarios related to horizontal and vertical gene transfer.

2 Which DNA Marker at Which Scale and for Which Purpose?

Since their usage in forest genetics, molecular markers have shown a great capacity for tracing the distribution of genetic diversity at many different scales reflecting different evolutionary processes. These are macro- and micro-evolutionary processes based on different mutations. Likewise, the respective molecular markers have to be carefully selected and validated: there is no universal marker which can be applied throughout all scales of

Table 1. Which neutral plant DNA marker at which scale for which purpose?

	Above species level	Within species level		
Power of discrimination	Differentiation of species	Differentiation of populations	Differentiation of individuals/clones	Differentiation of within-individual variation
Process	Hybridisation, introgression	Phylogeography, migration, gene flow, introgression of lineages	Mating, gene flow, drift, seed dispersal, vegetative propagation	Genomic stability: aneuploidy, chimera
DNA markers genomic origin	Organelle DNA markers, often chloroplast DNA markers, often cp genes (above genera) or introns (within genera)	Maternally inherited organelle DNA markers, often introns and intergenic spacer or organelle SSRs	Nuclear SSRs, AFLPs and RAPDs, SNPs in non-coding regions	Nuclear SSRs, ISSRs
Diagnostic techniques	Sequencing	Sequencing, PCR-RFLP, automated fragment length analysis	Automated fragment length analysis (SSRs, AFLPs, RAPDs) Sequencing, SSCP (SNPs)	Automated fragment length analysis
Mutation rate	Extremely low to low	Low	High	High
Mode of inheritance/gene action	Uniparental transmission	Uniparentally maternal inheritance	Biparental Codominant Dominant	Biparental Codominant
References/reviews	Chase et al. (1993); Gielly and Taberlet (1994); Wu et al. (1998); Hewitt (2000)	Petit et al. (2003); Petit and Vendramin (2003); Vendramin et al. (2003)	Gillet (1999); Vendramin et al. (2003)	Gomez et al. (2001); Hristoforoghu et al. (2000); Leroy et al. (2000); Deutsch et al. (submitted)

SSR Simple sequence repeat, *AFLP* amplified fragment length polymorphism, *RAPD* random amplified polymorphic DNA, *SNP* single nucleotide polymorphism, *SSCP* single-strand confirmation polymorphism, *ISSR* inter-sequence simple repeat

interest. Before starting an experiment or study, the appropriate marker category, therefore, has to be determined (Gillet 1999). Table 1 gives a survey of the available neutral DNA markers used to trace different processes related to issues of genetic resources. It is a rough orientation and should not be considered complete. The references are also incomplete and serve for orientation only.

3 Case Studies with Fir (*Abies* sp.), Norway Spruce [*Picea abies* (Karst.) L.] and Poplar (*Populus* sp.)

From our own fields of research, we selected case studies to demonstrate the power of DNA markers for identifying gene pools, maternal and paternal lineages, as well as past genetic bottlenecks at a broader geographical scale. At a regional or local scale, factors affecting the level of genetic diversity are a matter of concern. DNA markers are tools to understand the processes of local gene flow and their disturbance. In natural stands, used as certified seed lots, it is of interest to identify the mother trees from subsets of their seeds. Furthermore, spatial patterns of intron and exon nucleotide variation in a candidate gene were analysed in order to determine whether this could be an approach to develop adaptively relevant genetic markers. Forest trees are not only grown in more or less natural forest stands, but are cultivated in plantations as well. While fir and Norway spruce grow in natural forest sites of more or less continuous distribution throughout their natural range, poplar is mainly used in plantation forestry and is among those tree species that are routinely genetically transformed.

3.1 DNA Markers in Natural Populations

Abies is one of the forest trees where we have conducted a multi-scale analysis using DNA markers. *Abies* is a genus with nine species and found in central Europe and the Mediterranean area. These species are allopatric with the exception of *A. alba* and *A. cephalonica*, which are parapatric (Parducci 2000). The central European silver fir (*Abies alba* Mill.) is the most common European fir. It is the conifer species of mountainous areas where it predominantly occurs in mixed forest stands together with Norway spruce and beech (*Fagus sylvatica*). It is supposed to be natural in most of its stands and an ecologically important member of the mountainous ecosystems due to its extensive root systems. Conservation programmes have been initiated after the forest decline of the 1970s. Such programmes necessarily rely on the knowledge of large-scale spatial distribution and patterns of genetic resources, as well as on critical levels of genetic diversity within populations.

3.1.1 Identification of Maternal and Paternal Lineages and Historical Gene Flow

From isozyme gene studies, it is known that *Abies alba* Mill. survived in at least two southern refugia during the last glaciation (reviews in Konnert and Bergmann 1995 and Hewitt 1999). The development of organelle DNA markers with contrasting modes of inheritance allowed a new, differentiated perspective on the range-wide distribution of maternal and paternal lineages (Liepelt et al. 2002). Two DNA markers with contrasting modes of inheritance were applied to 100 populations covering the entire range of silver fir in Europe. The markers each exhibited two highly conserved alleles based on an insertion/deletion of 80 bp in the fourth intron of the mitochondrial *nad*5 gene and on a synonymous substitution in the chloroplast *psb*C gene. The geographical distribution of the maternally inherited mitochondrial variation confirmed the existence of at least two refugia with two recolonizing maternal lineages remaining largely separated throughout the range. Silver fir [*Abies alba* (Mill.)] turned out to be an excellent model to test whether pollen-mediated gene flow may eliminate the genetic imprints of Pleistocene refugial isolation. As the cline calculated from the *psb*C allele frequencies was as wide as the whole range, our results provided first striking evidence that even a species with very long generation times and heavy pollen grains was able to establish a highly efficient pollen-mediated gene flow between refugia (Liepelt et al. 2002). The same markers (*nad*5-4, *psb*C) and one additional cpDNA marker (intergenic region of *trn*C and *trn*D, primers designed by Parducci and Szmidt 1999) were applied at the next higher taxonomic level, namely throughout the Mediterranean *Abies* species (Liepelt et al., unpubl.). Similar phenomena were observed: the maternally inherited mtDNA marker *nad*5-4 distinguished five European maternal lineages, which were strictly separated in most cases. Introgression between maternal lineages was only observed among the parapatric species *A. alba* and *A. cephalonica*. Using the paternally inherited markers, a highly effective pollen-mediated interspecific gene flow was suggested (Liepelt et al., unpubl.), confirming a former hypothesis on weak interspecific reproductive barriers (Kormuták et al. 2002; Kormuták, pers. comm.). The study provided a new possibility of comparing the genetic consequences of forest genetic resources using DNA markers with a contrasting mode of inheritance. Particularly, the postglacial range-wide exchange of the paternally inherited variation poses the question: what is autochthony in terms of time and space and what is the evolutionary background of the delineations of common seed zones?

3.1.2 Identification of Genetic Bottlenecks and Genetic Erosion

In the same species, highly polymorphic chloroplast microsatellite (SSR simple sequence repeat) markers have been developed and proved to be uniparentally, i.e. paternally, inherited (Vendramin and Ziegenhagen 1997). The marker was supposed to be highly sensitive towards those stochastic population, genetic or demographic processes that may cause loss of genetic variation (Ziegenhagen et al. 2001). This sensitivity is based on the combination of three features characteristic for this type of marker: (1) selective neutrality, (2) high degree of polymorphism, and (3) uniparental inheritance. The latter reduces the reproductively effective population size to half the size of that when biparentally inherited markers are used (Birky et al. 1989). The loss of genetic diversity as depicted by this kind of marker is expected to be highest when isolation and drift have both become effective. To estimate the degree of such loss in conservation genetic studies, the measure 'allelic richness' was shown to be highly suited (Petit et al. 1998).

A range-wide analysis of a total of 714 individuals at two such loci resulted in the detection of 90 different two-locus-haplotypes (Vendramin et al. 1999). A Pyrenean population was found to exhibit the lowest values of allelic richness. Therefore, it was concluded that it suffered from a past bottleneck due to isolation and drift (see also Konnert and Bergmann 1995).

The same marker was tested for its usability in paternity analyses in a relic Ore Mountain silver fir stand (Ziegenhagen et al. 1998). In such relic stands, silver fir is supposed to suffer from inbreeding deficiency (Llamas-Gomez and Braun 1994). As demonstrated by our study, chloroplast microsatellites may be useful to estimate pollen trapping or outcrossing rates of isolated trees or groups of trees in relic stands.

3.1.3 Seed Source Identification

A current concern is how to improve control methods for verifying the origin and identity of traded forest seeds (Konnert et al. 2002; Cremer et al. 2003; Ziegenhagen et al. 2003). Such a control should ensure the goal of maintaining an appropriate level of genetic diversity in offspring from seed lots. A methodological breakthrough was obtained in *Abies alba*. It was possible to extract and analyse DNA from single dry wings of the seeds (Ziegenhagen et al. 2003). Using three highly polymorphic chloroplast microsatellite loci (Vendramin and Ziegenhagen 1997; Liepelt et al. 2001), the wings were haplotyped. A comparison with the haplotypes of endosperm and embryos of the same seeds revealed the unambiguous maternal origin of the wing tissue (Fig. 1). It was also demonstrated that individual mother trees could be identified from the haplotypes of the single wings. Thus, a tool has become available for direct molecular identification of mother trees by simply genotyping or haplotyping maternal tissues of their fruits or seeds (Ziegenhagen et al. 2003). Such tools are currently being introduced in certification initiatives to control the trade of forest seeds in southern Germany (Konnert et al. 2002).

3.1.4 Nucleotide Variation Within Introns and Exons of a Gene – a Source of Markers for Adaptively Relevant Variation?

So far, we have introduced case studies on DNA markers originating from more or less non-coding genomic regions. These markers can be regarded as selectively neutral as there has been no evidence of any selective relevance up to now. It is of great future importance to develop markers for tracing vari-

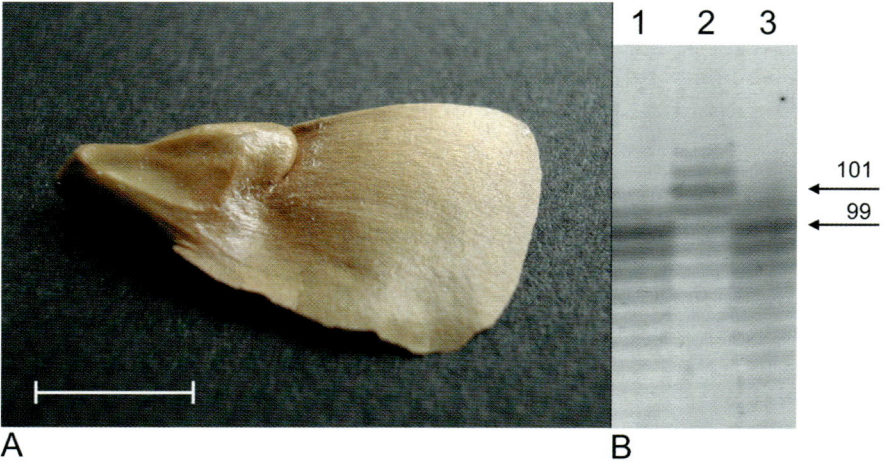

Fig. 1. Morphology of an *Abies alba* (Mill.) seed and haplotypes of its endosperm, embryo and wing. **A** Seed of *A. alba* including the dry membranous wing, *scale bar* 1 cm. **B** Polyacrylamide gel electrophoresis of the endosperm (*lane 1*), the embryo (*lane 2*) and the wing (*lane 3*) of the same seed of an *A. alba* individual analysed at the chloroplast microsatellite locus Pt 30249 (Liepelt et al. 2001). The gel exhibits typical patterns with prominent target alleles and slippage bands in the same lanes. *Arrows* indicate the target alleles, the sizes of which are given in base pairs

able genomic regions or regulatory domains with adaptive relevance (Purugganan 2000). This would enable us to identify populations and individuals with the capacity to adapt to certain environmental conditions, which is an urgent topic in conservation biology.

In a pilot study, we analysed a candidate isozyme gene system that had been argued to be involved in adaptive processes in Norway spruce (Scholz and Bergmann 1994; Rothe and Bergmann 1995). Two genomic full-length alleles of a phosphoenolpyruvate carboxylase (PEPC, EC4.1.1.31) were isolated and sequenced in the gymnosperm species Norway spruce [*Picea abies* (L.) Karst.]. Homology of the two full-length PEPC-1 alleles was as high as 99.8%. From exon variation, several synonymous and one exchange of an amino acid were deduced. The introns harboured various polymorphic regions that combined or recombined into a considerable number of 'within-gene genotypes', when sampled throughout a whole population (Ipsen and Ziegenhagen 2001; Ziegenhagen et al. 2002a). The occurrence of within-gene genotypes was analysed for spatial autocorrelation in a large natural population of Norway spruce (Ziegenhagen et al. 2002a). As far as could be concluded from the selected intron/exon regions analysed, however, there was no evidence for any adaptive relevance of this variation. This does not exclude the actual existence of such variation in PEPC. Either the within-gene candidate regions for adaptation remained undetected or the spatial scale analysed was too small to find a spatial structure deviating from stochastic distributions of genotypes.

Currently, worldwide research is progressing in this field of generating a new marker generation. For example, the detection and monitoring of SNPs (single nucleotide polymorphism) will be automated and enhanced in full-length candidate genes or in ESTs in the future (expressed sequence tagged sites; e.g. Schubert et al. 2001).

3.2 DNA Markers in Plantation Forestry

Populus species are among the few forest trees that are extensively studied, bred and cultured worldwide. In addition to *Eucalyptus*, the genus *Populus* has developed into a model for hardwood trees (dicot angiosperms) for forest tree biotechnology and genetic studies (Chaffey 1999; Chaffey et al. 2002; Taylor 2002; Wullschleger et al. 2002; Campbell et al. 2003). Advantages of *Populus* are: rapid growth, prolific sexual reproduction, ease of cloning, small genome, facile transgenesis, and tight coupling between physiological traits and biomass productivity.

3.2.1 Clone Identification

For identification and regular use of *Populus* clones used in plantation forestry, markers are required that differentiate between the clones. In principle, all marker types that generate a clone-specific pattern are useful. Few poplar clones are commercially traded (e.g. clone 'Astria'). For these clones, it could be of future interest to establish fingerprint patterns as a reference such as, e.g. microsatellite, RAPD (random amplified polymorphic DNA) or AFLP (amplified fragment length polymorphism) fingerprint patterns. Due to their codominant Mendelian inheritance, nuclear microsatellite markers are the markers of choice for individual identification in natural populations. For this purpose, they need to be validated by inheritance analysis. In addition, various individuals need to be genotyped to determine how many loci are necessary to obtain a maximum exclusion percentage. Such a laborious effort is not useful when only clones need to be differentiated. Instead, it was possible to use highly polymorphic markers that could be universally applied without prior sequence information of the multiple loci. A laborious and cost-intensive development of microsatellite markers can thus be avoided. For example, M13-fingerprinting can reliably distinguish between individuals (Degen et al. 1995; Fladung and Ziegenhagen 1998). Due to their more complex multi-banding patterns, AFLPs may be even more powerful for differentiation purposes. The use of just one single primer-enzyme combination enabled us to distinguish clones of poplar that originated from a half-sib relationship (Ziegenhagen et al. 2002b).

In a recent pilot study, leaf samples harvested from different *Populus* trees had to be analysed with respect to clone identity to a reference sample (Mar-

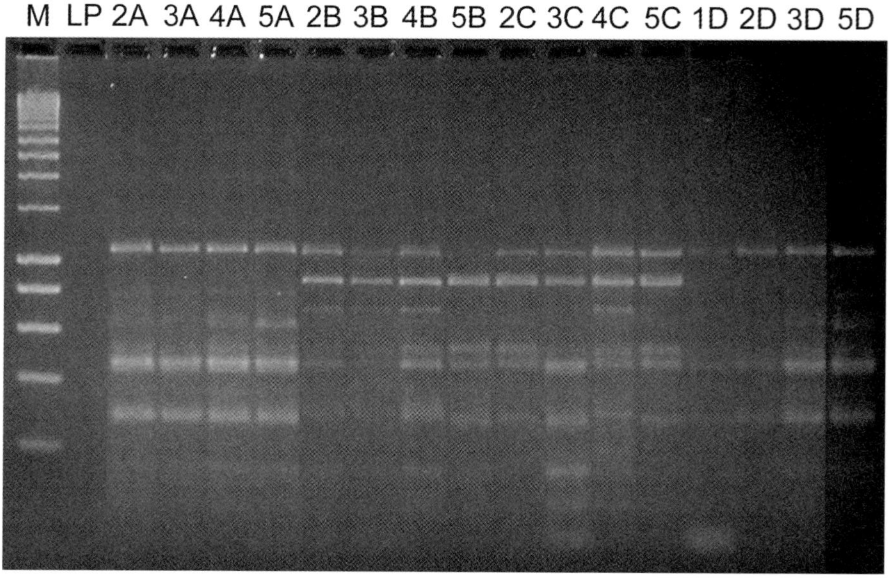

M LP 2A 3A 4A 5A 2B 3B 4B 5B 2C 3C 4C 5C 1D 2D 3D 5D

Fig. 2. M13-PCR fingerprint pattern of four *Populus* leaf samples with four repetitions. *A* Reference sample, *B* "Abt. :433a1-4", *C* "Südrand Nochten/Fürst Pückler", and *D* Südrand Nochten/ Hann. Münden". *M* molecular weight marker, *LP* water control

kussen and Fladung, unpubl.). The M13-PCR fingerprint analysis was used to compare the reference sample (A) with three samples doubted for their belonging to clone A: B = "Abt.:433a1-4", C = "Südrand Nochten/Fürst Pückler", and D = "Südrand Nochten/Hann. Münden". Four different ramets were analysed for each sample to exclude possible PCR-based contaminations (Fig. 2). The results indicate that (1) the patterns of A and D clearly differ from the similar patterns of B and C, and (2) the suspected ramets are mainly identical (Markussen and Fladung, unpubl.). In the case of B and C, we cannot exclude a mislabelling or mixing up of two ramets at any stage of the performance, from the field to the laboratory. The power of the marker was demonstrated, however, the verification that all four clones are different and that ramets have possibly been mislabelled is still required.

3.2.2 Control of Ploidy Level

Conservation issues may also be addressed at the molecular level when a status needs to be conserved that has been obtained by tissue or cell culture.

In a recent study, haploid, double- and triple-haploid plants from microspore or immature pollen culture from a *Populus nigra* × hybrid were obtained (Deutsch et al., submitted). The long generation cycles of trees are an obstacle in the search for mutants, and in *Populus* self-pollination is not even feasible, since they are dioecious. Therefore, haploid as well as double-haploid plant material and its long-term maintenance would be highly desirable for molecular mapping, mutagenesis, and gene-tagging approaches in trees.

For the *Populus* line mentioned, the successful isolation, culture and plant regeneration from isolated, immature pollen of poplar are reported (Deutsch et al., submitted). Important factors like storage of donor material, stress pretreatment, exposure to growth regulators and other culture conditions influencing the success of plant regeneration from microspores are considered. A large number of calli obtained from microspores were subjected to further analysis. Their ploidy level and haploid origin were investigated with flow cytometry and microsatellite markers, respectively. Six regenerative callus lines have maintained their haploid status for a period of 12–24 months to date.

Populus nigra L. has a number of 2n=2x=38 chromosomes (Gallego Martin et al. 1987). This chromosome number was confirmed in the diploid metaphase plates of root tips for the donor trees used in the experiments. Analysis of the tissues by flow cytometry produced clear peaks of DNA content at channel 50 using pollen as haploid standard and at channel 100 using tissue of the male donor tree as diploid control.

Using flow cytometry, however, only the ploidy level can be assessed. Therefore, microsatellite markers were applied to confirm the true haploid origin or multiple haploidy of the regenerates. Before microsatellites were available, isozyme gene markers were commonly used to test double haploidy in the regenerants of anther cultures in trees. Microsatellites have been developed as an elegant tool for individual identification, paternity analysis, genome mapping and many more applications due to a much higher degree of heterozygosity than allozymes have ever exhibited (review in Vendramin et al. 2003).

Five SSR markers developed from *P. nigra* (van der Schoot et al. 2000; Smulders et al. 2001) turned out to be excellent tools for verifying the haploid origin and/or homozygosity at each locus. The use of these SSRs demonstrated that in these immature pollen cultures even some heterozygous tissues were regenerated that might be caused by chimeric tissue, genetic reconstitutions, or irregular meiotic divisions. Somatic cells or unreduced microspores having exactly the same genome and regeneration potential as the donor plant did not seem to be the cause, since the three calli showing the same allelic combination as the donor plants did not regenerate to plantlets. In total, 77 lines were analysed at the five microsatellite loci for checking their true haploid origin. The pollen donor trees 'Aue 1' and 'Aue 2' were heterozygous at all loci, this being a prerequisite for investigating the haploid origin of the regenerates. All haploid as well as 57 diploid and three tetraploid lines exhibited one of the paternal alleles at all five SSR loci. For these lines, a haploid origin was unambiguously confirmed. In the diploid and tetraploid lines, homozygosity and thus multiple haploidy were determined. Three diploids as measured by flow cytometry revealed the same heterozygous allelic combination as the donor trees. They regenerated unequivocally worse than the control and did not develop into plantlets in vitro. In this case, a haploid origin was rejected.

In the due course of regeneration or further cultivation of haploid lines, SSR markers are very suitable to check for the stability of a ploidy level referring to the original alleles. We included the haploids in the SSR analysis to confirm their "true" haploid status, since chimeras, genetic reconstitutions, and aneuploids having one chromosome more or less than the 19 chromosomes of the haploid level could not be excluded by flow-cytometric analysis. The test for aneuploidy with five markers is a small, but conceivable option, not all 19 chromosomes were covered. A missing chromosome or an additional chromosome of the other chromosome set might be detected. A doubled chromosome would not be detected.

3.2.3 Risk Assessment of Transgene Flow

3.2.3.1 Risk Assessment of Vegetative Dispersal of Transgenic Poplar

Besides horizontal transfer, vertical transgene transfer is a matter of great concern and regarded as the main risk in field release of transgenic plants. In Germany, the first release experiment with genetically transformed trees, transgenic poplar, was initiated in 1996 (Fladung and Muhs 2000). Transgenic *Populus* carries the risk of both vegetative and generative dispersal of the transgene. To ensure that the trees remain in the vegetative phase, duration of the field trial was limited to 5 years. The field release experiment served as a basis to monitor a putative occurrence of transgenic root suckers without provoking bio-safety problems, as wood suckers can be completely deleted. In total, 444 1-year-old trees, including eight transgenic aspen lines carrying either 35S-*rolC* or rbcS-*rolC* gene construct (Fladung et al. 1996) and three control lines, were transferred to the field. After 3 years of growth, an increasing number of root suckers were detected that carried the phenotype of the wild-type or the rbcS-*rolC*-like phenotype (Fladung et al. 2003). In total, 234 root suckers were harvested in 2000 and 2001 and analysed for their transgenic status. This was done by using the sequence of the gene construct for designing diagnostic primers. By means of this marker, more than half of the root suckers were shown to carry the *rolC* gene (Fladung et al. 2003). We concluded that the vegetative dispersal capacity of transgenic perennial plants is important and should be urgently included in risk assessment studies.

3.2.3.2 Risk Assessment of Pollen-Mediated Transgene Flow into Natural Poplar Populations

Poplar is known to be obligatory outcrossing and also to have weak reproductive barriers among different species (Guries and Stettler 1976). Due to this tremendous risk potential, pollen-mediated vertical transgene flow cannot be analysed experimentally in field release trials. For example, in the due course of the mentioned field trial with 35S-*rolC* and rbcS-*rolC* transgenic aspen trees, two flower buds were detected on one single, 3-year-old tree of the female transgenic aspen clone Esch5:35S-*rolC*-1 in 1998. The next year, no flowering in any tree was observed, while in the year 2000 11 additional trees from three different transgenic lines all transformed with the 35S-*rolC* construct were detected with flower buds. All trees carrying female flower buds were removed from the field a long time before flower maturation.

Risk assessment is only possible by means of simulation models (Bialozyt et al. 2002). For parametrising these models natural gene flow via pollen needs to be understood. Microsatellite markers are therefore applied to analyse pollen-mediated gene flow at a local scale (Fladung, unpubl.). The data will be used to up-scale this process by means of computer models. Furthermore, it is necessary to understand whether such processes can be generalised or whether they are dependent on landscape and/or climatological conditions. The meteorological model METRAS, operating in a real northern German landscape, is currently being parameterised for pollen flow (in cooperation with Dr. Heinke Schlünzen, Meteorological Institute, Hamburg). The output matrices will be linked with genetic models and pollen-mediated transgene scenarios will be simulated for the real landscape (Bialozyt et al. 2002).

4 Conclusions

The presently available molecular markers harbour great potential for enhancing their usage, e.g. in the control of illegally traded timber or forest reproductive material including transgenics, since it is methodologically feasible to fingerprint wood tissues of forest tree species (Deguilloux et al. 2002; Ziegenhagen et al. 2003; Fladung et al., submitted). In addition to an enhanced usage of available markers, new marker generations are needed. The presently available DNA markers are supposed to be selectively neutral. Currently, much effort is being put into the development of a new marker generation, which should target genomic or regulatory regions of adaptive relevance, e.g. SNP detection and their validation in ESTs or candidate genes. Furthermore, knock-out (RNAi, RNA interference) and activation-tagging technologies may be promising. The availability of such markers would greatly facilitate the evaluation of genetic resources in terms of their evolutionary adaptive capacity.

Acknowledgements. For providing us with a tremendous amount of plant material, especially for range-wide investigations, we are indebted to many colleagues whose names are explicitly mentioned in the original articles cited. For their substantial input of conceptional and laboratory work, we would like to thank our colleagues and coworkers Ronald Bialozyt, Frank Deutsch, Anja Ipsen, Sandeep Kumar, Sascha Liepelt and Torsten Markussen. We highly appreciate the technical assistance of Susanne Jelkmann, Vivian Kuhlenkamp and Olaf Nowitzki. The data presented were obtained in various national and international projects (EU, German Federal Ministries of Education and Research and Environment, Federal Agency of Environment, and Deutsche Forschungsgemeinschaft). We are indebted to many fruitful discussions with our friends and colleagues Giovanni G. Vendramin (CNR, Florence, Italy) and Rémy J. Petit (INRA, Bordeaux, France).

References

Bartels H (1971a) Genetic control of multiple esterases from needles and macro-gametophytes of *Picea abies*. Planta 99:283–289

Bartels H (1971b) Isoenzymes and their significance for forest tree breeding and genetics. Allg Forstz 3:50–52

Bergmann F (1973) Genetische Untersuchungen bei *Picea abies* mit Hilfe der Isoenzym-Identifizierung. III. Geographische Variation an 2 Esterase und 2 Leucin-aminopeptidase-Loci in der schwedischen Fichtenpopulation. Silvae Genet 22:63–66

Bergmann F (1974a) Genetischer Abstand zwischen Populationen. II. Die Bestimmung des genetischen Abstands zwischen europäischen Fichtenpopulationen (*Picea abies*) auf der Basis von Isoenzym-Genhäufigkeiten. Silvae Genet 23:28–32

Bergmann F (1974b) The genetics of some isoenzyme systems in spruce endosperm (*Picea abies*). Genetika 6:353–360

Bialozyt R, Ziegenhagen B, Fladung M (2002) Modellierung des Genflusses für die Risikoabschätzung gentechnisch veränderter Bäume. In: Peschel T, Mrzljak J, Wiegleb G (eds) Verhandlungen der Gesellschaft für Ökologie, vol 32. Landschaft im Wandel – Ökologie im Wandel. Verlag Die Werkstatt, Göttingen, p 424

Birky CW Jr, Fuerst P, Maruyama T (1989) Organelle gene diversity under migration, mutation, and drift: equilibrium expectations, approach to equilibrium, effects of heteroplasmatic cells, and comparison to nuclear genes. Genetics 121:613–627

Campbell MM, Brunner AM, Jones HM, Strauss SH (2003) Forestry's fertile crescent: the application of biotechnology to forest trees. Plant Biotech J 1:141–154

Chaffey N (1999) Wood formation in forest trees: from *Arabidopsis* to *Zinnia*. Trends Plant Sci 4:203–204

Chaffey N, Cholewa E, Regan S, Sundberg B (2002) Secondary xylem development in *Arabidopsis*: a model for wood formation. Physiol Plant 114:594–600

Chase MW, Soltis DE, Olmstaed RG, Morgan D et al. (1993) Phylogenetics of seed plants: an analysis of nucleotide sequences from the plastid gene *rbc*L. Ann Mo Bot Garden 80:528–580

Cremer E, Liepelt S, Ziegenhagen B, Hussendörfer E (2003) Combined use of chloroplast DNA-microsatellite and isozyme gene markers for seed source identification in silver fir. For Genet 10(3):165–171

Degen B, Ziegenhagen B, Gillet E, Scholz F (1995) Computer-aided search for codominant markers in complex DNA banding patterns – a case study in *Abies alba* Mill. Silvae Genet 44:274–282

Degouilloux M-F, Pemonge M-H, Petit RJ (2002) Novel perspectives in wood certification and forensics: dry wood as a source of DNA. Proc R Soc Lond B 269:1039–1046

Fladung M, Ziegenhagen B (1998) M13 DNA fingerprinting can be used in studies on phenotypical revisions of forest tree mutants. Trees 12:310–314

Fladung M, Muhs HJ (2000) Field release with *Populus tremula* (*rolC*-gene) in Großhansdorf. Umweltbundesamt (ed). Humboldt University, Berlin, pp 40–45

Fladung M, Muhs HJ, Ahuja MR (1996) Morphological changes observed in transgenic *Populus* carrying the *rolC* gene from *Agrobacterium rhizogenes*. Silvae Genet 45:349–354

Fladung M, Nowitzki O, Ziegenhagen B, Kumar S (2003) Vegetative dispersal capacity of field released transgenic aspen trees is besides generative propagation also an important component in risk assessment. Trees Struct Funct 17:412–416

FSaatG (1994) Gesetz über forstliches Saat- und Pflanzgut (FSaatG) in der derzeit anzuwendenden Fassung der Bekanntmachung von 26. Juli 1979 Bundesgesetzblatt (BGBl I:1242), zuletzt geändert durch Artikel 22 des Gesetzes vom 2. August 1994 (BGBl I:2018)

Gallego Martin F, Sánchez Anta MA, Navarro Andrés F (1987) Datos cariológicos de algunas Salicaceas. Stud Bot 6:163–167

Gielly L, Taberlet P (1994) The use of chloroplast DNA to resolve plant phylogenies: noncoding versus *rbc*L sequences. Mol Biol Evol 11(5):769–777

Gillet EM (ed) (1999) 'Which marker for which purpose?' Final compendium of the research project 'Development, optimization and validation of molecular tools for assessment of biodiversity in forest trees' in the European Union DGXII Biotechnology FW IV Research Programme 'Molecular Tools for Biodiversity'. URL: http://www.sub.gwdg.de/ebook/y/1999/whichmarker/index.htm

Gladstone WT, Ledig FT (1990) Reducing pressure on natural forests through high-yield forestry. For Ecol Manage 35:69–78

Glaubitz JC, Moran GF (2000) Genetic tools: the use of biochemical and molecular markers. In: Young A, Boshier D, Boyle T (eds) Forest conservation genetics. CSIRO Publ, Collingwood, Australia pp 39–59

Gomez A, Pintos B, Aguirinao E, Manzanera JA, Bueno MA (2001) SSR markers for *Quercus suber* tree identification and embryo analysis. J Hered 92:292–295

Guries RP, Stettler RF (1976) Pre-fertilization barriers to hybridisation in the poplars. Silvae Genet 25:37–44

Hamrick JL, Godt MJW (1990) Allozyme diversity in plant species. In: Brown AD, Kahler AL, Sunderlands (eds) Plant population genetics, breeding, and genetic resources. Sinauer Associates, Sunderland, Mass, pp 43–63

Hewitt GM (1999) Post-glacial re-colonization of European biota. Biol J Linnean Soc 68:87–112

Hewitt GM (2000) Speciation, hybrid zones and phylogeographie – or seeing genes in space and time. Mol Ecol 10:537–549

Hristoforoghu K, Endemann M, Wilhelm E (2000) Monitoring of genetic stability in somatic embryo clones of *Quercus robur* L. with flow cytometry and microsatellites. In: Espinel S, Ritter E (eds) Proceedings of the international congress "Applications of biotechnology to forest genetics" (Biofor 99), Vitoria-Gasteiz, 22–25 Sept 1999, Vitoria-Gasteiz, Spain

Ipsen A, Ziegenhagen B (2001) New insights into allelic diversity of a phosphoenolpyruvate carboxylase in the conifer *Picea abies* (L.) Karst. Planta 214:265–273

Konnert M, Bergmann F (1995) The geographical distribution of genetic variation of silver fir (*Abies alba*, Pinaceae) in relation to its migration history. Plant Syst Evol 195:19–30

Konnert M, Fromm M, Hussendörfer E (2002) Referenzproben zur Identitätssicherung von forstlichem Vermehrungsgut. AFZ Der Wald 5:214–215

Kormuták A, Vookova B, Ziegenhagen B (2002) Reproductive isolation between Colorado white fir (*Abies concolor*) and the Mediterranean firs. Biologia (Bratislava) 57:527–532

Langner W (1953) Eine Mendelspaltung bei Aurea Formen von *Picea abies* (L.) Karst. als Mittel zur Klärung der Befruchtungsverhältnisse im Walde. Z Forstgenet Forstpflanzenzücht 2:49–51

Leroy XL, Leo K, Branchard M (2000) Plant genomic instability variation detected by microsatellite-primers. EJB Electr J Biotechnol 3

Liepelt S, Kuhlenkamp V, Anzidei M, Vendramin GG, Ziegenhagen B (2001) Pitfalls in determining size homoplasy of microsatellite loci. Mol Ecol Notes 1:332–335

Liepelt S, Bialozyt R, Ziegenhagen B (2002) Wind-dispersed pollen mediates postglacial gene flow among refugia. Proc Natl Acad Sci USA 99:14590–14594

Llamas-Gómez L, Braun H (1994) Part A: Untersuchungen über ökologisch-genetische Anpassungsvorgänge bei der Tanne (*Abies alba* Mill.) in unterschiedlich immissionsbelasteten Regionen unter besonderer Berücksichtigung des Erzgebirges. In: Sächsische Landesanstalt für Forsten (ed) Schriftenreihe der Sächsischen Landesanstalt für Forsten – Genetik und Waldbau der Weißtanne, Graupa, pp 1–64

Müller-Starck G, Ziehe M (eds) (1992) Genetic variation in forest tree populations in Europe. Sauerlaender's Verlag, Frankfurt am Main

Neale DB, Devey ME, Jermstad KD, Ahuja MR, Alosi MC, Marshall KA (1992) Use of DNA markers in forest tree improvement research. New For 6:391–407

Parducci L (2000) Genetics and evolution of the Mediterranean *Abies* species. PhD Thesis, Umea, Sweden

Parducci L, Szmidt AE (1999) PCR-RFLP analysis of cpDNA in the genus *Abies*. Theor Appl Genet 98:802–808

Petit RJ, Vendramin GG (2003) Plant phylogeography based on organelle genes: an introduction. In: Weiss S, Ferrand N (eds) Phylogeography of southern European refugia. Kluwer, Dordrecht (in press)

Petit RJ, El Mousadik A, Pons O (1998) Identifying populations for conservation on the basis of genetic markers. Conserv Biol 12:844–855

Petit RJ, Aguinagalde I, Beaulieu J-L, Bittkau C, Brewer S, Cheddadi R, Ennos R, Fineschi S, Grivet D, Lascoux M, Mohanty A, Müller-Starck G, Demesure-Musch B, Palmé A, Martin JP, Rendell S, Vendramin GG (2003) Glacial refugia: hotspots but not melting pots of genetic diversity. Science 300:1563–1565

Purugganan MD (2000) The molecular population genetics of regulatory genes. Mol Ecol 9:451–1461

Rothe GM, Bergmann F (1995) Increased efficiency of Norway spruce heterozygous phosphoenolpyruvate carboxylase phenotype in response to heavy air pollution. Angew Bot 69:27–30

Scholz F, Bergmann F (1994) Genetic effects of environmental pollution on tree populations. In: Kim ZS, Hattemer HH (eds) Conservation and manipulation of genetic resources in forestry. Kwangmungak Publ, Seoul, Korea, pp 34–50

Schubert R, Müller-Starck G, Riegel R (2001) Development of EST-PCR markers and monitoring their intrapopulational genetic variation in *Picea abies* (L.) Karst. Theor Appl Genet 103:1223–1231

Smulders MJM, van der Schoot J, Arens P, Vosman B (2001) Trinucleotide repeat microsatellite markers for black poplar (*Populus nigra* L.). Mol Ecol Notes 1:188–190

Taylor G (2002) *Populus*: Arabidopsis for forestry. Do we need a model tree? Ann Bot 90:681–689. URL: http://www.ejb.org.content/vol3/issue2/full2

Van der Schoot J, Pospiskova M, Vosman B, Smulders MJM (2000) Development and characterization of microsatellite markers in black poplar (*Populus nigra* L.). Theor Appl Genet 101:317–322

Vendramin GG, Ziegenhagen B (1997) Characterization and inheritance of polymorphic plastid microsatellites in *Abies*. Genome 40:857–864

Vendramin GG, Degen B, Petit RJ, Anzidei M, Madaghiele A, Ziegenhagen B (1999) High level of variation at *Abies alba* chloroplast microsatellite loci in Europe. Mol Ecol 8:1117–1126

Vendramin GG, Scotti I, Ziegenhagen B (2003) Microsatellites in forest trees: characteristics, identification and applications. In: Kumar S, Fladung M (eds) Molecular genetics and breeding of forest trees. Haworth Press, Bringhamton, NY (in press)

Wu J, Krutovskii KV, Strauss SH (1998) Abundant mitochondrial genome diversity, population differentiation and applications. Genetics 150:1605–1614

Wullschleger SD, Jansson S, Taylor G (2002) Genomics and forest biology: *Populus* emerges as the perennial favorite. Plant Cell 14:2651–2655

Young A, Boshier D, Boyle T (eds) (2000) Forest conservation genetics. CSIRO Publ, Collingwood, Australia

Ziegenhagen B, Fladung M (1997a) Molekulare Methoden zur Erfassung von Biodiversität bei Waldbäumen. In: Welling M (ed) Biologische Vielfalt in Ökosystemen – Konflikt zwischen Nutzung und Erhaltung, Senatsarbeitsgruppe "Ökosysteme/Ressourcen", Braunschweig, 22–24 Apr 1997. Braunschweig. Köllen Druck + Verlag = Schriftenr. d. BML. Reihe A: Angew Wiss 465:397–399

Ziegenhagen B, Fladung M (1997b) Variation in psbC gene region of gymnosperms and angiosperms as detected by a single restriction site polymorphism. Theor Appl Genet 94:1065–1071

Ziegenhagen B, Scholz F, Madaghiele A, Vendramin GG (1998) Chloroplast microsatellites as markers for paternity analysis in Abies alba. Can J For Res 28:317–321

Ziegenhagen B, Degen B, Petit RJ, Anzidei M, Madaghiele A, Scholz F, Vendramin GG (2001) Highly polymorphic uniparentally inherited DNA markers for spatial genetic analysis of silver fir (Abies alba Mill.) populations. In: Müller-Starck G, Schubert R (eds) Genetic response of forest systems to changing environmental conditions. Forest tree sciences, vol 70. Kluwer, Dordrecht, pp 139–149

Ziegenhagen B, Ipsen A, Kuhlenkamp V, Vendramin GG (2002a) Nucleotide diversity of a nuclear gene in gymnosperms – DNA marker development and application in a large natural population of Norway spruce [Picea abies (L.) KARST]. Symposium of population and evolutionary genetics of forest trees. IUFRO Research Group 2.04.00. Stara Lesna, Slovakia, 25–29 Aug 2002

Ziegenhagen B, Brettschneider R, Kuhlenkamp V, Fladung M. (2002b) Non-radioactive DIG-labelled AFLPs for application in forest trees. In: Van Dyke K, Van Dyke C, Woodfork K (eds) Luminescence biotechnology: instruments and applications. CRC Press, Boca Raton, pp 211–222

Ziegenhagen B, Liepelt S, Kuhlenkamp V, Fladung M (2003) Molecular identification of individual oak and fir trees from maternal tissues of their seeds. Trees 17:345–350

Section IV Legal Aspects

IV.1 Intellectual Property Rights in the Field of Molecular Marker Analysis

P. Jorasch[1]

1 Introduction

Intellectual property rights – and especially patents – become more and more important in biotechnology as there are many industrial applications with high economic value. The economic value of biotechnological inventions, especially in the field of agrobiotechnology, is increasing with the worldwide expansion of the cultivation of transgenic plants (Herrlinger et al. 2003). Beyond this, patents on plant-related inventions can influence the funding that is available for research, in particular in the private sector of biotechnology (Fleck and Baldock 2003). Conventional (non-transgenic) plant breeding and plant breeding research is also strongly influenced by biotechnological processes and methods. One tool that has found its way into conventional plant breeding is molecular marker analysis of significant traits (e.g. resistance against pathogens, yield) or DNA fingerprinting with the help of molecular markers to obtain information on the relationship between individual plants. Moreover, with the improvements in our understanding of the genomes and in our knowledge of the relations between genotype and phenotype in economically important crops, this tool will become more significant in practical plant breeding. The importance of molecular marker analysis for the different applications was recognized very early so that many patents have been filed in the last 10–15 years. For scientists in all institutions, public or private sector, an understanding of intellectual property rights (IPRs) is fundamental in both research and development (Kowalski et al. 2002), but many scientists are still not aware of the rising number of patents in this field. This chapter will give an overview of patents for methods and applications in the field of microsatellite markers or simple sequence repeat (SSR) markers.

---

[1] Gesellschaft für Erwerb und Verwertung von Schutzrechten GVS mbH, Kaufmannstr. 71–73, 53115 Bonn, Germany

2 What Is a Patent?

Patents are granted for inventions that are new, involve a creative step and can be applied in industry. A patent gives its owner the exclusive monopoly to use his invention, preventing others from using it without permission for a certain time (Shear and Kelley 2003). This means that if someone wants to use a patent-protected invention, the permission for use (license) must be obtained from the patent owner. One exception where the user of a protected invention does not need the permission of the patent owner is the so-called research exemption [e.g. §11 (2) of the German Patent Law]. It allows license-free activities concerning the improvement or testing of patent-protected inventions. Use of patent-protected inventions in research and development under the provisions of the patent, however, does not fall under the research exemption and is dependent on the permission of the patent owner. In most countries, patent specifications are published 18 months after the application date. The different patent offices, but also private companies, provide online patent databases that can be searched by different keywords. The results of such a patent investigation, in which different keywords concerning microsatellite marker analysis were used, are provided here.

3 Microsatellite or Simple Sequence Repeat Markers

When considering a typical experiment regarding molecular marker analysis with microsatellite markers, one can divide this experiment into different steps (Fig. 1). Starting with the plant and the extraction of its DNA, in some experiments, the DNA is cut by restriction enzymes. After this step, specific primers are used to perform a PCR reaction. The resulting PCR fragments can be analysed by different methods like gel electrophoreses, mass spectrometry or micro-array analysis. This analysis will provide information on specific traits of different plants for marker-assisted selection or on their genetic relationship to each other (fingerprinting).

An investigation of the patent specifications that have been filed in this field shows that there are many patents claiming different steps of this typical marker experiment. Figure 1 shows some of these patents and indicates which step of the experiment is claimed.

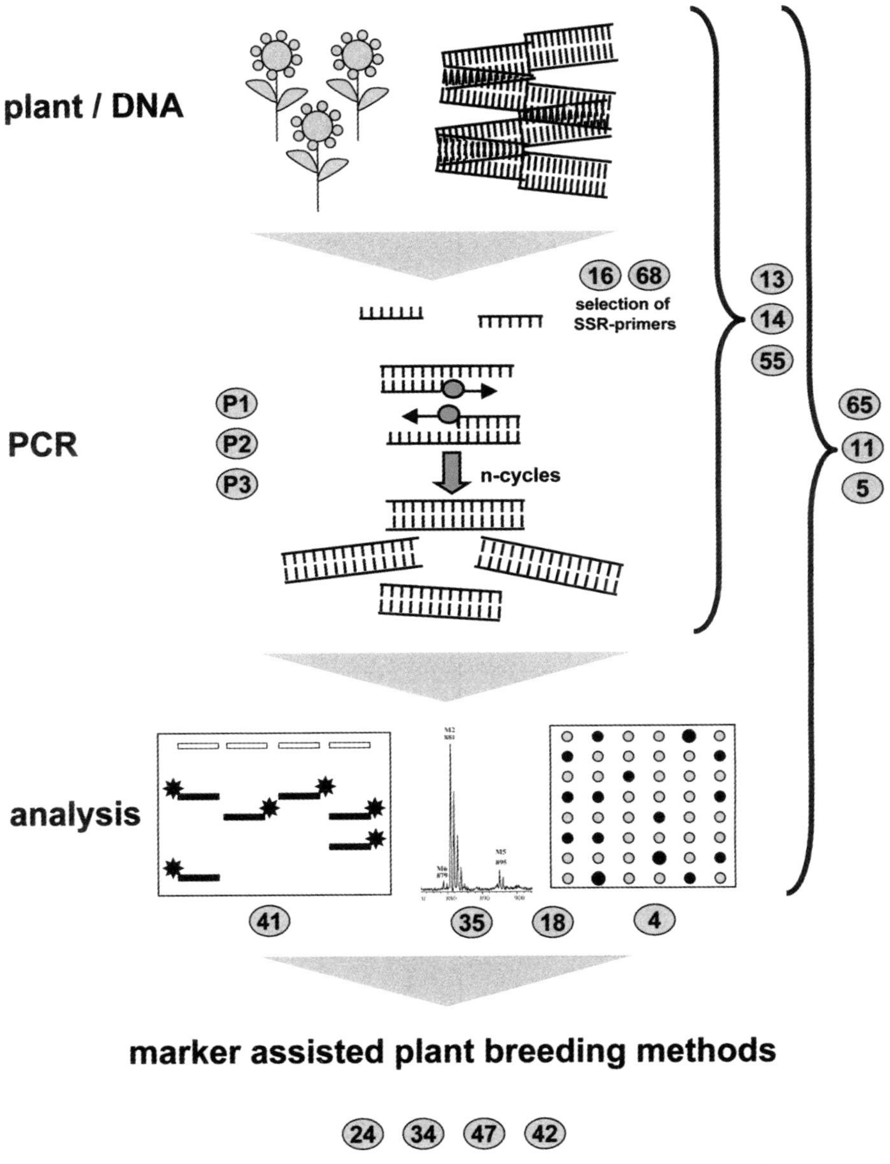

Intellectual Property Rights in the Field of Molecular Marker Analysis 435

marker assisted plant breeding methods

Fig. 1. An overview of a typical microsatellite marker experiment and some sample patents that are relevant for the different steps of such an experiment. Patents are indicated by *numbers*. The experiment is divided into different steps. Starting with the isolation of DNA of a plant, the DNA is sometimes cut by restriction enzymes. After the selection of specific SSR primers, a PCR reaction is carried out. There are different possible methods for the analysis of the resulting PCR fragments, here exemplified by gel electrophoresis (the fluorescent label of the PCR product is indicated by ✱), mass spectrometry and microarray analysis. Molecular marker analysis results in marker-assisted plant breeding. The *numbers* indicate patents that are relevant for the different steps of the experiment. Numbering of the patents is in accordance with the consecutive numbering of the patents in Table 1

4 The Selection of Microsatellite Primers and the PCR Reaction

Figure 1 shows two typical patents claiming primers for microsatellite marker analysis. Patent No. 16 (Röder et al. 1997) claims specific microsatellite markers from *Triticum aestivum*. Patents claiming specific primer sequences for marker analysis have become rare in the last few years. The problem is that patents are published 18 months after their registration (Art. 93, European Patent Convention). After publication, the owner of a patent has difficulty controlling whether someone unauthorized is using the patented primer sequences illegally because the plant that was analyzed by the primers does not show which primer was used for the analysis. As a consequence, primer sequences as specific as that are normally not patented and thereby published, but rather treated as a business secret that is licensed to users. This gives the inventor a controlling mechanism for the use of his invention. In contrast, patent No. 68 (Nagaraju 2003) claims a certain class of SSR primers, the inter-simple sequence repeat-PCR primers. Here, the scope of protection of the claim is broader, making it easier for the patent owner to control who is using the invention.

After selecting the primers, the PCR experiment follows. Most researchers are aware of patents concerning PCR methods. The basic patents on PCR were registered in 1985. Patents No. P1–P3 (Mullis 1992; Mullis et al. 1992, 1993) in Fig. 1 indicate these basic patents owned by Hoffmann La Roche. A license for PCR can be relatively easily obtained by buying a licensed polymerase and a licensed thermocycler. As there are also many cheaper non-licensed polymerases and thermocyclers on the market, the manufacturers indicate in their instructions for use that the product is not licensed for performing PCR reactions. Meanwhile, there are many other patents concerning registered PCR methods. These patents claim special polymerases or methods like RT-PCR and quantitative PCR. They are not listed in this context because their discussion would go beyond the scope of this chapter.

In the previous paragraphs, patent specifications claiming the primer sequences, on the one hand, and patent specifications claiming the PCR method, on the other, were discussed. However, if one has a closer look, one can also find patents claiming both steps, like patents No. 13 (Morgante and Vogel 1997), 14 (Kuiper et al. 1997) and 55 (van Eijk et al. 2001; Fig. 1).These patent specifications claim processes for detecting polymorphisms between different samples of DNA. The processes comprise the amplification of nucleic acid segments using defined primer sequences, sometimes starting with the previous restriction of the DNA sample by restriction endonucleases, and the ligation of certain adaptor sequences similar to the AFLP (amplified fragment length polymorphism) approach. Patent No. 55 (van Eijk et al. 2001) even claims a combined method between microsatellite and AFLP marker analysis using special RAMP primers (random amplified microsatellite polymorphism primers) for the analysis of microsatellite sequences.

5 Analysis of PCR Products

Figure 1 shows three different methods for the analysis of the resulting PCR products. The most common one, the analysis by gel electrophoresis, can also be claimed by patents, if for example special fluorescent labels for detection are used. Such a method is claimed by patent No. 41 (Shuber and Pierceall 2002). The claimed method comprises the PCR reaction with fluorescent primers, the detection of the labelled extension products and the comparison of the PCR product size.

A second method of analysis, mass spectrometry, is claimed by patent No. 35 (Hillenkamp and Köster 1999). This patent generally claims the analysis of nucleic acids by mass spectrometry in general, and not just for microsatellite marker analysis. For high-throughput analysis of probes, the microarray technique is preferred. This method is protected by a patent of Affymetrix, patent No 4 (Fodor et al. 1998). This specification not only protects the detection of microsatellites by microarray analysis, but also the detection of nucleic acid sequences in general which comprises microsatellites. Meanwhile, there are other patents claiming further developments of this technique, but the discussion of these would also go beyond the scope of this chapter.

Another high-throughput technique described in patent No. 18 (Olek 1996) combines the method of mass spectrometric and microarray analysis of microsatellite markers.

Patent specifications No. 5 (Caskey and Edwards 1992), No. 11 (Perlin 1995) and No. 65 (Saint-Louis and Paquin 2003; Fig. 1) summarize the complete experimental process from DNA extraction to the analysis of the PCR products, in which different PCR methods are combined, for example, use of certain labelled nucleotide triphosphates and different analytical tools like mass spectrometry or computer analytical tools.

6 Marker-Assisted Breeding Methods

The most comprehensive patent specifications claim complete plant breeding methods in which molecular marker analysis is used. Examples are patent specifications No. 24 (Byrum and Reiter 1998), No. 34 (Beavis 1999), No. 47 (Jansen and Beavis 2001) and No. 42 (Openshaw and Bruce 2001; Fig. 1). These comprise the previously mentioned experimental steps in that they claim the association of the genotype with phenotypic traits of interest by molecular marker analysis. The patents differ in the selection of plant populations that are the basis for the analysis, the statistical methods applied in the analysis and the integration of molecular biological techniques like expression profiling of genes. The claims of these patents are not restricted to microsatellite markers. They also comprise other well-known marker techniques like AFLPs, RFLPs (restriction fragment length polymorphisms) or RAPDs (random amplified polymorphic DNA). These patent specifications were filed in the late 1990s and are still in the process of examination in Europe. Details concerning the legal status of these patent applications are shown in Table 1.

7 Conclusions

Molecular marker analysis is one of the most powerful tools in modern plant breeding. However, as for most innovative applications, IPRs play an important role. As shown above, the implementation of microsatellite marker analysis for plants is also strongly dependent on IPRs. To identify these rights, a biotechnological process or method has to be dissected into its essential components and processes, with each part to be analysed under the IP microscope (Kowalski et al. 2002). This means that scientists must educate themselves on these issues so that they can make informed decisions regarding their research practices (Kimpel 1999). Beyond this, patents describe the latest inventions made by innovative researchers and companies and the publication of these patents guarantees their public availability. This, in turn, allows the further development and improvement of these innovative techniques.

Table 1. Results of a patent investigation concerning SSR-marker technology. The results are from a patent investigation in the Delphion (www.delphion.com) and Epoline (www.epoline.org) patent databases. Thought has been given to patent specifications in the field of microsatellite marker analysis. Column one comprises European (*EP*) and international (*WO*) patent publication numbers. Publication numbers of *US*, New Zealand (*NZ*) and Japanese (*JP*) patents are included when no EP or WO publications were available. The column "*main claim*" includes the first claim of a patent specification, if the respective patent is still in force. Comments on the legal status of the patent specifications were derived from the patent databases

Publication number and title	Publication date (V) and priority date (P)	Assignee	Main claims	Comment on legal status
1. EP0237362: Process for detecting specific nucleotide variations and genetic polymorphisms present in nucleic acids and kits thereof	V-16.09.87 P-13.03.86 P-22.06.86	Hoffmann La Roche	1. A process for detecting the presence of a specific nucleotide sequence in nucleic acid in a sample, which includes: (a) Treating the sample, together or sequentially, with four different nucleoside triphosphates, an agent for polymerization of the nucleoside triphosphates, and two oligonucleotide primers for said nucleic acid under hybridizing conditions such that a primer will hybridize to said nucleic acid and an extension product of the primer be synthesized which is complementary to said nucleic acid, wherein said primers are selected such that the extension product synthesized from one primer, when separated from its complement, can serve as a template for synthesis of the extension product of the other primer (b) Treating the sample under denaturing conditions to separate the primer extension products from their templates (c) Treating the sample, together or sequentially, with said four nucleoside triphosphates, an agent for polymerization of the nucleoside triphosphates, and oligonucleotide primers such that a primer extension product is synthesized using each of the single strands produced in step (b) as a template, wherein steps (b) and (c) are repeated a sufficient number of times exponentially to increase the amount of said nucleic acid and to result in detectable amplification thereof (d) Directly transferring, without gel fractionation, product derived from step (c) to a membrane	Granted EP-patent

Table 1. (Continue)

Publication number and title	Publication date (V) and priority date (P)	Assignee	Main claims	Comment on legal status
			(e) Treating the membrane from (d) under hybridization conditions with a labeled sequence-specific oligonucleotide probe capable of hybridizing with the amplified nucleic acid only if the sequence of the probe is complementary to a region of the amplified nucleic acid (f) Detecting whether the probe has hybridized to an amplified nucleic acid in the sample.	
2. WO9004651: Mapping quantitative traits using genetic markers	V-03.05.90 P-19.10.88	Whitehead Institute for biomedical research Cornell research foundation, Inc.		WO-application withdrawn
3. EP 0561796: Oligonucleotide constructs and methods for the generation of sequence signatures from nucleic acids	V-05.03.92 P-24.08.90	The University of Tennessee research corporation		Granted in EP, but lapsed, data supplied by contracting states AT, BE, CH, DE, FR, GB, IT, LI, NL, SE
4. EP 0834576: Detection of nucleic acid sequences	V-08.04.98 P-06.12.90	Affymetrix, Inc.	1. A method for detecting nucleic acid sequences in two or more collections of nucleic acids, comprising: (a) Providing an array comprising more than 100 different polynucleotide probes bound to a solid surface (b) Contacting said array of probes under hybridisation conditions with:	Granted in EP, opposition filed against

Table 1. (Continue)

Publication number and title	Publication date (V) and priority date (P)	Assignee	Main claims	Comment on legal status
			(i) a first collection of nucleic acids comprised of first-labelled nucleic acids having at least some sequences complementary to probes of said array (ii) at least a second collection of nucleic acids comprised of second-labelled nucleic acids having at least some sequences complementary to probes of said array, wherein said first and second labels are distinguishable from each other (c) Detecting hybridisation of first and second labelled complementary nucleic acids to probes of said array	
5. EP 0639228: DNA typing with short tandem repeat polymorphisms and identification of polymorphic short tandem repeats	V-20.08.92 P-31.01.91	Baylor College of Medicine	A DNA profiling assay for detecting polymorphisms in at least one short tandem repeat, comprising the steps of: extracting DNA from a sample to be tested; amplifying said at least one short tandem repeat in the extracted DNA, wherein the short tandem repeat sequence is characterized by the formula (Aw Gx Ty Cz)n wherein A,G,T, and C represent the nucleotides; w, x, y and z represent the number of each nucleotide and range from 0 to 7; the sum of w+x+y+z ranges from 4 to 7; and n represents the repeat number and ranges from about 5 to 50; and detecting said polymorphisms by identifying said amplified extension products for each different sequence, wherein each different sequence is differentially labelled.	In EP still under examination
6. US6455758: Process predicting the value of a phenotypic trait in a plant breeding program	V-24.09.02 P-19.02.91	Dekalb Genetics Corp.		US only, just maize

Table 1. (Continue)

Publication number and title	Assignee	Publication date (V) and priority date (P)	Main claims	Comment on legal status
7. EP 0552545: Detection of polymorphisms in simple sequence repeats using oligonucleotide ligation	Pioneer Hi-Bred International, Inc.	V-28.07.93 P-17.01.92		Refusal of application 13.11 1996
8. US 5746023: Method to identify genetic markers that are linked to agronomically important genes	Du Pont	V-05.05.98 P-07.07.92 P-23.07.93 P-18.07.95	1. A method for identifying alleles associated with agronomic fitness of crop plants, comprising: (a) Selecting a sample of current-day elite lines of a given crop to form an elite population; (b) Selecting the predominant and earliest known ancestral lines of said elite lines by considering the pedigrees of said elite lines (c) Conducting a genetic marker survey to determine the genotype of said elite lines and said ancestral lines (d) Using the pedigrees of said elite lines and genotypes of said ancestral lines to calculate the probability of each elite line inheriting each allele from said ancestral lines (e) Calculating the expected allele frequency of each allele within said elite population by averaging the probabilities calculated in step d) for each elite line (f) Calculating the observed allele frequency within said elite population; (g) Comparing said observed allele frequency with said expected allele frequency for each said allele in said elite population to identify alleles at each locus that have been inherited more frequently than expected	US only

Intellectual Property Rights in the Field of Molecular Marker Analysis 443

Table 1. (Continue)

Publication number and title	Publication date (V) and priority date (P)	Assignee	Main claims	Comment on legal status
			(h) Producing crop plants with superior agronomic fitness; such that new crop plants with superior agronomic fitness can be efficiently identified with said genetic markers that are diagnostic of said alleles that have been inherited more frequently than expected	
9. EP 0733126: Immobilized mismatch binding protein for detection of mutations and polymorphisms, and allele identification	V-11.05.95 P-04.11.93	ValiGene Corporation	1. A method of detecting a mutation from a non-mutated sequence of a double stranded target DNA in a sample, the method comprising; (a) Denaturing any double stranded DNA in the sample into single strands and allowing the single strands to reanneal into duplexes (b) Incubating the denatured and reannealed duplexes of step (a) with a mismatch-binding protein immobilized by adsorption on a solid support, either (i) In the presence of a detectably labelled DNA having a mismatch and capable of binding to the mismatch binding protein; or (ii) Wherein the mismatch-binding protein was preincubated with and allowed to bind a detectably labelled DNA having a mismatch (c) Detecting the amount of detectably labelled DNA having a mismatch bound to the mismatch-binding protein, wherein the presence of a mutation in the double stranded target DNA of the sample results in a decrease in the binding of the detectably labelled DNA to the mismatch-binding protein	Granted in EP, but lapsed, data supplied by contracting states AT/ 24–04–2002 GR/ 24–04–2002 NL/ 24–04–2002 PT/ 24–07–2002 SE/ 24–07–2002 in DE in force
10. WO 9515400: Genotyping by simultaneous analysis of multiple microsatellite loci	V-08.06.95 P-03.12.93	The Johns Hopkins University		WO-application deemed to be withdrawn

Table 1. (Continue)

Publication number and title	Publication date (V) and priority date (P)	Assignee	Main claims	Comment on legal status
11. EP 0714537: Method and system for genotyping	V-28.12.95 P-17.06.94	Perlin, Mark W	1. A method for genotyping comprised of the steps: (a) Obtaining DNA or RNA material from a genome (b) Amplifying a location of the material, with the length of the location not exceeding 50 kb and the location containing a multinucleotide repeat region (c) Labelling the amplified material with labels (d) Converting the labels with a sensing device which produces a first electrical signa; (e) Removing a reproducible pattern of the amplification from the first electrical signal using a program residing in the memory of a computer to form a third electrical signal (f) Producing from the third electrical signal a genotype of the material at the location	In EP still under examination
12. EP0828853: Method for nucleotide sequence amplification	V-25.04.96 P-18.10.94	Genzyme Corporation		24.01.01 EP-application deemed to be withdrawn
13. EP0804618: Compound microsatellite primers for the detection of genetic polymorphisms	V-06.06.96 P-28.11.94	E.I. Du Pont de Nemours and Company	1. An improved method of detecting polymorphisms between two individual nucleic acid samples comprising amplifying segments of nucleic acid from each sample using primer-directed amplification and comparing said amplified segments to detect differences, the improvement comprising wherein at least one of the primers used in said amplification consists of a perfect compound simple sequence repeat in which two different repeating sequences are either directly adjacent or are separated by no more than three intervening bases	Granted in EP

Table 1. (Continue)

Publication number and title	Publication date (V) and priority date (P)	Assignee	Main claims	Comment on legal status
14. EP0721987 EP0805875 WO9622388: Amplification of simple sequence repeats	V-17.07.96 P-16.01.95 V-12.11.97 P-16.01.96	Keygene, N.V.	1. A Process for the selective amplification of restriction fragments comprising simple sequence repeats, comprising the following: (a) Digesting a starting DNA with two or more different restriction enzymes, at least one of these enzymes cleaving at or near its recognition nucleotide sequence overlapping or flanking with the simple sequence repeat (referred to as first restriction enzyme) and at least one of these enzymes cleaving the restriction fragments into amplifiable restriction fragments, in a preferably four-base sequence (referred to as second restriction enzyme, to obtain restriction fragments (b) Ligating an appropriate double stranded oligonucleotide adaptor to each of the ends of the restriction fragments produced by said restriction enzymes (c) Amplifying the restriction fragments of step (b) using two or more different amplification primers with the following general structure: one primer having a sequence at the V end matching the common sequence of the restriction fragments produced with the first restriction enzyme, or part thereof, and at the V end at least five nucleotides matching the sequence of the simple sequence repeat (referred to as primer one; one primer having a sequence at the V end matching the common sequence of the restriction fragments produced with the second restriction enzyme, or part thereof, and at its V end ranging from 0, 11 21 31 4 or more especially 0 to 3 randomly chosen nucleotides (referred to as primer two) (d) Recovering the amplified fragments	EP0805875 still under examination EP 0721987 deemed to be withdrawn

Table 1. (Continue)

Publication number and title	Publication date (V) and priority date (P)	Assignee	Main claims	Comment on legal status
15. EP 0815261: DNA diagnostics based on mass spectrometry	V-26.09.96 P-17.03.95	Sequenom, Inc.	1. A process for detecting one or more target nucleic acid sequences present in a biological sample, comprising the steps of: (a) Hybridizing one or more detector oligonucleotide with one or more nucleic acid molecules and removing unhybridized detector oligonucleotide (b) Ionizing and volatizing the product of step (a) (c) Analyzing the ionized and volatilized nucleic acid by mass spectrometry, wherein detection of the detector oligonucleotide by mass spectrometry indicates the presence of the target nucleic acid sequence in the biological sample	Granted in EP
16. EP 0833324: Microsatellite markers for plants of the species *Triticum aestivum* and tribe Triticeae and the use of said markers	V-16.01.97 P-28.06.95	Institut für Pflanzengenetik und Kulturpflanzenforschung	The patent claims 230 microsatellite markers for wheat	Granted in EP
17. WO9712059: Brown stern rot resistance in soybeans	V-03.04.97 P-26.09.95	Pioneer Hi-Bred International, Inc.		WO9712059 discontinued in Europe
18. EP 0870062: Genomic analysis process and agent	V-15.05.97 P-09.11.95	GAG Bioscience Zentrum für Umweltforschung und Technologie	1. Method for microsatellite analysis, wherein there exist – The fixation of microsatellite amplificates from genomic DNA samples separated into individual microsatellite markers before or after amplification to defined positions of a matrix – Evaporation of the individual positions in a mass spectrometer – Mass-spectrometry determination of the molecular weight	Granted in EP (PT, ES, DK. DE, AT)

Table 1. (Continue)

Publication number and title	Publication date (V) and priority date (P)	Assignee	Main claims	Comment on legal status
19. EP 1034307 EP 0815263 EP 1086247: Methods for the detection of nucleic acids	V-10.06.99 P-04.12.997 V-03.07.97 P-22.12.95 P-14.08.96 V-23.12.99 P-16.06.98	Exact Sciences Corporation	EP 0815263: 18. A method for detecting a nucleic acid sequence change in a target allele in a subpopulation of cells in a biological sample, comprising the steps of: (a) Determining (i) an amount of wild-type target allele in the biological sample (ii) an amount of a reference allele in the biological sample (b) Detecting a nucleic acid sequence change in the target allele in a subpopulation of cells in the biological sample, statistically significant difference in the amount wild-type target allele and the amount of reference allele obtained in said determining step being indicative of a nucleic acid sequence change	EP 1086247 application deemed to be withdrawn 07.08.02 EP1034307 claims fetal chromosomal abnormalities
20. EP 0912761: Detection of nucleic acid sequence differences using coupled ligase detection and polymerase chain reactions	V-04.12.97 P-29.05.96	Cornell Research Foundation, Inc.	1. A method for identifying one or more of a plurality of sequences differing by one or more single-base changes, insertions, deletions, or translocations in a plurality of target nucleotide sequences comprising: Providing a sample potentially containing one or more target nucleotide sequences with a plurality of sequence differences; providing one or more oligonucleotide probe sets, each set characterized by: (a) A first oligonucleotide probe, having a target-specific portion and a 5' upstream primer-specific portion (b) A second oligonucleotide probe, having a target-specific portion and a 3' downstream primer-specific portion, wherein the oligonucleotide probes in a particular set are suitable for ligation together when hybridized adjacent to one another on a corresponding target nucleotide sequence, but have a mismatch which interferes with such liga-	In EP still under examination

Table 1. (Continue)

Publication number and title	Publication date (V) and priority date (P)	Assignee	Main claims	Comment on legal status
			tion when hybridized to any other nucleotide sequence present in the sample; blending the sample, the plurality of oligonucleotide probe sets, and the ligase to form a ligase detection reaction mixture; subjecting the ligase detection reaction mixture to one or more ligase detection reaction cycles comprising a denaturation treatment, wherein any hybridized oligonucleotides are separated from the target nucleotide sequences, and a hybridization treatment, wherein the oligonucleotide probe sets hybridize at adjacent positions in a base-specific manner to their respective target nucleotide sequences, if present in the sample, and ligate to one another to form a ligation product sequence containing (a) the 5' upstream primer-specific portion, (b) the target-specific portions connected together, and (c) the 3' downstream primer specific portion with the ligation product sequence for each set being distinguishable from other nucleic acids in the ligase detection reaction mixture, and, wherein the oligonucleotide probe sets may hybridize to nucleotide sequences in the sample other than their respective target nucleotide sequences, but do not ligate together due to a presence of one or more mismatches and individually separate during the denaturation treatment; providing one or a plurality of oligonucleotide primer sets, each set characterized by (a) an upstream primer containing the same sequence as the 5' upstream primer-specific portion of the ligation product sequence and (b) a downstream primer complementary to the 3' downstream primer-specific portion of the ligation product sequence, wherein one of the primers has a detectable reporter label; providing a polymerase; blending the ligase detection reaction mix-	

Table 1. (Continue)

Publication number and title	Publication date (V) and priority date (P)	Assignee	Main claims	Comment on legal status
			ture with the one or a plurality of oligonucleotide primer sets, and the polymerase to form a polymerase chain reaction mixture; subjecting the polymerase chain reaction mixture to one or more polymerase chain reaction cycles comprising a denaturation treatment, wherein hybridized nucleic acid sequences are separated, a hybridization treatment, wherein the primers hybridize to their complementary primer-specific portions of the ligation product sequence, and an extension treatment, wherein the hybridized primers are extended to form extension products complementary to the sequences to which the primers are hybridized, wherein, in a first cycle, the downstream primer hybridizes to the 3' downstream primer-specific portion of the ligation product sequence and extends to form an extension product complementary to the ligation product sequence, and, in subsequent cycles, the upstream primer hybridizes to the 5' upstream primer-specific portion of the extension product complementary to the ligation product sequence and the 3' downstream primer hybridizes to the 3' downstream portion of the ligation product sequence; detecting the reporter labels; and distinguishing the extension products to indicate the presence of one or more target nucleotide sequences in the sample	
21. US 5811239: Method for single base-pair DNA sequence variation detection	V-22.09.98 P-13.05.96	Frayne Consultants		US only

Table 1. (Continue)

Publication number and title	Publication date (V) and priority date (P)	Assignee	Main claims	Comment on legal status
22. EP 0943019: Sets of labelled energy transfer fluorescent primers and their use in multi component analysis	V-23.07.98 P-15.01.97	Incyte Pharmaceuticals, Inc.	1. A set of four fluorescently labelled oligonucleotide primers, wherein three of said oligonucleotide primers have a common donor fluorophore "A" and acceptor fluorophore "B" in energy transfer relationship and are separated by different distances so as to provide three distinguishable fluorescent signals, and the fourth primer has two donor fluorophores "A" so as to provide a fluorescent signal different from said three distinguishable fluorescent signals	Granted in EP (BE, CH, DE, FR, GB, L,I NL, AT)
23. EP 0986651: Polymerases for analyzing or typing polymorphic nucleic acid fragments and uses thereof	V-22.03.00 P-07.02.97 V-08.07.99 P-07.02.97 V-13.08.98 P-06.01.98	Life Technologies, Inc.	1. A method of identifying, analyzing or typing a polymorphic DNA fragment in a sample of DNA, said method comprising contacting said sample of DNA with one or more DNA polymerases substantially reduced in the ability to add one or more non-templated nucleotides to the 3' terminus of a DNA molecule, amplifying said polymorphic DNA fragment within said sample and analyzing said amplified polymorphic DNA fragment	In EP still under examination
24. EP0972076: A method for identifying genetic marker loci associated with trait loci	V-24.09.98 P-27.03.97	Asgrow seed Inc, Du Pont De Nemours and Co	1. A method for identifying a genetic marker locus associated with a trait locus from a crop species, the method comprising: (a) Creating a genotypic survey for a crop species using germplasm of multiple ancestry, the survey created using genetic markers, wherein individual entries of the germplasm survey are not members of a segregating population created for the purposes of the analysis (b) Comparing the genotypic survey to phenotypic data collected on the same entries used to create the genotypic survey or their progeny (c) Estimating the association between genetic marker loci and trait loci (d) Identifying a genetic marker locus that is associated with the trait locus	In EP still under examination

Intellectual Property Rights in the Field of Molecular Marker Analysis 451

Table 1. (Continue)

Publication number and title	Publication date (V) and priority date (P)	Assignee	Main claims	Comment on legal status
25. EP1056528 EP1042503 EP1030933 EP1027121 EP1023463 EP1017841 EP1017466 EP1002137: Method of detecting mutant DNA by MIPC and PCR	V-14.05.99 V-15.04.99 V-22.04.99 V-05.11.98 V-22.04.99 V-25.02.99 V-05.11.98 V-18.02.99 P-25.04.97 P-23.09.97 P-17.10.97 P-27.10.97 P-30.10.97 P-05.12.97 P-05.01.98 P-13.03.98 P-10.04.98 P-04.08.98 P-06.10.98	Transgenomic Inc.	EP1056528: 1. A method for separating a mixture of polynucleotides comprising: (a) Flowing a mixture of polynucleotides having a target range of base pairs through a separation column containing a separation medium having a nonpolar separation surface (b) Separating said mixture by eluting said column using a mobile phase having a composition which remains essentially constant for the duration of the chromatographic separation EP1042503: 1. A method for detecting a putative mutant DNA in a sample of DNA, the method comprising the steps of: (a) Amplifying the sample of DNA using PCR (b) Hybridizing the amplified sample to form a mixture of homoduplexes and heteroduplexes (c) Separating the product of step (b) into fractions by denaturing matched ion polynucleotide chromatography (d) Blind collecting the fractions from step (c) at a retention time corresponding to the retention time of the heteroduplex EP1023463: 1. A method for enhancing the detection of a polynucleotide separated by matched ion polynucleotide chromatography comprising: (a) Covalently attaching a chemical tag to said polynucleotide to form a tagged polynucleotide (b) Applying said tagged polynucleotide to a separation medium having a non-polar surface	EP1030933 withdrawal of application 12–05–2003 EP1027121 Application deemed to be withdrawn 25–07–2003 EP1002137 withdrawal of application 09–07–2003 EP1056528, EP1042503, EP1023463, EP1017841 and EP1017466 in EP still under examination

Table 1. (Continue)

Publication number and title	Publication date (V) and priority date (P)	Assignee	Main claims	Comment on legal status
			(c) Eluting said tagged polynucleotide from said surface with a mobile phase containing a counterion agent and an organic solvent (d) Detecting said tagged polynucleotide, wherein said medium is characterized by having a DNA separation factor of at least 0.05 EP1017841: 1. A method for analyzing a sample of double stranded DNA to determine the presence of a mutation therein comprising: (a) Contacting said sample with a mutation site binding reagent (b) Chromatographically separating and detecting the product of step (a) EP1017466: 1. A method for separating a mixture of polynucleotides, comprising flowing a mixture of polynucleotides having up to 1500 bp through a separation column containing polymer beads having an average diameter of 0.5 to 100 microns, said beads having a surface composition essentially completely substituted with a moiety selected from the group consisting of unsubstituted, methyl, ethyl, hydrocarbon, and hydrocarbon polymer, and wherein said beads are characterized by having a DNA separation factor of at least 0.05; and separating said mixture of polynucleotides	
26. EP 0983383: Length determination of nucleic acid repeat sequences by discontinuous primer extension	V-03.12.98 P-27.05.97	PE Corporation (NY)	1. A method for determining the number of repeat units in a repeat region of a target nucleic acid comprising the steps of: (a) Annealing a primer-complementary portion of a target nucleic acid to a primer thereby forming a target-primer hybrid (b) Performing a first primer extension reaction using a first primer extension reagent	In EP still under examination

Intellectual Property Rights in the Field of Molecular Marker Analysis 453

Table 1. (Continue)

Publication number and title	Publication date (V) and priority date (P)	Assignee	Main claims	Comment on legal status
			(c) Separating the target-primer hybrid and unreacted first primer extension reagent (d) Performing a second primer extension reaction using a second primer extension reagent, wherein at least one of the first or second primer extension reagents includes an extendible nucleotide having a label attached thereto (e) Separating the target-primer hybrid from unreacted second primer extension reagent (f) Measuring a signal produced by the label (g) Treating the label so as to render the label undetectable (h) Repeating a cycle of steps (a) through (g) until the signal is substantially less than a signal detected in a previous cycle (i) Determining a number of repeat units in a repeat region of the target nucleic acid	
27. EP 1002127: Method of determining the genotype of an organism using an allele specific oligonucleotide probe which hybridises to microsatellite flanking sequences	V-14.01.99 P-02.07.97 P-27.03.98 P-01.04.98	University of Bristol		EP-application deemed to be withdrawn
28. EP 1025262: Sequence-based screening	V-04.03.99 P-26.08.97	Diversa Corporation	1. A method for identifying a desired activity encoded by a genomic DNA population comprising: (a) Obtaining a single-stranded genomic DNA population (b) Contacting the single-stranded DNA population of (a) with a DNA probe bound to a ligand under conditions and for sufficient time to	In EP still under examination

Table 1. (Continue)

Publication number and title	Publication date (V) and priority date (P)	Assignee	Main claims	Comment on legal status
			allow hybridization and to produce a double-stranded complex of probe and members of the genomic DNA population which hybridize thereto (c) Contacting the double-stranded complex of (b) with a solid phase specific binding partner for said ligand so as to produce a solid phase complex (d) Separating the solid phase complex from the single-stranded DNA population of (b) (e) Releasing from the probe the members of the genomic population which had bound to the solid phase-bound probe (f) Forming double-stranded DNA from the members of the genomic population of (e) (g) Introducing the double-stranded DNA of (f) into a suitable host cell to produce an expression library containing a plurality of clones containing the selected DNA (h) Screening the expression library for the desired activity	
29. WO 9914376: Detection of aneuploidy and gene deletion by PCR-based gene-dose co-amplification of chromosome specific sequences with synthetic sequences with synthetic internal controls	V-30.03.99 P-19.09.97	Genaco Biomedical Products, Inc.		WO-application deemed to be withdrawn

Table 1. (Continue)

Publication number and title	Publication date (V) and priority date (P)	Assignee	Main claims	Comment on legal status
30. WO 9914375: DNA typing by mass spectrometry with polymorphic DNA repeat markers	V-25.03.99 P-19.09.97	Genetrace Systems, Inc.		EP-application deemed to be withdrawn
31. EP 1045927: Method for identifying genes underlying defined phenotypes	V-22.07.99 P-15.01.98	ValiGen, Inc.		EP-application deemed to be withdrawn
32. EP 1058727: Materials and methods for identifying and analyzing intermediate tandem repeat DNA markers	V-12.08.99 P-04.02.98	Promega Corporation	1. A method for detecting a target intermediate tandem repeat DNA sequence having a low incidence of stutter artifacts, comprising the steps of: (a) Providing a sample of DNA having at least one target intermediate tandem repeat sequence, wherein the target intermediate tandem repeat sequence is a region of the DNA containing at least one repeat unit consisting of a sequence of five (5), six (6), or seven (7) base pairs repeated in tandem at least two (2) times (b) Detecting the target intermediate tandem repeat sequence in the sample of DNA, wherein an average stutter artifact of no more than 2.4% is observed	In EP still under examination
33. WO 9946404: DNA Sequences and their use for the selection of cereals	V-16.09.99 P-10.03.98	Scottish Crop Research Institute		EP-application deemed to be withdrawn
34. EP1042507: QTL mapping in plant breeding populations	V-01.07.99 P-04.05.98	Pioneer HiBred	A method of identifying quantitative trait loci in a mixed defined plant population comprising multiple plant families, the method comprising;	In EP still under examination

Table 1. (Continue)

Publication number and title	Publication date (V) and priority date (P)	Assignee	Main claims	Comment on legal status
35. EP 1075545: Infrared matrix-assisted laser desorption/ionization mass spectrometric analysis of macro-molecules	V-11.11.99 P-07.05.98	Sequenom, Inc.	(i) Quantifying a phenotypic trait across lines sampled from the mixed population, thereby providing a quantified population phenotype (ii) Identifying at least one genetic marker associated with the distribution of phenotypic trait by screening a set of markers for associations with the quantified population phenotype (iii) Identifying the quantitative trait loci based on the association of the phenotypic trait and genetic marker

1. A process for performing matrix assisted laser desorption/ionization (MALDI) of a nucleic acid in preparation for analysis by mass spectrometry, comprising the steps of: (a) Depositing a solution containing the nucleic acid and a liquid matrix on a substrate, thereby forming a homogeneous, thin layer of a nucleic acid/liquid matrix solution (b) Illuminating the substrate with infrared radiation, so that the nucleic acid in the solution is desorbed and ionized | In EP still under examination |
| 36. US 6074831: Partitioning of polymorhpic DNAs | V-13.06.00 P-09.07.98 | Agilent Technologies Inc. | | In US only |
| 37. EP1106687: Method for isolating satellite sequence | V-13.06.01 P-18.08.98 | Japan as represented by Director General of Ministry of Agriculture, Forestry and Fisheries National Institute of Agrobiologica | 1. An isolation method for satellite sequences, wherein a genomic DNA is cleaved by a nucleotide sequence-independent method, the isolation method comprising: (a) Obtaining randomly cleaved fragments of the genomic DNA (b) Selecting, from the fragments obtained in a), fragments comprising the satellite sequences | In EP still under examination |

Table 1. (Continue)

Publication number and title	Publication date (V) and priority date (P)	Assignee	Main claims	Comment on legal status
38. WO 0017341: Myrtaceae microsatellites	V-30.03.00 P-23.09.98 P-16.02.99	Business and Research Management PTY Ltd.		Myrtaceae only, WO-application deemed to be withdrawn
39. WO 0042210: Microsatellite DNA markers and uses thereof	V-20.07.00 P-15.01.99	USA, the Secretary of Agriculture		Tree, forest only, WO-application deemed to be withdrawn
40. US 6573047: Detection of nucleotide sequence variation trough fluorescence resonance energy transfer label generation	V-03.06.03 P-13.04.99 P-11.04.00	DNA Sciences, Inc.		US only
41. EP 1203100: Methods for detecting nucleotide insertion or deletion using primer extension	V-15.02.01 P-11.08.99	Exact Sciences Corporation	1. A method for detecting a nucleic acid insertion or deletion the method comprising the steps of: (a) Selecting a nucleic acid having a known wild-type sequence and having a target region comprising a repeat sequence having at most three different types of nucleotide bases selected from the group consisting of dGTP, dATP, dTTP, and dCTP (b) Contacting a sample with an oligonucleotide primer that is complementary to a portion of said nucleic acid immediately upstream of said target region (c) Extending said primer in the presence of nucleotide bases that are complementary to the nucleotide bases of the target region, thereby to form a primer extension product	In EP still under examination

Table 1. (Continue)

Publication number and title	Publication date (V) and priority date (P)	Assignee	Main claims	Comment on legal status
			(d) Extending the primer extension product in the presence of a labelled nucleotide complementary to a nucleotide base downstream from the target region in said nucleic acid, wherein said labelled nucleotide is not complementary to any of the nucleotide bases of the target region, thereby to produce a labelled extension product comprising a sequence that is complementary to the entire target region (e) Detecting the labelled extension product (f) Comparing the size of the labelled extension product detected in step (e) to a standard, wherein a labelled extension product smaller than the standard is indicative of the presence of a deletion in the target region and a labelled extension product larger than the standard is indicative of the presence of an insertion in the target region Further comprising the step of terminating the primer extension product by incorporating a terminator nucleotide in said product that is complementary to a nucleotide downstream from the target region in a wild-type nucleic acid, wherein said terminator nucleotide is not complementary to any of the nucleotides of the target region, said step of terminating the primer extension product being performed simultaneously with or immediately after step (d).	
42. EP1230385: Marker-assisted identification of a gene associated with a phenotypic trait	V-19.04.01 P-08.10.99	Pioneer Hi Bred	1. A method of associating a gene with a phenotypic trait of interest, comprising: (a) Segregating members of a biological population by the presence or absence of at least one genetic marker in linkage disequilibrium with said phenotypic trait, wherein said phenotypic trait is statistically associated with more than one genetic locus (b) Expression profiling segregated members of (a) (c) Determining from expression profiles of (b) said gene associated with said phenotypic trait	In EP still under examination

Intellectual Property Rights in the Field of Molecular Marker Analysis 459

Table 1. (Continue)

Publication number and title	Publication date (V) and priority date (P)	Assignee	Main claims	Comment on legal status
43. WO 0140512: Resistance gene	V-07.06.01 P-29.11.99	Plant Bioscience Limited		WO-application deemed to be withdrawn
44. EP 1250452: Methods for determining single nucleotide variations and genotyping	V-07.06.01 P-02.12.99	DNA Sciences, Inc.		EP-application withdrawn
45. NZ 0509194: Simple sequence repeats (microsatellites) in clover	V-25.05.01 P-24.12.99	Agriculture Victoria Services PTY Ltd.		AUH, NZ only
46. NZ 0509193: Molecular markers in ryegrass and fescues	V-25.05.01 P-24.12.99	State of South Australia as represented by South Australian Research and Development Institute		NZ only
47. EP 1265476: MQM mapping using haplotyped putative QTL-alleles: a simple approach for mapping QTLs in plant breeding populations	V-12.07.01 P-30.12.99	Pioneer HiBred	1. A method of mapping a phenotypic trait to a corresponding chromosomal location or region, the progeny descending from a plurality of families resulting from related or unrelated crosses (ii) Assigning phenotypic values to at least one phenotypic trait segregating in the population of progeny (iii) Determining a genotype for at least one haplotype in the population of progeny, which at least one haplotype comprises a plurality of genetic markers	In EP still under examination

Table 1. (Continue)

Publication number and title	Publication date (V) and priority date (P)	Assignee	Main claims	Comment on legal status
			(iv) Applying a statistical model which evaluates correspondence between the haplotype and the assigned phenotypic value, thereby identifying a chromosomal location corresponding to the phenotypic trait	
48. WO 0151627: Soybean SSRs and methods of genotyping	V-19.07.01 P-07.01.00	–		EP-application deemed to be withdrawn
49. WO 0162967: A method that compares genomic sequences	V-30.08.01 P-22.02.00	Genena Ltd.		WO-application is deemed to be withdrawn
50. WO 0162966: Methods for characterizing polymorphisms	V-30.08.01 P-24.02.00	Gemini Inc		WO-application has not entered EP-Phase
51. JP 2000060559: Isolation of satellite sequence	V-29.02.00	Natl. Inst. of Agrobiological Resources		JP only
52. WO 0179482: Gene mapping method	V-25.10.01 P-13.04.00	Inoko, Hidetoshi		JP, US only

Table 1. (Continue)

Publication number and title	Publication date (V) and priority date (P)	Assignee	Main claims	Comment on legal status
53. EP 1278894: Identification of genetic markers	V-29.01.02 P-02.05.00	Centre National de la Recherche Scientifique, Institut National de la Santé et de la Recherche Médicale	1. A method for the identification of the presence of a genetic marker in a DNA sample comprising the following steps: (a) Selection of sequences specific of said genetic marker (b) Fixation of oligonucleotides comprising said specific sequences or the complementary sequences on a solid support (c) Addition of a mixture of DNA fragments representing the said DNA sample to the solid support in a way that hybridization is possible (d) Detection of the presence of the genetic marker in the DNA sample by the presence of a signal corresponding to the hybridization of a fragment of the DNA sample to the specific oligonucleotide, wherein said specific sequences are flanking sequences of said genetic marker and said DNA sample has been reduced in complexity	In EP still under examination
54. WO 0185988: Methods for detecting nucleic acid molecules having particular nucleotide sequences	V-15.11.01 P-09.05.00	Diatech Pty. Ltd.		WO-application is deemed to be withdrawn
55. EP 1282729: Microsatellite-AFLP	P-15.05.00 P-12.01.01 V-22.11.01	Keygene N.V.	1. Use of a RAMP primer and an AFLP primer in analysing a nucleic acid sequence, in particular in analysing a nucleic acid sequence for the presence polymorphisms associated with microsatellites	In EP still under examination
56. EP 1297181A2: Sample generation for genotyping by mass spectrometry	V-03.01.02 P-30.06.00	Centre National de Genotypage	1. A method for DNA genotyping by mass spectrometry, comprising the steps of: (a) Reduction of the complexity of the DNA sample (b) Generation of allele-specific products on the products generated in step (a), wherein the generation of allele specific products in step (b) is achieved by at least one method that uses (an) allele-specific oligonucleotide(s)	In EP still under examination

Table 1. (Continue)

Publication number and title	Publication date (V) and priority date (P)	Assignee	Main claims	Comment on legal status
57. WO 0205628: Pollen polymyx plant breeding method utilizing molecular pedigree analysis	V-24.01.02 P-18.07.00	Weyerhaeuser Company	(c) Mass spectrometric analysis of the products generated in step b, 1 0 wherein the mass spectrometric analysis in step c. is performed on the products generated in step b. without purification or separation from the reaction mixture	WO-application has not entered EP-Phase
58. WO 0238804: Method for marking samples containing DNA by means of oligonucleotides	V-16.05.02 P-08.11.00	Agrobiogen GmbH Biotechnologie	No English translation of the claims available 1. Verfahren zur Kennzeichnung von DNA-enthaltenden Proben, bei dem mindestens ein Kennzeichnungs-Oligonukleotid mit einer zu kennzeichnenden Probe zusammengebracht wird und die Probe zusammen mit dem Kennzeichnungs-Oligonukleotid einer Untersuchung unterworfen wird, wobei das Kennzeichnungs-Oligonukleotid ausgewählt ist aus der Gruppe bestehend aus artifiziellen Mikrosatelliten-Oligonukleotiden oder artifiziellen Single- Nukleotide-Polymorphismus-Oligonukleotiden	In EP still under examination
59. EP 1207210: Method for melting curve analysis of repetitive PCR products	V-09.07.02 P-15.11.00	F. Hoffmann La Roche AG	1. Method for analysis of a target nucleic acid consisting of repetitive and non repetitive sequences comprising; (a) Hybridization of at least one polynucleotide hybridization probe comprising a first segment which is complementary to a non repetitive region and a second segment which is complementary to an adjacent repetitive region, said second segment consisting of a defined number of repeats (b) Determination of the melting point temperature of the hybrid which has been formed between the target nucleic acid and the at least one hybridization probe	In EP still under examination

Table 1. (Continue)

Publication number and title	Publication date (V) and priority date (P)	Assignee	Main claims	Comment on legal status
60. WO 02086159: Method for genotyping Microsatellite DNA markers	V-31.10.02 P-23.04.01	Galileo Genomics	1. A method for genotyping different alleles of a microsatellite DNA locus by using combinations of at least three oligonucleotides for each allele on the locus comprising: (a) Providing a sample containing the microsatellite DNA (b) Selecting at least three oligonucleotides comprising: (i) A 5' primer which comprises at least a 5-base pair sequence that is complementary a flanking region of a repeat region of the microsatellite (ii) A central primer which is complementary to a repeated region of the microsatellite DNA (iii) A plurality of 3' primers which comprises: (a) A sequence that is complementary to the 5' flanking sequence of the repeated region of the microsatellite (b) A number (n) of repeat units at the 5' end of the plurality of 3' primers (c) Mixing the sample and primers such that the primers and microsatellite DNA hybridize (d) Adding a ligating reagent (e) Detecting the presence of ligation products that consist of all the oligonucleotide primers	
61. WO 0209563A1: Methods for the reduction of stutter in microsatellite amplification using sorbitol	V-14.11.02 P-07.05.01	PE Corporation (NY)	1. A method for reducing stutter in the amplification of a microsatellite comprising the steps of: (a) Providing a sample comprising a microsatellite of interest, said microsatellite having a G+C content of greater than 50% (b) Contacting said sample with at least one enzyme having nucleic acid polymerase activity	

Table 1. (Continue)

Publication number and title	Publication date (V) and priority date (P)	Assignee	Main claims	Comment on legal status
62. WO 02090562: Methods for the reduction of stutter in microsatellite amplification	V-14.11.02 P-07.05.01	Applied Biosystems Inc.	(c) Incubating said sample with said enzyme for a time and under conditions sufficient to amplify said microsatellite; wherein said incubation is performed in the presence of an amount of sorbitol effective to reduce said stutter relative to the amount of stutter observed in the absence of sorbitol	
63. WO 0185988: Methods for detecting nucleic acid molecules having particular nucleotide sequences	V-15.11.01 P-09.05.01	Diatech Pty. Ltd.		Belonging to WO02090562
64. WO 03023055: Method for detecting mutations, insertions, deletions and polymorphisms on DNA and the use thereof	V-20.03.03 P-12.09.01	Max-Delbrück-Centrum für Molekulare Medizin	No English translation of the claims available
1. Methode zum Nachweis von Mutationen, Insertionen, Deletionen und Polymorphismen auf der DNA, dadurch gekennzeichnet, dass – eine erste spezifische PCR-Reaktion durchgeführt wird mit einem Primerpaar, dass die zu untersuchende Stelle im Genom flankiert, wobei am Y-Ende jedes Primers zusätzlich eine jeweils unterschiedliche universelle Oligonukleotidsequenz hängt, die mit der zu untersuchenden DNA-Sequenz nicht komplementär ist; – eine zweite universelle PCR-Reaktion durchgeführt wird mit einem markierten, vorrangig fluoreszenzmarkierten oder biotinmarkierten, Primerpaar, dass komplementär zu den in der ersten PCR verwendeten 15 universellen Oligonukleotidsequenzen ist; – die PCR-Produkte durch Erhitzung denatu-	WO-application deemed to be withdrawn

Table 1. (Continue)

Publication number and title	Publication date (V) and priority date (P)	Assignee	Main claims	Comment on legal status
			riert und danach schnell wieder abgekühlt werden; – eine Auftrennung der PCR-Produkte erfolgt; und – die Detektion der Laufeigenschaften der PCR-Produkte, wobei die Laufeigenschaften abhängig sind von der speziellen Faltung und der Konformation der DNA-Einzelstränge und die spezielle Faltung wiederum von der DNA-Sequenz	
65. WO 03035906: Method for genotyping Microsatellite DNA markers by mass spectrometry	V-01.05.03 P-26.10.01	Galileo Genomics	1. A method for genotyping different alleles of a microsatellite DNA locus by using enzymatic and/or chemical agents that produce short, single stranded DNA fragments of a size suitable for mass spectrometry analysis, said method comprising: (a) Providing a genomic DNA sample containing the microsatellite DNA (b) Performing PCR amplification of a microsatellite DNA marker locus, using: (i) An appropriate combinations of oligonucleotides (ii) A dNTP mix in which the 2'-thymidine 5'-triphosphate is replaced by 2'-uridine 5'-triphosphate (iii) A thermostable DNA polymerase that is capable of incorporating uridine nucleotides at positions where thymidine nucleotides are usually incorporated (iv) An appropriate buffer (c) Treating the PCR fragment with uracyl-DNA-glycosylase (d) Treating further the UDG-treated DNA with an enzymatic or chemical agent that cleaves DNA at a basic site to yield single-stranded DNA products	

Table 1. (Continue)

Publication number and title	Publication date (V) and priority date (P)	Assignee	Main claims	Comment on legal status
66. WO 03040395: Universal nucleotides for nucleic acid analysis	V-26.06.03 P-07.11.01	Applera Corporation	1. A method of sequencing at least one target nucleic acid template comprising: (a) Forming a reaction composition comprising at least one target nucleic acid template, at least one primer, at least one polymerase, at least one universal nucleotide, and at least one specific terminator (b) Incubating the reaction composition under appropriate conditions to generate at least one primer extension product comprising at least one or more of the universal nucleotides and at least one or more of the specific terminators (c) Separating at least one or more of the primer extension products, wherein the separating comprises at least one mobility-dependent analysis technique (MDAT) (d) Detecting at least one or more of the primer extension products	
67. EP 1217079: Microsatellite markers from *triticum tauschii*	V-26.06.02 P-?	Institut National de la Recherche Agronomique (INRA)		EP-application deemed to be withdrawn
68. WO03085133: Novel FISSR-PCR primers and method of identifying diverse genomes of plant and animal systems including rice varieties, a kit thereof	V-16.10.03 P-08.04.02	Centre for DNA fingerprinting and diagnostics, India	1. A set of inter-simple sequence repeats (ISSR)-PCR primers of SEQ ID Nos. 1 to 37 for genotyping eukaryotes, (for example: SEQ ID NO. 1. GATGCTGATACACACACACACACA)	

Table 1. (Continue)

Publication number and title	Publication date (V) and priority date (P)	Assignee	Main claims	Comment on legal status
P1 EP 0509612: Process for amplifying and detecting nucleic acid sequences	V-21.10.92 P-28.03.85 P-25.10.85 P-07-02–86	F. Hoffmann La Roche Inc.	1. A first and second single-stranded oligonucleotide allowing amplification of a specific template nucleic acid sequence contained in a single- or double-stranded nucleic acid or in a mixture of such nucleic acids, wherein (a) One oligonucleotide of said oligonucleotides contains a part which is substantially complementary to said template nucleic acid sequence in said single-stranded nucleic acid or in one strand of said double-stranded nucleic acid (b) The other oligonucleotide of said oligonucleotides contains a part which is substantially complementary to a complement of said template nucleic acid sequence in said single-stranded nucleic acid or in said strand of said double-stranded nucleic acid (c) Said parts of oligonucleotides (a) and (b) have attached to their 5'-end a nucleotide sequence which is non-complementary to said template nucleic acid sequence and which comprises a restriction site; and wherein (d) The parts of said oligonucleotides of (a) and (b) that have substantial complementarity are different and define the termini of the specific template nucleic acid sequence to be amplified	Granted EP-patent, no opposition
P2 EP 0201184: Process for amplifying nucleic acid sequences	V-21.10.92 P-28.03.85 P-25.10.85	F. Hoffmann La Roche Inc.	1. A process for exponentially amplifying at least one specific double-stranded nucleic acid sequence contained in a nucleic acid or a mixture of nucleic acids wherein each nucleic acid consists of two complementary strands, of equal or unequal length, which process comprises:	Granted EP-patent, but opposition filed against

Table 1. (Continue)

Publication number and title	Publication date (V) and priority date (P)	Assignee	Main claims	Comment on legal status
			(a) Treating the strands with a molar excess of two oligonucleotide primers, one for each of the strands, under hybridizing conditions and in the presence of an inducing agent for polymerization and the different nucleotides, such that for each strand an extension product of the respective primer is synthesized which is complementary to the nucleic acid strand, wherein said primers are selected so that each is substantially complementary to one end of the sequence to be amplified on one of the strands such that an extension product can be synthesized from one primer which, when it is separated from its complement, can serve as a template for synthesis of an extension product of the other primer (b) Separating the primer extension products from the templates on which they were synthesized to produce single-stranded molecules (c) Treating the single-stranded molecules generated from step (b) with the primers of step (a) under hybridizing conditions and in the presence of an inducing agent for polymerisation and the different nucleotides such that a primer extension product is synthesized using each of the single-strands produced in step (b) as a template; and, if desired (d) Repeating steps (b) and (c) at least once; whereby the amount of the sequence to be amplified increases exponentially relative to the number of steps in which primer extension products are synthesized	

Table 1. (Continue)

Publication number and title	Publication date (V) and priority date (P)	Assignee	Main claims	Comment on legal status
P3 EP 0200362: Process for amplifying, detecting, and/or cloning nucleic acid sequences	V-20.01.93 P-28-03.85 P-25.10.85 P-07.02.86	F. Hoffmann La Roche Inc.	1. A process for detecting the presence or absence of at least one specific double-stranded nucleic acid sequence in a sample, or distinguishing between two different double-stranded nucleic acid sequences in said sample, which process comprises first exponentially amplifying the specific sequence or sequences (if present) by the following steps, and then detecting the thus-amplified sequence or sequences (if present): (a) Separating the nucleic acid strands in the sample and treating the sample with a molar excess of a pair of oligonucleotide primers for each different specific sequence being detected, one primer for each strand, under hybridizing conditions and in the presence of an inducing agent for polymerization and the different nucleoside triphosphates such that for each of said strands an extension product of the respective primer is synthesized which is complementary to the strand, wherein said primers are selected so that each is substantially complementary to one end of the sequence to be amplified on one of the strands such that the extension product synthesized from one primer, when it is separated from its complement, can serve as a template for synthesis of an extension product of the other primer of the pair (b) Treating the sample resulting from (a) under denaturing conditions to separate the primer extension products from their templates (c) Treating as in (a) the sample resulting from (b) with oligonucleotide primers such that a primer extension product is synthesized using each of the single strands produced in step (b) as a template; and, if desired,	Granted, but opposition filed against

Table 1. (Continue)

Publication number and title	Publication date (V) and priority date (P)	Assignee	Main claims	Comment on legal status
			(d) Repeating steps (b) and (c) at least once; whereby exponential amplification of the nucleic acid sequence or sequences, if present, results thus permitting detection thereof; and, if desired, (e) Adding to the product of step (c) or (d) a labelled oligonucleotide probe capable of hybridizing to said sequence to be detected; and (f) Determining whether said hybridization has occurred	
P4 EP 0497784: Quantitation of nucleic acids using the polymerase chain reaction	V-19.12.95 P-21.08.89 P-28.09.89	F. Hoffmann La Roche Inc.	1. Use of an internal standard for the quantitation of at least one target nucleic acid segment contained within a sample in an amplification method, said internal standard comprising on one strand a nucleic acid segment comprising a 5' sequence and a 3' sequence, which sequences provide an upstream primer hybridization site and the complement of a downstream primer hybridization site which are identical to an upstream primer hybridization site and the complement of a downstream primer hybridization site within said target nucleic acid segment, wherein said internal standard nucleic acid segment and said target nucleic acid segment are co-amplified using the same set of primers and wherein upon amplification said internal standard nucleic acid segment and said target nucleic acid segment can be distinguished	Granted EP-patent, but opposition filed against

References

Beavis WD (1999) QTL mapping in plant breeding populations. Patent EP 1042507
Byrum J, Reiter R (1998) A method for identifying genetic marker loci associated with trait loci. Patent EP 0972076
Caskey T, Edwards A (1992) DNA typing with short tandem repeat polymorphisms and identification of polymorphic short tandem repeats. EP 0639228
Fleck B, Baldock C (2003) Intellectual property protection for plant-related inventions in Europe. Nat Rev Gen 4:834–838
Fodor S, Dower W, Solas D (1998) Detection of nucleic acid sequences. Patent EP 0834576
Herrlinger C, Jorasch P, Wolter FP (2003) Biopatentierung – eine Beurteilung aus Sicht der Pflanzenzüchtung. In: Baumgartner C, Mieth D (eds) Patente am Leben? Ethische, rechtliche und politische Aspekte der Biopatentierung. Mentis, Paderborn, pp 245–258
Hillenkamp F, Köster H (1999) Infrared matrix-assisted laser desorption/ionization mass spectrometric analysis of macro-molecules. Patent EP1075545
Jansen RC, Beavis WD (2001) MQM mapping using haplotyped putative QTL-alleles: a simple approach for mapping QTL's in plant breeding populations. Patent EP 1265476
Kimpel JA (1999) Freedom to operate: intellectual property protection in plant biology and its implications for the conduct of research. Annu Rev Phytopathol 37:29–51
Kowalski SP, Ebora RV, Kryder RD, Potter RH (2002) Transgenic drops, biotechnology and ownership rights: what scientists need to know. Plant J 31:407–421
Kuiper M, Zabeau M, Vos P (1997) Amplification of simple sequence repeats. Patent EP 0805875
Morgante M, Vogel J (1997) Compound microsatellite primers for the detection of genetic polymorphisms. Patent EP 0804618
Mullis K (1992) Process for amplifying nucleic acid sequences. Patent EP 0201184B1
Mullis K, Arnheim N, Saiki R, Erlich H, Horn G, Scharf S (1992) Process for amplifying and detecting nucleic acid sequences. Patent EP 0509612B1
Mullis K, Arnheim N, Saiki R, Erlich H, Horn G, Scharf S (1993) Process for amplifying, detecting, and/or cloning nucleic acid sequences. EP 0200362B1
Nagaraju J (2003) Novel FISSR-PCR primers and method of identifying genotyping diverse genomes of plant and animal systems including rice varieties, a kit thereof. Patent WO 03085133
Olek A (1996) Amplification of simple sequence repeats. Patent EP 0870062
Openshaw S, Bruce WB (2001) Marker assisted identification of a gene associated with a phenotypic trait. Patent EP 1230385
Perlin M (1995) Method and system for genotyping. Patent EP0714537
Röder M, Plaschke J, Ganal M (1997) Microsatellite markers for plants of the species *Triticum aestivum* and tribe Triticeae and the use of said markers. Patent EP 0835324B1
Saint-Louis D, Paquin B (2003) Method for genotyping microsatellite DNA markers by mass spectrometry. Patent WO03035906
Shear RH, Kelley TE (2003) A researcher's guide to patents. Plant Phys 132:1127–1130
Shuber A, Pierceall W (2002) Methods for detecting nucleotide insertion or deletion using primer extension. Patent EP1203100
Van Eijk M, Peleman J, De Ruiter-Bleeker M (2001) Microsatellite-AFLP. Patent EP1282729

Subject Index

α-Amylase inhibitor 181
Abies spec. 413 ff
advanced backcross lines 58, 64, 312
alfalfa 139 ff
allele
– mining 49
– specific hybridisation (ASH) 32
– specific associated primer (ASAP) 161
– specific PCR (AS-PCR) 29
amplified fragment length polymorphism (AFLP) 26, 66, 94, 108, 158, 248, 337, 360, 436 f
anchor
– locus 163
– marker 199
Aphanomyces root rot 164
Arabidopsis thaliana 55, 70 ff, 89 ff, 116
array
– micro 45
– oligonucleotide 46
Ascochyta 164
association studies 10, 146, 209, 362
autopolyploid 143
azuki bean 172

background selection 208, 319
Bacillus thuringiensis (Bt) 343
bacterial artificial chromosome (BAC) 69, 116, 207, 234
barley 77, 229
bean yellow mosaic virus (BYMV) 160
beech 387 ff
beet cyst nematode 130
Beta spec. 77, 121 ff
betalain 131
BIN map 232
bolting 123, 131
Brassica spec. 89, 92
– Rf line 100
– Rfk1 region 100
breeding
– backcross 12, 15, 311, 319
–, – recurrent 15, 319

– by design 313
– population 90
– whole genome 19
bruchid beetle 178
bulked segregant analysis 9, 57, 65, 159, 220

candidate gene 207
– mapping 306, 405, 408
Capsicum spec. 189 ff
carbohydrate 124
centromere 198
Cercospora leaf spot 132
certation 65
chromosome
– assignment to 198
– carrier 320
– intact donor 323, 331
– non-carrier 324
– paracentric inversion 217
– rearrangement 217
– translocation 192, 198
cleaved amplified polymorphic sequence (CAPS) 28, 116, 247
cold sweetening 221
complementation 62, 73
conifer 269 ff
conserved ortholog set 24
core collection 48
cost analysis 348
cowpea 172
cultivar
– development 157
– ideotype 166
– protection 190
– specific marker 192
cysteine proteinase inhibitor 182

deletion 374
distorted segregation 193
diversity array technology (DarT) 46
doubled haploid line 189, 194, 338
drought tolerance 347

earliness   344, 407
economics of MAS   348
embryo rescue   189
epistasis   204, 346
Eucalyptus spec.   399 ff
- reticulate evolution   401
expressed sequence tag (EST)   115, 207, 235, 247, 270, 273, 274
- database   336
- SSR   148

Fagus spec.   387 ff
fine-scale genetic structure   62, 390
fingerprinting   110, 335, 434
fir   413 ff
flowering genetics   161, 341
fluorescence assay   382
forest   413 f
- conservation   393
FPC database   372
functional genomics   47, 115, 336
Fusarium wilt   159

gene
- cluster   161, 204
- common ancestral   206
- density   67
- flow   403
- major effect   405
- major resistance   197
- orthologous   199, 206
- pool   400
- pyramiding   310
- stacking   338
genetic
- distance   378
- diversity   140
- erosion   418
- gain ΔG 339
- linkage map   9, 142, 155, 230
- management   393
- maps, comparative   24, 96, 209, 229
-, - consensus   157, 163, 231
genome sequence   40
- rice project   245
genomics
- comparative   267, 269
genotype
- construction   338
- environment interaction   346
Glycine max   171 f, 371 ff
grain yield   344

H^2 339
haplotype map   372
Helianthus spec.   107 ff
heritability   339
heterochromatin   200
heteroduplex analysis   32
heterosis   142, 149
heterotic group   111
high polymorphic locus   95
Hordeum vulgare   77, 229
hybrid inviability   191, 402
hybridisation   402
- in situ (FISH)   126
- somatic   220

in vitro regeneration   113
inbreeding   149, 394
insertion   374
intellectual property rights   101, 433
interval mapping   162
introgression line   209, 220
isozyme   190

karoytype   192

Lablab purpureus   177
lignin   408
linkage
- block analysis and selection   16
- disequilibrium (LD)   10, 224, 391
- drag   345
- equilibrium   224
- group   93, 109, 143, 216
Lycopersicon spec.   76, 78

Maize   335 ff, 344
male sterility   191
- cytoplasmic (CMS)   97
map
- based cloning   43
- comparative,   24, 96, 209, 229, 267 ff
- consensus,   157, 163, 231
- genus-wide   406
- high-density,   66, 94
- linkage,   216
- low-resolution,   64
- physical,   218, 233, 372
- soybean,   371
- tomato,   199
- transcript,   235
marker
- AFLP   26, 66, 94, 108, 158, 248, 337, 360, 436, 438
- ASAP   161

Subject Index 475

- anchored clone   373
- assisted selection (MAS)   4, 156, 208, 224, 305, 335, 353, 362, 434
-, - SLS-MAS   349
- biochemical   156, 354
- classical   246
- colinearity   355
- data point   325
- density   350
- diagnostic   257
- DNA   56, 90 ff, 100 f, 306
- index score   339
- microsatellite   25, 110, 247, 255, 306, 337, 360, 388, 433
- molecular   3 ff, 67, 155, 230, 335
- morphological   156, 354
- PCR-based   93
- positions   323
- RAPD   25 ff, 158, 246, 337, 360, 438
- RFLP   23 ff, 92, 108, 140, 190, 246, 336, 355, 438
- SAMPL   374
- SNP   28 ff, 47, 73, 115, 124, 190, 235, 306, 337, 361, 374
- SSR   25 ff, 41, 108, 205, 247, 306, 337, 360, 373, 388, 433
- trait association   6 ff
Medicago spec.   139 ff
metaphore agarose   382
micro-colinearity   96
microsatellites   25, 110, 247, 255, 306, 337, 360, 388, 433
micro-synteny   238
miniature inverted repeat transposable elements (MITE)   248
minisatellite   110
minisequencing   29
MITE   248
molecular beacons   382
mung bean   172
mutation   72

nearly isogenic line (NIL)   56 f, 251, 312, 338
nitrogen metabolism   124
nuclear-encode Rf gene   97
nutraceutical   380

oleic acid, high   114
ortholog   267 ff
Oryza sativa   39 ff, 55, 245 ff

Phaseolus vulgaris   176
paralogous   267, 268, 273
patent   433

paternity analysis   392, 403
pathogenesis   221
pea   155 ff
- disease   159
- enation mosaic virus (PEMV)   161
- germplasm   163
- powdery mildew   158
- seed-borne mosaic virus (PSbMV)   160
peach   279 ff
pepper   189 ff
- fruit pungency   205
Phaseolus spec.   171
phenotypic
- selection (PS)   337
- value   339
- variation   114
Phoma macdonaldii   111
photosynthesis   113
PIF   251
Picea spec.   413 ff
Pinus spec.   267 ff
Pinaceae   267 ff
Pisum sativum   155 ff
pleiotropy   221
polygenic trait   337, 335, 379
polymerase chain reaction (PCR)   336
polymorphic loci   108 f
polyploid   371
population
- minimum size   320, 331
- structure   147, 322, 400
Populus spec.   423 f
positional cloning   42 f, 55
potato   215 ff
Powdery mildew   178
probability distribution   332
Prunus spec.   279 ff
pyramidization   209, 354, 361

quality
- control among seeds   100
- protein maize (QPM)   348
- trait   344

quantitative trait locus (QTL)   41, 56, 78, 107, 145, 197, 200, 218, 245, 251, 260, 267, 273, 307, 335, 405 ff
- mapping   58, 64 f, 338

random amplified polymorphic DNA (RAPD)   25 f, 158, 246, 337, 360, 438
rapeseed   89, 92, 100
recombinant inbred line (RIL)   56 ff, 108, 251, 338

recombination   60, 66, 100
– rate of   223
recombinational mapping   55
recurrent parent genome   325, 346, 378
reserve genetics   336
resistance   197, 200, 353 ff
– *Ascochyta*   164
– *Fusarium* wilt   159
– gene analogs (RGA)   125, 361
– gene homologues   207
– genes   113, 158, 355
– insect   343
– nematode   123, 130
– plant virus   159 f,
– powdery mildew   178
– quantitative   353 ff
– race-specific gene   353
– single locus   160
restorer   123, 131
restriction fragment length polymorphism (RFLP)   23, 92, 108, 140, 190, 246, 267, 336, 355, 438
retrotransposon   127, 200
Rhizomania   123, 129
rice   39 ff, 76 ff, 245 ff
– heading-date1 (Hd1)   251
– Onaga I   250
– stowaway OS-1   250
– Tos 17   250
rice bean   178
risk assessment   424

satellite DNA   127 f
seed characteristic   165, 344
seed orchard   392, 403
selection
– foreground   327
– four-stage   326
– in early generations   14
– intensity   323
– marker assisted (MAS)   4, 156, 208, 224, 305, 335, 353, 362, 434
– three-stage   326
– trait-based   12
– two-stage   326
– whole genome   15
selectively amplified microsatellite polymorphic locus (SAMPL)   374
self-incompatibility   215
sequence
– characterized amplified region (SCAR)   158

– polymorphism   55
– tagged site (STS)   97, 247, 360
simple sequence repeat (SSR)   25, 41, 108, 205, 247, 306, 337, 360, 373, 388, 433
simulation studies   342
simultaneous introgression   330
single dose restriction fragment (SDRF)   143
single nucleotide polymorphism (SNP)   28, 47, 73, 115, 124, 190, 235, 306, 337, 361, 374
single, large-scale, marker-assisted selection (SLS-MAS)   349
snaPshot assay   74
solanaceae   189, 215 ff
Solanum spec.   215 ff
soybean   177, 371 ff
– cyst nematode (SCN)   380
– map   371
species delimitation   191
spruce   413 ff
stearic acid   111
Striga gesnerioides   178
sudden death syndrome (SDS)   380
sugar beet   77, 121 ff
sunflower   107 ff
synteny   205, 216, 236, 267, 270

targeted local lesions in genomes (TILLING)   48
tandem base repeats   95
tetraploid   139
tetrasomic inheritance   148
tomato   76 ff
– map   199
transformation   182
transgressive segregant   379
transposon   248
– miniature inverted repeat (MITE)   248 f
– ping   251
– pong   251
– tabito   250
trisomics   121, 124, 193
Triticum spec.   255 ff, 353 ff

Vigna spec.   171 ff

wheat   255 ff, 353 ff

yeast artificial chromosome (YAC)   234

Zea mays   335 ff

Printing: Mercedes-Druck, Berlin
Binding: Stein+Lehmann, Berlin